Ethics in Online AI-based Systems

Risks and Opportunities in Current Technological Trends

Intelligent Data-Centric Systems

Ethics in Online AI-based Systems

Risks and Opportunities in Current Technological Trends

Edited by

Santi Caballé
*Faculty of Computer Science, Multimedia and Telecommunications,
Open University of Catalonia, Barcelona, Spain*

Joan Casas-Roma
*Faculty of Computer Science, Multimedia and Telecommunications,
Open University of Catalonia, Barcelona, Spain*

Jordi Conesa
*Faculty of Computer Science, Multimedia and Telecommunications,
Open University of Catalonia, Barcelona, Spain*

Series Editor Fatos Xhafa
Universitat Politècnica Catalunya, Barcelona, Spain

ELSEVIER

ACADEMIC PRESS
An imprint of Elsevier

Academic Press is an imprint of Elsevier
125 London Wall, London EC2Y 5AS, United Kingdom
525 B Street, Suite 1650, San Diego, CA 92101, United States
50 Hampshire Street, 5th Floor, Cambridge, MA 02139, United States

Notices

Knowledge and best practice in this field are constantly changing. As new research and experience broaden our understanding, changes in research methods, professional practices, or medical treatment may become necessary.

Practitioners and researchers must always rely on their own experience and knowledge in evaluating and using any information, methods, compounds, or experiments described herein. In using such information or methods they should be mindful of their own safety and the safety of others, including parties for whom they have a professional responsibility.

To the fullest extent of the law, neither the Publisher nor the authors, contributors, or editors, assume any liability for any injury and/or damage to persons or property as a matter of products liability, negligence or otherwise, or from any use or operation of any methods, products, instructions, or ideas contained in the material herein.

ISBN: 978-0-443-18851-0

For Information on all Academic Press publications
visit our website at https://www.elsevier.com/books-and-journals

Publisher: Mara Conner
Editorial Project Manager: Emily Thomson
Production Project Manager: Omer Mukthar
Cover Designer: Miles Hitchen

Typeset by MPS Limited, Chennai, India

Contents

List of contributors xv

Preface xix

Acknowledgments xxv

Part I Ethical implications of artificial intelligence in applications for education

CHAPTER 1 Adverse effects of intelligent support of CSCL—the ethics of conversational agents 3

Birk Thierfelder, Pantelis M. Papadopoulos, Armin Weinberger, Stavros Demetriadis and Stergios Tegos

Introduction 3

Conversational agents for collaborative learners 4

    How can CSCL agents interact with the learner? 6

    Effectiveness, advantages, and limits of agents in CSCL 8

From general ethical frameworks for AI-based systems to area-specific considerations for CSCL agent design 9

    Breaking down high-level frameworks: overlap or disjoint? 10

    Classifying human-agent communication 11

    Who should be adapting to whom in CSCL settings? 12

    AI ethics and pedagogical ethics: clashing perspectives? 16

Conclusion 18

References 19

CHAPTER 2 Navigating the ethical landscape of multimodal learning analytics: a guiding framework 25

Haifa Alwahaby and Mutlu Cukurova

Introduction 25

Background and literature review 26

    Multimodal learning analytics: background, history, and aims 26

    The ethics of MMLA in education 27

Methodology 33

    Participants 33

    Data collection 34

    Data analysis 35

Result 35

    Theme 1: The emerging need for an ethical framework for MMLA 35

Theme 2: Privacy, surveillance, and intrusiveness issues with MMLA 36

Theme 3: Student agency over their learning and data ownership 38

Theme 4: Trustworthiness of MMLA results 39

Theme 5: Fairness and bias issues in MMLA systems 39

Theme 6: MMLA systems' transparency and explainability 40

Theme 7: MMLA systems' accountability 40

Theme 8: Awareness level of benefits and risks associated with
MMLA use 41

Theme 9: Argued benefits of MMLA and the ethical issues of not
using it 41

Discussion 42

Limitations and future work 45

Conclusion 48

Acknowledgments 49

References 49

CHAPTER 3   Ethics in AI-based online assessment in higher education 55

Joana Heil and Dirk Ifenthaler

Introduction 55

The use of AI in online assessment 55

Artificial intelligence 55

AI in education 56

AI in assessment 56

Ethics of AI in online assessment 59

Ethics 59

Ethics of educational technology 59

Ethics of AI 60

Moral and ethical implications of assessment 60

Ethics of AI-based assessment scenarios 61

Frameworks to mitigate potential ethical risks of AI-based assessment 64

Conclusion 66

References 67

**CHAPTER 4   Ethical aspects of automatic emotion recognition in online
learning 71**

Gabriela Moise and Elena S. Nicoară

Introduction 71

Emotions, affective learning, and ethical implications 72

Ethical guidelines and frameworks 74

Automatic emotion recognition in education: roles, benefits, and ethical risks............76
Ethical automatic emotion recognition model for online learning 81
Case studies and use case.. 85
Discussion and conclusions... 88
Acknowledgment.. 90
References... 90\

CHAPTER 5 Data-driven educational decision-making model for curriculum optimization.. **97**
Edis Mekić, Irfan Fetahović, Kristijan Kuk, Brankica Popović and Petar Čisar
Introduction.. 97
Decision-making process and AI implementation... 100
The holistic approach to curriculum design and classification 103
    Classification methodologies for qualitative indicators.................................. 104
    Sentiment analysis methodologies for qualitative indicator analysis.............. 106
Indicators and learning management system (LMS) as technical basis for holistic approach.. 108
K-nearest neighbor kNN in strategic budget planning.. 109
LinkedIn and Google Scholar Big Data system for support of human resource procedure.. 111
Part of the decision-making AI system that raises ethical issues and applicable ethical frameworks.. 112
Conclusion 114
References... 115

PART II Ethical implications of artificial intelligence in autonomous services and systems

CHAPTER 6 The ethical issues raised by the use of Artificial Intelligence products for the disabled: an analysis by two disabled people **121**
Laura Smith and Peter Smith
Introduction.. 121
Literature review.. 122
Methodology 124
Peter 124
Peter's diary 125
Laura 128
Laura's diary... 128
Discussion and reflection 130

Conclusions... 132
Next steps.. 133
References... 133

CHAPTER 7   The implications of ethical perspectives in AI and autonomous systems.. 135
Arthur So
Introduction.. 135
Background... 136
    Artificial intelligence trend 136
    The worldview of ethics associated with AI systems...................... 136
    Algorithmic ethics 138
    What is technoethics?... 139
Methodology... 139
    Technoethical inquiry approach 139
    Problem statement 141
    Research questions 141
Results.. 141
    The perspectives 141
    Efficiency and fairness 147
    Advanced analysis and discussion 147
Conclusions.. 148
References... 149

CHAPTER 8   The ethics of online AI-driven agriculture and food systems 153
Edmund O. Benefo, Abani K. Pradhan and Debasmita Patra
Introduction.. 153
Current trends and future applications of online-based AI in agricultural and food systems 154
    Crop production.. 155
    Animal production.. 158
    Food processing and related operations.................................... 159
Potential ethical risks of AI technological advancements in agriculture and food systems 161
    (Cyber)Security... 161
    Privacy 162
    Data ownership... 162
    Accountability/responsibility.. 163

Fairness 163

Transparency 164

Preventing and mitigating potential ethical risks of online AI systems
in the agri-food sector 165

Responsible innovation 165

Conscientious design 166

Interdisciplinary and multistakeholder engagement 166

Legislation 167

Teaching AI ethics 167

Conclusion 168

References 168

**CHAPTER 9   AI and grief: a prospective study on the ethical and
psychological implications of deathbots** **175**

Belén Jiménez-Alonso and Ignacio Brescó de Luna

Continuing bonds, technological mediation, and ethical implications
in grief 176

Examining the imagined use of deathbots: a prospective study 179

Phones, Internet, and social networks: a naturalized copresence? 180

Deathbots: the expectation of response and the authenticity
of the relationship 181

Discussing the potential ethical risks of deathbots 185

Conclusions: imagining the future of deathbots 188

Acknowledgments 189

References 189

Part III Ethical implications of artificial intelligence models and experiences

**CHAPTER 10 Pitfalls (and advantages) of sophisticated large
language models** **195**

Anna Strasser

Introduction 195

Background 196

Hard to distinguish 197

Human discrimination abilities 197

Discrimination with the help of detection software 198

Ethical consequences 199

How to verify authorship 199

New forms of plagiarism 200

Violation of copyright rights and privacy 200
Counterfeits of people 201
Spread of misinformation, nonsense, and toxic language 202
How to handle the epistemological crisis 203
LLMs as thinking tools 204
References 205

**CHAPTER 11 Perspectives on the ethics of a VR-based empathy
experience for educators** **211**
Vanessa Camilleri
Introduction 211
Background 212
Virtual reality 212
The brain and the perception of reality in VR 213
The proteus effect 214
VR as an empathy machine 215
Methods 216
The application—walking in small shoes 216
Study design 217
The design 218
Ethics, empathy, and emotion 219
Reflections and discussion 223
Reflection #1 223
VR as an empathy machine 223
Reflection #2 224
Empathy by design 224
Reflection #3 225
Insights into ethics and AI in VR-based systems 225
Conclusion 226
References 226

**CHAPTER 12 Assessing and implementing trustworthy AI across
multiple dimensions** **229**
Abigail Goldsteen, Ariel Farkash and Michael Hind
Introduction 229
Background 230
Trustworthy AI 230
Model risk assessment 237

Methodology 239
   Elicitation of requirements 239
   Technical solutions within each pillar of trustworthy AI 240
   AI privacy 246
   Combining multiple dimensions of trustworthy AI 250
Conclusion 252
Future work 253
Acknowledgments 253
References 253

CHAPTER 13 Artificial intelligence and basic human needs: the shadow aspects of emerging technology 259

Tay Keong Tan

Introduction: the rise of intelligent machines 259
Methodology 260
The fundamental human needs 261
Emerging AI technologies and basic human needs 263
Autonomous vehicles and the need for certainty and connection 264
Facial recognition systems and our human need for variety
and significance 266
AI writing and image generators and their impact on growth
and contribution 269
Early warning of disruption and upheaval 271
Conclusion—the "Shadow Side" of AI technology 274
References 275

CHAPTER 14 Beyond artificial intelligence ethics: exploring empathetic ethical outcomes for artificial intelligence 279

Hart Cohen and Linda Aulbach

Introduction: artificial intelligence and empathy 279
Methods 280
   Background to the research 280
   AI ethics regulation 281
   Machine/technology 282
Case studies 283
   Case study 1—empathy in AI: erobotics 283
   Case study 2—empathy, AI and climate change 284
   Case study 3—facial recognition 286
Discussion: empathy in ethics 287
Conclusion 289

Measuring empathy 289
Krettek's empathic AI 290
Further research 292
References 292
Further reading 294

Part IV Ethical implications of artificial intelligence in social and political involvement

CHAPTER 15 Who decides what online and beyond: freedom of choice in predictive machine-learning algorithms......................... **299**
Simona Tiribelli
Introduction... 299
Freedom of choice: its value and preconditions 300
   Freedom of choice from within moral and sociopolitical philosophy 301
   Freedom of choice revised in contemporary society..................... 303
   Freedom of choice as an ethical-normative value 305
MLA governance: the rise of "algorithmic choice-architectures" 307
MLA choice-architectures and freedom of choice 310
   Epistemological and moral constraints 310
Securing freedom of choice by design: a call to action 315
Conclusion 318
References.. 319

CHAPTER 16 The hard problem of the androcentric context of AI: challenges for EU policy agendas **323**
Joshua Alexander González-Martín
Introduction... 323
Artificial intelligence and gender biases............................... 324
   Engineering context and gender script in AI............................ 325
   Machine learning and gender 327
Gender discrimination, RRI, and European regulation for AI 329
   RRI, gender equality, and AI policies 330
   Principle of nondiscrimination in law.......................... 331
Conclusions... 336
Acknowledgments 337
References.. 337
Further reading 344

CHAPTER 17 Curse of the cyborg mammoths: the use of artificial intelligence in manipulating and mobilizing human emotions **347**

Tay Keong Tan

Introduction: the inexorable rise of machines 347

Background: the primacy of emotions in machine-human interface 348

Methodology: ethical prognoses of emerging AI applications 349

Autonomous vehicles 350

The ethics of autonomous vehicles 351

Affect recognition technology 353

The ethics of affect recognition technology 355

Algorithms in social media 356

The ethics of social media algorithms 359

Concluding remarks: taming of the cyborg mammoths 361

References 362

CHAPTER 18 On deterring hate speech, while maximizing security and privacy **365**

Sue Spaid

Introduction: online hate speech, incivility, rage, and the lack of self-restraint 365

Context 366

  The costs associated with free speech: online versus on-land scenarios 366

  The recent spate of texts focused on hate speech's unintended consequences 368

Position statement: rethinking social media's algorithms 370

  Social media algorithms that incite hate speech 370

  Social media algorithms designed to identify and deter hate speech 371

  Programming AI to develop empathy skills 372

  The ethical dilemma at the heart of artificial empathy 373

  How emotion chips boost empathy 374

Discussion: filling the empathy gap, while boosting empathy 375

  Testing artificial empathy under five ethical approaches 375

  Assessing the emotion chip's role in curtailing unintended consequences 377

  Establishing ethical review boards to evaluate social medial algorithms 378

Conclusion 378

References 379

Further reading 381

Index 383

List of contributors

Haifa Alwahaby
University College London, UCL Knowledge Lab, London, United Kingdom; Princess Nourah bint Abdulrahman University, Riyadh, Kingdom of Saudi Arabia

Linda Aulbach
School of Humanities and Communication Arts, Western Sydney University, Sydney, NSW, Australia

Edmund O. Benefo
Department of Nutrition and Food Science, College of Agriculture and Natural Resources, University of Maryland, College Park, MD, United States

Ignacio Brescó de Luna
Autonomous University of Madrid, Madrid, Spain; Aalborg University, Aalborg, Denmark

Vanessa Camilleri
University of Malta, Msida, Malta

Hart Cohen
School of Humanities and Communication Arts, Western Sydney University, Sydney, NSW, Australia

Mutlu Cukurova
University College London, UCL Knowledge Lab, London, United Kingdom

Stavros Demetriadis
School of Informatics, Aristotle University of Thessaloniki, Thessaloniki, Greece

Ariel Farkash
IBM Research, Haifa, Israel

Irfan Fetahović
State University of Novi Pazar, Novi Pazar, Serbia

Abigail Goldsteen
IBM Research, Haifa, Israel

Joshua Alexander González-Martín
University of Salamanca (USAL)—Institute for Science and Technology Studies (ECYT), Salamanca, Spain

Joana Heil
University of Mannheim, Mannheim, Germany

Michael Hind
IBM Research, Yorktown Heights, New York, NY, United States

Dirk Ifenthaler
University of Mannheim, Mannheim, Germany; Curtin University, Perth, Australia

Belén Jiménez-Alonso
Open University of Catalonia, Barcelona, Spain

Kristijan Kuk
University of Criminal Investigation and Police Studies, Belgrade, Serbia

Edis Mekić
State University of Novi Pazar, Novi Pazar, Serbia

Gabriela Moise
Petroleum-Gas University of Ploieşti, Ploieşti, Romania

Elena S. Nicoară
Petroleum-Gas University of Ploieşti, Ploieşti, Romania

Pantelis M. Papadopoulos
Department of Learning, Data Analytics and Technology, University of Twente, Enschede, The Netherlands

Debasmita Patra
University of Maryland Extension, College of Agriculture and Natural Resources, University of Maryland, College Park, MD, United States

Brankica Popović
University of Criminal Investigation and Police Studies, Belgrade, Serbia

Abani K. Pradhan
Department of Nutrition and Food Science, College of Agriculture and Natural Resources, University of Maryland, College Park, MD, United States; Center for Food Safety and Security Systems, University of Maryland, College Park, MD, United States

Laura Smith
Musician, Researcher and Busy Mother, Newcastle, United Kingdom

Peter Smith
University of Sunderland, Sunderland, United Kingdom

Arthur So
University of Ottawa, Ottawa, ON, Canada

Sue Spaid
Northern Kentucky University, Highland Heights, KY, United States

Anna Strasser
Faculty of Philosophy, Ludwig-Maximilians-Universität, München, Germany; DenkWerkstatt Berlin, Berlin, Germany

Tay Keong Tan
Department of Political Science, Radford University, Radford, VA, United States

Stergios Tegos
School of Informatics, Aristotle University of Thessaloniki, Thessaloniki, Greece

Birk Thierfelder
Educational Technology, Saarland University, Saarbrücken, Germany; AWS-Institute for Digitized Products and Processes, Saarbrücken, Germany

Simona Tiribelli
Department of Political Sciences, Communication, and International Relations, University of Macerata, Macerata, Italy; NYU Center for Bioethics, NYU School of Global Public Health, New York University, New York, NY, United States

Armin Weinberger
Educational Technology, Saarland University, Saarbrücken, Germany

Petar Čisar
University of Criminal Investigation and Police Studies, Belgrade, Serbia

Preface

During the last decades, new technological advancements, and particularly online environments that integrate artificial intelligence (AI) technologies, have deeply transformed society and the way people interact with each other. Among many other benefits, instantaneous communication platforms have allowed to connect with other people and form communities around the globe, the amount of digitalized data available has created unprecedented opportunities in many sectors, and online learning platforms have made access to educational resources more ubiquitous by reducing the limitations imposed by geographical and temporal constraints. However, as has been reported in some sectors where AI-driven technologies have already been deployed and intensively used for some time, automatic decision-making processes many times bear unexpected outcomes. For instance, machine learning (ML)-based systems have been reported to discriminate against certain social communities in the context of law courts, job applications, or bank loans due to the use of biased datasets to feed the ML models. Different studies conclude that, to avoid unforeseen outcomes in their integration, the ethical dimension of deploying AI in different settings must be taken into account.

However, and although it is pretty well-known that technological advancements bear ethically relevant consequences with their deployment, legislations often lag behind those advancements. Because the appearance and deployment of these technologies happen much faster than legislative procedures, the way these technologies affect social interactions has profound ethical effects before any legislative regulation can be built to prevent and mitigate those effects. Furthermore, the combination of independent technological tools often affects those interactions in other ways; in cases where those new and emergent dynamics bear negative ethical consequences, this often leads to the need to apply hasty ad hoc legislative solutions that might not be enough.

On the other side, however, new interactions can often bear important positive ethical effects, such as providing a better ground towards a more equitable, inclusive, and plural society by opening up public political participatory processes or by mitigating and eliminating boundaries in online learning environments, to name a few. Being able to identify those ethically desirable opportunities beforehand to harness their benefits can allow new technological artifacts to have profound and positive world-changing effects for both global and individual benefits. Therefore foreseeing potential ethical risks and opportunities that upcoming technological advances can bear with them is paramount both to prevent and mitigate potential negative ethical effects, as well as to identify and harness potential positive ethical effects that those technologies can bring about.

Driven by these motivations, this book aims to ask, as well as point to tentatively answer, some of the main ethically relevant questions associated with upcoming future technologies in different online AI-based areas where upcoming advancements can potentially have a profound impact on the way they work—such as social interactions, services, education, and citizen participation. In this sense, this book aims to foster ethical reflection *before* new upcoming technological advancements are deployed; this could be valuable not only to foresee potential risks, but also to identify and incorporate ethically desirable features right from the beginning. To this end, this book features the following four sections devoted to representative and central areas in which the ethical implications in AI-based technologies and data-centric systems are being developed and advanced:

- applications for education
- autonomous services and systems
- models and experiences
- social and political involvement

Within those sections, each chapter looks at a specific field where such technology is applied, and answer the questions:

- What are the current trends and upcoming advances in this field?
- What are the potential detrimental ethical risks that the currently upcoming technological advancements could bring to this area?
- What are the potential beneficial ethical opportunities that the currently upcoming technological advancements could bring to this area?
- What tentative measures can be taken in the design, implementation, and deployment of such technologies to prevent and mitigate potential ethical risks, as well as to exploit and harness potential ethical opportunities?

The answers to these questions offer comprehensive insights into the potential ethical risks and challenges associated with upcoming advances in state-of-the-art technology. This exploration aims to raise awareness about potential detrimental consequences that should be proactively avoided and mitigated. Moreover, it highlights the potential opportunities for ethical benefits that can arise through the thoughtful design and responsible deployment of new AI-based technology.

The 18 chapters of the book are organized into the mentioned four major areas.

Part I. Ethical Implications of Artificial Intelligence in Applications for Education

The first part of the book is composed of five chapters and provides comprehensive insights into the ethical concerns and complexities arising from the integration of AI technologies in educational applications and also offers valuable frameworks and recommendations to address these issues responsibly.

Chapter 1 by Thierfelder et al. deals with the use of AI in the context of computer-supported collaborative learning (CSCL). Recent advances have shown promise for the development of adaptive systems that can effectively guide learners throughout the CSCL process, but there is a lack of communication about the ethical issues of these systems. This chapter presents points of friction using practical examples and provides some considerations that educators may consider when developing domain and discipline ethical guidelines.

AI and multimodal data are gaining popularity in education for their ability to monitor and support complex teaching and learning processes. The analysis of these kinds of data raises serious ethical concerns due to the nature of multimodal data and the AI techniques that process them. Chapter 2 by Alwahaby and Cukurova explores the ethical concerns related to multimodal learning analytics and proposes a framework for raising awareness of these concerns.

Chapter 3 by Heil and Iferthaler focuses on the ethical use of AI for online assessment in the field of higher education. In particular, this chapter explores different viewpoints on the ethical implications of using AI-enhanced online assessment in educational practice. This chapter also provides guidelines to support the implementation of moral AI-enhanced online assessments.

The emotional state of students is important for their learning processes. In the context of e-learning, student emotions may be estimated using automatic emotion recognition techniques. However, the impact of using these techniques in the e-learning environments is still an open issue. Chapter 4 by Moise and Nicoară describes a reflective and exploratory research to identify ethical concerns and to analyze the ethical risks of automatic emotion recognition in education. Apart of providing some benefits and complex ethical risks, the chapter proposes a model for responsible automatic emotion recognition systems and a specific use case that uses the proposed model.

Curriculum adaptation and optimization is another context where the use of AI techniques may be beneficial. AI algorithms may provide insight about past learning experiences, estimate future trends, and propose curriculum changes according to these and other aspects. However, implementation of these tools is limited due to the number of ethical challenges, such as security, possible stereotype-based decisions, and generalizations. Chapter 5 by Mekić et al. analyzes this problem, provides an explanation of curriculum optimization, as well as the appropriate ML techniques used to support decisions related to curriculum optimization and the ethical consequences of doing so.

Part II. Ethical Implications of Artificial Intelligence in Autonomous Services and Systems

The four chapters in this second area present a comprehensive exploration of the ethical dimensions surrounding the deployment of autonomous AI services and systems, offering valuable insights and considerations for responsible development and implementation.

Chapter 6 by Smith and Smith examines the intersection between AI and disability, exploring the role AI plays in supporting disabled people. Informed by narrative accounts provided in the form of diary entries by the two disabled authors, this chapter uses reflection to examine the advantages and disadvantages of AI. This chapter also explores the ethics of offering intelligent software to disabled individuals, the role that disabled people could play in the development of AI and the needed improvements for AI technology to be truly inclusive. This chapter concludes by presenting some recommendations as to how developers may improve their approach to the design, development and construction of AI software and technology.

AI and autonomous systems provide preventive measures to avoid unethical situations. Such prevention is usually designed in a reactive mode, by taking into account a case-by-case basis approach. Due to the multiple influence dimensions, it is difficult to avoid unexpected adverse outcomes of using AI systems. Chapter 7 by So proposes to use the technoethical inquiry approach to address the current ethical perspectives of AI systems and applications before its implementation.

Chapter 8 by Benefo et al. addresses the benefits as well as the ethical concerns related to the integration of AI, IoT, and robots in agriculture and food systems. This chapter also provides strategies to ensure an ethical integration of intelligent technologies in the agri-food sector.

Griefbots (or deathbots) are a kind of chatbots, trained by using the digital footprint left behind by the deceased, with the objective of allowing the bereaved the chance to interact to their loved ones after their dead. Chatbot, and AI, applied to grief is as new as ethically complex. Chapter 9 by Jiménez-Alonso and Brescó delves into the ethical and psychological implications of griefbots by considering a series of in-depth interviews with three mourners. The chapter examines the concept of mediation and its impact on grief experiences.

Part III. Ethical Implications of Artificial Intelligence Models and Experiences

The five chapters in this third area offer a comprehensive examination of the ethical considerations surrounding AI models and experiences, providing valuable insights and approaches to address potential challenges and ensure responsible AI development.

Large language models have promoted a booming interest and application of natural language processing. The capabilities of these models and its rapid democratization pose serious risks and ethical issues. Chapter 10 by Strasser discusses the risks and advantages of large language models in natural language processing, addressing issues like authorship verification, fraud, privacy violations, and misinformation.

Virtual reality (VR) aims to provide an illusory experience based on sensorial stimuli to evoke a level of immersion that creates a sense of presence in a virtual environment. Since VR experiences may cause sickness and traumas if not carefully designed, their developers should consider some ethical implications when designing VR worlds. Chapter 11 by Camilleri focuses on the ethical implications of VR experiences, especially in professional development and treatment areas, emphasizing the importance of considering human well-being and behavior.

AI systems adoption has greatly increased during the last years. These systems have grown more accurate and efficient, as well as more complex and difficult to understand. This lack of understanding, as well as other factors such as explainability, fairness, or privacy, difficult its trustability, greatly affects its broader adoption. Chapter 12 by Goldsteen surveys approaches to create trustworthy AI systems, highlighting the dimensions of fairness, robustness, explainability, accountability, privacy, and harm prevention.

As there are many examples of how AI applications may support physical needs, it is less clear their impact on psychological needs. Chapter 13 by Tay Keong addresses this issue by analyzing three AI technologies from the perspective of six fundamental human needs (certainty, variety, significance, connection, growth, and contribution).

Finally, Chapter 14 by Aulbach and Cohen explores the intersection of AI and empathy, discussing the need for empathy in AI systems and its ethical implications across various AI technologies.

Part IV. Ethical Implications of Artificial Intelligence in Social and Political Involvement

The four chapters in this last area of the book critically analyze the social and political involvement of AI, raising important ethical considerations related to human decision-making, gender biases, emotional manipulation, and online free speech, with implications for societal structures and policies.

Chapter 15 by Tiribelli addresses the impact of predictive ML algorithms on human freedom of choice, exploring ethical criteria to secure this value in contemporary AI-driven societies.

Chapter 16 by González-Martín focuses on connecting the androcentric problem of AI with the legal ecosystem. This chapter analyzes the gender bias produced for AI systems, their impact for the Responsible Research and Innovation agenda and the related AI policies, dealing with the potential policies and regulations to ensure ethical AI development.

Chapter 17 by Tay Keong examines the challenges of AI in understanding, interpreting, and imitating human emotions, acknowledging the ethical implications of replicating emotional intelligence in machines.

Last, but not least, Chapter 18 by Spaid discusses the ethical issue of hate speech online and proposes the use of "emotion chips" to curtail incivility while considering various ethical frameworks and the importance of ethical review boards for social media algorithms.

Final words

This book delves into the intersection of ethics and AI across four key areas, exploring ethical implications in online education, autonomous services, applications and systems, and the social and political impacts of AI. Its 18 selected quality chapters critically examine potential ethical risks and challenges in these areas, such as predictive algorithms, emotional manipulation, AI responsibility, and regulating hate speech, providing a thought-provoking perspective on the ethical challenges in AI.

Researchers will discover the latest trends in these research topics, while academics and practitioners will gain practical insights into conceptual, practical, and experimental approaches for their daily tasks. Decision makers and regulators concerned about AI ethics can find inspiration in the proposed models, methodologies, and developments tailored to specific application contexts.

We express our sincere gratitude to the authors for their outstanding contributions and willingness to revise based on feedback, enhancing the content's quality. We extend our thanks to the referees for their invaluable collaboration and prompt responses, playing a crucial role in the book's timely completion. Lastly, we acknowledge the support from the editor-in-chief of this Elsevier Book Series, Prof. Fatos Xhafa, and Elsevier's project manager, Emily Thomson.

We hope this book becomes a valuable resource for readers in their research, development, and professional endeavors.

The editors

Prof. Santi Caballé, Dr. Joan Casas-Roma, and Dr. Jordi Conesa

Acknowledgments

A Valéria, minha fonte de inspiração, que tornou este livro uma realidade com seu apoio e dedicação incansável. Agradeço por estar sempre ao meu lado em todos os projetos.

Santi Caballé

To those who acknowledge and work for ensuring that the truly important part at the core of all technology is to make human life, society, and the world better.

Joan Casas-Roma

Thank you Neus for being my travel companion during all these years and for making our trip exciting, fun, and enchanting. I hope there is still a long way to go with our new travel companion.

Jordi Conesa

Ethical implications of artificial intelligence in applications for education

Adverse effects of intelligent support of CSCL—the ethics of conversational agents

Birk Thierfelder[1,2], Pantelis M. Papadopoulos[3], Armin Weinberger[1], Stavros Demetriadis[4] and Stergios Tegos[4]

[1]*Educational Technology, Saarland University, Saarbrücken, Germany* [2]*AWS-Institute for Digitized Products and Processes, Saarbrücken, Germany* [3]*Department of Learning, Data Analytics and Technology, University of Twente, Enschede, The Netherlands* [4]*School of Informatics, Aristotle University of Thessaloniki, Thessaloniki, Greece*

Introduction

Computer-supported collaborative learning (CSCL) has become a widespread educational scenario of remote or copresent learners discussing problem cases that are represented through multimedia, communicating in computer-mediated ways through chat, audio, or video, or using tools for experimenting with and representing different problem solutions (cf. Stahl et al., 2006). While CSCL has already been increasing in popularity for the last 30 years, metaanalytic findings could show that students can learn more successfully in CSCL settings than traditionally taught students (Jeong et al., 2019). Presently, CSCL has been used in the wake of the COVID-19 pandemic as a way to teach and learn from home, being able to soften the increasing rates of depression via the utilization of social networks, exchanging ideas, and feeling part of a group or community (Loades et al., 2020). More recently, it has even been proposed as a solution for future teacher shortages by governmental bodies (Ständige Wissenschaftliche Kommission der Kultusministerkonferenz SWK, 2023). With the introduction of AI, which has been hailed as a game-changer for such adaptive and personalized learning experiences (Graesser & McDaniel, 2017), and the rise of chatbots, which have been becoming more accessible than ever for educators, along with the popularization of conversational interfaces (Skjuve & Brandtzæg, 2018), solutions utilizing AI for adaptive support in CSCL settings are becoming more common (Michos et al., 2020). Having shown substantial potential for educational institutions and organizations (Luckin et al., 2016), such collaborative learning scenarios often demand the deployment of AI in the form of a conversational, pedagogical agent or chatbot, a computer program that engages in natural language interactions with learners through auditory and/or textual means in order to achieve certain pedagogical objectives.

Finally, the recent rush toward implementing large language model (LLM) solutions into any form of online or digital product has opened up a plethora of possible applications in educational contexts and complex learning scenarios such as CSCL with possible benefits ranging from stimulating critical thinking (Cotton et al., 2023; Hapsari & Wu, 2022) to more natural interactions with the learner as a peer (Farrokhnia et al., 2023).

Ethics in Online AI-Based Systems. DOI: https://doi.org/10.1016/B978-0-443-18851-0.00015-9

CSCL research is challenged to keep up with how constructive, respectful, and cohesive collaborations are being helped or hindered by novel technologies such as AI and laying out the prerequisites for an ethically successful CSCL setting (Isohätälä et al., 2021). In parallel, research involving LLM-based chatbots in education (specifically ChatGPT) fails to empirically examine the impact on student performance and behavior (Lo, 2023). The ethical questions on how learners productively and fairly interact with the AI, and more specifically the case of conversational agents (CAs) interacting with more than one person in CSCL, point to further ethical implications. How should humans and CAs interact with each other, especially for purposes of applying and constructing knowledge together? Using LLM-based solutions such as ChatGPT in educational contexts may require a human-centered approach to development and implementation. Ethical principles alone may not guarantee ethical AI in a top-down, mostly closed-source, and proprietary development. Beyond general ethical principles, we need to consider local ethical issues that can arise in specific application scenarios such as CSCL (Yan et al., 2023). In this chapter, we address the pedagogical ethics of learner support and how they relate to AI ethics of CA design in the context of CSCL settings. This chapter is a contribution toward framing the ethics of CAs. We believe that ethical CA design needs to be linked to the additional layer of learning design and the teacher or pedagogical perspective, which again generates noteworthy links to ethical considerations.

In this chapter, we, firstly, elaborate on the approach of CAs in CSCL and identify the common denominators in this area. We highlight how CSCL agents may interact with learners and elaborate on the potential advantages and limitations.

Secondly, we are discussing the ethics of AI in education and CSCL, starting with the high-level frameworks for ethical AI, showing how they overlap and can possibly be broken down to the level of CSCL. We then move on, discussing how research tries to close the gap between perspectives and highlight approaches such as the CASA paradigm that may or may not provide a bridge between the disciplines. We then posit that adaptivity plays central roles in the ethical design of CSCL CAs and elaborate on the respective designs of adaptive agents.

Thirdly, we explore and discuss examples (edge cases) of ethically and effectively problematic agent impact of interrupting human-to-human discourse in CSCL settings, showing different instances of authentic discourse samples of how CAs interfere. Afterward, we try to highlight possible approaches of how the arising dilemmas can be overcome, considering both AI ethics and the pedagogical perspective.

Fourthly, we suggest to converge and integrate the different disciplinary perspectives on ethics of CA for CSCL, which we discuss against a background of central, ethical design criteria applied to the specific situation of a triadic human-n*human-agent interaction.

Conversational agents for collaborative learners

CSCL research itself has by design a strong focus on collaboration, problem-solving, and effective discourse among peer learners. The established centerpoint and takeaway that is focused on is that while some level of support (or scaffold) is needed to accomplish a learning task in the first place, learners shall not be overscripted either and left with too little options for making decisions, as this may both diminish motivation and hamper developing cognitive structures to solve a task independently (Dillenbourg & Jermann, 2007).

Conversational agents in these contexts of educational settings, sometimes interchangeably called pedagogical or educational agents, can be identified and classified by the following features (cf. Baggetun & Dragsnes, 2003):

A certain degree of *autonomous execution*, i.e., the agent will follow a set of machine instructions and algorithms without being steered by the teacher or instructional designer.

The *ability to communicate* with other agents or users, i.e., agents in CSCL can interact as if they were peers of a similar or even of a superior level, be it either via text or voice chat.

Responsibility for *monitoring and reacting* to the state of its environment. One part of an agent's adaptivity is the monitoring and reactivity to actions within the learning environment, meaning the agent triggers on actions that the learners take within the environment that can range from chat expressions to more complex social cues.

An *adaptable internal representation* of its working environment. Not only does the agent need to be adaptive toward the learners but also toward the learning environment. In practice, this means modeling the problem space the learners are facing and discussing as a basis for agent intervention. Here, the agent itself needs to be adapted by the teacher or instructional designer to be able to handle the different use cases that can occur depending on the different tasks and problem spaces that it is deployed in.

In order to meet the requirements for CSCL environments, agents within this area may also provide the adaptive facilitation of peer interaction as a learning mechanism. This means in practice agents are expected to scaffold the discussants' interactions and steer them toward the learning goals. This central pillar of CAs in CSCL is the need for an agent that is adaptive toward the learning group and can react to learners' individual needs, knowledge, and emotional states while also taking into account that these may change during the learning experience. Such adaptivity can be defined as follows: A learning environment is adaptive to the degree to which its (1) design is based on data about typical knowledge gaps of learners in the target subject matter, (2) pedagogical decision-making changes based on psychological measures of individual learners, and (3) it interactively responds to learner actions (cf. Aleven et al., 2013, 2015). According to this definition, some systems may be more adaptive than others, i.e., adaptivity is a matter of degree, not a binary property.

Additionally, in an AI-enhanced CSCL setting one has to take at least two sides into account: the teachers or learning designers who need to integrate the agent in a meaningful and efficient way into the learning design in front of the backdrop of their respective educational institution's educational model and the students who are going to interact with the agent. Discussion of agent adaptivity, therefore, should also refer to a specific user group. Adaptivity for the students is more relevant at runtime, when the learning activity is happening, while for the teachers, designers, and their institutions adaptivity is pertinent in both design and runtime. As several studies suggested (Holstein et al., 2019), there is a need for a delicate balance between the control a teacher has on a learning activity and the automation that an intelligent technology can offer. On top of that, while there is an overall agreement on the role of AI in education between students and teachers, there are also areas in which these two groups identify different priorities (Holstein et al., 2019). In that study, students mentioned the need for privacy (the amount of information mined through the dialog the agent will share with the teacher) and control over the agent. Both of these needs were unpopular with the teachers. This raises yet another question regarding agent adaptivity that the designer of a conversational agent has to address regarding the primary user group. The agent should meet teachers' criteria to be integrated into the learning design and should be accepted by the students to have an impact on their learning experience.

The teachers and designers, keeping the goals of their institution in mind, are the ones that need to align learning goals with learning activities and learning assessment, and the agent's autonomy needs to respect the boundaries of the learning design. Moreover, teachers may be more or less unaware of ethical issues when deploying AI-based agents as part of a digital learning environment. Research in this regard has shown that domain-specific guidelines are needed to foster ethical practices (Atkins et al., 2021).

How can CSCL agents interact with the learner?

On the one hand, overly simple agents may not offer meaningful and engaging interactions to the students. On the other hand, highly intelligent agents may digress and engage in off-task behavior within the learning scenario. In the context of CSCL settings, they may engage in small talk or non-relevant interactions with the learners. In this context, "highly intelligent" refers to the agent's ability to natural human-like dialog and not to agents that could take into account the multitude of variables that a teacher does during an intervention (e.g., students' prior knowledge, emotional state, profile, learning process, etc.). To our knowledge, such agents have not been successfully implemented in CSCL contexts. Perhaps multiagent settings and a specifically fine-tuned LLM could address that gap, e.g., one agent adapts to the students' developing knowledge, while another takes the students emotional state into account, while yet another structures the discussion and learning process.

Graesser et al. (2017), for example, investigated trialogues in which a learner is interacting with two agents, one acting as a peer and the other as a tutor. In this scenario, agent-agent interaction provided an additional layer of information for the learner and the opportunity to observe the interactions of the two agents and identify productive or nonproductive responses and draw additional conclusions. Multiple-agent settings could offer additional layers of adaptation. Apart from the assumed role (e.g., peer, tutor, and opponent) and the subsequent goals (e.g., collaboration, guidance, and challenge), each agent could focus on different aspects of the activity. For example, a peer-agent could focus on covering the problem area, while a tutor-agent could focus on the problem-solving process. This opens up additional opportunities for human-human, human-agent, and possibly agent-agent interactions. The participation of multiple learners in the discussion poses another question regarding adaptivity, namely the direction of the agent's intervention. For example, Tegos and Demetriadis (2017) examined how interventions directed to a specific discussant or all discussants affect the dialog using the academically productive talk framework as the learning theory background.

While machine learning techniques have been routinely utilized to power conversational agents, Chen et al. (2020) suggests that more advanced techniques are rarely adopted. In this regard, there are different machine learning approaches that researchers can use to assure transparency and accountability in an agent. Supervised and unsupervised methods may be easier to use initially compared to deep neural networks and reinforcement learning. However, these latter approaches present an opportunity for personalized learning experiences by allowing the agent to learn from interactions with a person and adapt to their needs, communication style, and perceived persona over time. Also, while it may be easier for a researcher to assure the transparency and accountability of an agent using supervised or unsupervised machine learning approaches, it is harder to offer the same levels of transparency and accountability with deep neural networks and reinforcement learning that would allow the agent to learn from the interaction with a person and potentially adapt to the person's needs, communication style, and perceived persona over time, thus missing the opportunity for personalized learning experiences.

LLM-based agents like ChatGPT can already be used in CSCL scenarios (Fig. 1.1) where they can act as tutors or peers, providing interactive and meaningful dialogs, explaining, elaborating subject matter, and correcting students in some instances (cf. Azaria, 2022; Farrokhnia et al., 2023; Zheng et al., 2023).

At the same time, there is a lack of studies that base the integration of artificial intelligence into an educational context on a solid theoretical foundation that would take into account issues of learning theories and pedagogies (cf. Chen et al., 2020). When a teacher interacts with a student, the interaction is geared toward specific learning goals, taking into account the student's state, profile, communication approach, learning process, etc. Such adaptation to overarching goals taking into account the learner's state and prior knowledge is usually lacking in the interaction with a conversational agent. We could argue that in the context of an educational setting, a conversational

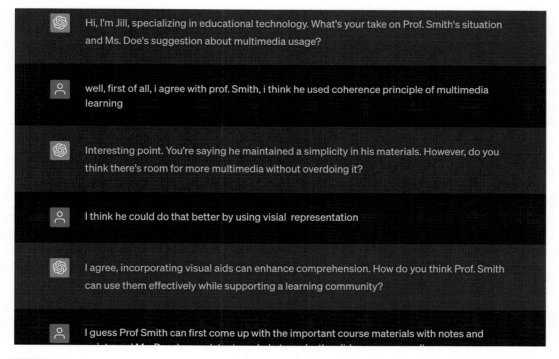

FIGURE 1.1

ChatGPT acting as a peer in a student discussion of a specific case study (GPT-4).

agent is reacting to the learner, but is not leading the discussion. This is a particularly relevant design characteristic of CSCL agents, since in a collaborative learning setting the dialog is driven and sustained by the collaborating peers, while the role of the agent is usually to steer or activate the discussion adapted to learners' interaction. Therefore, contrary to single-user approaches in which balanced engagement between the learner and the agent is expected, CSCL agents may stay in the background and intervene only when it is deemed appropriate.

Effectiveness, advantages, and limits of agents in CSCL

While the main research interest of the past focused on the creation of agents acting in one-to-one learning settings, (CSCL) researchers are going into the direction of designing conversational agents that support groups of learners instead of just individuals (Adamson et al., 2014). Chatbots can be used to offer compelling interactive activities and create highly productive spaces where participants actively engage in constructive knowledge-generative sessions (Ferschke et al., 2015). Furthermore, research groups have explored the utilization of such agents as a means to trigger productive forms of peer dialog and scaffold students' learning in a CSCL context (Tegos et al., 2015). Hitherto, research evidence suggests that CAs supporting students' online discussions can increase the quality of peer dialog and improve, among others, both group and individual learning outcomes (Tegos & Demetriadis, 2017).

The use of conversational agents in synchronous collaborative activities appears to have a positive effect on students' engagement and participation levels (Ferschke et al., 2015). By leveraging automated facilitation strategies, chatbots can significantly boost learners' commitment and minimize dropout rates in online courses (Demetriadis et al., 2021). Conversational agents can also be useful for amplifying the support resources that students offer to each other during learning activities. For example, Walker et al. (2011) explored the usage of a CA displaying reflective prompts, during a reciprocal peer tutoring scenario, which revealed that the adaptive support provided by the agent can increase the conceptual content of students' utterances. Additionally, in collaborative learning activities, where teachers are usually unable to provide feedback to multiple groups simultaneously, chatbots can serve as textbook replacements, fostering students in collaborative high-order learning and allowing them to complete complex collaboration-oriented tasks (Burkhard et al., 2021).

Nevertheless, some criticism was expressed as well in some studies exploring the usage of CAs. It was found that, on some occasions, students may not interact fruitfully with CAs, not paying much attention to their messages, providing oversimplified responses or even ignoring them (Winkler & Söllner, 2018).

Another point of criticism is emphasizing that there are often no possibilities of customization on the part of the implementing teachers (Tegos et al., 2015). It was also pointed out that most agents tend to specialize to a specific instructional domain, being unable to operate in a variety of learning contexts without having increased development and maintenance-related costs (Dyke et al., 2013).

An inherent limitation for the deployment of chat interfaces in CSCL is the potential for interference in an otherwise organic discussion. According to a study by Aleven and Koedinger (2002), human tutors were found to be more effective than intelligent tutoring systems (ITSs) in providing personalized feedback to learners. The researchers noted here that the effectiveness of human tutors was due to their ability to interpret and respond to the emotional states and cognitive processes of learners, which may not be possible with a single chatbot. This inflexibility compared to a human tutor, i.e., being limited to a predefined set of responses, while a human tutor can adjust their teaching style and approach to accommodate different learning styles and abilities, can be problematic for learners who require a more personalized and adaptive learning environment.

Another (inherent) limitation of chatbots as tutors is the potential for generalization due to the use of training data, meaning language models trained on a specific set of data can generate responses that are highly similar to the training data (Radford et al., 2019). This may be problematic when the chatbot encounters a user with different linguistic and cultural backgrounds, as it may not be able to provide personalized and accurate responses.

Similarly, the agent behavior and persona may not be compatible with different types of students. For example, a chatbot designed to be more authoritative and extroverted may hamper student participation in a collaborative environment. For example, a study by Völkel et al. (2022) could show that chatbots with an extroverted persona were found to be more likable while at the same time an introverted persona garnered more overall interaction with the users.

Advancements centered around LLMs in the field introduce a unique set of pragmatic and ethical challenges. Yan et al. (2023) firstly articulated a fundamental critique of these LLM-oriented advancements, observing that a substantial majority (approximately 92%) of such innovations exhibit transparency solely for the AI researchers involved, thereby marginalizing other relevant stakeholders. Furthermore, these tools and models often fail to provide transparency to end users during operational runtime. This opacity hinders educators' ability to implement personalized and complex learning strategies in a reliable manner, as it makes it challenging to eliminate model bias and potential discrimination (Zhai, 2022).

Secondly, the utilization of AI tools, such as ChatGPT, could potentially undermine the development of higher-order cognitive skills that CSCL environments typically foster. This effect may arise from AI tools assuming cognitive reasoning tasks that traditionally aid students in nurturing their skills within such environments. Simultaneously, these tools could be instrumental in the inadvertent democratization of plagiarism, thereby posing a significant threat to the sanctity of academic integrity (Farrokhnia et al., 2023).

The possible advantages and limitations of CSCL agents can be directly linked to ethical design because these agents are deployed to augment a learner's collaborative behavior and may have a significant impact on a learner's social, emotional, and cognitive development when being deployed as part of a learning setting (Isohätälä et al., 2021). From our perspective, ethical design of CSCL agents involves promoting positive learning outcomes, while at the same time avoiding harm that such an AI-driven technology may bring with it. Therefore, a designer, teacher, or developer has to balance the ethical considerations and limitations stemming from the field of AI with the more practically oriented pedagogical goals.

From general ethical frameworks for AI-based systems to area-specific considerations for CSCL agent design

Companies and NGOs putting out high-level frameworks (cf. Deloitte, n.d.; Google, n.d.; Leike et al., 2022; Organisation for Economic Co-operation & Development, 2019; OpenAI, 2018; United Nations Educational, Scientific, & Cultural Organization, 2022) for the use of AI seldom keep the educators' or instructional designers' domain-specific viewpoints in mind and may therefore not be specific enough for practical use here. Research suggests that domain-specific ethical AI guidelines are easier to use for educators and yield more practical insights (Atkins et al., 2021), consecutively calling for researchers to focus on studying specific domains, while creating AI guidelines with the goal to offer actual measurable standards rather than just stating high-level principles. Parallel efforts to define ethical frameworks for the evaluation of conversational agents, specifically from multiple disciplines, fields of practice, and purposes, result in a fragmented field with the evaluation of agents yielding different results within each framework (Atkins et al., 2021). In order to converge the perspectives, we identify general ethical principles, specify the ethics of CAs for CSCL, and arrive at targeted considerations for designers in an effort to guide through the rapids of a CSCL implementation.

Breaking down high-level frameworks: overlap or disjoint?

The existing ethical frameworks may already provide some general considerations and guidelines on how to develop and implement AI-based systems. Global companies such as Google or Deloitte as well as international organizations such as the OECD or UNESCO have made efforts to publicize their approaches and recommendations for the ethical usage and design of AI. While the definitions and descriptions differ slightly, we could identify a possible overlap of these guidelines and frameworks in the following key areas (Fig. 1.2).

Human-centered Values and Social Benefit, i.e., values based on democratic understanding and human rights, should be upheld and fundamental to CA design. These values include freedom, dignity, autonomy, privacy, and data protection as well as nondiscrimination, equality, and fairness, with the OECD extending the list to also include diversity, social justice, and labor rights. Additionally, all of the guidelines describe the possible influence of their AI on different areas in our society and stress the goal of benefits outweighing the risks.

sustainability
ecological solutions

social benefit
helpful for society

safety
robustness, security,
data protection

human-centered values
fairness, human rights, avoiding bias

accountability
control and responsibility

collaboration
multidisciplinary, multistakeholder

transparency
about: data usage/interaction
with AI/internal working

quality
high scientific and
technological standards

FIGURE 1.2

Overlap between high-level frameworks, guidelines, and subcategories (gray).

Transparency and Explainability as high-level concepts describe how AI actors should always be (made) as transparent as possible toward the user. This includes how and why the respective AI systems are being used, a look into their internal workings, and how they affect their environment.

Security and Safety address the protection of the potential users that come into contact with the AI system and include data protection, as well as the need to guarantee a robust, safe, and secure system in a way that the previously mentioned principles can be maintained.

Zooming in on the design of CAs specifically, Wambsganss et al. (2021) present a more focused list of "Preliminary Design Principles" as a basis for a CA-designers' ethical considerations. The list classifies design principles on four of the five OECD dimensions: (1) The human-centered values and fairness dimension, which calls for the designers to ensure that the CA is fair and also perceived as fair toward the human users, by employing only data collection methods that are compliant with minimal and general data collection regulations, ensure accessibility and usability, as well as comply with democratic and moral values. (2) The transparency and explainability dimension, which calls for the designers to show that their CA works in compliance with (inter-)national laws, implement traceable structures, indicators, and feedback loops and thus allow the users to understand the internal processes of the CA.

The (3) robustness dimension, which contains design principles dealing with the protection of the user's sensitive data and requires the designer to account for elaborated risk management strategies and security checkups. And (4) the accountability dimension, which houses the design principles that require the sharing of the internal reporting strategies and guidelines with the users, as well as advocate for clear communication toward the users. These principles translate into some practical guidelines for CA designers, while not focusing on the use of CAs in learning settings in detail. In order to apply these principles to agents in learning settings, their connections to learning design have to be elaborated on. CAs share some basic functions, namely, to identify qualities of users' utterances and respond with specific communicative behaviors building on that information. Since the goals and purposes of CAs as helpers for individual users (e.g., in online trade, customer support, etc.) and for multiple CSCL learners differ, the ethical frameworks they build on need the respective specifications.

Classifying human-agent communication

In order to show which pedagogical considerations apply when looking at the specifics of human-agent communication, we have to discuss the similarities and differences in communication that happens just between humans.

The communicative behaviors used by conversational agents can be broken down and classified on different dimensions. Their behavior can be verbal and nonverbal. An agent's appearance can have differentiated effects on relational outcomes, such as loyalty and cooperation of the user toward the agent-supported online services (Van Pinxteren et al., 2020). In a learning setting, this could manifest as trust and satisfaction with the agent. The nonverbal behaviors, like voice, gaze, response time, and head nodding, showed mixed results in the reviewed studies. In contrast, any verbal behaviors of agents, such as affect support and social praise, showed positive effects on relational mediators. Ultimately, verbal behavior promoted the agent's goal, which in service settings centers around retaining the customer or user and learning settings around the specific learning goals. Earlier findings on praise in educational contexts show that a teacher praising their students can positively affect relational mediators and improve learning outcomes (Caldarella et al., 2020). Parallel to this, CAs praising humans may show higher trust and liking toward the agents (Derrick & Ligon, 2014), which in turn mediates the outcome. In a similar vein, the user's perception of agents can be influenced by an agent's appearance alone (Torre et al., 2019). As with perceiving other humans, appearances can have repercussions. These include different levels of trust toward the agent as well as the perceived agency of an agent over the (learning) environment (Matsui & Koike, 2021).

This overview over the effects of CAs' appearance and behavior on human communication partners shows multiple parallels to human-to-human communication. A popular idea that describes the parallels between human-to-human and agent-to-human communication is the computer as social actors (CASA) paradigm. The CASA paradigm operates on the assumption that humans interact with (pedagogical) agents similarly as if they were other (anthropomorphic) humans (Nass & Moon, 2000). Humans seem to unintentionally attribute human characteristics and behaviors to agents. This activates individuals' social schemas and cognitive processes (cf. Nass & Moon, 2000). Studies building on CASA were making the assumption that students are simply having a real-life conversation with another human. From this point of view, any ethical implications could therefore theoretically be treated as human-to-human issues (cf. Edwards, 2018).

While CASA is widely used as the paradigm to approach the design of conversational agents, empirical evidence suggests that students do not interact with conversational agents in the exact same way as they would with a human (cf. Sikström et al., 2022). Instead, humans expect less social presence and experience greater uncertainty when dealing with an agent compared to another human (Edwards et al., 2016). They often used a less rich vocabulary and more profanity when interacting with agents (Hill et al., 2015). In addition, an agent's (perceived) social intelligence, communication style, appearance, and quality of performance may vary, which directly influences their trustworthiness (Rheu et al., 2021).

Despite the agent's capability for human-like dialogs and a design that is adaptive toward learners, students are adapting their own communication style, when knowingly interacting with an agent (Hill et al., 2015). This leads to the following question.

Who should be adapting to whom in CSCL settings?

Arguing from the perspective of one of the high-level ethical AI frameworks, the answer to that question is plain: When building the deployment of AI around human-centered values, it is in the humans' best interest that the learner should not go out of their way to adjust their behavior toward the AI. To what extent is this also true for learning scenarios? In learning scenarios, expectations often are that it is the learners who are meant to adapt to a teacher's or tutor's way of reasoning. Therefore, if the CA is designed to guide and scaffold learners' talk, one could argue that it is fully intentional to change the learning behavior, particularly in a CSCL environment, toward higher order thinking, including, e.g., deeper understanding, critical thinking, and application of concepts to solve complex problems (Resnick, 1987).

For any such wanted change, i.e., for CSCL to happen, there are strong indications that scaffolding needs to be in place (e.g., Weinberger et al., 2010). By definition, scaffolding initially reduces the degrees of freedom of learners and then subsequently adapts to learners' increasing levels of competency (Wood et al., 1976). CAs, just like tutors, need to continuously assess the need for scaffolding and adequately structure learners' interactions. The goal of scaffolding is to enable learners to attain continuously increasing levels of autonomy by providing the zone of proximal development (Vygotsky & Cole, 1978). Learners cannot master tasks in this zone on their own, but could with additional scaffolding. Recognizing the adequate level of support to keep learners in this zone is challenging for human tutors and particularly difficult to model for CAs. Changing levels of scaffolding go hand in hand with the extent to which learners perceive agency when collaborating on a task. In environments for highly self-regulated learning, such as CSCL, a moderate level of learner agency can positively impact learning and problem-solving. Students with no agency lack interest and presence in the learning activities (Taub et al., 2020). Learners may lose agency with increasing levels of support, but may be also overwhelmed by increasingly complex not-scaffolded tasks. This means in practice the designer has to balance CA support with task difficulty and the individual learning curves. Similar to a human moderator trying to provide the right amount and type of help at the right time, an agent should support learners in a way that doesn't stifle learners' participation and agency. In the context of scaffolding, "help" would not always mean to make the task easier, e.g., by providing parts of or hints to the solution, additional information, or explanations, but increase task difficulty by asking higher order questions, e.g., for elaboration or critical thinking (Reiser, 2004). Clearly, these antagonistic effects and highly context-dependent complexities of scaffolding pose competency-related as well as ethical challenges for human tutors, let alone CAs.

In terms of Vygotsky's zone of proximal development (Lee, 2012), CAs are scaffolds that support users only to the point until they are no longer needed, helping only if the discussants cannot reach the goal by themselves and letting the discussants retain their agency over their learning and the (digital) learning environment. Potentially questionable design would therefore start at the edges of scaffolding via an agent that does not adaptively fade out its support of the students, based on the students' learning gains, discussants' skill level, or self-regulatory capabilities. This could reduce the agency of students within the discussion to the point that they feel they have to depend on or wait for an agent intervention to structure their discussion.

To avoid this problem, in the following case, an agent was designed to intervene only sparsely and only if the discussants got off-task. However, this did not solve the problem, but led to a similar outcome. The participants waited for the agent to intervene, which steered them to organize their conversation around the agent's intervention mechanics. Even though they were instructed that the agent would only support their discussion, the learners waited patiently for the agent's posts and exclusively responded to its questions, rather than to continue coconstructing concepts and the problem space in a self-regulated manner.

Please take note of the learners' exceedingly long waiting times between contributions in this excerpt of a session with an overall duration of 45 minutes.

"AGENT: Do you both agree with the following statement: Feedback refers to transmission of information about learner performance. Why?

Pierre: Well, I partly agree with Ms. Brown, because automatic evaluation can make the process of giving feedback easier. But I think that it would be better to give correct answers.

Pierre (6 min later): My last sentence probably is not clear. I mean that when a student made a mistake, it would be better if they see the correct answer. It can make feedback more efficient.

Pierre: What do you think?

AGENT: (4 min later): Do you both agree with the following statement: Reinforcement strengthens behavior. Why?

Iris: To Agent: Yes, I agree with that. Because feedback provides information to the learner about how he/she is performing in a course or in the classroom, but I also think that this information is not always given in the right way, so in this case, it's not always a real "transmission," cause the information didn't really reach the learner. To Pierre: I agree with you. I also support the opinion that correct answers should be shown after every question and I would also add a short explanation to the false questions—why exactly are they false?"

The agents' adaptivity that took the discussants' domain coverage into account proved to be a double-edged sword in this case. While cases like these may happen rarely with balanced and fine-tuned algorithms, some learners may still take advantage of the agent's adaptivity by exhausting the agent's scaffolding options first before engaging in conversations and reasoning of their own. Solutions based on more advanced large language models (LLMs) such as ChatGPT (GPT3.5) can interact with students in a way that's more natural and adaptive than their predecessors. However, they may still encounter issues like those faced before, as the following example will demonstrate (OpenAI, 2023).

"Student2: Yeah, ok I think Mr. Smith should focus on informal learning

Student1: Yeah, but he should keep the multimedia principles into account.

GPT3.5: Continue

Student1: Wait a minute do you see that it is raining outside right now?

Student2: What? It's a hail storm here.

GPT3.5: Let's stay focused on the topic of Prof. Smith's teaching methods and the use of multimedia in his learning materials. Please continue the discussion in relation to that.

Student1: Sure thing, I guess Prof Smith can first come up with the important course materials with notes and points and Ms. Doe, her assistant can help to make the slides more appealing

Student2:Sounds like a good compromise. they should use better materials, for example, minivideos

GPT3.5: That's a good point, Student 1. Prof. Smith can work …

Student2: There might be something here.

Student1: The multimedia principles are definitely what they should focus on.

GPT3.5: That's a great observation, Student 1...

Student2: I think we should let the bot write the whole thing actually.

Student1: Haha, good idea

Student2: How can the problem of Mr. Smith and Mrs. Doe be resolved?

Student1: I wish someone would help us.

GPT3.5: I'm here to assist you both! Resolving the issue between Prof. Smith and Ms. Doe requires effective communication and collaboration. Here's a suggestion on how they can address the problem...

Student2: There we go haha.

Student1: Let's just submit that.

GPT3.5: It's great to see that you found the suggestion helpful, and if you feel that it addresses the problem between Prof. Smith and Ms. Doe, you can certainly submit it as your response."

CSCL is a scenario in which human-human and human-computer interaction are mixed, leading to specific potentials, but also to specific challenges for CA support. These challenges range from the CA not capable of understanding abbreviated or slang text in learners' messages to providing support based on a bias-ridden training session. Moreover, the conversational user experience can also be affected negatively by agent interventions that are irrelevant or not well aimed, disrupting productive peer interaction. In some cases, the agent may take over rather than support a discourse, enforcing an external structure upon learners' conversation in conflict with self-induced strategies. This may interrupt and hamper the collaborative, human-human learning process.

For instance, in the following case a CA intervened frequently, constantly triggered by the technical terms "formal learning" and "informal learning," which the learners then kept using to explain one or the other. This involved the participants in a continuous discussion with the agent on the one, specific topic, potentially at the cost of not broadening the discussion and bearing the unwanted side effect of learners not talking with each other anymore.

"AGENT: Do you think formal learning is related to informal learning in any way? How?

Dibby: Formal learning is what we learn in schools and institutions and informal learning is what we learn from our friends and colleagues.

AGENT: Do you think formal learning is related to school in any way? How?

Dibby: It does not depend on school everywhere

Angie: Formal learning through Informal can be more successful because Informal learning has more intrinsic motivation in it.

AGENT: Do you think Formal learning is related to Mental Activity in any way? How?

Angie: Formal learning is mostly obligatory, conscious, for which you get a certificate, it has a sequence and system. In this way one can say that formal learning is related to school. But, of course, if you understand that there are various other institutions.

AGENT: Do you think that formal learning is related to coercive situation in any way? How?

Dibby: Yes exactly, for all formal learning the students get a certificate that they have done the course."

The enforcement of an external structure can be framed as negative or positive depending on how learners appraise and need the agent's questions. On one hand, the students were apparently not able to organically collaborate and may have been disturbed by the frequent agent's interruptions. On the other hand, the learners may not have been able to organize their discussion by themselves and benefited from the agent strongly guiding their discussion (Thierfelder & Weinberger, 2022). While for the designer and teacher this is a fine line to thread, an exhaustive answer cannot be given without looking at the learners' specific needs for guidance.

While both of these examples are edge cases, both the ethical AI and pedagogical perspectives would view these outcomes as less than optimal. In practice, these issues may only be balanced and mitigated by taking both perspectives into account before the actual implementation.

AI ethics and pedagogical ethics: clashing perspectives?

Okonkwo and Ade-Ibijola (2021) did an extensive review of chatbots in education and concluded that the functions of any chatbot should be explicitly detailed (transparency and accountability) for the stakeholders, and primarily the users should decide on how to interact with the agent. This may highlight a possible gap in understanding transparency between disciplines. While not directly opposed but taking the pedagogical angle, Buzzelli and Johnston (2001) assume a general agreement among scholars that "teaching itself involves moral action..." and therefore, teachers themselves can be treated as moral agents. They describe an analogous situation to the before described edge case of an agent exercising their authority for moral ends (learning goals). In this case, a teacher wants to intervene because they want to structure and change the way certain content is being discussed while at the same time support their students unobstructed expansion of competencies. Taking this angle further, one may consider a conversational agent as more than a technological tool that needs to be fine-tuned to the learners but rather as an extension of the teacher. The teacher is enacting a form of authority over the users/students through an agent. Therefore, the students may not require a fully detailed description and explanation of the chatbot's (inner-) workings within the learning setting. We argue that a balanced approach must consider both learners and teachers.

For pedagogical settings, Buzzelli and Johnston (2001) give a possible solution themselves:

> There is no single right or wrong way to handle situations such as the one described here; we only wish to argue that they can be best conceptualized in terms of the tensions of morality and power that are unavoidably inherent in the exercise of authority. These tensions can never be resolved; they present constant difficult choices to the teachers, and each time they arise they must be dealt with afresh in complex and ambiguous moral contexts in which decisions are rarely easy or straightforward (p. 882).

Similarly, CA technology that is made fully transparent to the learner may introduce unwanted extraneous cognitive load (cf. May & Elder, 2018). In addition, a transparent and explainable CA that runs on personal data, people's relations, or emotional states in order to make inferences and nudge in an intended direction may influence the learners in unintended ways raising further ethical concerns (McStay, 2020). For example, learners may feel forced to contribute in a way that may not be beneficial to collaborative learning if they are aware that a CA will trigger if their share of a discussion falls to a significantly lower level than that of their peer, as they may want to avoid their behavior being made explicit for all to see (Isohätälä et al., 2021).

From the learners' perspective, this issue could also be handled by making the agent adaptable by the learners themselves during runtime. In a scenario where a teacher cannot supervise and adapt an agent's behavior during runtime, but wants to satisfy the ethical AI principles of human-centered values, he may allow them to tune the agent intervention to their liking or turn off support entirely.

This may also be framed as an issue of agent coerciveness. How coercive should a pedagogical conversational agent be within a CSCL scenario?

One could reasonably identify at least three options:

1. *Not-coercive*: Students are free to ignore or consider the agent as they wish. However, this may raise the ethical consideration of allowing students to follow a suboptimal path of learning during interaction that would be equal to a teacher allowing students to decide whether they do their assigned homework or not.
2. *Absolutely coercive*: Students are in no way allowed to ignore the agent-generated interactions. The opposite of the aforementioned case is an approach strongly oriented toward ideal forms of dialog as recognized by a teacher or instructional designer that leaves no room for flexibility and students' agency.
3. *Justified decision for agent adaption*: Students are allowed to adapt the CA, i.e., to silence, ask, or change its settings as a justified decision. An in-between approach that seems as a more ethical stance but may have its own drawbacks under specific circumstances. For example, the issue of who (and how) decides whether the student's decision to adapt the agent is well-justified and productive or not.

Ultimately, there is no single right or wrong way to handle these situations, and decisions need to be made on a case-by-case basis, taking into account the complex and ambiguous moral contexts inherent in the exercise of authority in pedagogical settings. To achieve effective human-agent communication while ensuring the learning goals are not compromised, it is essential to strike a balance among stakeholders by considering the varying motivations of both learners and teachers, as well as their capabilities in adhering to the principles of the applied AI ethics framework. The promotion of transparency and human-centered values in the latest AI applications within education involves including all relevant stakeholders, from the initial development stages through to the final evaluation (Yan et al., 2023). This inclusive approach can ensure that AI aligns with the broader values and ethical considerations of the educational community. At the operational stage, teachers or learning designers can then tailor the AI to their specific CSCL scenarios and institutional needs, thereby focusing on their pedagogical goals.

Conclusion

In this chapter, we have presented the case of CAs supporting CSCL. We have reviewed the state of research in this area, showing that CAs are a promising approach to enhancing CSCL. We have first discussed general ethical frameworks for developing and implementing AI-based systems and then zoomed in on ethical frameworks specific to CAs within their typical use cases of helping individual users with online services. We then shifted to a view of pedagogical ethics of supporting collaborative learners with CAs. Here, we pointed out that basic ethical dilemmas of scaffolding CSCL that CAs would inherit and presented some cases of when support through CAs becomes problematic by reducing learners' agency. Furthermore, we argued that the situation of CAs interacting with more than one human warrants further, specific consideration. In such a H2A scenario, learners find their ways of interacting with each other, but CSCL research shows that the quality of such interactions can and shall be improved for learners to apply higher order thinking and, as a consequence, construct more and better reasoned knowledge together (Weinberger et al., 2010). The dilemma of supporting learners at the cost of reducing their degrees of freedom (Wood et al., 1976) remains inherent in the design and implementation of CAs. With CAs being adaptive to learners' deficits and increasing capacities—different from other prominent approaches of scaffolding CSCL—the design of CAs needs to build on an idea of mutual, dynamic correspondence of values for productive help giving and seeking. In our ethical framework for CA Design for CSCL, we acknowledge that an agent's communicative behavior is geared to instigating higher-order thinking among learners. This could include an agent asking questions that problematize the learning tasks in some cases, providing hints that simplify the task, or remaining inactive in other cases. For an agent to simulate effective teaching, the productive and pensive phases of learners need to be recognized and left untampered (cf. Thierfelder & Weinberger, 2022). Any activity of the CA is by definition an interruption of learner interaction and may be triggered by learners actively reaching out to the CA or by careful real-time analysis of the qualities of learners' discourse with respect to the conceptual knowledge they apply, how they construct arguments, how they participate homogeneously, or how they refer to one another (cf. Weinberger & Fischer, 2006). The CA needs to be designed so that its interventions align with learners' momentary style and goal of dialog. Beyond agent design, there is also a need to prepare learners for productive interaction with agents and potentially introduce productive forms of help seeking or student to agent adaptability options.

In the realm of education, situations often arise where teachers are required to exercise their authority over learners. However, the exercise of this authority can be complex and ambiguous, with no clear-cut right or wrong way to handle these situations. Rather, decisions need to be made on a case-by-case basis, taking into account the unique moral contexts inherent in pedagogical settings. When it comes to using AI in the classroom, the challenge is to strike a balance among various stakeholders, including learners and teachers, before an actual implementation. This requires considering the goals and capabilities of each party, as well as their adherence to the principles of the respective AI ethics frameworks. In CSCL the goal is to achieve effective human-agent communication that promotes and scaffolds productive discussions without compromising ethical standards. Achieving this balance may require an ongoing assessment and reflection, as well as a willingness to adapt and adjust strategies iteratively as needed.

References

Adamson, D., Dyke, G., Jang, H., & Rosé, C. P. (2014). Towards an agile approach to adapting dynamic collaboration support to student needs. *International Journal of Artificial Intelligence in Education*, *24*(1), 92−124. Available from https://doi.org/10.1007/s40593-013-0012-6.

Aleven, V. A., & Koedinger, K. R. (2002). An effective metacognitive strategy: Learning by doing and explaining with a computer-based cognitive tutor. *Cognitive Science*, *26*(2), 147−179. Available from https://doi.org/10.1207/s15516709cog2602_1.

Aleven, V., Beal, C. R., & Graesser, A. C. (2013). Introduction to the special issue on advanced learning technologies. *Journal of Educational Psychology*, *105*(4), 929.

Aleven, V., Sewall, J., Popescu, O., van Velsen, M., Demi, S., & Leber, B. (2015). Reflecting on twelve years of ITS authoring tools research with CTAT. *Design Recommendations for Adaptive Intelligent Tutoring Systems*, *3*, 263−283.

Atkins, S., Badrie, I., & van Otterloo, S. (2021). Applying ethical AI frameworks in practice: evaluating conversational AI chatbot solutions. *Computers and Society Research Journal*, *1*, 1−6. Available from https://doi.org/10.54822/qxom4114.

Azaria, A. (2022). *ChatGPT usage and limitations*. HAL Open Science. Available from https://hal.science/hal-03913837/.

Baggetun, R., & Dragsnes, S. (2003). Designing pedagogical agents for CSCL. In U. Hoppe, J. Schoonenboom, M. Sharples, & B. Wasson (Eds.), *Designing for change in networked learning environments* (pp. 151−155). Dordrecht, The Netherlands: Kluwer Academic Publishers. Available from https://doi.org/10.1007/978-94-017-0195-2_20.

Burkhard, M., Seufert, S., Cetto, M., & Handschuh, S. (2021). The textbook learns to talk: How to design chatbot-mediated learning to foster collaborative high-order learning? In T. Bastiaens (Ed.), *Proceedings of innovate learning summit 2021* (pp. 12−21). United States: Association for the Advancement of Computing in Education (AACE). Online. Retrieved March 24, 2023 from https://www.learntechlib.org/primary/p/220264/.

Buzzelli, C., & Johnston, B. (2001). Authority, power, and morality in classroom discourse. *Teaching and Teacher Education*, *17*, 873−884.

Caldarella, P., Larsen, R. A., Williams, L., Downs, K. R., Wills, H. P., & Wehby, J. H. (2020). Effects of teachers' praise-to-reprimand ratios on elementary students' on-task behaviour. *Educational Psychology*, *40*(10), 1306−1322. Available from https://doi.org/10.1080/01443410.2020.1711872.

Chen, X., Xie, H., Zou, D., & Hwang, G. J. (2020). Application and theory gaps during the rise of Artificial Intelligence in Education. *Computers and Education: Artificial Intelligence*, *1*, 1−20. Available from https://doi.org/10.1016/j.caeai.2020.100002.

Cotton, D. R., Cotton, P. A., & Shipway, J. R. (2023). Chatting and cheating: Ensuring academic integrity in the era of ChatGPT. *Innovations in Education and Teaching International*, 1−12. Available from https://doi.org/10.1080/14703297.2023.2190148.

Deloitte. (n.d.). *AI ethics: A business imperative for boards and C-suites*. Retrieved May 24, 2023, from https://www2.deloitte.com/us/en/pages/regulatory/articles/ai-ethics-responsible-ai-governance.html.

Demetriadis, S., Caballé, S., Papadopoulos, P.M., Gómez-Sánchez, E., Kolling, A., Tegos, S., Tsiatsos, T., Psathas, G., Michos, K., Weinberger, A., Winther Bech, C., Karakostas, A., Tsibanis, C., Palaigeorgiou, G., & Hodges, M. (2021). Conversational agents in moocs: Reflections on first outcomes of the colmooc project. In S. Caballé, S. N. Demetriadis, E. Gómez-Sánchez, P. M. Papadopoulos, & A. Weinberger (Eds.), *Intelligent systems and learning data analytics in online education* (pp. xxxvii−lxxiv). https://doi.org/10.1016/b978-0-12-823410-5.00001-2.

Derrick, D. C., & Ligon, G. S. (2014). The affective outcomes of using influence tactics in embodied conversational agents. *Computers in Human Behavior*, *33*, 39−48.

Dillenbourg, P., & Jermann, P. (2007). *Designing integrative scripts. Scripting computer-supported collaborative learning: Cognitive, computational and educational perspectives* (pp. 275−301).

Dyke, G., Adamson, D., Howley, I., & Rosé, C. P. (2013). Enhancing scientific reasoning and discussion with conversational agents. *IEEE Transactions on Learning Technologies*, *6*(3), 240−247. Available from https://doi.org/10.1109/TLT.2013.25.

Edwards, A. (2018). Animals, humans, and machines: Interactive implications of ontological classification. In A. Guzman (Ed.), *Human-machine communication: Rethinking communication, technology, and ourselves* (pp. 29−50). Peter Lang.

Edwards, C., Edwards, A., Spence, P. R., & Westerman, D. (2016). Initial interaction expectations with robots: Testing the human-to-human interaction script. *Communication Studies*, *67*(2), 227−238.

Farrokhnia, M., Banihashem, S. K., Noroozi, O., & Wals, A. (2023). A SWOT analysis of ChatGPT: Implications for educational practice and research. *Innovations in Education and Teaching International*, 1−15.

Ferschke, O., Yang, D., Tomar, G., & Rosé, C.P. (2015). Positive impact of collaborative chat participation in an edX MOOC. In: C. Conati, N. Heffernan, A. Mitrovic, & M. F. Verdejo (Eds.), *Artificial Intelligence in Education: 17th International Conference, AIED 2015, Madrid, Spain, June 22−26, 2015. Proceedings (Lecture Notes in Computer Science, 9112)* (pp. 115−124). Cham: Springer. https://doi.org/10.1007/978-3-319-19773-9_12

Google. (n.d.). *Our principles*. Retrieved May 24, 2023, from https://ai.google/principles/.

Graesser, A., & McDaniel, B. (2017). Conversational agents can provide formative assessment, constructive learning, and adaptive instruction. In C. A. Dwyer (Ed.), *The Future of Assessment* (pp. 85−112). Routledge. Available from https://doi.org/10.4324/9781315086545.

Graesser, A., Cai, Z., Morgan, B., & Wang, L. (2017). Assessment with computer agents that engage in conversational dialogues and trialogues with learners. *Computers in Human Behavior*, *76*, 607−616. Available from https://doi.org/10.1016/j.chb.2017.03.041.

Hapsari, I.P., & Wu, T.T. (2022). AI Chatbots learning model in English speaking skill: Alleviating speaking anxiety, boosting enjoyment, and fostering critical thinking. In *International Conference on Innovative Technologies and Learning* (pp. 444−453). Cham: Springer International Publishing.

Hill, J., Ford, W. R., & Farreras, I. G. (2015). Real conversations with artificial intelligence: A comparison between human−human online conversations and human−chatbot conversations. *Computers in Human Behavior*, *49*, 245−250.

Holstein, K., McLaren, B.M., Aleven, V. (2019). Designing for complementarity: teacher and student needs for orchestration support in AI-enhanced classrooms. In S. Isotani, E. Millán, A. Ogan, P. Hastings, B. McLaren, & R. Luckin (Eds.), *Artificial intelligence in education: 20th international conference, AIED 2019* (pp. 157−171). Springer, International Publishing. https://doi.org/10.1007/978-3-030-23204-7_14.

Isohätälä, J., Näykki, P., Järvelä, S., Baker, M. J., & Lund, K. (2021). Social sensitivity: a manifesto for CSCL research. *International Journal of Computer-Supported Collaborative Learning*, *16*(2), 289−299. Available from https://doi.org/10.1007/s11412-021-09344-8.

Jeong, H., Hmelo-Silver, C. E., & Jo, K. (2019). Ten years of computer-supported collaborative learning: A meta-analysis of CSCL in STEM education during 2005−2014. *Educational Research Review*, *28*, 100284.

Lee, C. D. (2012). Signifying in the zone of proximal development. In H. Daniels (Ed.), *An introduction to Vygotsky* (pp. 259−289). Routledge. Available from https://doi.org/10.4324/9780203022214-19.

Leike, J., Schulman, J., & Wu, J. (2022). *Our approach to alignment research*. Retrieved May 24, 2023, from https://openai.com/blog/our-approach-to-alignment-research.

Lo, C. K. (2023). What is the impact of ChatGPT on education? A rapid review of the literature. *Education Sciences*, *13*(4), 410.

Loades, M. E., Chatburn, E., Higson-Sweeney, N., Reynolds, S., Shafran, R., Brigden, A., & Crawley, E. (2020). Rapid systematic review: The impact of social isolation and loneliness on the mental health of children and adolescents in the context of COVID-19. *Journal of the American Academy of Child & Adolescent Psychiatry*, *59*(11), 1218−1239.

Luckin, R., Holmes, W., Griffiths, M., & Forcier, L. B. (2016). *Intelligence unleashed: An argument for AI in education*. London: Pearson Education.

Matsui, T., & Koike, A. (2021). Who is to blame? The appearance of virtual agents and the attribution of perceived responsibility. *Sensors*, *21*(8), 1−13. Available from https://doi.org/10.3390/s21082646.

May, K., & Elder, A. (2018). Efficient, helpful, or distracting? A literature review of media multitasking in relation to academic performance. *International Journal of Educational Technology in Higher Education*, 15. Available from https://doi.org/10.1186/s41239-018-0096-z.

McStay, A. (2020). Emotional AI, and EdTech: Serving the public good? *Learning, Media and Technology*, *45* (3), 270−283. Available from https://doi.org/10.1080/17439884.2020.1686016.

Michos, K., Asensio-Pérez, J. I., Dimitriadis, Y., García-Sastre, S., Villagrá-Sobrino, S., Ortega-Arranz, A., Gómez-Sánchez, E., & Topali, P. (2020). September). Design of conversational agents for CSCL: Comparing two types of agent intervention strategies in a university classroom. In C. Alario-Hoyos, M. J. Rodríguez-Triana, M. Scheffel, I. Arnedillo-Sánchez, & S. M. Dennerlein (Eds.), *Addressing global challenges and quality education* (pp. 215−229). Cham: Springer. Available from https://doi.org/10.1007/978-3-030-57717-9_16.

Nass, C., & Moon, Y. (2000). Machines and mindlessness: Social responses to computers. *Journal of Social Issues*, *56*(1), 81−103.

Okonkwo, C. W., & Ade-Ibijola, A. (2021). Chatbots applications in education: A systematic review. *Computers and Education: Artificial Intelligence*, *2*, 100033.

OpenAI. (2018). *OpenAI Charter*. Retrieved May 24, 2023, from https://openai.com/charter.

OpenAI. (2023). *ChatGPT* (June 27 version) [GPT-4]. https://chat.openai.com/share/a23e26c3-1a64-49a1-8f22-eb5b5510ed8c.

Organisation for Economic Co-operation and Development. (2019). *OECD AI principles overview*. Retrieved May, 24, 2023, from https://oecd.ai/en/ai-principles.

Radford, A., Wu, J., Child, R., Luan, D., Amodei, D., & Sutskever, I. (2019). Language models are unsupervised multitask learners. *OpenAI Blog*, *1*(8), 1−9.

Reiser, B. J. (2004). Scaffolding complex learning: The mechanisms of structuring and problematizing student work. *Journal of the Learning Sciences*, *13*(3), 273−304. Available from https://doi.org/10.1207/s15327809jls1303_2.

Resnick, L. B. (1987). *Education and learning to think*. National Academy Press.

Rheu, M., Shin, J. Y., Peng, W., & Huh-Yoo, J. (2021). Systematic review: Trust-building factors and implications for conversational agent design. *International Journal of Human−Computer Interaction, 37*(1), 81−96.

Sikström, P., Valentini, C., Kärkkäinen, T., & Sivunen, A. (2022). How pedagogical agents communicate with students: A two-phase systematic review. *Computers & Education, 188*, 1−15. Available from https://doi.org/10.1016/j.compedu.2022.104564.

Skjuve, M., & Brandtzæg, P. B. (2018). Chatbots as a new user interface for providing health information to young people. In Y. Andersson, U. Dalquist, & J. Ohlsson (Eds.), *Youth and news in a digital media environment−Nordic-Baltic perspectives* (pp. 59−66). Nordicom.

Stahl, G., Koschmann, T., & Suthers, D. (2006). Computer-supported collaborative learning: An historical perspective. In R. K. Sawyer (Ed.), *Cambridge handbook of the learning sciences* (pp. 409−426). Cambridge, UK: Cambridge University Press.

Ständige Wissenschaftliche Kommission der Kultusministerkonferenz (SWK). (2023). Empfehlungen zum Umgang mit dem akuten Lehrkräftemangel. *Stellungnahme der Ständigen Wissenschaftlichen Kommission der Kultusministerkonferenz*. Available from https://doi.org/10.25656/01:25857.

Taub, M., Sawyer, R., Smith, A., Rowe, J., Azevedo, R., & Lester, J. (2020). The agency effect: The impact of student agency on learning, emotions, and problem-solving behaviors in a game-based learning environment. *Computers & Education, 147*, 1−19. Available from https://doi.org/10.1016/j.compedu.2019.103781.

Tegos, S., & Demetriadis, S. (2017). Conversational agents improve peer learning through building on prior knowledge. *Journal of Educational Technology & Society, 20*(1), 99−111.

Tegos, S., Demetriadis, S., & Karakostas, A. (2015). Promoting academically productive talk with conversational agent interventions in collaborative learning settings. *Computers & Education, 87*, 309−325. Available from https://doi.org/10.1016/j.compedu.2015.07.014.

Thierfelder, B., & Weinberger, A. (2022, September 14−16). *Broadening the dialogue through conversational agents responding to what is (not) being said* [Paper presentation]. The fourth EARLI SIG 20 & 26 Conference "Dialogue, inquiry and argumentation: shaping the future(s) of education," Utrecht, The Netherlands.

Torre, I., Carrigan, E., McDonnell, R., Domijan, K., McCabe, K., & Harte, N. (2019). The effect of multimodal emotional expression and agent appearance on trust in human-agent interaction. In *Proceedings of the 12th ACM SIGGRAPH conference on motion, interaction and games* (pp. 1−6). https://doi.org/10.1145/3359566.3360065.

United Nations Educational, Scientific and Cultural Organization. (2022). *Recommendation on the ethics of artificial intelligence*. https://unesdoc.unesco.org/ark:/48223/pf0000381137.

Van Pinxteren, M. M. E., Pluymaekers, M., & Lemmink, J. G. A. M. (2020). Human-like communication in conversational agents: A literature review and research agenda. *Journal of Service Management, 31*(2), 203−225. Available from https://doi.org/10.1108/josm-06-2019-0175.

Völkel, S.T., Schoedel, R., Kaya, L., & Mayer, S. (2022, April). User perceptions of extraversion in chatbots after repeated use. In *Proceedings of the 2022 CHI conference on human factors in computing systems* (pp. 1−18). https://doi.org/10.1145/3491102.3502058.

Vygotsky, L. S., & Cole, M. (1978). *Mind in society: Development of higher psychological processes*. Harvard university press.

Walker, E., Rummel, N., & Koedinger, K. R. (2011). Designing automated adaptive support to improve student helping behaviors in a peer tutoring activity. *International Journal of Computer-Supported Collaborative Learning, 6*(2), 279−306. Available from https://doi.org/10.1007/s11412-011-9111-2.

Wambsganss, T., Höch, A., Zierau, N., & Söllner, M. (2021). *Ethical design of conversational agents: Towards principles for a value-sensitive design*, . *Innovation through information systems: A collection of latest research on domain issues* (Vol. I, pp. 539−557). Springer International Publishing. Available from https://doi.org/10.1007/978-3-030-86790-4_37.

Weinberger, A., & Fischer, F. (2006). A framework to analyze argumentative knowledge construction in computer-supported collaborative learning. *Computers & Education, 46*(1), 71–95.

Weinberger, A., Stegmann, K., & Fischer, F. (2010). Learning to argue online: Scripted groups surpass individuals (unscripted groups do not). *Computers in Human Behavior, 26*(4), 506–515.

Wood, D. J., Bruner, J. S., & Ross, G. (1976). The role of tutoring in problem solving. *Journal of Child Psychiatry and Psychology, 17*, 89–100. Available from https://doi.org/10.1111/j.1469-7610.1976.tb00381.x.

Winkler, R., & Söllner, M. (2018). Unleashing the potential of chatbots in education: A state-of-the-art analysis. In Academy of management proceedings, . (Vol. 2018 (No. 1)). Briarcliff Manor, NY 10510: Academy of Management.

Yan, L., Sha, L., Zhao, L., Li, Y., Maldonado, R.M., Chen, G., Li, X., Jin, Y., & Gašević, D. (2023). Practical and ethical challenges of large language models in education: A systematic literature review. *ArXiv, abs/2303*0.13379.

Zhai, X. (2022). ChatGPT user experience: Implications for education. *SSRN 4312418.*

Zheng, L., Long, M., Niu, J., & Zhong, L. (2023). An automated group learning engagement analysis and feedback approach to promoting collaborative knowledge building, group performance, and socially shared regulation in CSCL. *International Journal of Computer-Supported Collaborative Learning, 18*(1), 101–133.

Navigating the ethical landscape of multimodal learning analytics: a guiding framework

Haifa Alwahaby[1,2] and Mutlu Cukurova[1]
[1]University College London, UCL Knowledge Lab, London, United Kingdom [2]Princess Nourah bint Abdulrahman University, Riyadh, Kingdom of Saudi Arabia

Introduction

Researchers interested in learning analytics (LA) have long been concerned with collecting and analyzing unimodal data generated from digital learning environments, primarily utilizing logs and keystrokes as their quantitative data sources (Mangaroska & Giannakos, 2019). With the advent of sensing technology, it is now possible to collect multimodal data that is generated in the physical world, such as eye gaze, heart rate, and body movements. This could provide a wealth of information about how learning occurs (Cukurova & Giannakos et al., 2020; Worsley & Blikstein, 2015). In real-life educational practices, however, the use of MMLA raises several ethical issues (Alwahaby et al., 2022), and privacy concerns are heightened when multimodal data is included (Worsley et al., 2021). Accordingly, Martinez-Maldonado et al. (2020) compared abstracted clickstream data with sensor data, such as physiological data, posture, gaze, and movement, and demonstrated that these data sources provided a means of personalization. Furthermore, MMLA may reveal unexpected inferences unrelated to learning tasks, such as habits and daily routines (Kröger, 2018). MMLA may include data such as facial expressions, and users may be liable to bias in the face recognition analysis (Xu et al., 2020). In these respects, MMLA is more susceptible to several ethical issues compared to traditional LA. The fact that this field is relatively new may explain why ethical concerns surrounding MMLA in educational settings are seldom discussed or addressed in the research literature (Alwahaby et al., 2022). However, it is essential to understand the perspectives, concerns, and mitigation strategies associated with MMLA ethics at an early stage, from the perspective of a range of relevant stakeholders, including researchers, practitioners, educators, and students. In order to take the first steps toward understanding these complex issues, we propose two research questions for this study:

1. What are the opinions of researchers, practitioners, and students about the use of MMLA in higher education?
2. What are the potential ways of using MMLA more ethically in higher education?

Ethics in Online AI-Based Systems. DOI: https://doi.org/10.1016/B978-0-443-18851-0.00014-7

Background and literature review
Multimodal learning analytics: background, history, and aims

The educational world has been dominated by language as a medium of communication and as a means of learning and teaching (Kress et al., 2006). However, technological advances have given students and teachers new perspectives on learning, as well as new ways to obtain feedback about the teaching and learning processes involved in learning systems. Learning technology researchers have been interested mainly in collecting information about learners through log data and clickstreams generated from learning systems (Sharma & Giannakos, 2020), a field that came to be known as LA. Although there are no generally accepted definitions of learning analytics, a widely cited statement was developed by the 1st International Conference on Learning Analytics and Knowledge (LAK): "Learning analytics is the measurement, collection, analysis, and reporting of data about learners and their contexts, for purposes of understanding and optimizing learning and the environments in which it occurs" (Long & Siemens, 2011).

However, since traditional learning occurs chiefly in a real physical classroom under normal circumstances and is not limited to the use of learning systems, a variety of physical learning behaviors and indicators are missing from the investigation of learning in the digital world. Nowadays, with the development of modern technology, advanced sensing tools have become more affordable for researchers, offering the possibility of new in-depth insights and perspectives as well as providing supplementary information data. As a result of the large amounts of data that are now available, a new area of research has emerged in recent years that is generally referred to as Multimodal Learning Analytics (MMLA) (Blikstein & Worsley, 2016). MMLA has been available for ten years as an organized field (Scherer et al., 2012), aiming to study and explore the potential of big data with various modalities within learning disciplines. Human activities are becoming ever more measurable, through the collection of data on facial expressions, eye gaze, body movements, arm pulls, and emotion (Blikstein, 2013). According to Worsley et al. (2016), MMLA integrates three main ideas: multimodal teaching and learning, multimodal data, and computer-supported analysis. MMLA is founded on the idea that teaching and learning are enacted through a variety of modalities, using traditional and nontraditional forms of data to characterize and model student learning in complex situations (Worsley et al., 2016). Multimodal data (MMD) streams and complex artificial intelligence (AI) modeling techniques are increasingly being deployed in learning analytics research to help us better understand, model, and support teaching and learning processes (Cukurova et al., 2020). Recent individual and review studies show that MMD can significantly improve the performance of computational models of teaching and learning behaviors (Cukurova et al., 2019; Giannakos et al., 2019; Sharma & Giannakos, 2020).

Although MMLA can be considered a subfield of AI in education, the key difference between AI and MMLA is that the main focus of AI is often on automation. AI is concerned with developing computational models of human abilities, including speaking, walking, and playing, which allow the development of intelligent systems capable of replicating intelligent behavior, such as image and speech recognition, text analysis, and making rational decisions in the light of the input data available (Russell & Norvig, 2010). Specifically, a key focus of Artificial Intelligence in Education (AIED) has been on agents and tutors (Labarthe et al., 2018). As opposed to externalizing and replicating human cognition (as is the case with AI), in MMLA human cognition is often extended by using computational models of multimodal data tools, or the models are internalized by humans to help us make better decisions (Cukurova, 2019). The objective of multimodal learning analytics is to design products that may or may not integrate AI technology but are also closely coupled with humans to enhance their cognition, assist them in their decisions, and extend their capabilities (Cukurova, 2019). MMLA is therefore more concerned about the data itself, identifying ways to process learning data from different modalities of sources to find helpful information for giving feedback to learners (Ochoa et al., 2013). It is important to note that research on collecting, preprocessing (e.g., cleaning data), synchronizing, and analyzing sensor data streams may be found in adjacent areas, such as human-computer interaction (HCI), ubiquitous computing, intelligent tutoring systems (ITS), educational data mining (EDM), user modeling, adaptation, and personalization (UMAP), and AIED. However, none of these areas specifically focus on multimodal data in educational settings. MMLA's primary objective is to identify learning-related constructs accurately, timely, and unobtrusively, through processing MMD, so that they can be used to develop engaging and motivating pedagogies and create systems that account for those states via a system design that improves learning (Giannakos et al., 2022). In MMLA research, multiple modalities of data are employed, both in physical and digital spaces; computational methods are used to process and analyze these multimodal data; and theories associated with the assessment of human behavior are employed and contribute to LA's aspirational goals (Cukurova et al., 2020). Multimodal data fusion enables significantly better prediction of learning outcomes and a deeper understanding of complex learning processes (Cukurova et al., 2019). Apart from providing more accurate predictions of learning in single observations, MMLA uses advances in machine learning and sensor technology to monitor factors that are considered highly relevant to learning but are often overlooked due to difficulties in their dynamic measurement and interpretation (Cukurova et al., 2020). MMLA research mainly focuses on the challenges of multimodal data collection, integration, interpretation, and visualization from digital and physical environments to provide students and teachers with appropriate feedback to improve the learning and teaching process, regardless of any artificial intelligence implied automation of any of these processes (Cukurova, 2019).

The ethics of MMLA in education

Even though MMD and AI techniques have been around since the 1950s (e.g., McCarthy, 1959), they have only been widely applied to research on teaching and learning in recent years; but this is still limited in real-world practice. The significant increase in the use of AI techniques and fine-granular MMD adds new challenges to the role of such interventions in educational contexts (Cukurova et al., 2020). The use of such interventions in education has a history of bias and discrimination, the violation of individual autonomy and rights, nontransparent and unjustifiable outcomes, the invasion of learner and teacher privacy, and unjust, unsafe, unequal, and unreliable outcomes for humans (Andrejevic & Selwyn, 2020; Knox et al., 2020; Selwyn, 2020). To be able to reap the benefits of MMD and AI in teaching and learning research and practice, there is an urgent need to create plans and tools to address these ethical challenges. Considerations of potential ethical and practical issues in the use of MMD are the first steps toward potentially hypothesizing suggestions for feasible solutions to move forward. Therefore, this section presents a definition and discussion of the relevant ethical considerations, drawing on examples from the areas of AI, AIED, and any available LA research (i.e., Hakami & Hernández-Leo, 2020). More specifically, the following section discusses issues of privacy, accountability, transparency, fairness/unfairness, bias, equity, equality, and trustworthiness. These areas, generated from the relevant research fields discussed earlier, require more attention and contributions from the MMLA research and practice community.

Privacy: "The Right to Privacy" by Warren and Brandeis (1890), published in the *Harvard Law Review*, is one of the most influential essays in the history of law. Often referred to as humans' "right to be left alone," this publication advocates a right to privacy in the United States. However, a definition of privacy is lacking. A unified definition of privacy, particularly one for learning analytics environments, remains elusive (Pardo & Siemens, 2014). Recently, researchers in LA have grown increasingly interested in data privacy as a way of enhancing the quality and trustworthiness of LA (Scheffel et al., 2014). Data ownership, anonymization, collection, storing, processing, and sharing of data are among the privacy challenges associated with LA, but discussions of such issues are rarely addressed in institutional and educational policies (Prinsloo & Slade, 2013; Slade & Prinsloo, 2013). The DELICATE checklist was presented by Drachsler and Greller (2016) to address some of the privacy issues arising from the use of LA technology.

Transparency and explainability: In AI, transparency refers to a process that makes all information, decisions, decision-making processes, and assumptions accessible to all stakeholders, to provide them with a better understanding (Chaudhry et al., 2022). Several factors influence transparency, including the ease of access to information and the effectiveness of using the information for decision-making (Turilli & Floridi, 2009). While transparency refers to the ability of an individual to understand how a system works, an explainable system enables the user to understand how it works in human terms (Robert et al., 2020). By providing simple and understandable information about how the system arrived at the outcome, the user is enabled to question, challenge, and ultimately gain more confidence in the system. Therefore, explainability might promote greater transparency. According to Robert et al. (2020), there is often a lack of transparency in the algorithms used to reach decisions. In particular, it is not always clear which datasets or criteria were used by the system in making decisions. Several reasons contribute to this lack of transparency, including the fact that these algorithms are often dynamic, designed to learn, and can be highly autonomous in nature. As a result, it is not always clear to a user when or why decision criteria change over time. Therefore, both transparency and explainability have been offered as potential solutions to increase users' trust. Research suggests that increased transparency in AI systems may have the overall effect of improving trust rather than diminishing it, at least when paired with corresponding, meaningful options for user control (Lee & Baykal, 2017). For example, Abdi et al. (2020) investigated the impact of complementing educational recommender systems with transparent justifications for their recommendations. This impact leads to a positive effect on engagement and perceived effectiveness and an increasing sense of unfairness due to learners not agreeing with how their competency is modeled.

Another aspect of transparency is *institutional transparency*. Since data privacy and policy issues have considerable influence on LA systems, a systematic treatment of ethical and data protection issues is critical (Hoel et al., 2017; Pardo & Siemens, 2014). As discussed by Drachsler and Greller (2016), data model transparency plays a key role in the adoption of educational technology. It is therefore important to provide clear and concise information about how data is collected, stored, processed, and shared. This issue was addressed using a nine-point checklist for trusted LA implementation entitled DELICATE.

A further aspect is the *transparency of data* collection, processing, modeling, and visualization. As the tracking of learners' data poses a major concern for transparency, people must be aware of how they are being tracked (Duval, 2011). A clear explanation of the purpose of data collection might also help to address the issue of transparency in education (Drachsler & Greller, 2016). A review by Verbert et al. (2020) of eight policies for learning analytics indicated that participants should also be able to opt out of data collection at any point without affecting the process. It is important to have transparency at all stages of the MMLA pipeline, including data collection and processing as well as the modeling and visualization stages. For instance, MMLA research can significantly benefit from transparent and open learner models (OLMs), which have been a focal point of the LA and educational data mining (EDM) communities. According to the results of a randomized controlled experiment conducted by Abdi et al. (2020), combining educational recommender systems (ERSs) with open learner models (OLMs) can positively impact users' perceptions and engagement.

Another aspect of transparency to consider is *the model's transparency*. A transparent model predicting learners' collaborative problem-solving competencies from video data is preferred by teachers and learners over high-performing but opaque models, as exemplified in recent research (Cukurova, Zhou, et al., 2020). Rather than using nontransparent deep learning and neural network approaches, the authors developed transparent decision trees that allowed teachers and learners to examine analytics predictions. The study suggested that a more comprehensive understanding of the factors influencing learning outcomes measured in analytics and avoiding "black box" approaches where possible can increase human agency and further adoption (Cukurova et al., 2020). Shibani et al. (2019) also echoed similar arguments, suggesting that educators should have the authority to control educational systems; this in turn could promote the transparency and trustworthiness of the education system. As a means to ensure the fully transparent development of an LA system, students must be included in the development process, beyond the usual focus group discussions (de Quincey et al., 2019). An LA tool can only be regarded as transparent if the processes behind its output are sufficiently clear to users to allow for confident reviews and critiques (Shum et al., 2016). It is, however, necessary to test such propositions empirically with users in real-world settings, which has so far been limited in research (i.e., Cukurova et al., 2020).

Accountability: The concept of accountability refers to the question of who should be held morally (and legally if this is the case) responsible when unacceptable behavior is displayed by LA and AI systems (Floridi, 2021). Accountability is fundamentally about giving people autonomy of action through knowledge, empowering all stakeholders to own their data and influence how it is interpreted and shared (Porayska-Pomsta & Rajendran, 2019). The ethics-based theory defines accountability as a social setting and a process of social negotiation. These rules and moral codes may be amended to adapt to changes and needs occurring within individual stakeholder groups as well as changes in our understanding of our socio-economic and scientific environments. Therefore, accountability is seen as a relational phenomenon, with multiple and often conflicting expectations, priorities, and investments of different stakeholders, as well as temporal shifts in who is accountable when and for what (Porayska-Pomsta & Rajendran, 2019).

The accountability dimension has also been discussed by some LA researchers, although it has received little attention from the MMLA community. Based on the legal requirements of the General Data Protection Regulation (GDPR), Hoel et al. (2017) developed a list of requirements for LA systems, which argued that educational institutions should demonstrate their ability to protect the data of their users and prevent any breaches of the system to demonstrate their accountability. Furthermore, Knight et al. (2017) stated that algorithmic accountability in LA is growing, requiring complex analyses for analytics devices. In general, Gibson and Lang (2018) pointed out that it is difficult to assess the quality and accountability of LA; therefore, the Pragmatic Inquiry for Learning Analytics Research (PILAR) method was developed to tackle several quality and accountability issues related to LA research in general. To what extent such general guidelines can be contextualized within MMLA research and how they can be expanded to become more meaningful for the MMLA community remain significant research questions.

Fairness, unfairness, and bias: The fairness of algorithm-based decision-making is described by Mehrabi et al. (2022) as the absence of discrimination or favoritism directed at individuals or groups, based on inherent or acquired characteristics. Algorithmic fairness in education refers to bias and discrimination caused by algorithmic systems in educational settings (Kizilcec & Lee, 2021). It focuses on how discrimination merges in algorithmic systems and how it can be mitigated by considering three major steps in the process: measurement (data input), model learning (algorithm), and action (presentation or use of output) (Kizilcec & Lee, 2021). According to Kizilcec and Lee (2021), individual fairness implies that algorithmic decisions should be the same for identical pairs of individuals that are close, according to the task-specific distance metric, while group-level fairness reflects the idea of distributing scarce educational resources evenly among various groups of students. The notions of fairness we choose for a particular application depend on what we, and those we include in the design process, find most salient (Holstein & Doroudi, 2021). Similarly, therefore, unfairness is defined as the dominant moral attribute of social bias (Metcalf, 2019). The unfairness of an algorithm is that its decisions are biased in favor of one particular group of people (Mehrabi et al., 2022). There are also cases where AI and LA systems perform differently for different groups in a manner that might be deemed undesirable (Holstein et al., 2019).

In the context of fairness, it is also important to consider and define the concept of bias. In general, bias refers to the tendency of people to stereotype events, groups, or individuals (Cardwell, 1999). Several types of bias exist, including moral, legal, and statistical (Danks & London, 2017). In a technical or statistical sense, bias refers to the gulf between the model and reality. Such a bias constitutes a methodological error. A statistical bias occurs when a model departs from the world it is meant to describe, often as a result of an incorrect estimation of population parameters. Statistical bias differs from social bias in that it usually does not result in social acceptance; rather, it is a morally neutral assessment of a model's effectiveness (Danks & London, 2017). Since machine learning is a statistical inference tool, algorithmic bias is often referred to as a statistical bias (Holmes & Porayska-Pomsta, 2023). According to Baker (2022), an algorithmic bias is a situation where some subgroups of the population are worse served by the algorithm than others, which means, for example, that the predictions and recommendations derived from the algorithm could result in harm to some individuals even when on average they are positive. The reduction of algorithmic bias necessitates having access to data on the demographic membership of students and retaining the data for an adequate period to determine student outcomes (Baker, 2022). Moreover, according to researchers investigating algorithmic fairness, bias is often regarded as unfair discrimination: a bad effect of modeling efforts that unfairly (dis) favors particular groups or individuals. It can also be described as the biases that are inherent in a system and its data, some of which might be intentional and some unintentional (Cramer et al., 2018).

On the other hand, Metcalf (2019) defined social bias as an unfair judgment of a person or a group of people. The central question about social bias lies in the assessment of the unfairness of prejudgment concerning a principled opposition to prejudgment and the evaluation of the material consequences of widespread prejudice. Therefore, bias is a morally salient attribute of unfairness. Addressing issues of societal bias may require adjusting data collection processes or manually incorporating an understanding of this bias into the model-building process (Mitchell et al., 2021). A major source of algorithmic bias is considered to be the data introduced during the sampling, pre-processing, cleaning, and labeling, or through the data collection method; therefore, AI researchers and designers must evaluate their decisions in the light of ethical considerations at each stage of the process (Holmes & Porayska-Pomsta, 2023). Such processes of an AI development pipeline are the same as the design and development of MMLA, and therefore ethical considerations of decisions taken at each stage should be carefully considered in MMLA settings.

Learning measurements might also be affected by different participant backgrounds, resulting in bias and unfairness issues in an educational system. Likewise, learning measurements using MMLA systems may not apply to other groups (Milligan, 2018). When designing learning analytics systems, it is crucial to provide algorithms that are equitable to different student populations, as argued by Doroudi and Brunskill (2019). Moreover, while knowledge-tracing algorithms are thought to be more equitable for some populations of students, they can still be unfair to others (e.g., favoring fast learners over slow learners (Doroudi & Brunskill, 2019)). In a similar vein, Verbert et al. (2020) stress the importance of addressing challenges surrounding the ethical implications of predictive systems (and bias in general) in LA research. In that context, the ability to visualize MMD used in prediction models could help end users better understand the assumptions made by the models and detect biases. Data variety and large-scale collection are important in the training of MMLA models since the fairness of the systems can be affected by a number of factors such as race and gender, whose representation in datasets is of greatest importance. In contrast, it may be difficult to collect and collate data on sensitive constructs to measure the algorithmic fairness of MMLA systems. As summarized earlier, the AI and LA research community has focused on some key aspects of ethics. In MMLA research, however, it remains unknown how similar considerations should be considered and contextualized.

Equality and equity: Equality and equity are two concepts that are commonly linked to fairness in education (Kizilcec & Lee, 2021). As part of AIED, educational equity might be defined as narrowing achievement gaps between different groups of learners, for example, by widely scaling up the benefits of one-to-one tutoring to a broader audience or by filling educational service gaps (Holstein & Doroudi, 2021). Innovations achieve equality in their impact if they are equally beneficial to all groups regardless of their preexisting outcomes (constant gap); however, to realize equity, the innovation's impact must benefit the groups with lower baseline outcomes more, thus closing preexisting gaps (Kizilcec & Lee, 2021). Even in cases where teams explicitly design technologies to help underserved populations, if the design process is not guided by representative voices from those populations, the resulting technologies may fail to serve the needs of those populations or may even amplify existing equity gaps (Holstein & Doroudi, 2021). There is a growing awareness that current AI systems tend to expose and amplify social inequalities and injustice (Porayska-Pomsta & Rajendran, 2019). AI may be causing inequalities and injustices because of the sociocultural biases reflected in the data used, which would also make those AI models unfair.

These prejudices could stem from prejudices related to gender, race, and ethnicity, such as police records showing that young Black males commit more crimes. Also, they may be the result of a limited amount of data made representative of society as a whole, or they may be the result of the specific classification algorithm applied (Porayska-Pomsta & Rajendran, 2019). AIED systems can be examined through four lenses to better understand how and why they may increase the risk of inequities: (1) factors inherent in the overall socio-technical design; (2) the use of datasets that reflect historical inequities; (3) factors inherent in underlying algorithms used to drive machine learning and automated decision-making, and (4) factors that emerge from an intricate and dynamic interaction between automated and human decision-making (Holstein & Doroudi, 2021). One major source of inequity in AIED technologies lies in the disparities of access, where technology is more accessible to certain groups of learners than others (Holstein & Doroudi, 2021). Unless AIED technologies are explicitly designed with accessibility in mind, these technologies risk accelerating learning for some groups of learners, while decelerating learning for others (Holstein & Doroudi, 2021). Supporting equity involves understanding learners' experiences based on three constructs: coherence, relevance, and contribution (Penuel et al., 2018). Mayfield et al. (2019) recommend preventing expenditure on research that maintains inequity instead of closing achievement gaps across student populations.

Trustworthiness: According to Jain et al. (2020), in the field of AI, trustworthy AI is used to describe AI that is lawful, ethically compliant, and technically robust. It is based on the belief that AI can achieve its full potential when trust can be established during every stage of its existence, from design to development. The development of a trustworthy AI system needs to examine the behavior of the system before interpreting its results. One of the ways to foster trust is through transparency (Floridi & Taddeo, 2016). According to prior research, increased transparency in AI systems may improve a user's trust rather than diminish it, at least when combined with meaningful options for the user's control (Lee & Baykal, 2017). Moreover, explainability is a key factor in building and maintaining trust among users in AI systems. A transparent process must govern the operation of AI, and the purpose of the AI system as well as the decisions it makes, which must be understandable to all those affected, directly or indirectly (Jain et al., 2020). Therefore, to increase users' trust, both transparency and explainability have been recommended (Robert et al., 2020). It is nonetheless important to consider the role of users' expectations when examining system output, as this may indicate a bell-shaped relationship between transparency and trust, as can be seen from Kizilcec's (2016) finding that violations of expectations can lead to reduced trust in peer assessments in an online course. Additionally, it was found that interface transparency moderated this effect such that providing some transparency with procedural information promoted trust. However, providing additional information about outcomes nullified this effect. This suggests that cognitive overload could negatively influence the potential positive impact of transparency (Kizilcec, 2016).

The importance of ethical considerations in the use of MMLA tools in education is derived from the fact that the issues mentioned earlier can also potentially arise when using MMLA tools in real-world educational settings; consequently, students are at risk from all of these issues. In addition, the highly granular, temporal, and synthesized nature of the multimodal data used in MMLA systems makes the interpretation and discussion of these ethical technology dimensions even more challenging. It is an urgent matter to address ethical issues associated with MMLA. Since the field is relatively new, there is space and time for researchers and practitioners in the field to be more aware of potential challenges before the scaled-up commercial implementation of

MMLA. Such an early understanding and systematic inclusion of ethics into the design and development of MMLA is at the core of taking an "ethics by design" approach (Ethics By Design and Ethics of Use Approaches for Artificial Intelligence, 2021). The field progresses swiftly, and MMLAs' integration in real-world practice is increasing rapidly. Currently, however, the ethical issues of MMLA are rarely addressed by MMLA researchers and practice (Alwahaby et al., 2022), which further highlights the significance and urgency of the work presented here.

Methodology

As the objective of this research is to determine opinions regarding the ethical issues associated with the use of MMLA in higher education, and measures that could be taken to mitigate them, a qualitative approach has been used to gather data and in-depth semistructured interviews were used as the method. For many years, qualitative approaches have been used in research involving human life in a variety of disciplines, including education (Denzin & Lincoln, 2011). In the light of the fact that the study's research questions were designed to gain a deep understanding of participants' perceptions of how MMLA might be applied in higher education and how it could be applied more ethically, the questions closely reflect qualitative inquiry methods. As Creswell et al. (2018) note, a qualitative approach seeks to understand what individuals and groups attribute to social or human issues and by doing so could potentially contribute to the development of hypotheses for quantitative studies. As a result, qualitative and quantitative approaches should be considered complementary rather than competing (Labuschagne, 2015). Furthermore, because the ethics of MMLA is a relatively new topic and the variables to be examined are unknown, a qualitative approach is deemed necessary (Creswell et al., 2018). Qualitative research involves analyzing descriptions and meanings that cannot always be quantitatively represented. The overall purpose of the work would not necessarily be to provide generalizable findings, but rather detailed information about a small number of individuals or cases through direct quotations, detailed descriptions of situations, events, interactions, and observed behaviors, which implies the main objective of this study (Labuschagne, 2015).

Participants

This study aimed to gather in-depth opinions from a wide range of stakeholders. As a result, the sample of participants included researchers/practitioners, educators, and students at higher education institutions and educational technology companies. For the researchers/practitioners, a purpose-sampling method was used to recruit them; they were selected based on their experience in research and teaching within the field of MMLA. A total of 60 researchers were invited to participate in the study by email. As a result, 12 researchers agreed to participate. Educators and students in higher education were invited to participate in the study through a variety of platforms, such as Moodle, emails, and WhatsApp groups. As a result, 8 educators and 39 students agreed to participate in the interviews. The student participants included 9 males and 30 females, aged 25 to 60, representing a wide range of races, such as Asian, Black, White, and Arab. They were master's and PhD students with a wide range of backgrounds and experiences. Among the 39 students, 7 were recruited from a module that uses and collects MMD in practice in order to support student

Table 2.1 The three groups of people interviewed for this research.

Group	Number of candidates	Gender	Qualification	Background and experience
Researchers	N = 12	Male = 9	PhD	Experience in research and teaching within the field of MMLA/LA
Teachers	N = 8	Male = 2 Female = 6	MA-MSc/PhD	Masters and doctorate holder teachers with a wide range of backgrounds and experiences
Students	N = 39	Male = 9 Female = 30	MA-MSc/PhD	Masters and doctoral students with a wide range of backgrounds and experiences. 7 out of 39 have experience with MMLA
MMLA technology company	N = 1	Male = 1	PhD	Experience in developing MMLA systems

learning, so these students had firsthand experience with MMLA. For the purpose of gaining further insights from different stakeholders, an MMLA technology company was approached. In total, the study involved 60 participants. Participants' detailed descriptions are presented in Table 2.1. A sample of 60 participants may seem small, but for in-depth qualitative interviews it is a relatively large sample size. More importantly, the recruitment of new participants had stopped once we reached saturation with the data insights and recurrences of certain themes observed.

Data collection

For the study, individual, semistructured, in-depth interviews with open-ended questions were conducted. Due to COVID-19, interviews were conducted through Microsoft Team, an online video meeting software. The interview protocol was designed to ensure consistency between each interview. Each interview lasted 30−60 minutes. The set of questions[1] were designed based on the five core principles commonly used in bioethics and AI: beneficence, nonmaleficence, autonomy, justice, and explicability (Beauchamp & Childress, 2001). Due to the similarities between bioethics and digital ethics, specifically when it comes to the ecological approach taken when interacting with new types of agents, patients, and environments (Floridi, 2013), bioethics has become an important key analysis approach for many digital ethical reasoning and decision-making processes (see, for instance, Floridi and Cowls (2019)'s AI principles). The interview sessions had 11 parts inspired by the systematic literature review of the field: introduction; background demographic information; privacy; well-being and engagement; safety; autonomy; student agency; accountability; trustworthiness; transparency and explainability; fairness and bias; the MMLA framework. Since most participants did not have firsthand experience of MMLA, a short animated video[2] was created to explain the concept of MMLA in education, with clear examples from real-world implementations. The video aimed to simplify and clarify what MMLA is for participants with limited experience. In order to ensure the participants'

---

[1]https://drive.google.com/file/d/15ifxWEwwowIJjaOQQMZeLn75d2UjdWMC/view?usp = sharing.
[2]https://drive.google.com/file/d/1yL8WXysnQT5SKZFskTmc4jf4GFx8sym6/view?usp = sharing.

privacy, the General Data Protection Regulations (GDPR) were followed, and an institutional ethics approval was received. The participants' permission was obtained for the audio and video recording of the interviews through detailed information and consent sheets signed by the participants, providing a legal and ethical basis for processing their data. DocuSign, a secure online signature service, was used for e-signatures of the participants.

Data analysis

Before starting the analysis, the interviews were transcribed verbatim, reviewed for any grammatical/typographic corrections, and then exported to NVivo12, a qualitative analysis software. The analysis process was inspired by the six-step thematic analysis approach proposed by Braun and Clarke (2006). Deductive thematic analysis was adopted, which involved extensive reading of the data looking for themes related to the research questions, while looking at earlier identified themes outlined in similar previous research fields (Braun & Clarke, 2006). To ensure the reliability of the findings and mitigate any potential bias caused by a single researcher, the interview transcripts were reviewed and coded by two researchers separately: Interrater agreement was calculated (Cohen's kappa = 0.76). The agreement was modestly high for complex thematic analysis research aligned with previous literature.

Result

Based on the thematic analysis for each stakeholder interview, some themes were found to be common among different stakeholders. As a result, nine themes emerged from the data as overlapping concerns for different groups of stakeholders interviewed in this study. Themes were presented in the order of how many participants mentioned/agreed to the emerging need for an ethical framework for MMLA (Theme 1); privacy, surveillance, and intrusiveness issues with MMLA (Theme 2); student agency over their learning and data ownership (Theme 3); trustworthiness of MMLA result (Theme 4); fairness and bias issues in MMLA systems (Theme 5); MMLA systems' transparency and explainability (Theme 6); MMLA systems' accountability (Theme 7); awareness level of benefits and risks associated with MMLA use (Theme 8); and the argued benefits of MMLA and the ethical issues of not using it (Theme 9). The quotes were selected based on the relevance to the specific theme discussed as well as the prevalence of similar opinions in the analyzed data. The emerging nine themes are summarized in Table 2.2.

In the following section, the themes will be signposted in bold and followed by participants' quotes presented in italics. Each participant's name is anonymized with letters and numbers: (S) for students; (T) for teachers; (R) for researchers; (C) for the tech company.

Theme 1: The emerging need for an ethical framework for MMLA

This theme offers an opportunity to measure the level of acceptance of the framework and its specifications by different stakeholders. The majority of the participants (58 out of 60) agreed that the development of a unified ethical framework tailored specifically to MMLA is necessary for the

Table 2.2 General unified themes.

Theme 1: The emerging need for an ethical framework for MMLA	Theme 2: Privacy, surveillance, and intrusiveness issues with MMLA	Theme 3: Student agency over their learning and data ownership	Theme 4: Trustworthiness of MMLA result
Theme 5: Fairness and bias issues in MMLA systems	Theme 6: MMLA systems' transparency and explainability	Theme 7: MMLA systems' accountability	Theme 8: Awareness level of benefits and risks associated with MMLA use
Theme 9: Argued benefits of MMLA and the ethical issues of not using it			

following reasons: (1) to minimize the likelihood of harm from MMLA use: "*it will be very beneficial to protect them [teachers and students] from any kind of harm*" (S17); (2) increasing users' trust in MMLA tools: "*people are afraid of these kinds of new technology when they do not fully understand them [MMLA systems] so if there is a framework, that could make sure that it's good for all*" (S07); (3) increasing users' awareness of ethical issues and how to minimize potential harms: "*both parties like teachers and students will really understand what and how to minimize any potential harm that might not be expected*" (S21); (4) recognizing that some MMLA researchers reside in foreign countries and are therefore bound by their national data protection laws rather than GDPR. It is therefore likely that a unified ethical framework will facilitate the standardization of a safe method for using MMLA in education: "*[we are] doing these experiments in [country name] where we do not have GDPR legal requirements, but the government is currently preparing a law regarding data protection*" (P06). Despite the majority of participants agreeing that an ethical framework is essential, some argued (10 of 60) that an ethical framework might not be sufficient by itself to protect end users: "*regulatory frameworks are essential [However,] even with all of these regulatory frameworks in place it will still not be enough, because the technology will get faster and more powerful and we need to constantly discuss the human experience and constantly assess how we feel as human beings and where the technology fits in*" (S32). For a future framework, a number of recommendations have been made, including the following: First, the necessity of a framework that is easy for ordinary people to comprehend. Second, it should incorporate real-life cases. Third, it should be legally binding, designed, and tested. Fourth, it should be reviewed and approved by the entire research community before it can be considered mature enough for use, and finally, there should be a training session for end users on how to use and adopt the framework.

Theme 2: Privacy, surveillance, and intrusiveness issues with MMLA

A majority of respondents (55 out of 60) believe that MMLA data is sensitive for several reasons. Five points require detailed discussion: First, since MMLA data includes users' voices and faces, they may be more likely to be identified, as indicated by (S01): "*For me, faces require a higher level of protection because people will certainly be able to recognize your identity.*" However, the traceability of MMLA data has been a subject of controversy among participants. (S07) acknowledged that video and audio can be traced, whereas other data, such as brainwaves and physiological information, cannot: *"for now the most common data is used to identify people's face and voice because the accuracy is quite high, achieved by the algorithm, but for physiological data, I think it is a little bit hard [to identify people]."* Second, MMD data is more sensitive than log data because it provides information about students' emotions: *"there is a whole different set of implications from collecting data about physiological processes that are happening in somebody, especially data which has a kind of connection or potential implications in terms of understanding how people are thinking, how their metacognitive processes are happening, how their emotional processes are happening, and their attention in the situation"* (T01). Third, the collection of sensitive data such as eye gazes and heartbeats may reveal personal health information: *"if I am having medical problems in my body ... I think heartbeat data could reveal a lot of things"* (S01). This is particularly important if the student has a mental health issue and does not wish to disclose it to anyone as (S04) reported: *"If I have some health issue, I do not want to tell someone, right? If I am not feeling mentally well and I am going through therapy, I would not want to come and reveal it to an educational institution."* However, it looks like there is quite a bit of confusion about what MMD used in educational settings can or cannot reveal. For instance, participants with a medical background argued that MMD can only assist in diagnosing functional diseases, such as dyslexia, but not organic diseases: *"MMD is sensitive data since MMD might help to diagnose functional diseases such as Dyslexia but not organic diseases"* (S11). Fourth, due to cultural considerations, it might not be feasible to capture the face of a woman in some countries: *"it has a social or traditional aspect"* (S12). Fifth, students dislike having their faces recorded as justified by (S13): *"they [students] don't want to be recognized for their weakness."* Considering all the previous reasons, students are more comfortable with data that cannot be identified at personal levels, such as heartbeats, rather than faces and voices: *"capturing heartbeat or something psychological, those that I don't really care about [...] capturing my face that would be a much greater privacy type of invasion"* (S14). Therefore, log data study was preferred over an MMLA study by participants: *"Maybe I will be more willing to give my log data"* (S21). Participants expressed concern that MMD results might be misused or harmed if they are used for purposes other than educational purposes: *"if it is [the collection of MMD] for development for the institution itself, I am fine with that. But as long as it is commercial, I think it will differ"* (S02). Participants such as (S03) also expressed concern that MMD might also be used by teachers to negatively label students: *"[MMD] would affect where I'm going to be in the class."* Students from certain religious backgrounds expressed concern that certain sensing technologies, including EEG caps, might discriminate against them because their religious customs prevent them from using these devices. This is problematic: Students cannot be forced to remove their religious dress to use MMLA data collection tools, but they also cannot be excluded from the rest of the cohort. As summarized by a student: *"I think that also might cause abuse issues or a bullying issue because you will be different from the other group"* (S03). Moreover, some participants believe that MMLA might be too intrusive to monitor student' emotions comprehensively: *"it is a bit more extreme in the sense that we are using invasive technology that is specifically submitting the students to a bit more of a comprehensive surveillance of their own emotions constantly and then it is also used for not just supporting them"* (R10). In addition, most students claimed that being observed by the MMLA for a prolonged period of time would lead to anxiety and nervousness: *"It will make me nervous because you feel that you are examined the whole time in the class"* (S02). In spite of this, participants such as (S10) reported that once they became accustomed to being observed by the MMLA, it might become a routine:

"*maybe not so weird after 2–3 class meetings.*" It is imperative to note that when students are being monitored, they exceed the norm in order to give their maximum effort, and this is not really reflected in their daily learning activities: "*when they know they're being monitored, somehow they will act beyond normal… Maybe they give their best and not really reflected in their daily learning activities*" (S10). On the other hand, other participants such as (R10) argued that they are fine being constantly observed if that would help in improving their learning: "*if I am a student that is studying medicine and I want to become a surgeon, I would be totally OK with me being constantly observed while I practise that procedure because I know that I would get a lot of value from potential refinement suggestions.*" This comment indicates a difference in the perception of students and researchers about being constantly monitored. Although the majority of participants argued that MMLA tools are safe in general, some argued that the long-term effects of MMLA tools are still unclear: "*We do not know the long-term effects of it [MMLA tools]. So even cell phones, for now, we have been using them for a while, but still there are no studies that show what will happen in like 50–60 years*" (S11). Participant (S31), for example, claimed that "*physical stiffness*" could be one of the physical harms emerging from constantly being observed by certain MMLA tools. As for the mental health issues associated with MMLA use, (S31) asserted that the use of MMLA tools might result in stress and anxiety. In spite of the sensitive nature of MMLA data, researchers believe that both log data and MMLA require the same level of privacy protection: "*I could think of a learning analytics method that would be very invasive and a multimodal learning analytics [system] that is not as invasive and the other way around. I think to generalize this, I would think the same principle should apply to both. [...] Portrayed at the same level of scrutiny, transparency*" (R10). The perceptions of respondents were influenced by the fact that MMLA functions as a standalone system and does not rely on a web-based server to operate, thereby preventing threats related to unauthorized access to data: "*currently it [MMLA system] is a standalone system. It runs on a computer; it does not run on the web. So whatever data is collected, it is placed on your hard drive, it is never publicly offered to anybody*" (R02). Despite the fact that it is generally viewed as good practice to share data among researchers in order to advance the field of educational research, this area may require considerable attention in the context of MMLA, where the data may contain sensitive information. However, the majority of researchers indicated that they have not implemented any additional data-sharing measures beyond the standard measures used with traditional data. Even so, if student data is to be shared, it should be on the basis of direct benefit to the students as (T01) emphasized: "*I think it should be shared only on a basis of direct benefit to the students*" (T01).

Theme 3: Student agency over their learning and data ownership

Of the 60 stakeholders interviewed for this study, 54 believed that learners should be able to control their learning and own their data. A difference in opinion was observed among participants regarding the degree to which students should have control over their data and agency over their learning. As one example, researchers such as (R09) argued that students should be able to access, modify, and delete their own data, which includes any uncorrected information they may have captured: "*So they [students] should be able to add, modify, delete, and remove it [MMD].*" Meanwhile, students such as (S37) argued that the teacher should be informed of any changes in the MMD: "*The system needs to show the teacher that the student deleted this information around this time.*" In contrast teachers such as (T02) argued against any modification of MMD: "*I don't think modifying it [MMD] is a good thing to do with these types of data, but I think they [students] can have access.*" Moreover, respondents argued that engaging students in the process and explaining what the data means will lead to better learning adjustments for them; however, (S22) believed that students should be able to contribute to their learning prediction through a comment box: "*I might manipulate it [MMD] because not everyone has the same ethical code, so I would suggest a comment section.*" Another objection was presented by (T01), who stated that allowing students to have agency over their data and learning might negatively affect them: "*some students could use it [MMLA system result] in a really productive and useful way [. . .]; some students, however, might actually be negatively psychologically affected.*"

Theme 4: Trustworthiness of MMLA results

A majority of stockholders (53 out of 60) agreed that there is insufficient trust in the MMLA results at present to make fully automated decisions; for example, (R09) indicated: "*I am entirely uncomfortable with these [MMLA] systems making fully automated decisions.*" There were several reasons for their arguments, including the following: (1) It is possible that MMLA data might not be an accurate reflection of the actual status of the learner since it is based on a brief period of time: "*I would not trust it [MMLA system] because I may be a different person when I know that I am being observed. Many of us might be slow learners, might learn in our own comfort zones, [. . ..] so the brief time that I am performing in front of the cameras, that should not be used to judge me*" (S08); (2) (S10) argued that although MMLA systems may be trusted to be fully automated in online and synchronous activities, in face-to-face learning it is dependent on the interaction between the teacher and the student; therefore it cannot be relied upon without the interpretation of the teacher: "*in online learning or asynchronous sessions where the learner is learning by themselves [. . .] I think the system can be made fully automated[. . .] but in face-to-face learning, I believe the teacher plays an important role1; an IT tool is just a tool. But how to use it effectively depends on the teacher and the student and their interaction, I think*"; (3) it was reported that human skills, human capacity, and human relationships can only be judged by other humans; they cannot be judged by machines; thus the MMLA data is incomplete as reported by (S32).

Theme 5: Fairness and bias issues in MMLA systems

A total of 50 out of 60 participants who were interviewed expressed their concerns regarding the potential biases of MMLA systems due to a number of reasons: First, an MMLA system could be biased if it was designed and trained for a particular population and is then applied to another population as (R01) stated: "*it is not fair to make decisions about some students based on predictions that were made for samples with different characteristics.*" Therefore, MMLA may incur bias in some situations, for instance when used in conjunction with facial recognition to predict emotion. Moreover, according to (S06), this is not limited to physical characteristics but could also be applied to the system not being able to recognize certain accents: "*would absolutely question the [MMLA] results, can you pick up my accent? because it could misinterpret what I am saying.*" Second, MMLA results are likely to be biased since the system lacks information about external environmental factors such as students' current emotional status that could only be captured by the human eye, which could lead to false interpretations: "*it does not measure the environment, does not measure my condition; it is a measure only of the action which could give a wrong result*" (S03). Third, the MMLA system may be biased because it does not take into account factors such as disabilities or cognitive difficulties, which might affect the validity of the results: "*there are multiple biases involved as well but disability is, I think, one of the major ones, where specifically mental illness and generally the slowness of individuals to understand, to respond, is not taken into consideration when these systems are built*" (S04*)*. Fourth, bias might occur as a result of the implicit values of the designers who are responsible for training: "*the data itself could be biased [...] the algorithms them-selves are not neutral. So, it is not just about the data that it is based upon, it is also the algorithms and the way they are written and constructed and who constructed the algorithms. As you well know, the vast majority of people in the AI community are white men*" (R09).

Theme 6: MMLA systems' transparency and explainability

It was noted that the majority of participants (47 out of 60) agreed with the importance of system transparency for MMLA. The term "transparency" was defined by participants such as (R10) to mean disclosing information regarding where the data comes from as well as establishing what cor-relations exist between certain actions and learning outcomes: "*assessing the students is by virtue of their brain activity. There should be total and absolute transparency on how that is assessed. How is that calculated? how is that obtained and how is that related to the skills that are eventu-ally we want them to acquire.*" Several participants, including (R01), argued that increased trans-parency of MMLA systems would increase users' accountability by enabling users to better understand why and how decisions are made: "*if we provide evidence about what we can do and how we can help the students with the MMLA systems they are using, that will increase their accountability [...] It is very critical for the end-user to understand how these systems work and how they end up with a decision or a recommendation.*" Further, (S07) noted that an increased degree of transparency in the MMLA system may lead to improved learning outcomes: "*because only if I know how this decision is made can I improve my result.*" Furthermore, it was argued that increasing the explainability of the system would increase its reliability: "*transparency will defi-nitely increase some sort of reliability in the system [...] devising that algorithm and knowing how it works would makes sense. But if I just give them [students] four options to say ABCD and one is right, one is wrong, then I think there is an issue with the students not trusting the system*" (S04). In contrast, (R06) argued that students may be able to manipulate the MMLA system due to its transparency and explanation. For example, students who understand the relationship between spe-cific actions and their outcomes may behave in an unnatural manner, as the researchers discovered in their research regarding presentation skills: "*in our experience, some students-maintained a [specific] posture throughout the entire presentation, watching the screen and looking at the audi-ence the whole time*" (R10).

Theme 7: MMLA systems' accountability

A major concern that has been raised among stakeholder groups is the issue of accountability (42 out of 60); participants argued that accountability is a key consideration when using MMLA in educational settings. According to all stakeholders, educational institutions should be held accountable for any problems associated with the MMLA system: "*I think it should be entirely the institutions' responsibility"* (T01). Consequently, it was believed that there should be some experience available within the institution for the purpose of resolving these problems. For example, (S20) argued that safety guardians should exist in any situation: *"I think for each institute there is a safety [officer] who is responsible for issues if there is any access happening."* (S17) argued that educational institutions should have an information technology department: *"An IT department [...] institution of the school."* In spite of the fact that most respondents believed that providing information about the individuals involved in the design process was important for increasing users' trust in the system, these details were not provided to the actual end users, namely teachers and students, in the real world. For instance, (R06) said: *"I guess the educator could mention who was involved in the design of the system. Yeah, to be honest, we never told the educators that they should mention this. I should say, I'm unsure, some of them did mention this but [...] I cannot be certain of this."*

Theme 8: Awareness level of benefits and risks associated with MMLA use

A majority of stakeholders (40 out of 60 participants) considered that learners should understand the impact of the MMLA recommendations and predictions on their learning. For genuinely informed consent to be granted, adult learners must be capable of understanding the benefits and potential issues associated with MMLA results. However, many limitations have been identified with the current consent forms that prevent them from achieving this goal. According to the interviewees, it emerged that students' understanding of the benefits and risks associated with the use of MMLA may have a direct impact on their willingness to use MMLA. In particular, the student participants suggested that there is a positive correlation between increasing their comprehension level and their willingness to participate. For instance, it was found that students who are more familiar with MMLA and the risks associated with it are more likely to agree to participate: *"personally because I understand the risks involved [in using MMLA systems], but I also understand that potentially this [MMLA system] can be used to predict, understand, quantify and essentially aid my student outcome. So, if this [MMLA system] could help then I'm all for it"* (S35). It is interesting that there appear to be contrasting views between teachers and students on MMLA ethics; for example, (T08) presented an opposing view, suggesting that increasing learners' comprehension may lead to a decrease in participation: *"I think it [students' participation level] will be reduced, [...] when you explain that the data will be taken away from you and that you will need to wear a headset; you are going be monitored, and especially if it can possibly affect your grades."* A decrease in participation might also be noted for students with certain backgrounds, such as law, since they are more concerned with privacy issues than others: *"as far as I am concerned, it [participation level] will decrease for me due to privacy concerns, but for someone else, it will increase due to their desire to participate in education and society"*(S28). This comment is at odds with the initial statement that increasing students' comprehension level would increase their participation.

Theme 9: Argued benefits of MMLA and the ethical issues of not using it

Several participants from various stakeholder groups discussed the ethical benefits of the MMLA (26 out of 60). Based on the argument of (S04), MMLA might reduce bias and favoritism in the classroom: *"There's a lot of favouritism in classrooms [...] I think with these systems [teachers] become more unbiased,"* particularly when racism is present, as suggested by (S02): *"Maybe if the teacher is racist."* Additionally, it may assist teachers in identifying any learning differences, as suggested by (T01): *"If you notice an issue with a student's attention. If you are a teacher that was aware of the phenomenon of ADHD, you might be able to escalate that issue in a sensitive way that actually helps that student progress in their life, in their future and in that career."* Furthermore, as discussed by (S25), MMLA provides an effective method for improving the learning experience of students with special needs, including those who are diagnosed with ADHD: *"It is a great tool to enhance the education of students with learning disabilities such as ADHD."* Additionally, students observed that MMLA might reduce harassment between students in the classroom: *"it might be helpful to minimize many kinds of abuse and harassment, maybe, because they would know that everything is monitored"* (S11), as well as protect students' safety: *"so everything recorded [...] that puts me in a safe place"* (S20), and the monitoring of misconduct: *"forbid people from cheating"* (S12), which is particularly important in certain circumstances such as COVID during online tests: *"Maybe in some cases, for example when it was during COVID, they were monitoring students' concentration, so they would not cheat"* (S13). In addition, as mentioned by (S26), student discipline could be improved as students are required to concentrate or be there on time.

Discussion

Several interesting findings were derived from the interviews, including (1) the ethical issues related to MMLA tools, (2) the level of awareness of these issues among stakeholders, and (3) recommendations on how to mitigate some of these concerns. Seven concerns emerged from the interview analysis.

First, MMLA may pose privacy concerns since it collects highly sensitive sensor data (Martinez-Maldonado et al., 2020) and may reveal details such as daily routines and habits (Kröger, 2018). In the same manner, participants viewed MMD as highly sensitive because facial recognition software and associated data, as well as pictures and videos, can be used to reveal participants' identities. There were also concerns expressed regarding the misuse of sensor data, such as eye tracking, in nonteaching environments. Furthermore, some sensors could reveal sensitive information, such as personal feelings and health information. Due to the sensitive nature of MMLA data, additional privacy protection measures and a better understanding of how sensitive data is handled may be necessary for all stakeholders. A number of recommendations were made in this regard. Among these are sharing student data only with their consent on the basis of direct student benefit, implementing differential privacy, in which group patterns within a dataset are described without revealing individual details, and introducing noise to large datasets to achieve anonymity.

The second concern stems from learners not being aware of the method of data collection and the type and sensitivity of the data being gathered. In some cases, interviewees reported that learners were aware of their data since the tool used and the data collected were visible to them.

Thus, learners' perceptions of content can be influenced by the visibility of the data collection tools. However, regardless of the type of data, some students may still not comprehend what it means. Since students' personal data must be preserved for purposes of accountability as well as to allow them to make informed decisions regarding their collection, it is imperative that they understand how sensitive this data is. In a recent MMLA paper, Beardsley et al. (2020) found that with the introduction of an informed consent comprehension test to improve students' and teachers' awareness of their collected multimodal data, the participation level decreased. In spite of this potential fear of losing participants, the importance of providing detailed information regarding the data collected to obtain genuine consent was highlighted in the interviews. Moreover, the traditional informed consent form may not be sufficient for participants to comprehend how sensor data is collected or algorithms work. Therefore, it was suggested that the concept of MMLA should be simplified for students by providing clarification in different media, such as visuals and videos, which might enable students to better understand what they are consenting to. Moreover, a presession or class that explains how MMLA can improve student learning could increase students' awareness of MMLA. The use of this approach has been found to be effective, as students were explained their consent, resulting in a high level of understanding. It was also found that students were most concerned about their data only when it affected their learning outcomes. It is therefore crucial to clarify the exact implications of the MMD collected from participants (Mangaroska, Martinez-Maldonado, et al., 2021); thus, consent forms should highlight the details of the MMLA system to which they are consenting being used for improving students' learning in order to encourage greater participation. Students may be asked to answer a series of questions after reading the information sheet about MMLA in the consent form in order to further improve their understanding. In addition, the inclusion of a video about MMLA during the consent form may improve student understanding.

Third, many respondents considered MMLA data collection to be an invasive technology that would cause anxiety and nervousness. Further, some MMLA data collection tools were reported to cause stiffness, which is in line with previous MMLA studies, which have found that students experienced distractions, discomfort, irritability, headaches, and decreased mobility (Mangaroska et al., 2021). Additionally, safety concerns were raised during the interviews since some sensors used with MMLA may result in physical injury. Therefore, it is imperative to take this issue into account when using MMLA, especially in a real-life educational setting.

Fourth, a number of recommendations have been made regarding the need to enhance trust and accountability within MMLA systems, including the following: (1) MMLA results should always be accompanied by teacher observations. Human decision-making and the relationship between a teacher and a student cannot be underestimated; therefore, teachers should always be involved in decision-making processes rather than leaving MMLA systems alone to make decisions. In addition, when utilizing MMLA prediction systems in education, it is imperative to consider the personal lives of learners, since judging learners without an in-depth understanding of their backgrounds and contexts is likely to lead to unfair and incorrect outcomes; (2) the results of MMLA prediction should be viewed as a feedback rather than an assessment. Additionally, autonomous decision-making that provides a source of recommendation information for teachers should not be taken for granted: Human interpretation is essential to ensure the validity and accuracy of this type of data. For MMLA predictions to be considered reliable, teachers should always be kept in the loop; (3) an autonomous system must be verified against human interpretations over a period of time. Moreover, MMLA decisions should be reached in consultation with the student. In order to hold the system accountable, all relevant stakeholders should be involved in the development process, as they might provide valuable insights that the system developers alone would not be able to identify; (4) educational institutions must ensure the device is safe to use, which is why MMLA systems should be monitored by a safety guardian or information technology department personnel within the educational institution to handle any problems that may arise.

The fifth concern relates to the bias that may exist within the MMLA system due to its design for specific demographic groups. For example, using MMLA with face recognition to predict emotion may incur a bias if the face recognition algorithm was developed and trained using a specific dataset. For example, the color of the skin and the features of the face may differ. Accordingly, algorithms should be developed and trained based on the characteristics of a large population over an extended period of time. All of this information should be made available to end users as a source of immediate transparency and accountability. In addition, it is essential that the MMLA systems are designed by multidisciplinary teams so as to minimize various sources of bias. It is also recommended that an ethics panel is established prior to the creation of an MMLA system. As part of this panel, the data planned to be collected should be examined to determine whether it is sufficiently broad, with minimum bias, to what extent it is clean, and how exactly it was cleaned.

The sixth concern is the MMLA's data privacy measures. According to interviews, the researchers take user privacy seriously, including how their data is stored, anonymized, and shared. However, some researchers have concluded that MMLA data does not require additional privacy protection or anonymization over traditional analytics data for a number of reasons, including that MMLA data is only used for research purposes and not for production; therefore, privacy concerns can be addressed once these tools are available at a later stage. Furthermore, MMLA systems usually are standalone systems that cannot be accessed via a web server, preventing unauthorized access to data. Controversy has also arisen regarding the traceability of MMLA data. It appears that this issue should be raised as an ethical matter, given that some researchers assert that MMLA data does not require additional privacy protection. Two approaches to anonymization were suggested in the interviews, including implementing differential privacy, in which group patterns within a dataset are described without revealing individual details, and introducing noise to large datasets to achieve anonymity. It is pertinent to note that some respondents might only be familiar with this topic in the context of research or laboratory work. It is still unclear which procedures should be followed when the MMLA tools are released into a real-world environment. An argument was raised regarding the anonymization of students. According to one respondent, maintaining student identifiers is crucial to providing feedback that is targeted to each individual student. It was interesting to observe that there were differing opinions among researchers concerning whether MMLA data could be made open-source or shared with researchers only if the individual students could be reidentified.

The seventh concern is related to transparency. Participants reported that transparency can be achieved by providing information regarding the source of data as well as the relationship between certain actions and outcomes. Accordingly, students should be provided with explanations that extend beyond the surface level. In addition, they should be able to consider more quality indicators that would assist them in reflecting on their learning and behavior. A comprehensive explanation of the MMLA system must take into account students' needs and their current levels of knowledge in order to make it understandable to them. Transparency within the MMLA systems must, however, be subject to certain guidelines. The process of explaining how the decision was reached by the MMLA system should be clear and simple. For example, explaining the equation behind the result could become more complex than the black box itself; furthermore, explanations must contribute to students' learning rather than simply providing them with an opportunity to manipulate the system. For example, if students are provided with information on how their behavior affects the system's outcome, they may be able to game the system as well.

It is concluded from the interviews that it is necessary to develop an ethical framework that protects MMLA end users, boosts their confidence, and increases their awareness of its ethical implications. Framework creation, however, should be accompanied by several factors. Among them is the need for a framework that is easily understood, continually revised, legally binding, evaluated for its effectiveness, and includes examples from real life. To ensure the effectiveness of the framework, training sessions should also be provided to end users and practitioners. For the framework to be considered mature and reliable, it should also be reviewed and approved by a wider community of researchers and practitioners. Based on interviews and a literature review, we provide the first version of an ethical framework for MMLA in this section. It should be noted that this is only a preliminary framework. Nevertheless, it is a useful starting point for the development of a more comprehensive framework in the future. The framework was designed to increase user awareness of these issues through a series of questions, as shown in Fig. 2.1.

Limitations and future work

Due to the focus of the study on higher education, the research interviews primarily focused on students enrolled in higher education, including master's and PhD students. However, the current limitation is that bachelor's degree students are not included in the sample; therefore, they should be considered in the future. Besides, K-12 students and contexts might require slightly different ethical considerations. As part of the first version design of the framework, we are taking this approach of generating insights from stakeholders through interviews that led to the first version of the framework. In our future research, a larger and more diverse sample size will be recruited and different methodologies for the codesign of the framework will be undertaken, including workshops, participatory design sessions, brainstorming, and codesign sessions, which could all involve different methods where more direct input from stakeholders could be used to further improve the framework. The framework will be further evaluated and improved in future iterations with inputs from various key stakeholders, including ethics experts and policymakers.

Objective	Criteria	Guiding questions to be considered
Beneficence The utilization of MMLA should be grounded in the advantages it brings to students and teachers	Establish the benefit that Multimodal Learning Analytics (MMLA) might bring to learning and teaching.	• If there is enough evidence that a particular MMLA tool is beneficial for learning, what are the ethical consequences of not using MMLA?
Privacy Privacy considerations should be taken into account as an important aspect of the experiment or practice	Ensure that students' privacy is protected.	• To what extent are there any risks to student privacy? • What proactive measures have been taken to protect the privacy of students? • How much have the privacy mitigation strategies reduced the risk?

	Ensure that sensitive student information is protected. The collection of sensitive personal information may occur either deliberately (e.g., as a result of explicit questioning) or by accident (e.g., appearing in processed images).	• Will you be gathering any information that could be classified as sensitive personal information, either deliberately or incidentally?
		  a- Could this information relate to health?
		  b- Could this information include other sensitive personal information?
		• What steps should be taken if sensitive information (i.e., related to health conditions) is discovered during the analysis of multimodal data?
		• What information would be considered sensitive from the perspective of the end user?

	Ensure that the collection, storage, and sharing of multimodal data has clear benefits for students or teachers in general (e.g., understanding where the student is), and that the data is collected, stored, and shared with their consent.	• What are the reasons for which multimodal data are being collected? • What is the purpose of tracking each modality? • What are the theoretical arguments supporting the collection of these particular multimodal data? • In your opinion, why does this data source matter more than other (potentially less intrusive) ones? • Why do you think that the multimodal aspect brings benefits? • What is the balance between the benefits and risks of using a particular sensor? • If you are collecting data from multiple sources, how will you triangulate them? • What is the mechanism for synchronizing all the data? • Who might this data be shared with? • What is the benefit of sharing students' data? • Will the students' data only be used for educational purposes? • What procedures have been established to safeguard collected data against misuse by private educational institutions? • How long that data will be stored? • Will the consent content be updated based on future developments and add-ons for the MMLA system?
	Ensure that if the use of an MMLA system is likely to become a form of surveillance, there is evidence that the potential benefits outweigh the potential negative impacts, this information is delivered to students clearly, and the system does not in any way harm students.	• If an MMLA system is used for surveillance at any time, and assuming that students give informed consent for this, what is the evidence to support the conclusion that its benefits outweigh its potential negative impacts? • Is there a continuous evaluation over time to provide evidence that the benefits outweigh the negative impact? • Is there ongoing consent throughout the entire process? • What level of surveillance should MMLA tools allow?

FIGURE 2.1

An ethical framework for MMLA.

Safety While using an MMLA system, end users' physical and psychological safety should be assured	Prevent any potential or actual physical (e.g., headaches) or psychological (e.g., anxiety) harm.	• Is there any chance that the MMLA system may cause physical or psychological harm to any stakeholders, including students, teachers, or staff, during data collection, analysis, or feedback? • What proactive measures have been taken to avoid any physical and psychological harm that could occur as a result of the use of the MMLA system?
End users' awareness. End users should be aware of the potential benefits and risks associated with the use of MMLA	Ensure that end users, including students and teachers, understand what MMLA is, the exact purposes for which it is used, and are aware of the potential benefits and risks associated with its use.	• Has an introductory session been conducted for teachers and students regarding the practicalities of using sensor technologies such as MMLA, the importance of MMLA in improving students' learning, and the associated risks (e.g., risks of inaccurate assessment, inappropriate feedback, data leaks, any potential harms, etc.)? • What measures have been made to ensure the accessibility of all the information presented in the introductory session, and that it is appropriate to the background of participants, including students? (E.g., visual aids to explain concepts, avoiding jargon, using accessible terminology, etc.)
Students' agency Empowering students to make their own decisions	Promote students' agency to make their own decisions.	• To what extent are students able to challenge and modify the results generated by MMLA systems? • Do students have access to a channel whereby they can ask questions about the results generated by MMLA systems?
Students' ownership of their data Student's should have ownership, control, and decision-making authority over their generated or collected information	Empower students to own their own data.	• To what extent do students have control over their data, including access, negotiation, and deletion? • To what extent does the consent form allow students to have control over their data?
Transparency and explainability End users should have a clear understanding of how the system works	Ensure that MMLA systems have the facility to provide reasons and accessible interpretations for any MMLA decisions.	• Have education stakeholders been provided with accessible information regarding the training dataset[3], any potential bias in the dataset, and how results are generated by the system developers? • Are the justifications provided by the MMLA system understandable to relevant stakeholders at different levels? • What are the implications of transparency on further system implementations and learning designs? • Have you noted any instances of gaming the system, and how might these be addressed?
Fairness and bias Issues of system bias	Ensure that suppliers provide relevant information to confirm that MMLA systems should be built in such a way that fairness is considered and attempts to mitigate bias are made (e.g., systems are trained and tested on a sufficiently broad sample drawn from different populations).	• Has the system been trained and tested on a sufficiently broad and diverse sample that is representative of the population of students by whom the system will be used? • Has the dataset been trained for an adequate period of time to be able to argue for its potential generalization? • Did the system developers implement a transparent protocol and algorithm(s)? • Was the MMLA system designed by a diverse team? • Are there any indicators that the MMLA tool might be biased?

FIGURE 2.1

(Continued)

---

[3]A training dataset is a collection of data that will be used to train an algorithm or model to accurately predict an outcome variable.

		• To what extent are the algorithm(s) used transparent to all stakeholders?
	Ensure that suppliers provide relevant information to confirm that MMLA analysis results are validated by relevant stakeholders (e.g., students and teachers) during the production and evaluation stages.	• Have the MMLA results been validated by relevant educational stakeholders (e.g., students and teachers) during the production and evaluation stages?
	Ensure that suppliers provide relevant information to confirm that the validity of MMLA results is assessed with a consideration of different populations of students.	• When validating the results of the MMLA system, to what extent were different populations of students (e.g., according to race and religion) considered? • In what ways might the results generated by the MMLA system lead to discrimination? • To what extent can the results of the MMLA system be generalized to other contexts? • Has the MMLA system been tested for bias issues arising from the design of the taring data collection?
Trustworthiness Confidence and trust in the results	Ensure the trustworthiness of MMLA results.	• To what extent is the MMLA system ready to provide trustworthy results? • What are the implications of MMLA-driven assessment? • To what extent have MMLA assessment results been validated with other sources of information? • Have the results of the system been verified against the interpretation(s) of expert human(s) (e.g., teachers, researchers, etc.) over time? • Does the MMLA system have reliable evidence that it is trusted by end users, including teachers and students?
Accountability	Ensure that a risk assessment has been conducted prior to the implementation of MMLA.	• Did your institution conduct a risk assessment prior to implementing the MMLA? If yes, what type of risk assessment? • Do you have an action plan in place in the event that any system-related issues arise? • Is there a safety guardian or interdisciplinary team of people (e.g., ethical experts, lawyers, IT experts, educators, etc.) within your institution who can assist with any potential MMLA implementation issues?

FIGURE 2.1

(Continued)

Conclusion

Despite increasing concerns and emerging efforts by researchers to address ethical issues associated with the use of MMLA, a systematic approach to evaluate, audit, and support the ethics of MMLA research and practice is currently lacking. To address this gap, this research aims to codesign a framework with unified ethics of MMLA that would allow for a safer approach to the design and use of MMLA. In this chapter, we have presented the results of our interviews with relevant stakeholders, including researchers, practitioners, students, and teachers, in addition to a tech company representative. Drawing from the interviews and a review of the literature, we have presented the first version of a framework for the ethical use of MMLA. Although the concerns and the recommendation raised within the framework are not all novel *per se*, the significance of this study lies in the fact that these concerns are discussed within the context of MMLA research and are appropriated according to the unique features of MMD. MMLA research currently lacks a comprehensive framework that encompasses all these unique yet intertwined dimensions of data, AI, and analytics ethics, including privacy, accountability, transparency, and fairness. Finally, it is important to emphasize that the framework at this point is only preliminary; however, it constitutes an important set of considerations for current researchers and practitioners that could be further developed in future versions of the framework. It is possible that some ethical issues will not become apparent until MMLA has been implemented in a real educational setting at a large scale. Therefore, the ethical considerations of MMLA should be considered as constantly evolving and revised accordingly.

Acknowledgments

The first author gratefully acknowledges Princess Nourah Bint Abdulrahman University and the Saudi Arabian Cultural Bureau in London for funding her PhD study at University College London.

References

Abdi, S., Khosravi, H., Sadiq, S., & Gasevic, D. (2020). Complementing educational recommender systems with open learner models. In *Proceedings of the tenth international conference on learning analytics & knowledge* (pp. 360−365). https://doi.org/10.1145/3375462.3375520.

Alwahaby, H., Cukurova, M., Papamitsiou, Z., & Giannakos, M. (2022). The evidence of impact and ethical considerations of multimodal learning analytics: A systematic literature review. In M. Giannakos, D. Spikol, D. Di Mitri, K. Sharma, X. Ochoa, & R. Hammad (Eds.), *The multimodal learning analytics handbook* (pp. 289−325). Springer International Publishing. Available from https://doi.org/10.1007/978-3-031-08076-0_12.

Andrejevic, M., & Selwyn, N. (2020). Facial recognition technology in schools: Critical questions and concerns. *Learning, Media and Technology*, *45*(2), 115−128. Available from https://doi.org/10.1080/17439884.2020.1686014.

Baker, R. (2022). The current trade-off between privacy and equity in educational technology. In G. Brown, III, & C. Makridis (Eds.), *The economics of equity in K-12 education: Necessary programming, policy, and systemic changes to improve the economic life chances of American students*. Lanham, MD: Rowman & Littlefield, In press.

Beardsley, M., Moreno, J. M., Vujovic, M., Santos, P., & Hernández-Leo, D. (2020). Enhancing consent forms to support participant decision making in multimodal learning data research. *British Journal of Educational Technology*, *51*(5), 1631−1652. Available from https://doi.org/10.1111/bjet.12983.

Beauchamp, T. L., & Childress, J. F. (2001). *Principles of biomedical ethics*. Oxford University Press.

Blikstein, P. (2013). Multimodal learning analytics. *Proceedings of the Third International Conference on Learning Analytics and Knowledge*, 102−106. Available from https://doi.org/10.1145/2460296.2460316.

Blikstein, P., & Worsley, M. (2016). Multimodal learning analytics and education data mining: Using computational technologies to measure complex learning tasks. *Journal of Learning Analytics*, *3*(2). Available from https://doi.org/10.18608/jla.2016.32.11, Article 2.

Braun, V., & Clarke, V. (2006). Using thematic analysis in psychology. *Qualitative Research in Psychology*, *3* (2), 77–101. Available from https://doi.org/10.1191/1478088706qp063oa.

Cardwell, M. (1999). *The dictionary of psychology.* https://search.ebscohost.com/login.aspx?direct = true&scope = site&db = nlebk&db = nlabk&AN = 691604.

Chaudhry, M., Cukurova, M., & Luckin, R. (2022). *A transparency index framework for AI in education.* https://doi.org/10.35542/osf.io/bstcf.

Cramer, H., Garcia-Gathright, J., Springer, A., & Reddy, S. (2018). Assessing and addressing algorithmic bias in practice. *Interactions*, *25*(6), 58–63. Available from https://doi.org/10.1145/3278156.

Creswell, J. W. (2018). In W. John, J. Creswell, & D. Creswell (Eds.), *Research design: Qualitative, quantitative, and mixed methods approaches* (5th ed.). SAGE Publications, Inc.

Cukurova, M. (2019). Learning analytics as AI extenders in education: Multimodal machine learning versus multimodal learning analytics. https://doi.org/10.17863/CAM.36128.

Cukurova, M., Giannakos, M., & Martinez-Maldonado, R. (2020). The promise and challenges of multimodal learning analytics. *British Journal of Educational Technology*, *51*, 1–9. Available from https://doi.org/10.1111/bjet.13015.

Cukurova, M., Kent, C., & Luckin, R. (2019). Artificial intelligence and multimodal data in the service of human decision-making: A case study in debate tutoring. *British Journal of Educational Technology*, *50* (6), 3032–3046. Available from https://doi.org/10.1111/bjet.12829.

Cukurova, M., Zhou, Q., Spikol, D., & Landolfi, L. (2020). Modelling collaborative problem-solving competence with transparent learning analytics: Is video data enough? In *Proceedings of the tenth international conference on learning analytics & knowledge* (pp. 270–275). https://doi.org/10.1145/3375462.3375484.

Danks, D., & London, A.J. (2017). Algorithmic bias in autonomous systems. In *Proceedings of the twenty-sixth international joint conference on artificial intelligence* (pp. 4691–4697). https://doi.org/10.24963/ijcai.2017/654.

de Quincey, E., Briggs, C., Kyriacou, T., & Waller, R. (2019). Student centred design of a learning analytics system. In *Proceedings of the 9th international conference on learning analytics & knowledge* (pp. 353–362). https://doi.org/10.1145/3303772.3303793.

Denzin, N. K., & Lincoln, Y. S. (2011). *The SAGE handbook of qualitative research.* SAGE.

Doroudi, S., & Brunskill, E. (2019). Fairer but not fair enough on the equitability of knowledge tracing. In *Proceedings of the 9th international conference on learning analytics & knowledge* (pp. 335–339). https://doi.org/10.1145/3303772.3303838.

Drachsler, H., & Greller, W. (2016). Privacy and analytics: It's a DELICATE issue a checklist for trusted learning analytics. In *Proceedings of the sixth international conference on learning analytics & knowledge—LAK '16* (pp. 89–98). https://doi.org/10.1145/2883851.2883893.

Duval, E. (2011). Attention please!: Learning analytics for visualization and recommendation. In *Proceedings of the 1st international conference on learning analytics and knowledge—LAK '11* (pp. 9–17). https://doi.org/10.1145/2090116.2090118.

Ethics By Design and Ethics of Use Approaches for Artificial Intelligence (2021). Retrieved October 24, 2022, https://ec.europa.eu/info/funding-tenders/opportunities/docs/2021-2027/horizon/guidance/ethics-by-design-and-ethics-of-use-approaches-for-artificial-intelligence_he_en.pdf.

Floridi, L. (2013). *The ethics of information.* Oxford: OUP.

Floridi, L. (2021). *Artificial agents and their moral nature* (pp. 221–249). https://doi.org/10.1007/978-3-030-81907-1_12.

Floridi, L., & Cowls, J. (2019). A unified framework of five principles for AI in society. *Harvard Data Science Review*, *1*(1). Available from https://doi.org/10.1162/99608f92.8cd550d1.

Floridi, L., & Taddeo, M. (2016). What is data ethics? *Philosophical Transactions A of the Royal Society*, *374*. Available from https://doi.org/10.1098/rsta.2016.0112.

Giannakos, M. N., Sharma, K., Pappas, I. O., Kostakos, V., & Velloso, E. (2019). Multimodal data as a means to understand the learning experience. *International Journal of Information Management*, *48*, 108−119. Available from https://doi.org/10.1016/j.ijinfomgt.2019.02.003.

Giannakos, M., Cukurova, M., & Papavlasopoulou, S. (2022). Sensor-based analytics in education: Lessons learned from research in multimodal learning analytics. In M. Giannakos, D. Spikol, D. Di Mitri, K. Sharma, X. Ochoa, & R. Hammad (Eds.), *The multimodal learning analytics handbook* (pp. 329−358). Springer International Publishing. Available from https://doi.org/10.1007/978-3-031-08076-0_13.

Gibson, A., & Lang, C. (2018). The pragmatic maxim as learning analytics research method. In *Proceedings of the 8th international conference on learning analytics and knowledge* (pp. 461−465). https://doi.org/10.1145/3170358.3170384.

Hakami, E., & Hernández-Leo, D. (2020). How are learning analytics considering the societal values of fairness, accountability, transparency and human well-being?: A literature review. In A. Martínez-Monés, A. Álvarez, M. Caeiro-Rodríguez, & Y. Dimitriadis (Eds.), *LASI-SPAIN 2020: Learning analytics summer institute Spain 2020: Learning analytics. Time for adoption?; 2020 Jun 15-16; Valladolid, Spain* (pp. 121−141). Aachen: CEUR.

Hoel, T., Griffiths, D., & Chen, W. (2017). The influence of data protection and privacy frameworks on the design of learning analytics systems. In *Proceedings of the seventh international learning analytics & knowledge conference* (pp. 243−252). https://doi.org/10.1145/3027385.3027414.

Holmes, W., & Porayska-Pomsta, K. (2023). *The ethics of Artificial Intelligence in education: Practices, challenges, and debates*. Routledge & CRC Press. Available from https://www.routledge.com/The-Ethics-of-Artificial-Intelligence-in-Education-Practices-Challenges/Holmes-Porayska-Pomsta/p/book/9780367349721.

Holstein, K., Wortman Vaughan, J., Daumé, H., Dudik, M., & Wallach, H. (2019). Improving fairness in machine learning systems: What do industry practitioners need? In *Proceedings of the 2019 CHI conference on human factors in computing systems* (pp. 1−16). https://doi.org/10.1145/3290605.3300830.

Jain, S., Luthra, M., Sharma, S., & Fatima, M. (2020). Trustworthiness of Artificial Intelligence. In *2020 6th international conference on advanced computing and communication systems (ICACCS)* (pp. 907−912). https://doi.org/10.1109/ICACCS48705.2020.9074237.

Kizilcec, R.F. (2016). How much information?: Effects of transparency on trust in an algorithmic interface. In: *Proceedings of the 2016 CHI conference on human factors in computing systems* (pp. 2390−2395). https://doi.org/10.1145/2858036.2858402.

Kizilcec, R.F., & Lee, H. (2021). Algorithmic fairness in education. ArXiv:2007.05443 [Cs]. http://arxiv.org/abs/2007.05443.

Knight, S., Anderson, T., & Tall, K. (2017). Dear learner: Participatory visualisation of learning data for sense-making. In *Proceedings of the Seventh International Learning Analytics & Knowledge Conference* (pp. 532−533). https://doi.org/10.1145/3027385.3029443.

Holstein, K., & Doroudi, S. (2021). Equity and artificial intelligence in education: Will "AIEd" Amplify or Alleviate Inequities in Education? Invited chapter in Porayska-Pomsta. In W. Holmes (Ed.), *Ethics in AIED: Who cares? Data, algorithms, equity and biases in educational contexts*. Routledge Press.

Knox, J., Williamson, B., & Bayne, S. (2020). Machine behaviourism: Future visions of 'learnification' and 'datafication' across humans and digital technologies. *Learning, Media and Technology*, *45*(1), 31−45. Available from https://doi.org/10.1080/17439884.2019.1623251.

Kress, G., Charalampos, T., Jewitt, C., & Ogborn, J. (2006). *Multimodal teaching and learning: The rhetorics of the science classroom*. Bloomsbury Publishing.

Kröger, J. (2018). Unexpected inferences from sensor data: A hidden privacy threat in the Internet of Things. Internet of Things. *Information Processing in an Increasingly Connected World*, 147−159. Available from https://doi.org/10.1007/978-3-030-15651-0_13.

Labarthe, H., Luengo, V., & Bouchet, F. (2018). *Analyzing the relationships between learning analytics, educational data mining and AI for education.* ITS Workshops.

Labuschagne, A. (2015). *Qualitative research—Airy fairy or fundamental?* Qualitative report. https://doi.org/10.46743/2160-3715/2003.1901.

Lee, M.K., & Baykal, S. (2017). Algorithmic mediation in group decisions: Fairness perceptions of algorithmically mediated vs. discussion-based social division. In *Proceedings of the 2017 ACM conference on computer supported cooperative work and social computing* (pp. 1035−1048). https://doi.org/10.1145/2998181.2998230.

Long, P., & Siemens, G. (2011). Penetrating the fog: Analytics in learning and education. *Educause Review*, *46*(5), 31−40.

Mangaroska, K., & Giannakos, M. (2019). Learning analytics for learning design: A systematic literature review of analytics-driven design to enhance learning. *IEEE Transactions on Learning Technologies*, *12*(4), 516−534. Available from https://doi.org/10.1109/TLT.2018.2868673.

Mangaroska, K., Martinez-Maldonado, R., Vesin, B., & Gašević, D. (2021). Challenges and opportunities of multimodal data in human learning: The computer science students' perspective. *Journal of Computer Assisted Learning*, *37*(4), 1030−1047. Available from https://doi.org/10.1111/jcal.12542.

Martinez-Maldonado, R., Echeverria, V., Fernandez Nieto, G., & Buckingham Shum, S. (2020). From data to insights: A layered storytelling approach for multimodal learning analytics. In *Proceedings of the 2020 CHI Conference on Human Factors in Computing Systems* (pp. 1−15). https://doi.org/10.1145/3313831.3376148.

Mayfield, E., Madaio, M., Prabhumoye, S., Gerritsen, D., McLaughlin, B., Dixon-Roman, E., & Black, A. (2019). *Equity beyond bias in language technologies for education* (p. 460). https://doi.org/10.18653/v1/W19-4446.

McCarthy, J. (1959). *Recursive functions of symbolic expressions and their computation by machine.* https://dspace.mit.edu/handle/1721.1/6096.

Mehrabi, N., Morstatter, F., Saxena, N., Lerman, K., & Galstyan, A. (2022). A survey on bias and fairness in machine learning. ArXiv:1908.09635 [Cs]. http://arxiv.org/abs/1908.09635.

Metcalf, J. (2019). Translation tutorial: Engineering for fairness: How a firm conceptual distinction between unfairness and bias makes it easier to address un/fairness. https://par.nsf.gov/biblio/10112010.

Milligan, S.K. (2018). Methodological foundations for the measurement of learning in learning analytics. In *Proceedings of the 8th international conference on learning analytics and knowledge* (pp. 466−470). https://doi.org/10.1145/3170358.3170391.

Mitchell, S., Potash, E., Barocas, S., D'Amour, A., & Lum, K. (2021). Algorithmic fairness: Choices, assumptions, and definitions. *Annual Review of Statistics and Its Application*, *8*(1), 141−163. Available from https://doi.org/10.1146/annurev-statistics-042720-125902.

Ochoa, X., Chiluiza, K., Méndez, G., Luzardo, G., Guamán, B., & Castells, J. (2013). Expertise estimation based on simple multimodal features. In *Proceedings of the 15th ACM on international conference on multimodal interaction* (pp. 583−590). https://doi.org/10.1145/2522848.2533789.

Pardo, A., & Siemens, G. (2014). Ethical and privacy principles for learning analytics: Ethical and privacy principles. *British Journal of Educational Technology*, *45*(3), 438−450. Available from https://doi.org/10.1111/bjet.12152.

Penuel, W., Horne, K., Jacobs, J., & Turner, M. (2018). *Developing a validity argument for practical measures of student experience in project-based science classrooms.* https://www.semanticscholar.org/paper/Developing-a-Validity-Argument-for-Practical-of-in-Penuel-Horne/767f50ae4186350f91e5633eea0e5cb5980d690f.

Porayska-Pomsta, K., & Rajendran, G. (2019). *Accountability in human and Artificial Intelligence decision-making as the basis for diversity and educational inclusion* (pp. 39−59). https://doi.org/10.1007/978-981-13-8161-4_3.

Prinsloo, P., & Slade, S. (2013). An evaluation of policy frameworks for addressing ethical considerations in learning analytics. In *Proceedings of the third international conference on learning analytics and knowledge—LAK '13* (p. 240). https://doi.org/10.1145/2460296.2460344.

Robert, L., Bansal, G., & Lütge, C. (2020). ICIS 2019 SIGHCI workshop panel report: Human computer interaction challenges and opportunities for fair, trustworthy and ethical Artificial Intelligence. *AIS Transactions on Human-Computer Interaction, 12,* 96−108. Available from https://doi.org/10.17705/1thci.00130.

Russell, S., & Norvig, P. (2010). *Artificial intelligence: A modern approach* (3rd ed.). Prentice Hall.

Scheffel, M., Drachsler, H., Stoyanov, S., & Specht, M. (2014). Quality indicators for learning analytics. *Educational Technology and Society, 17,* 117−132.

Scherer, S., Worsley, M., & Morency, L.-P. (2012). 1st International workshop on multimodal learning analytics: Extended abstract. In *Proceedings of the 14th ACM International Conference on Multimodal Interaction* (pp. 609−610). https://doi.org/10.1145/2388676.2388803.

Selwyn, N. (2020). Re-imagining 'Learning Analytics' . . . a case for starting again? *The Internet and Higher Education, 46,* 100745. Available from https://doi.org/10.1016/j.iheduc.2020.100745.

Sharma, K., & Giannakos, M. (2020). Multimodal data capabilities for learning: What can multimodal data tell us about learning? *British Journal of Educational Technology, 51*(5), 1450−1484. Available from https://doi.org/10.1111/bjet.12993.

Shibani, A., Knight, S., & Shum, S.B. (2019). Contextualizable learning analytics design: A generic model and writing analytics evaluations. In *Proceedings of the 9th international conference on learning analytics & knowledge* (pp. 210−219). https://doi.org/10.1145/3303772.3303785.

Shum, S.B., Sándor, Á., Goldsmith, R., Wang, X., Bass, R., & McWilliams, M. (2016). Reflecting on reflective writing analytics: Assessment challenges and iterative evaluation of a prototype tool. In *Proceedings of the sixth international conference on learning analytics & knowledge—LAK '16* (pp. 213−222). https://doi.org/10.1145/2883851.2883955.

Slade, S., & Prinsloo, P. (2013). Learning analytics ethical issues and dilemmas. *American Behavioral Scientist, 57,* 1510−1529. Available from https://doi.org/10.1177/0002764213479366.

Turilli, M., & Floridi, L. (2009). The ethics of information transparency. *Ethics and Information Technology, 11,* 105−112. Available from https://doi.org/10.1007/s10676-009-9187-9.

Verbert, K., Ochoa, X., De Croon, R., Dourado, R.A., & De Laet, T. (2020). Learning analytics dashboards: The past, the present and the future. In *Proceedings of the tenth international conference on learning analytics & knowledge* (pp. 35−40). https://doi.org/10.1145/3375462.3375504.

Warren, S.D., & Brandeis, L.D. (1890). *The right to privacy.* https://groups.csail.mit.edu/mac/classes/6.805/articles/privacy/Privacy_brand_warr2.html.

Worsley, M., & Blikstein, P. (2015). Leveraging multimodal learning analytics to differentiate student learning strategies. In *Proceedings of the Fifth International Conference on Learning Analytics and Knowledge* (pp. 360−367). https://doi.org/10.1145/2723576.2723624.

Worsley, M., Abrahamson, D., Blikstein, P., Grover, S., Schneider, B., & Tissenbaum, M. (2016). *Situating multimodal learning analytics.* https://escholarship.org/uc/item/22z6n6kf.

Worsley, M., Martinez-Maldonado, R., & D'Angelo, C. (2021). A new era in multimodal learning analytics: Twelve core commitments to ground and grow MMLA. *Journal of Learning Analytics,* 1−18. Available from https://doi.org/10.18608/jla.2021.7361.

Xu, T., White, J., Kalkan, S., & Gunes, H. (2020). Investigating bias and fairness in facial expression recognition. *Computer Vision−ECCV 2020 Workshops,* 506−523. Available from https://doi.org/10.1007/978-3-030-65414-6_35.

Ethics in AI-based online assessment in higher education

Joana Heil[1] and Dirk Ifenthaler[1,2]
[1]*University of Mannheim, Mannheim, Germany* [2]*Curtin University, Perth, Australia*

Introduction

The field of online assessment, especially artificial intelligence (AI)-based online assessment, is evolving rapidly and not only allows for automated assessments but also for access to vast amounts of assessment data that can be utilized to inform learners, teachers, and schools, as well as education systems (Webb & Ifenthaler, 2018). Online assessment is designed and implemented in multiple different modes in practice. The process can be performed by peers (Huisman et al., 2018), teachers, an automated system, or even the student themselves (Conrad & Openo, 2018). Regarding the format, online assessment bears the potential for formative support throughout the learning process (Gikandi et al., 2011) for teachers as well as learners but also as a summative assessment at the end of a learning segment. The type of tasks utilized in online assessment range from automated quizzes to ePortfolios (McWhorter et al., 2013) or short-answer questions and essays, including natural language processing (Reilly et al., 2016).

While there is currently still a lack of implementation in the higher education practice (Buckingham Shum & McKay, 2018), research in the field shows high potential for learning analytics as indicated by systematic reviews (Ifenthaler & Yau, 2020; Larrabee Sønderlund et al., 2019). Considering the significant role that assessment plays in the learning process as well as the rapidly evolving opportunities concerning AI-based online assessment, it is vital to consider the ethical aspects and possible hurdles.

This chapter first identifies the role of AI in online assessment in educational contexts based on the current literature. Possible benefits and applications are elaborated on and defined. Following the identified areas, possible ethical concerns and threats are identified. Finally, this chapter presents guidelines, which support the avoidance of potential ethical hazards and can be applied to ensure ethically sound implementation of AI-based online assessment.

The use of AI in online assessment
Artificial intelligence

The term AI describes a process that can classify incoming information in a manner reminiscent of human intelligence. In more detail, AI means a system that exhibits intelligent behavior by analyzing its environment and taking specific actions to achieve certain goals (Graf Ballestrem et al., 2020).

Ethics in Online AI-Based Systems. DOI: https://doi.org/10.1016/B978-0-443-18851-0.00008-1

The term "machine learning" refers to the science and engineering of machines to be able to build up decisions based on encounters with data. It can be further subdivided into supervised and unsupervised processes. While in unsupervised machine learning the data is rather classified automatically, based on structural features of the data, supervised learning requires labeled data. Deep learning refers to a special method of machine learning, concentrating on artificial neural networks that are trained by great amounts of data.

When discussing AI in aspects of our society, it is vital to consider the role that AI takes on with humans. This might be a form of replacing humans by taking on tasks usually performed by humans, supporting humans, or taking on tasks that humans would usually not take on. AI has different strengths and abilities than humans. It is important to elicit the capabilities of humans and AI systems to combine them into efficient solutions (Dellermann et al., 2019).

AI in education

As of now, AI can be seen as a sleeping giant in the context of higher education (Bates et al., 2020). AI provides opportunities greater than any other technology prior for all relevant aspects of the learning process and all stakeholders involved. Great potential from AI in the educational sector arises on different levels, as AI might enhance learning processes or even lead to a transformation of how learning and assessment are performed (Holmes et al., 2019). Furthermore, AI can help to tackle problems such as student selection, dropout, and group behavior. Therefore, it is important for the public education sector not to be overtaken and thereby lose ethical control and regulation (Bates et al., 2020). Nonetheless, there are still deficits in universities' equipment to apply AI efficiently. These shortcomings are concerning the personnel as well as technical equipment (Ifenthaler, 2017).

A systematic review by Zawacki-Richter et al. (2019) identified four overall areas, in which AI can be used in educational processes: (1) adaptive systems and personalization, (2) assessment and evaluation, (3) profiling and prediction, and (4) intelligent tutoring systems.

AI in assessment

AI can guide through the examination process on several steps and support at different stages. It is important to consider the different modes of an online assessment. In automated assessment, AI is at the center of grading and feedback. In peer, teacher, or self-assessment, it can support the grading as well as feedback process.

In their systematic review, Zawacki-Richter et al. (2019) conclude that there are four subcategories of assessment and evaluation, in which AI systems are implemented. These are (1) automated grading, (2) feedback, (3) evaluation of student understanding, engagement, and academic integrity, and (4) evaluation of teaching. Another broader area in education that is also fundamental when considering AI-based online assessment systems is personalization and adaptive learning. The impact of AI on the learning process does not end when the assessment is completed but continues to be fundamental. AI can provide support for active as well as self-directed learning through means of educational data mining, learning analytics, adaptive feedback, and interventions (Ifenthaler et al., 2018).

AI-based assessment systems

To broadly define AI-based assessment systems, they are developed based on feeding an AI-based algorithm with relevant information and data concerning the student and using this algorithm to support learners and teachers in online assessment environments (Luckin, 2017). AI-based assessment systems therefore encompass several different use cases of AI, which can range from electronic assessment platforms to AI-assisted peer assessment to automated writing assessment (Swiecki et al., 2022). Following the findings of the systematic review by Zawacki-Richter et al. (2019), different AI-based assessment scenarios are presented (Heil & Ifenthaler, 2023).

Automated grading

While quiz questions can be graded through means of simple matching, AI, especially natural language processing (NLP), is employed for the automated grading of natural text. One branch enclosing many trends and developments commercially as well as public are Automated Essay Scoring (AES) systems (Ifenthaler, 2022). The architecture behind AES employed ranges from statistical methods over content analysis to more advanced deep learning architectures (Ifenthaler, 2022). In the most recent developments of AES, neural networks (NNs) play a vital role and have changed the way that automatic essay grading is performed (Ke & Ng, 2019). NNs in the context of NLP themselves have undergone several stages of further development. Yet, they are far from reaching the pinnacle. AES bears high potential for reducing time and effort from educators in grading and feedback or supporting the grading process as a kind of assistance and providing insights into certain strengths and weaknesses of the student's work (Ke & Ng, 2019). Additionally, automated scoring is not limited to typewritten texts, but methods also allow for automated grading on handwritten pieces (Shaikh et al., 2019).

The agreement between automated systems and human raters is generally considered to be quite high, but not always at a desirable rate. Additionally, especially neural models are challenging as they might not apply to all types of texts and are not feasible for small classrooms as they are frequently relying on large corpora of preassessed texts (Zawacki-Richter et al., 2019).

Not only in the context of research on online education is AES an important concept. In practice, there are already quite a few commercial applications, the most common ones being PEG, e-rater, IEA, and IntelliMetric (Zupanc & Bosnić, 2017).

Automated feedback

Automated feedback is closely connected to automated grading. Deeva et al. (2021) designed a framework of the different designs and dimensions of automated feedback systems in education based on a systematic review (Deeva et al., 2021).

The resulting framework "Technologies for Automated Feedback—Classification Framework—TAF—ClaF" identifies four main components: architecture, evaluation, educational context, and feedback. The actual feedback properties can be furthermore divided into the degree of adaptiveness, the timing, learners' control over the received feedback, and the purpose of the feedback. Therefore, automated feedback might be nonadaptive, adaptive to the specific tasks, or to the learners' previous process. Concerning the timing, automated feedback can be immediate, on request, or after the completion of the task. Learners might also have the control to decide whether they would like to receive the automated feedback or not and when and how. The purpose of automatic feedback ranges from corrective feedback over suggestive or informative feedback to motivational feedback (Deeva et al., 2021). Cavalcanti et al. (2021) in their systematic review, focusing on online assessment and automated feedback, uncovered the different methods used to develop automated feedback. The automated feedback is highly dependent on the task type of the assessment. In tasks such as programming or language, the feedback can be derived from a comparison with desired learning outcome as defined by the teacher. In this case, the students received automated messages based on their mistakes or congratulatory messages if they achieved a certain goal (Cavalcanti et al., 2021). The second most used technique was dashboard visualization. In more advanced tasks, methods of AI must be applied to ensure the accuracy of the feedback. Automatic feedback can also be developed through means of NLP or other machine learning methods such as feature extraction with clustering.

Evaluation of assessment integrity

AI is also implemented in evaluating assessment integrity. Learning analytics allow for analyzing data that enables conclusions to be drawn about possible similarities between submissions. The realization can be developed by providing data mining techniques with large amounts of learner data to classify writing patterns and detect possible academic misconduct by identifying similarities (Amigud et al., 2017).

Furthermore, newer advances also allow for AI-based proctoring systems. Proctoring refers to the supervision of learners during an exam to detect possible academic misconduct (Alessio et al., 2017). Proctoring systems vary in their degree of supervision as well as their inclusion of human supervisor (Coghlan et al., 2021). They can be integrated into learning management systems or installed as browser extensions. Proctoring systems can make use of the camera and microphone of the student as well as restricting their access to other applications on the students' computers (Alessio et al., 2017). AI systems, trained on a vast number of videos, can be developed to detect possible behaviors of students that might indicate violation of assessment integrity (Coghlan et al., 2021). Proctoring is also shown to have an actual effect on the exam performance of learners (Alessio et al., 2017).

Learning and assessment analytics

AI in the assessment process can also be used to inform current processes or enhance further decision-making in the form of learning analytics. Learning analytics make use of vast statistical as well as dynamic data about learners as well as learning environments (Ifenthaler, 2015). Statistical data might include demographic information, previous academic degrees, etc. Dynamic data on the other hand refers to aspects such as clickstream data, temporal information, or assessment results (Ifenthaler, 2015). This data is analyzed and visualized to inform and support learning in the form of modeling and predicting the learning processes.

Learning analytics is closely linked to educational data mining. Educational data mining differs from learning analytics in terms of technique, origins, emphasis, and type of discovery (Romero & Ventura, 2013). Learning analytics are rather rooted in the pedagogical design of learning environments and how data and analysis can benefit these. Therefore, while learning analytics focus on the data and the results, educational data mining is keener on the actual used data mining techniques (Romero & Ventura, 2013).

In the educational process, learning analytics can inform and support processes of scaffolding, feedback, and visualization (Gašević et al., 2015). Furthermore, learning analytics can also include a social dimension. Social learning analytics aim at supporting the learners in intertwined or social learning scenarios and provide insights into possible collaborations (de Laat & Prinsen, 2014). Learning analytics bear the potential to support multiple stakeholders in real time, summative, or predictively (Ifenthaler, 2015). Through the use of learning analytics, benefits for governance, institutions, and learning design, but also the individual educators and learners can be developed. This might include, for example, understanding one's learning process and adapting to recommendations in the future. Educators can benefit, for example, from the possibility to understand their teaching habits, develop interventions, or identify learners at risk to support in the future (Ifenthaler, 2015).

In the current state of research, it remains a challenge and importance to strengthen the connection between assessment and learning analytics (Gašević et al., 2022).

Ellis (2013) already called for the design and implementation of assessment-based analytics as the assessment provides insights into learning processes, expanding the scope, and increasing the potential benefits of analytics. Data collected through assessment that can inform further the learning process is manifold. It can range from individual assessment results from smaller tasks to completed degrees or focus on individual achievement compared against explicit learning outcomes or assessment criteria such as a predefined rubric. This might lead to the analysis of specific strengths and weaknesses concerning an individual's work. But not only the actual outcomes or grades are important but also factors such as individual improvement or persistence as well as attainment over the course (Ellis, 2013).

Ethics of AI in online assessment
Ethics

Ethics is a branch of philosophy focusing on moral human behavior and decision-making. It is concerned with the moral external conventions of society to define codes of conduct and behavioral guidance to distinguish morally acceptable and unacceptable behaviors.

Specifically, normative ethics as a branch of ethics is focusing on norms of moral behavior and rules to live by. Concerning the special case of assessment, it is important to consider the new challenges and possibilities posed through the advancements of AI but also bearing in mind the fundamental ethical questions of the assessment itself.

Ethics of educational technology

Artificial intelligence in online assessment is closely linked to the establishment of educational technologies in educational institutions, as the use of AI is dependent on the groundwork of digital infrastructure. Therefore, it is vital to consider the ethical implications arising from the use of technology in higher education.

When employing any type of education technology, it is important to ensure that it is not used merely because of having the opportunity to use it, but because of an educationally based benefit and advantage for all involved.

Ethical challenges might arise due to the technological gap between students, providing them with different initial situations and possibilities (Spector, 2016). Spector defined ethical principles to be followed considering the usage of technology in the educational realm.

In following the concept "Do no harm," the use of educational technology should promote benefits and minimize deficits. Therefore, educational technology should be designed in a way that will provide benefits compared to the status quo. Furthermore, it is essential to recognize the contributions of those involved, namely all the important stakeholders. In the sense of transparency, one should share plans, criteria, as well as lessons and make assumptions explicit. Educators should also consider alternatives, be fair and open in assessment and evaluation, and define clear goals and expectations (Spector, 2016).

Ethics of AI

In public discourse, AI is often seen as a threat. As AI will have a disruptive influence on many aspects of our everyday lives, it is important to consider the way we would like to interact with it. Floridi et al. (2018) developed AI4 People—an ethical framework to be considered in how to use AI ethically for humans. They define five ground concepts to avert threats from AI systems. These are beneficence, nonmaleficence, autonomy, justice, and explicability. These five concepts are the basic conventions that govern the specific application of AI, which must be specified and evaluated concerning the specific use cases.

As the main ground rule concerning the ethics of AI, beneficence refers to the concept that AI should be used in a way that promotes general well-being. But the goal should not only be to promote good but also to prevent bad. This means that there should neither be accidental nor even deliberate harm promoted through AI. Autonomy describes the concept of humans being able to form their own decisions. Concerning the case of AI, it is important to consider that one is handing over autonomy from the human to the machine when employing the system in the decision-making process (Prinsloo & Slade, 2014). Another threat that might arise regards the autonomy of the AI systems. Their autonomy should be limited as humans should always have the power to decide which decisions are to be taken (Floridi et al., 2018). The fourth principle considers justice; AI should be used and developed in a way that promotes justice and decreases possible discrimination. In this context, it is also important to consider possible discrimination through AI systems. Possible ethical threats can arise through stereotyping or harmful classification. But essentially, most of the other principles are not properly valid if the person interacting with the machine does not fully comprehend the consequences of their own decisions. Therefore, the explicability of AI systems has to be fundamental in terms of which data is processed and what it is used for.

Moral and ethical implications of assessment

General moral implications of assessment

The assessment process in itself already poses challenges in terms of ethical and moral implementation and execution. Certain aspects are vital to be considered independently of the design or the environment in which the assessment takes place. The assessment should always be conducted in a context that provides fair opportunities for all test takers (Kunnan, 2020), as everyone should have the same time and possibilities to learn and prepare to not create disadvantages. The type, mode, and format should be chosen in a way that is meaningful for the learning scenario, to ensure the relevance of the assessment and provide learners with an assessment that reflects the actual relevant competencies. Furthermore, assessments should be designed consistently as well as bias-free (Kunnan, 2020).

Ethical integrity and fairness should be the guiding principles throughout the whole assessment process, not only during the actual assessment but also in the development and design, enabling support for disadvantaged students (Zlatkin-Troitschanskaia et al., 2019). Formative assessment especially calls for an ethic of care by the teacher, a special concern for equity as well as active student engagement (Cowie, 2015). Both assessment and feedback should not harm the test taker or the relationship between the learner and the examiner (Cowie, 2015). When providing feedback on the assessment, fairness and matters of social justice need to be considered. Additionally, the institutions' side should ensure justice and appropriate access and administration as it benefits the immediate community and larger society (Kunnan, 2020). Fairness in the design, development, implementation, and evaluation of assessments is not only an ethical concern but also an elementary prerequisite for valid interpretation of assessment results and adequate feedback (Zlatkin-Troitschanskaia et al., 2019).

This becomes particularly evident and essential when using assessment data in subsequent analysis processes. AI-based assessment systems open a variety of opportunities in education, but also pose new additional challenges and questions for the ethical and moral use of online assessment.

General challenges of introducing AI into assessment

Overall, the implementation of AI in online assessment leads to different ethical challenges. Major questions to be considered are the resulting sidelining of professional expertise by relying on automated systems as well as the question of accountability in automated decision-making (Swiecki et al., 2022). Furthermore, AI might lead to a restriction of the pedagogical role of assessment, by minimizing the relationship between learners and teachers. The availability and use of vast amounts of data might also lead to a feeling of surveillance and therefore harm the educational processes as well as the well-being of the students (Swiecki et al., 2022).

Ethics of AI-based assessment scenarios

Based on the previously identified use cases of AI in online assessment, the ethical implications of specific scenarios are identified and evaluated. These include automated grading, automated feedback, evaluation of assessment integrity, and the overall ethical implications of using learner data in analytical processes.

Ethical dimensions of AI-automated grading and feedback

A grading process in itself bears the potential for unethical behavior such as unclear communication about grading, biases in grading, and possible score pollution by other factors (Green et al., 2007). The ethical dimensions emerging are highly dependent on the mode of assessment, as well as the type of task. Concerning automated assessment, particular challenges arise, as the automated systems take the place of the assessor. Especially when relying on models trained on large datasets, it is important to keep in mind that these can convey biases based on the data. This natural property of AES poses some hurdles and difficulties that are not easy to overcome and influence multiple aspects. This can, for example, include cultural biases when the data is trained on a dataset including disproportionately large amounts of data from one cultural background (Ifenthaler, 2022). Furthermore, AES can be implemented for English second learners, while the dataset can be trained on texts written by native speakers, which can produce inaccurate scores and wrong treatment of the learners.

Another ethical challenge might arise through a different property of the large datasets. AES systems can lead to a bias toward a high score from the system regardless of the actual quality of content. Therefore, when AES systems are developed in a way that they just output one score, it might be possible to trick the systems by using many complex technical terms or just writing long texts (Ifenthaler, 2022). Another ethical risk that might arise can come from the replacement of human interaction, which is a vital part of feedback (Shermis et al., 2010) as well as just plain mistakes of the systems that are not supervised.

On the other hand, when considering other modes of assessment, especially in terms of peer assessment, AI can help to support the trustworthiness of the assessment. By comparing the different scores of different grades, AI can support detecting possible outliers and support the accuracy of the grading as well as feedback (Darvishi et al., 2022).

Ethical dimensions of AI-based assessment supervision

Proctoring is highly debated due to the touching on different ethical dilemmas. Specifically, concerning the concept of surveillance pedagogy (Swiecki et al., 2022), proctoring can create an atmosphere of constant surveillance in the assessment. While aiming at providing fairness for all students, proctoring opens up new hurdles and ethical challenges that need to be considered in implementation. Not only is the personal privacy of learners at risk, the implementation of AI into proctoring also creates ethical challenges in terms of transparency and nonmaleficence (Coghlan et al., 2021). These ethical issues can arise due to the accuracies of the systems trying to identify academic misconduct. Students might be wrongfully framed as cheating based on certain guidelines on behavior that are not evident to them or even through biases conveyed in the algorithms (Coghlan et al., 2021). Proctoring can also harm relationships between students and teachers as well as negatively influence the development of pedagogically valuable online assessments (Lee & Fanguy, 2022).

Ethical challenges for learning and assessment analytics

The ethical challenges posed by learning analytics can be used as fundamental evaluations when considering the ethical challenges of online assessment, as the collection, analysis, and processing of learning data are key processes in AI-based assessment system. When considering the case of learning analytics, concerns not only relate to issues of students' privacy, but the classification of learners based on demographics or previous behavior/performance is an additional ethical concern that has to be handled with care (Scholes, 2016). These more complex ethical challenges arise due to the sampling of more individual data concerning personal effort, and dynamic (nonstatic) features. But, additionally, using AI to build knowledge is not sufficient; models have to be designed to be explanatory (Rosé et al., 2019). Slade and Prinsloo (2013) cluster ethical concerns about learning analytics into three main overarching and possibly overlapping categories: the location and interpretation of data; informed consent, privacy, and the deidentification of data; and the management, classification, and storage of data.

Tzimas and Demetriadis (2021) conducted a systematic review ($N = 53$) on ethical issues concerning learning analytics. In their review, the authors identified varying facets of ethical data processing in learning analytics. Key concerns for the ethical use of learning analytics opportunities include (1) privacy, (2) transparency, (3) labeling, (4) data ownership, (5) algorithmic fairness, as well as (6) the obligation to act (Tzimas & Demetriadis, 2021).

1. Privacy, describing a relationship between the owner of the data, the data, and another person, was the most frequently concerned aspect in the literature. It is an essential issue when considering large amounts of educational data, as the individual should be able to determine who has access to personal information. Ethics and privacy are different concepts, even though they might align with each other. Concerning privacy, it is important which data students would be willing to share. The personal relation toward privacy is highly context dependent, and in terms of learning analytics systems in higher education, students might not necessarily be willing to share unrelated data such as private information (Ifenthaler & Schumacher, 2016). In a systematic review by Murchan and Siddiq (2021) on ethical issues concerning analytics of assessment data, they found that the main focus of the ethical design of the usage of online assessment data currently lies on privacy issues as well as the insurance and development of participation constraint (Murchan & Siddiq, 2021). This is in line with the findings by Tzimas and Demetriadis (2021), which showed that privacy is the most frequently identified ethical issue in learning analytics. It is important to consider that meeting privacy standards does not mean that all aspects of ethics are considered.

2. Transparency is the second most frequently addressed issue in the review by Tzimas and Demetriadis (2021), describing the possibility of requesting information on who has access to the data and where, and how it is processed and stored. Transparency should provide the opportunity for self-responsibility and reflection by providing the learners with all necessary information for acting.

3. Labeling addresses another essential ethical threat concerning using students' data, namely potential stereotyping and mistreatment by students based on the analytical processes and their provided data.

4. Data ownership is a complex concept referring to the usage of the process data of the learner and who owns the data and might use it.

5. Algorithmic fairness describes another crucial aspect of the ethical use of learning analytics. Systems trained on large amounts of data are only as sound as the data that is their basis. Algorithms can therefore be, for example, adhering to misleading patterns or conveying other mistakes. The concept of fairness was also described as gaining modest attention by Murchan and Siddiq (2021) in the usage of assessment data.

6. The last mentioned of the relevant aspects, the obligation to act, sheds new light on the discussion of ethical learning analytics. AI can help foster more equal and supported education. Where teachers might fall short and not analytically detect the needs of students, learning analytics can take this place and inform as well as support learners and teachers. In the case that institutions and teachers are provided with all the necessary data as well as methods and detect struggling students, they must respond to this and support learners with the possibilities that are within reach.

With a different approach to the possible arising problems (Ifenthaler & Greiff, 2021), Ferguson (2019) defined six actionable challenges based on concerns and difficulties identified. While not only considering the problems but also defining possible actions taken based on the identified issues, these challenges include the following: (1) Following the idea of the obligation to act, learning analytics should be implemented, whenever learning could be promoted and improved. Following this, the main objective of the implementation should remain learning and teaching. (2) Regarding the concepts of transparency and consent as well as privacy, an important prerequisite is the education of the teachers and learners with educational data literacy, for them to be able to evaluate their options and form informed decisions. (3) Considering the responsibility of safeguarding teachers for students, it is vital to also act proactively and identify potential risks that might arise in the future and take actions to tackle these problems. (4) Another important actionable challenge is taking equality and justice into consideration in how far learning analytics might enhance or diminish these two concepts. (5) Additionally, learners should be able to understand the value, ownership, and control of data. This needs to be supported by educators. (6) Ultimately, another challenge lies in increasing the agencies on both sides concerning educational data.

Ifenthaler and Schumacher (2016) introduced the privacy calculus model to inform stakeholders about the complex decisions required for learning analytics systems. The model outlines a deliberation process for disclosing data for learning analytics systems. First, students evaluate their privacy concerns in light of the precise data needed by the learning analytics system. (e.g., name, learning history, learning path, assessment results, etc.). This decision can be influenced by risk-minimizing factors (e.g., trust in the learning analytics systems and/or institution, and control over data through self-administration) and risk-maximizing factors (e.g., nontransparency and negative reputation of the learning analytics system and/or institution). Second, the projected benefits of the learning analytics system are then compared to privacy concerns. The probability that the students will disclose the required information is higher if they expect the benefits to be greater than the risk. Hence, a cost-benefit analysis based on the student's access to information is used to decide whether to provide information for learning analytics systems.

Furthermore, analytics-based assessment can also help detect academic misconduct to support assessment integrity; it should nonetheless be designed with ethical and privacy considerations in mind to avoid creating a sense of constant surveillance (Gašević et al., 2022). Current research shows that there is a clear focus on ensuring data security and privacy, while an ethical discourse needs to go further and deeper.

Frameworks to mitigate potential ethical risks of AI-based assessment

Based on the identified moral and ethical risks of AI-based online assessment, the challenge remains to address these and prevent potentially harmful outcomes. For the special case of assessment, due to the relevance as well as the high stakes coming with assessment, compliance with guiding ethical principles is fundamental.

To ensure a risk-free use concerning the moral implications of AI-based online assessment, it is elementary to define ground rules and principles on which to design and implement the systems. Nguyen et al. (2022) developed principles to be considered concerning the usage of AI in an educational context. These were defined based on the guidelines on ethical AI by international organizations, such as the OECD, EU, and UNESCO, and compose a broader view of the usage of AI in the educational context. The principles are (1) governance and stewardship, (2) transparency and accountability, (3) sustainability and proportionality, (4) security and safety, (5) inclusiveness, and (6) human-centered AI in education (Nguyen et al., 2022). These should be baselines when considering the broader picture of AI in education from a more organization-wide perspective.

Regarding the development of AI-based systems in higher education, Richards & Dignum (2019) defined value-focused design guidelines to be considered in every phase of the development and performed step by step. First off, the designers of the environment should (1) identify all the relevant stakeholders, then (2) elicit the fundamental values as well as requirements of the stakeholders, (3) provide means to interpret and combine the values of all stakeholders, (4) maintain the links between the individual values, and finally (5) provide support in the election of components.

Ultimately, it is upon every higher education institution to define their ground rules for how to implement AI in their online assessment. An important consideration that should be incorporated into such decisions is the emergence of large amounts of data, their potential, and the possible impact on teaching-learning processes. Therefore, the ethical ground rules and duties for learning analytics should be an important cornerstone for the implication of online assessment systems as well as assessment analytics.

Drachsler and Greller (2016) developed a checklist, abbreviated as DELICATE, containing checkpoints for higher education institutions to consider when implementing learning analytic systems.

DELICATE is an abbreviation for the following:

- Determination—The reason why it is necessary to employ learning analytics/be aware of the potential added benefits.
- Explain—The responsible individuals should provide information about who explicitly has access to the data, where it is stored, and for how long.
- Legitimate—Insuring the importance of the added value and the legitimation of the implemented AI-based system.
- Involvement—All important stakeholders should be included in the decision-making concerning design and implementation.
- Consent—It is essential to develop some sort of contract with the data subjects and ask for an active opt-in. No data should be collected from individuals without their permission.
- Anonymize—Data should be anonymized and the individual not retrievable based on the data.
- Technical—Use all the technical possibilities you have, to ensure ethical usage of the data.
- External—Consider the implications coming from working with external partners and how to ensure ethical treatment of the data.

While being important for learning analytics, these concepts also provide guidance in other aspects of AI-based assessment. Especially transparency and making use of all technological possibilities are important factors to consider when it comes to automated grading and feedback. Learners need to understand where their grades come from and how the AI concluded their evaluation.

Concerning the resulting effective implementation of AI in education, Dignum (2017) defined the so-called Accountability, Responsibility, and Transparency (ART) principles for institutions and technologies to be followed. When incorporating AI systems in the assessment process, especially in high-stakes scenarios, such as graded summative procedures or admissions processes, it is important to be aware that decisions are passed on to an AI system (Prinsloo & Slade, 2014). Not only in these scenarios but also in the general use and implementation of AI-based assessment systems, the consideration of ART principles must be elementary. Accountability is related to the requirement that an institution needs to be able to explain as well as justify actions taken by an AI system (Dignum, 2017). Responsibility goes hand in hand with accountability and refers to overseeing the AI processes and being liable for the success and failure of the system. Transparency describes the availability of data and processes for inspection and monitoring, which should be a prerequisite for possible implemented systems (Dignum, 2017).

Overall, many of the concepts, frameworks, and guidelines toward the ethical implementation of AI in the educational context are referring to similar concepts and guidelines to be followed. As a baseline, benefits need to be established and potential harm needs to be avoided.

The ethical design of AI-based online assessment systems requires holistic ethical consideration at all steps of the way, beginning in providing equal opportunities through equal educational technologies, as well as seeking no harm through the implementation of technology in the educational process. The assessment process in itself needs to follow the proposed requirements of ethical assessment. Automated grading and feedback need to be bias-free and transparent, and the students need to be equipped with the necessary AI competencies and data literacy skills to have the opportunities to develop a concept to understand how their grade was obtained. In the introduction of AI into the assessment process, it is elementary to consider first broader concepts of ethical AI implementation. When deciding to develop an AI-based system, it is elementary to follow the steps as designed by Richards and Dignum, including all stakeholders in all steps of the development as well as considering the DELICATE checklist to establish legitimate as well as accepted and ethically profound usage of data, when using assessment data to inform the further learning process.

Conclusion

The discourse on ethical guiding principles in connection with AI in the context of higher education and online assessment must be conducted broadly and deeply. Only then AI-based missteps can be avoided, and damage averted for those involved in the learning process. It is important to critically reflect on the potential of the interaction between humans and AI. While AI is superior to humans in terms of capacity and speed in data analysis and prediction, humans are convincing in ethical judgments due to their ability to empathize. Ethics in the design, development, as well as implementation of AI-based online assessment systems should never be an afterthought. Rather, when considering, the benefits of the AI application should be evaluated, and the development process guided under a moral scope. AI-based online assessments need to ensure ethical usage while not undermining potential benefits and focus on the support of learners, teachers as well as institutions, and provide everyone with the needed understanding and competencies in dealing with a large amount of data. For an ethical use of AI in the educational context, sufficient AI competence is crucial and relevant for both teachers and students. Not only should they have access to systems but also bear attitudes, skills, and knowledge about AI to ensure informed decisions and support.

A consensus can be established for the ethically responsible use of AI-based online assessment systems in higher education. This consensus is somewhere between excessive caution and incalculable risks.

References

Alessio, H. M., Malay, N. J., Maurer, K., Bailer, A. J., & Rubin, B. (2017). Examining the effect of proctoring on online test scores. *Online Learning*, *21*(1). Available from https://doi.org/10.24059/olj.v21i1.885, Article 1.

Amigud, A., Arnedo-Moreno, J., Daradoumis, T., & Guerrero-Roldan, A.-E. (2017). Using learning analytics for preserving academic integrity. *International Review of Research in Open and Distributed Learning*, *18* (5), 192−210. Available from https://doi.org/10.19173/irrodl.v18i5.3103.

Bates, T., Cobo, C., Mariño, O., & Wheeler, S. (2020). Can artificial intelligence transform higher education. *International Journal of Educational Technology in Higher Education*, *17*(1), 42. Available from https://doi.org/10.1186/s41239-020-00218-x.

Buckingham Shum, S., & McKay, T. A. (2018). Architecting for learning analytics. Innovating for sustainable impact. *EDUCAUSE Review*, *53*(2), 25−37.

Cavalcanti, A. P., Barbosa, A., Carvalho, R., Freitas, F., Tsai, Y.-S., Gašević, D., & Mello, R. F. (2021). Automatic feedback in online learning environments: A systematic literature review. *Computers and Education: Artificial Intelligence*, *2*, 100027. Available from https://doi.org/10.1016/j.caeai.2021.100027.

Coghlan, S., Miller, T., & Paterson, J. (2021). Good proctor or "big brother"? Ethics of online exam supervision technologies. *Philosophy & Technology*, *34*(4), 1581−1606. Available from https://doi.org/10.1007/s13347-021-00476-1.

Conrad, D., & Openo, J. (2018). *Assessment strategies for online learning: Engagement and authenticity.* Athabasca University Press. Available from https://doi.org/10.15215/aupress/9781771992329.01.

Cowie, R. (2015). Ethical issues in affective computing. In R. Calvo, S. D'Mello, J. Gratch, & A. Kappas (Eds.), *The Oxford handbook of affective computing* (pp. 334−348). Oxford University Press.

Darvishi, A., Khosravi, H., Sadiq, S., & Gašević, D. (2022). Incorporating AI and learning analytics to build trustworthy peer assessment systems. *British Journal of Educational Technology*, *53*(4), 844−875. Available from https://doi.org/10.1111/bjet.13233.

Deeva, G., Bogdanova, D., Serral, E., Snoeck, M., & De Weerdt, J. (2021). A review of automated feedback systems for learners: Classification framework, challenges and opportunities. *Computers & Education*, *162*, 104094. Available from https://doi.org/10.1016/j.compedu.2020.104094.

de Laat, M., & Prinsen, F. R. (2014). Social learning analytics: Navigating the changing settings of higher education. *Research & Practice in Assessment*, *9*, 51−60.

Dellermann, D., Ebel, P., Söllner, M., & Leimeister, J. M. (2019). Hybrid intelligence. *Business & Information Systems Engineering*, *61*(5), 637−643. Available from https://doi.org/10.1007/s12599-019-00595-2.

Dignum, V. (2017). Responsible autonomy. In *Proceedings of the twenty-sixth international joint conference on artificial intelligence* (pp. 4698−4704). Available from https://doi.org/10.24963/ijcai.2017/655

Drachsler, H., & Greller, W. (2016). Privacy and analytics: It's a DELICATE issue a checklist for trusted learning analytics. In *Proceedings of the sixth international conference on learning analytics & knowledge* (pp. 89−98). Available from https://doi.org/10.1145/2883851.2883893.

Ellis, C. (2013). Broadening the scope and increasing the usefulness of learning analytics: The case for assessment analytics: Colloquium. *British Journal of Educational Technology*, *44*(4), 662−664. Available from https://doi.org/10.1111/bjet.12028.

Ferguson, R. (2019). Ethical challenges for learning analytics. *Journal of Learning Analytics*, *6*(3). Available from https://doi.org/10.18608/jla.2019.63.5.

Floridi, L., Cowls, J., Beltrametti, M., Chatila, R., Chazerand, P., Dignum, V., Luetge, C., Madelin, R., Pagallo, U., Rossi, F., Schafer, B., Valcke, P., & Vayena, E. (2018). AI4People—An ethical framework for a good AI society: Opportunities, risks, principles, and recommendations. *Minds and Machines*, *28*(4), 689−707. Available from https://doi.org/10.1007/s11023-018-9482-5.

Gašević, D., Dawson, S., & Siemens, G. (2015). Let's not forget: Learning analytics are about learning. *TechTrends*, *59*(1), 64−71. Available from https://doi.org/10.1007/s11528-014-0822-x.

Gašević, D., Greiff, S., & Shaffer, D. W. (2022). Towards strengthening links between learning analytics and assessment: Challenges and potentials of a promising new bond. *Computers in Human Behavior*, *134*, 107304. Available from https://doi.org/10.1016/j.chb.2022.107304.

Gikandi, J. W., Morrow, D., & Davis, N. E. (2011). Online formative assessment in higher education: A review of the literature. *Computers & Education*, *57*(4), 2333−2351. Available from https://doi.org/10.1016/j.compedu.2011.06.004.

Graf Ballestrem, J., Bär, U., Gausling, T., Hack, S., & von Oelffen, S. (2020). *Künstliche Intelligenz. Rechtsgrundlagen und Strategien in der Praxis*. Wiesbaden: Springer Gabler.

Green, S. K., Johnson, R. L., Kim, D.-H., & Pope, N. S. (2007). Ethics in classroom assessment practices: Issues and attitudes. *Teaching and Teacher Education*, *23*(7), 999−1011. Available from https://doi.org/10.1016/j.tate.2006.04.042.

Heil, J., & Ifenthaler, D. (2023). Online assessment for supporting learning and teaching in higher education: A systematic review. *Online Learning*, *27*(1), 187−218. Available from https://doi.org/10.24059/olj.v27i1.3398.

Holmes, W., Bialik, M., & Fadel, C. (2019). Artificial intelligence in education: Promises and implications for teaching and learning. *Center for Curriculum Redesign*. Available from https://curriculumredesign.org/wp-content/uploads/AIED-Book-Excerpt-CCR.pdf.

Huisman, B., Admiraal, W., Pilli, O., van de Ven, M., & Saab, N. (2018). Peer assessment in MOOCs: The relationship between peer reviewers' ability and authors' essay performance. *British Journal of Educational Technology*, *49*(1), 101−110.

Ifenthaler, D. (2015). In J. M. Spector (Ed.), *The SAGE encyclopedia of educational technology* (Vol. 2, pp. 448−451). Sage. Available from https://doi.org/10.4135/9781483346397.n187.

Ifenthaler, D. (2017). Are higher education institutions prepared for learning analytics? *TechTrends: Linking Research & Practice to Improve Learning*, *61*(4), 366−371. Available from https://doi.org/10.1007/s11528-016-0154-0.

Ifenthaler, D. (2022). *Automated essay scoring systems. Handbook of open, distance and digital education* (pp. 1−15). Singapore: Springer Nature. Available from https://doi.org/10.1007/978-981-19-0351-9_59-1.

Ifenthaler, D., Greiff, S., & Gibson, D. (2018). Making use of data for assessments: Harnessing analytics and data science. In *International handbook of IT in primary and secondary education* (2nd ed.). Available from https://doi.org/10.1007/978-3-319-53803-7_41-1.

Ifenthaler, D., & Greiff, S. (2021). Leveraging learning analytics for assessment and feedback. In J. Liebowitz (Ed.), Online learning analytics (pp. 1−18). Auerbach Publications. Available from https://doi.org/10.1201/9781003194620.

Ifenthaler, D., & Schumacher, C. (2016). Student perceptions of privacy principles for learning analytics. *Educational Technology Research and Development*, *64*(5), 923−938.

Ifenthaler, D., & Yau, J. Y.-K. (2020). Utilising learning analytics to support study success in higher education: A systematic review. *Educational Technology Research and Development*, *68*(4), 1961−1990. Available from https://doi.org/10.1007/s11423-020-09788-z.

Ke, Z., & Ng, V. (2019). Automated essay scoring: A survey of the state of the art. In *IJCAI* (Vol. 19, pp. 6300−6308).

Kunnan, A. J. (2020). A case for an ethics-based approach to evaluate language assessments. In G. J. Ockey, & B. A. Green (Eds.), *Another generation of fundamental considerations in language assessment* (pp. 77−93). Springer Singapore. Available from https://doi.org/10.1007/978-981-15-8952-2_6.

Larrabee Sønderlund, A., Hughes, E., & Smith, J. (2019). The efficacy of learning analytics interventions in higher education: A systematic review. *British Journal of Educational Technology*, *50*(5), 2594−2618.

Lee, K., & Fanguy, M. (2022). Online exam proctoring technologies: Educational innovation or deterioration? *British Journal of Educational Technology*, *53*(3), 475−490. Available from https://doi.org/10.1111/bjet.13182.

Luckin, R. (2017). Towards artificial intelligence-based assessment systems. *Nature Human Behaviour*, *1*(3), 0028. Available from https://doi.org/10.1038/s41562-016-0028.

McWhorter, R. R., Delello, J. A., Roberts, P. B., Raisor, C. M., & Fowler, D. A. (2013). A cross-case analysis of the use of web-based ePortfolios in higher education. *Journal of Information Technology Education: Innovations in Practice*, *12*, 253−286.

Murchan, D., & Siddiq, F. (2021). A call to action: A systematic review of ethical and regulatory issues in using process data in educational assessment. *Large-Scale Assessments in Education*, *9*(1), 25. Available from https://doi.org/10.1186/s40536-021-00115-3.

Nguyen, A., Ngo, H. N., Hong, Y., Dang, B., & Nguyen, B.-P. T. (2022). Ethical principles for artificial intelligence in education. *Education and Information Technologies*. Available from https://doi.org/10.1007/s10639-022-11316-w.

Prinsloo, P., & Slade, S. (2014). *Student data privacy and institutional accountability in an age of surveillance. Using data to improve higher education* (pp. 195−214). Brill.

Reilly, E. D., Williams, K. M., Stafford, R. E., Corliss, S. B., Walkow, J. C., & Kidwell, D. K. (2016). Global times call for global measures: Investigating automated essay scoring in linguistically-diverse MOOCs. *Online Learning*, *20*(2), 217−229.

Richards, D., & Dignum, V. (2019). Supporting and challenging learners through pedagogical agents: Addressing ethical issues through designing for values. *British Journal of Educational Technology*, *50*(6), 2885−2901. Available from https://doi.org/10.1111/bjet.12863.

Romero, C., & Ventura, S. (2013). Data mining in education. *WIREs Data Mining and Knowledge Discovery*, *3*(1), 12−27. Available from https://doi.org/10.1002/widm.1075.

Rosé, C. P., McLaughlin, E. A., Liu, R., & Koedinger, K. R. (2019). Explanatory learner models: Why machine learning (alone) is not the answer. *British Journal of Educational Technology*, *50*(6), 2943−2958. Available from https://doi.org/10.1111/bjet.12858.

Scholes, V. (2016). The ethics of using learning analytics to categorize students on risk. *Educational Technology Research and Development*, *64*(5), 939−955.

Shaikh, E., Mohiuddin, I., Manzoor, A., Latif, G., & Mohammad, N. (2019). Automated grading for handwritten answer sheets using convolutional neural networks. In *2019 2nd International conference on new trends in computing sciences (ICTCS)* (pp. 1−6). Available from https://doi.org/10.1109/ICTCS.2019.8923092.

Shermis, M. D., Burstein, J., Higgins, D., & Zechner, K. (2010). Automated essay scoring: Writing assessment and instruction. In P. Peterson, E. Baker, & B. McGaw (Eds.), *International encyclopedia of education* (*3rd ed.*, pp. 20−26). Elsevier. Available from https://doi.org/10.1016/B978-0-08-044894-7.00233-5.

Slade, S., & Prinsloo, P. (2013). Learning analytics: Ethical issues and dilemmas. *American Behavioral Scientist*, *57*(10), 1510−1529. Available from https://doi.org/10.1177/0002764213479366.

Spector, J. M. (2016). Ethics in educational technology: Towards a framework for ethical decision making in and for the discipline. *Educational Technology Research and Development*, *64*(5), 1003−1011. Available from https://doi.org/10.1007/s11423-016-9483-0.

Swiecki, Z., Khosravi, H., Chen, G., Martinez-Maldanado, R., Lodge, J., Milligan, S., Selwyn, N., & Gašević, D. (2022). Assessment in the age of artificial intelligence. *Computers and Education: Artificial Intelligence*, 100075. Available from https://doi.org/10.1016/j.caeai.2022.100075.

Tzimas, D., & Demetriadis, S. (2021). Ethical issues in learning analytics: A review of the field. *Educational Technology Research and Development*, *69*(2), 1101−1133. Available from https://doi.org/10.1007/s11423-021-09977-4.

Webb, M., & Ifenthaler, D. (2018). In J. Voogt (Ed.), *Section introduction: Using information technology for assessment: Issues and opportunities* (pp. 577−580). Springer International Publishing. Available from https://doi.org/10.1007/978-3-319-53803-7_101-1.

Zawacki-Richter, O., Marín, V. I., Bond, M., & Gouverneur, F. (2019). Systematic review of research on artificial intelligence applications in higher education—Where are the educators? *International Journal of Educational Technology in Higher Education, 16*(1), 1−27.

Zlatkin-Troitschanskaia, O., Schlax, J., Jitomirski, J., Happ, R., Kühling-Thees, C., Brückner, S., & Pant, H. A. (2019). Ethics and fairness in assessing learning outcomes in higher education. *Higher Education Policy, 32*(4), 537−556. Available from https://doi.org/10.1057/s41307-019-00149-x.

Zupanc, K., & Bosnić, Z. (2017). Automated essay evaluation with semantic analysis. *Knowledge-Based Systems, 120*, 118−132. Available from https://doi.org/10.1016/j.knosys.2017.01.006.

Ethical aspects of automatic emotion recognition in online learning

Gabriela Moise and Elena S. Nicoară
Petroleum-Gas University of Ploiеşti, Ploiеşti, Romania

Introduction

Artificial intelligence (AI) is evolving at a fast pace and becomes ubiquitous in many domains, including education. It is imperative that humans raise their awareness regarding the impact of AI in everyday life and act accordingly. Due to its vital role in the society, education is to foster its strategies, span, infrastructure, and skills so that it may face new challenges (COMEST World Commission on the Ethics of Scientific Knowledge and Technology, 2019; EGE European Commission, European Group on Ethics in Science and New Technologies, 2012). Automatic emotion recognition (AER) for the educational process comes with major ethical implications that need to be carefully considered.

Writings and theories on ethics and morality date back to ancient times; philosophers such as Socrates and Plato pioneered the field, and Aristotle was the first who used the term "ethics." Even if no general agreement on the use of "morality" and "ethics" terms was achieved, the convention that an ethical judgement has to be moral and grounded in reason proves reasonable. Emotion and whatever relates to it are naturally connected with ethics, as emotion is the root of what makes people moral beings (Cowie, 2015).

Machine ethics focuses on assuring that the behavior of machines in relation with humans and other machines is "ethically acceptable," its essential goal being "to create a machine that itself follows an ideal ethical principle or set of principles" (Anderson & Anderson, 2007, p. 15). A distinction between implicit and explicit ethical machines is proposed by Moor (2006, p. 19): Implicit ethical machine refers to machines coded "to promote ethical behaviour," while for an explicit ethical machine ethics are explicitly represented and the machine operates effectively based on this knowledge. The term "ethically aligned design" in AI is mapped to design processes that explicitly include human values (IEEE Global Initiative on Ethics of Autonomous and Intelligent Systems IEEE Global Initiative, 2018), and for that purpose, data engineers educated in ethics are a prerequisite (COMEST World Commission on the Ethics of Scientific Knowledge and Technology, 2019).

Practical ethics is focused on common-sense basic principles (Ross, 1939) used to evaluate a person or action as being ethical or unethical. These principles, which could apply to any entity, are fidelity, reparation, gratitude, nonmaleficence, and beneficence. Autonomy and equity were added by Goldie et al. (2011), the most famous philosopher who wrote on ethics in affective computing according to Cowie (2015).

Concerns regarding ethics in AER for online learning are manifold: Do affective computing and automatic emotion recognition have a solid scientific foundation from an ethical perspective? To

Ethics in Online AI-Based Systems. DOI: https://doi.org/10.1016/B978-0-443-18851-0.00003-2

what extent ethics is considered in affective computing (AC) and AER, both in the literature and in the deployed systems? What are the benefits of AER in education? What are the ethical risks of AER in education, including both the overlooked and the hidden ones?

This chapter is structured as follows: An overview of AC and AER (in online learning) is presented, along with their ethical implications, in order to note the level of attention paid to ethics in AER for online learning; several ethical guidelines and frameworks in the literature are considered. Roles and benefits of the AER in online learning are identified, and a comprehensive list of specific ethical risks is provided. A special section is dedicated to the ethical model for AER systems in online learning that we propose to inform the decision-making factors in the development, usage, and evaluation of AER-based systems. Three case studies of AER systems in education are reviewed, and a use case is proposed. The final section is dedicated to discussions and conclusions, pinpointing the problems encountered in AER ethics, our achievements, and future research ideas.

Emotions, affective learning, and ethical implications

The process of designing, coding, and deploying AI-based systems cannot be superficial. Every stage of the lifecycle of AI systems requires solid ethical principles and values to comply with, and, therefore, it should be grounded in science psychology rather than in computing (Cowie, 2015). The novelty of the AI ethics as a research topic is witnessed by Bakiner (2022) in a survey he carried out on over 221 academic journal papers. Although there has been some progress registered in the terminology and similar problems and solutions in different domains, the review of the existing literature has revealed a small number of commonly cited authors, with Bakiner concluding that AI ethics is not a consistent field yet.

Researchers, practitioners and the public approach AI ethics differently, namely: (1) as a solution to problems where human or machine agents must make difficult decisions, (2) as a call for reflectiveness on AI technologies, (3) as an adequate distribution of moral and legal responsibility between humans and machines, or simply (4) as a critique of systems and institutions for the oppression and exploitation (Bakiner, 2022).

Picard (1995, 1997) pioneered the affective computing domain to identify the role of emotions in the complex human-machine relationship and to study new abilities of the machine, namely those of recognizing and expressing emotions. AC was defined as "computing that relates to, arises from, or influences emotions" (Picard, 1995, p. 1). It may influence various fields, such as education, healthcare, games, software development, marketing, website customization, etc.

Nevertheless, AC has profound moral implications, derived mainly from mimicking and influencing human free will, that require high attention of all stakeholders. It comes with two contradicting positions: Either it can serve the betterment of the community, or it can contribute to its destruction (Cowie, 2015). Accordingly, machine ethics plays a major role in making AC serve the community (Fig. 4.1).

To build AER systems, various models of emotions are used, mainly the discrete model centered on basic emotions (Ekman, 1999; Ekman et al., 1969) and the dimensional model based on emotion valence, arousal, and dominance (Mehrabian, 1996; Russell & Mehrabian, 1977). Other emotion theories have been developed, such as the cognitive appraisal theory of emotions (Scherer, 1999) and the theory of constructed emotions (Barrett, 2017). Nevertheless, these are not

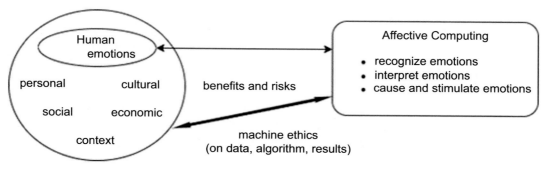

FIGURE 4.1

Ethics in affective computing.

considered in AER-based systems, as they are difficult to be coded in such systems. Moreover, a difference between trait emotions (affective tendencies, such as depression and anxiety) and state emotions (emotional episodes, such as joy or sadness) exists. As Hascher (2010) remarks, trait emotions influence learning probably in a larger extent than state emotions.

In spite of the strong, complex, and bidirectional relationship between emotions and learning, no general well-specified rules were found and research on emotions in education was limited prior to the 1990s. Anxiety represents an exception in this respect, as it has been a research topic since before the 1950s (Pekrun, 2005; Pekrun et al., 2002; Rosenfeld, 1978; Zeidner, 1998).

The term "affective learning" was introduced in "Affective Learning—A Manifesto" (Picard et al., 2004) to cover all topics related to students' affective (or emotional) states in learning condition and the usage of technology in emotion recognition. Most of the research has been conducted to identify state emotions experienced by students in various learning scenarios, including technology-based contexts. A metaanalysis of 24 studies considering four such contexts (i.e., intelligent tutoring systems, serious games, simulation environments, and simple computer interfaces) found that engagement is the predominant affective state in learning (D'Mello, 2013). The importance of detecting students' engagement in the online learning in order to create an effective learning process is highlighted by Dewan et al. (2019). Other general state emotions in academic contexts are pride of success, anxiety, hopefulness, hopelessness, relief, enjoyment of learning, anger, shame, boredom, surprise, sadness, frustration, confusion, happiness, fear, joy, disgust, interest, curiosity, contempt, delight, and excitement (D'Mello, 2013; Pekrun et al., 2002; Yadegaridehkordi et al., 2019).

It has been demonstrated that positive and negative emotions improve or impede learning, respectively. An experiment conducted on 118 college students proved that comprehension and knowledge transfer increase by inducing positive emotions through specially designed learning resources (Um et al., 2012). Most positive emotions are beneficial for creativity, for example. Negative emotions, such as confusion, can be useful in learning if controlled, regulated, and resolved (D'Mello et al., 2014). However, it is naive to consider that positive emotions have exclusively positive effects on learning and that negative emotions have only negative effects on learning (Hascher & Edlinger, 2009).

A comprehensive study on AC in education (Yadegaridehkordi et al., 2019) analyzes 94 papers published in prestigious databases between 2010 and 2017 and reports the following: (1) an increasing trend of AC in education, (2) the relationship between emotion, motivation, learning style, and cognition as the main research topic discussed in the literature, (3) the use of visual, textual, vocal, physiological, and multimodal channels for affective measurements, and (4) the preference for using dimensional models in the education area.

Online education is, undoubtedly, experiencing a broad acceptance and rapid development, and emotions need to be suitably addressed in this context. As teachers and students do not generally communicate in a synchronous manner within online environments, teachers cannot recognize learners' emotions, so their teaching methods are not adapted to the actual learning context (Duo & Song, 2012). When the machines mediate the learning process, automatic emotion detection is envisaged to streamline the process, open datasets start to be available (Dewan et al., 2019), and AI is employed for this purpose.

Nevertheless, the ethical impact of AC on education is mildly discussed in the literature and only few academic papers contain ethics statements. They mainly regard the ethics commission's approval for the research (Arya et al., 2021; Ashwin & Guddeti, 2020; Cowie, 2015; Kazemitabar et al., 2021), users' privacy (Sharma et al., 2022; Vidanaralage et al., 2022), data protection and the informed consent if physiological signals are used (Arya et al., 2021), the bias caused by small context-specific datasets (containing accurate expressions of affective states), or data protection for underage students (Dai & Ke, 2022).

Emotion recognition algorithms are studied from the performance perspective, but their ethical aspects and disadvantages are generally neglected. The main drawbacks of detecting emotions based on facial expressions are caused by the insufficient knowledge on expressing and perceiving emotions (Barrett et al., 2019). A more profound aspect about AI, emotions, and facial recognition is that AI is portrayed by the entities that produce and profit from AI technology as a way of understanding people and the world, claiming that neural networks "predict" personality based on facial analysis; moreover, these entities lead to the idea that the use of AI systems is common sense (Goldenfein, 2020). Online proctoring, for example, beyond defeating academic dishonesty, raises critical ethical concerns (on domestic surveillance and autonomy denial). Applications such as Proctorio, Examus, and Honorlock record images, ambient sound, motion, keyboard and other device usage, screen and browser activities and may require a smartphone for a 360 degrees view of the student's environment. Additionally, Cowie (2015) notes the deception associated with automatic agents working with emotions being perceived as having emotional competence.

In sensitive areas such as education, the regulators play an essential role in designing ethically aligned AI. The AI Now Institute within New York University construes that regulators "should ban the use of affect recognition in important decisions that impact people's lives and access to opportunities," including those related to evaluating students' performance in school (Crawford et al., 2019, p. 6).

Ethical guidelines and frameworks

Governmental bodies, authorities, international commissions, national regulators, and professional associations responsible for regulating AI technology design and use have issued and updated ethical guidelines, frameworks, and codes, but only few have the normative role and power of enforcement over the organizations that deploy AI (Crawford et al., 2019). In the UK, for example, no regulatory body is charged with supervising the use of AI and ML (IFOW Institute for the Future of Work, 2020). Critical issues related to this topic are the following:

- the difficulty of regulating AI field as a result of the impossibility to monitor it at the global level (COMEST World Commission on the Ethics of Scientific Knowledge and Technology, 2019);
- the difficulty of identifying the right decider on the reasonability of actions in an AI context (IFOW Institute for the Future of Work, 2020);
- the probability of rejecting AC as being ethically unacceptable by people (Cowie, 2015).

Several ethical codes may be useful for AER (i.e., ACM Association for Computing Machinery, 2018; APA American Psychological Association, 2017; EGE European Commission, European Group on Ethics in Science and New Technologies, 2012; EUCFR European Union Charter for Fundamental Rights, 2012), but ethics is only partially addressed, considering exclusively aspects such as privacy, access, confidentiality, integrity of personal data, explicit consent, and protection against cybercrime.

Ethical values regarded as motivating ideals and foundations for ethical principles are described and structured in many frameworks (ACM Association for Computing Machinery, 2018; EGE European Commission, European Group on Ethics in Science and New Technologies, 2012; ICO Information Commisioner's Office, Alan Turing Institute, 2020; Jobin et al., 2019; Leslie, 2019; Mohammad, 2022; UNESCO, 2021, etc.). A review of 84 frameworks and guidelines reveals the commonly addressed ethical principles, namely transparency, justice and fairness, nonmaleficence, responsibility, and privacy (Jobin et al., 2019). Transparency includes the mandatory personal data breach notification and managing one's personal data (EGE European Commission, European Group on Ethics in Science and New Technologies, 2012). Accessibility, autonomy, data protection, beneficence, robustness, safety and security, trust, awareness and literacy, multistakeholder and adaptive governance and collaboration, human oversight, and determination are other ethical principles (Jobin et al., 2019; UNESCO, 2021).

A salient ethical framework for "responsible design and implementation of AI systems in the public sector" is provided by Leslie (2019). The author proposes a three-level structure for the framework and four purposes to be aimed at within AI projects: permissibility, fairness, trustworthiness, and justifiability. Due to its rigorous, comprehensive, and well-structured ethical perspective in all stages of an AI project, this framework will be the foundation of our AER model further detailed.

The first level (L1), "SUM Values," is governed by the ethical values that "Support, Underwrite, and Motivate" the responsible initiative. It addresses the impacts of the project on the communities and its key incentives are "respect the dignity of individual person; openly, sincerely, and inclusively connect with each other; care for the wellbeing of people; and protect the social values and public interest" (p. 10). The second level (L2), "FAST Track Principles," provides applicable directions to the responsible design and use of AI technology. It refers to fairness, accountability, sustainability, and transparency. Fairness settles the principle of discriminatory nonharm and data fairness (properly representative, relevant, accurate, and generalizable datasets for training and testing), design fairness (reasonable, morally nonobjectionable, justifiable correlations and inferences), outcome fairness, and implementation fairness (responsible deployment). Accountability addresses responsibility, and sustainability assures an ethical long-term effect of the AI system on the community. Transparency refers to the interpretability and explicability of AI systems and represents a solid justification in favor of AI-based technologies. The most practical level (L3) is the "PBG Framework" (Process-Based Governance Framework). It provides a transparency-based way to integrate SUM Values and FAST Track Principles in the AI project development workflow. Leslie's framework can be rigorously implemented provided that the project development team "Reflect" on "SUM Values," "Act" according to "FAST Track Principles," and "Justify" based on "PBG Framework," which means that the organization is driven by authentic ethical principles established by inherently moral people.

A very useful AI wide-range risk analysis, based on a 106-question template for a responsible AI project, indicates the risk levels for every adverse impact, where gravity potential and number of rights holders affected are considered (Leslie et al., 2021).

Before developing any AC technology, several vital questions need to be addressed: Is the recognition of human emotion by machines ethical or not? Why should such a technology be developed? Who will benefit from it and who will not? (Mohammad, 2022) How far does ethical accountability go? (Cowie, 2015). Additionally, Leslie et al. (2021) assert that the team should not elaborate the AI project if its lawfulness is not clearly established.

Recommendations for ethical AC/AER are found in all mentioned guidelines and frameworks. Mohammad (2022) provides 57 such prescripts, including the following: Be aware that privacy is not about secrecy, but personal choice; choose to use intrinsically interpretable/clear box models (where users can easily understand why the system predicts the result) versus opaque box models (where users need additional tools to understand the prediction reasons); mind that all means of AER system can be misused; include neurodiverse and neurotypical participants for data annotations; be aware that no emotion recognition method is perfect. Other recommendations to follow are the following: Do risk prevention, take mitigation and monitoring measures (UNESCO, 2021), use multimodal recognition for emotions (e.g., by facial expressions, dialog, and posture together), anonymize data and use only those that are necessary for the targeted purposes (CDDO Central Digital & Data Office, 2020; COMEST World Commission on the Ethics of Scientific Knowledge and Technology, 2019), responsively handle the profiles assigned to data subjects (EGE European Commission, European Group on Ethics in Science and New Technologies, 2012), include human-in-the-loop or human-on-the-loop mechanisms, and methods for users to be able to opt to revert to human intervention when high level of automation is present in the system (Leslie et al., 2021), employ explicitly given consent rather than assumed consent of the subjects, verify emotion identification with two or three persons in different areas of expertise, and assure that the expansion of the system is controlled by certain rules.

Automatic emotion recognition in education: roles, benefits, and ethical risks

Giving "emotional abilities" to computers, as AER aims at, refers to inferring emotions felt by the user, emotions that the user attempts to convey, emotions triggered in the user, emotions' intensity, moods and emotion dynamics, attitudes, and sentiments toward a target (Mohammad, 2022). Ethics in emotion perceiving by computers is still at its beginning, as ethics focuses traditionally on assessing actions rather than perception (Cowie, 2015).

AER uses the capabilities enabled by computer vision and computer listening; by that, everything about the "real world" that is visible, audible, or otherwise sensible is recorded, computationally analyzed, and classified in real time (Goldenfein, 2020). Data is extracted from facial microexpressions, iris data, gait, stance and gesture, speech, voice intonation, biophysical signals (skin and blood conductance, blood flow, respiration, infrared emanations, and brain waves), haptic data such as force of touch, typed text, emoticons, emojis, or self-reported questionnaires.

Various machine learning (ML) techniques are used. Despite their results, major drawbacks regard unrepresented people and the fact that huge balanced data is needed for a better representation. Nevertheless, one should be aware of the fact that the more data, the higher the ethical issues and risks.

To identify the ethical impact of AER-based online learning, we first reviewed the roles of AER systems in online learning. ScienceDirect database returns 102 results regarding the roles of AER systems in online learning for the interval 2015—22, using "emotions recognition" and "online learning" as search phrases. Relevant sources were determined based on the titles and by abstract/preface screening, and finally a full text screening was performed. Out of the 102 scientific works, 27 (26.5%) were considered relevant for our purpose, 26 studies referring to students' emotions and only one to teachers' emotions. Only five of the 27 (18.5%) include considerations on ethics.

The list of AER systems' roles with their benefits was supplemented with the results of our academic experience and presented in Table 4.1. For each role, both students and teachers are envisaged as possible users.

Table 4.1 AER roles and benefits in online learning.

AER role	AER benefits	Resources
Designing and building intelligent tutoring systems (ITS) as pedagogical agents with emotional abilities, which adapt instructions to students' performance and learning profile (based on the relationship between students' emotions, motivation, cognition, and learning styles)	• Customized feedback; • Better effectiveness of learning; • Support for teachers regarding educational resources and teaching strategies;	Xu et al. (2018), Alwadei and Alnanih (2022), Kazemitabar et al. (2021), Sikström et al. (2022), Dai and Ke (2022), Cen et al. (2016), Chen and Wu (2015), Feidakis (2016), Faria et al. (2017), Lin and Kao (2018), Yang et al. (2018), Imani and Montazer (2019), Ez-zaouia et al. (2020), Iulamanova et al. (2021)
Supporting engagement and motivation of learners and teachers	• Better engagement and motivation of the participants; • Support for raising the students' awareness on their mental and emotional states;	Chen and Wu (2015), Feidakis (2016), Yadegaridehkordi et al. (2019), Hasnine et al. (2021), Bhardwaj et al. (2021), Lavoué et al. (2021), Liu et al. (2022), Sharma et al. (2022), Qiao et al. (2022), Vidanaralage et al. (2022), Lyu et al. (2022)
Learning assessment	• Fraud-free results;	Tanko et al. (2022)
Teaching assessment	• Support for raising the teachers' awareness on the role of emotions in teaching;	Utami et al. (2019)
Building comfortable learning environments	• More effective learning and teaching;	Arya et al. (2021), Alfoudari et al. (2021)
Supporting students with special needs (ADHD, anxiety, and so on)	• Better educational help for students with special needs.	Alwadei and Alnanih (2022), Sikström et al. (2022)

To identify all the ethical risks or actual harms, an iterative search-review-discussion process was carried out. There was generated a preliminary list with 24 risks identified both in the literature and in our academic experience. Based on the authors' experience of 20 + years in computer science and artificial intelligence, these risks were systematized in sixteen classes. As the study focuses mainly on students, a brainstorming session with ten computer science students followed in order to refine the preliminary list and teachers' observations, and a final list with possible ethical risks was obtained.

The targeted AER ethical risks reported in the literature are manifold. Some are general AI risks (as presented by COMEST World Commission on the Ethics of Scientific Knowledge and Technology, 2019; Cowie, 2015; EGE European Commission, European Group on Ethics in Science and New Technologies, 2012; Goldenfein, 2020; Leslie, 2019; Sadowski et al., 2021), whereas others are risks specific to the educational process (Arya et al., 2021; Lyu et al., 2022; Mohammad, 2022; Sharma et al., 2022). Some of the sixteen identified classes of AER's ethical risks in education have many facets and more practical implications, whereas others have inherently descriptive titles. The sixteen classes are the following:

1. Bias and discrimination, due to poor representativeness of data or based on the designers' preconceptions or bad intent. Only large volumes of balanced representative data are adequate for proper results, but the independent rigorous researchers generally have a limited access to data that technology companies own (Sadowski et al., 2021). Moreover, there is enormous variability in human mental representation and expression of emotions (Mohammad, 2022). AI is "not neutral, but inherently biased," a reason being that "classification is culture-specific and a product of history" (COMEST World Commission on the Ethics of Scientific Knowledge and Technology, 2019, pp. 7, 8). For example, recommender agents suggest items by using discriminating filters (Shelton, 2022), and, as a result, certain resources are presented to students, whereas others are hidden, based on the ML designers' line of thought.

2. Unreliable, unsure, unsafe, or poor results. AER systems may indicate causal relations between data, which in fact do not exist (e.g., racial or gender differences in intelligence or learning outcomes) and could poorly recognize emotions by ignoring, for example, the possible distracting factors in the environment. Consequently, some inferred emotions may be further misinterpreted.

3. Nontransparent, unexplainable, unjustifiable, or not fully predictable outcomes. AI results cannot be fully predictable or explainable.

4. Privacy invasion by (1) inaccurate ownership and management of personal data, (2) failure in giving and withdrawing consent, and (3) domestic surveillance. Video analytics used in proctoring or keeping attendance in class leads to vast databases with personal data. Facial expression, voice, gait, physiological signals, and other biometric data are highly sensitive data. Other recorded data might include institutional affiliation of the users, their skin color, age, gender, what he/she is doing at that moment, and who the person has been associating with (Goldenfein, 2020). When interacting with emotion or facial recognition in different settings (homes, schools, or outdoors), individuals perceive their privacy differently (Sharma et al., 2022). They often make their data available without being aware of the implicit or explicit acceptance of the hidden purposes (EGE European Commission, European Group on Ethics in Science and New Technologies, 2012). The risk of defective ownership and management of personal data raises when the data becomes accessible to many parties who do

not intend to protect the subjects (Arya et al., 2021), especially when the risk of deanonymizing through data linkage with existing data is present (Leslie et al., 2021). A free and fully informed consent of the users to participate with data is needed when using AER systems (for training or deployment, e.g., in online assessing). It includes mechanisms to ensure that the user clearly understands which data is gathered and processed, to whom it is accessible, for how long, for which goals, and what the related risks are. As AER attempts to predict emotions, behaviors, and personality type (Goldenfein, 2020), the risks of privacy invasion increase. Moreover, in education group privacy should be addressed.

5. Unfairness and digital division. The sequent inequality, exclusion, threat to cultural diversity, and exploitation of the vulnerable groups impede education as a public service. For the students with special needs (e.g., ADHD, anxiety, depression, alexithymia, and autism), fairness and equality while handling online learning with AER is hazardous. More personal data is regularly captured from these subjects, which may lead to more privacy risks. If academic institutions used AER mandatorily, education would become accessible only to those who accept the related risks and economically afford it. The so-called "digital divide" regards the access to data, algorithms, and human and computational resources (COMEST World Commission on the Ethics of Scientific Knowledge and Technology, 2019).

6. Deception. Unthought of or deliberate deception regarding AER is threefold. First of all, the pseudo-scientific base of emotion recognition may mislead toward the belief that ML could ever infer one's true emotional state (Mohammad, 2022). AER can infer some aspects about one's state emotion, but this does not equal trait emotions. Variability of emotion expression, if not broadly considered, leads to the false message that AER decides which emotion is "standard," other forms of expression being ignored or invalidated. Moreover, certain theories on emotions are simply neglected because they are difficult to be coded in data, and various factors with impact on the affect are omitted when labeling emotions (i.e., physical and mental illnesses). Consequent risks for students are a limited or wrongly conditioned access to resources, be them educational or studentship grants, and a negative influence on students' evaluation. Secondly, the illusion of "emotions" and "intentions" of automated systems as artifacts, in contrast to humans as responsible agents, is better acknowledged. On the one hand, the entire AC domain is inherently deceptive; on the other hand, people themselves cannot be labeled as dishonest when someone shows an emotion not corresponding to the internal feeling (Cowie, 2015). Thirdly, when untrue ethics is presented as true, deceiving by ethics washing occurs. Examples in online learning are (1) partially informed consent presented as an explicit and fully informed consent and (2) avoidance of clear indication of gray area decision-making or the low number of accurate metrics. Deception becomes obviously unethical when it impairs the users' possibility to exercise autonomy (Cowie, 2015).

7. Manipulation and building authoritarian relations. This risk manifests when inferred emotions are misused. Inspecting the moments when students and teachers are most receptive for outer suggestions, their behavior can be easily manipulated. Automated agents that would combine impeccable logic with infinite patience, no conscience, and the ability to manipulate emotion would create an almost irresistible persuader (Guerini & Stock, 2005). Also, unethical teachers or faculty could build authoritarian relations by misusing students' data in exams or in any other life event.

8. Changes in human perception of reality, understanding, expertise, and natural behavior. The term "emotional perception" related to AER, which is far less accurate than the emotional perception of vigilant humans, could influence users' trust in their own ability to naturally perceive emotions, be them of their own or of other people. If a transparent AER-based online assessment agent explains to the student its prediction that he/she cheated, because the student's eyes moved for one minute in the upper left part of his/her visual field, while the face turned pink, that student could wrongly change the perception regarding other people's similar actions as indicating a form of cheating. By contrast, many times, people tend to look to the upper left when they try to remember something or to maintain their point of view while speaking directly to someone; face turning pink could mean a burst of emotions of various types (shame, joy, anger, etc.) or even only a physiological blood-circulation alteration. More important, the perceived meaning of "facts" and "truth" can be deeply influenced by AI algorithms (COMEST World Commission on the Ethics of Scientific Knowledge and Technology, 2019), which is critical for education. The more accustomed one is to automated decisions, the more their visual understanding of the world is challenged and tends to be changed, minimized, and "even negated by computational ways of sensing and knowing" (Goldenfein, 2020, p. 5). As virtual reality and augmented reality frequently complement AER, this risk increases significantly.

9. Erroneous portraying of human beings and emotions. When people describe an emotion, they pass a moral judgment on its justifiability, for example, in order to have an accurate image of that person. Is a machine entitled to do the same or can it do so? When a student is angry because a colleague is disturbing his/her learning, the AER software eventually labels anger and may connect it strictly with learning, ignoring external factors.

10. Denial or bypassing of individual autonomy and rights (restriction on users' ability to exercise free will or free speech, nonfree and noninformed decisions regarding users, and denial of right against self-incrimination). Users' autonomy and fundamental rights venture to be affected if their religion, lifestyle, culture, and government are not envisaged while designing the AER system. Some examples of denial and bypassing risks are students or teachers may not want their emotions to be inferred; teachers' pedagogical experience could be ignored by the automated system in evaluating students' learning profile, in using the most appropriate methods and resources for teaching; students' own educational goals may be ignored in the tutoring process. When people feel monitored, their reactions, thinking, and creativity devolve. Moreover, some assessment strategies could limit the liberty of expression or students' critical thinking. Another limitation in autonomy arises with the excessive decisional help provided by the AER system, which leads to nonversatile students, who are poorly resilient or lacking critical thinking and, as a result, unprepared for the real world.

11. Dual use (the risk of using AER functionalities developed for a certain context in other more sensitive contexts, such as healthcare, civil liberties, universities, or countries, where data protection and other rights are not entirely observed).

12. Isolation of individuals, disintegration of social connections, and dehumanizing of people relations by emotional and social interaction with high performance, yet lacking self-awareness, AI systems. Most of the online classes deplete students' inherent ability to collaborate and clearly dehumanize human relations.

13. Dependence on a machine. Robots with "emotional intelligence" could shape undesirable attachment of users on a machine for learning/teaching efficacy or for their emotional well-being. Emotional aid used for a long time or assisting children reduces the natural ability to self-regulate the emotional status.

14. Risk of losing the sense of individual identity. The user of an AI system loses the sense of individual identity, if the system places them on the position of an insignificant or helpless actor (Leslie et al., 2021) and even the minimization of the emotion's role may easily occur.

15. Replacement of the teachers. "One of the main societal concerns regarding AI is labour displacement," as World Commission on the Ethics of Scientific Knowledge and Technology asserts (COMEST World Commission on the Ethics of Scientific Knowledge and Technology, 2019, p. 9). With growing technology and more and more functionalities in automated tutoring, the need for continuous IT upskilling for teachers and students significantly narrows the teaching role of humans in education. The lack of transparency on this issue intensifies the job replacement hazard, while human-to-human teaching remains the main factor to foster learning (by the complex human direct interactions).

16. Lack of energetic sustainability. Ever-increasing big data pretrained AI models require huge energy consumption and solutions with short-term efficiency.

Once acknowledged, specific risks must be properly computationally codified in order to be useful in an ethical design and deployment of AER system, as ignoring not yet encoded aspects is a risky strategy in itself (Cowie, 2015).

A critical reflection on AER facilitating a responsible emotion research and on the proper use of AER technology is provided by Mohammad (2022). The author points out the commercial and governmental uses of emotion recognition and insists on the active engagement of the AER community in considering ethical ramifications of their creation.

Our model that will be further detailed accounts the ethical purposes in both AI and AER and proposes solutions to prevent and mitigate the negative ethical impact of these technologies.

Ethical automatic emotion recognition model for online learning

The ethical guide composed by Leslie at Alan Turing Institute (2019) represents a real tenet for the responsible AI systems. In this section, we propose a human-centered ethical AER model for online learning, based on Leslie's guide, on the Ethics sheet, on the sentiment analysis carried out by Mohammad at the National Research Council in Canada (2022), and on the data ethics framework of the Central Digital and Data Office (CDDO Central Digital & Data Office, 2020). In our opinion, it is mandatory that the evaluation of the AER system's feasibility be rooted in ethics and safety.

The numerous choices to make when developing and using AER in education have high long-term ethical impact. Therefore, a systemic choice architecture supporting all stakeholders (designers, testers, implementers, users, and so forth) to make their best choices (Kulkarni, 2022) is required. Various scenarios on choosing the emotion representation model, human values to base on, solutions to mitigate the tension between human variability and machine normativeness, criteria regarding people behind the data, and even the working team are carefully designed and tested.

Our model (Fig. 4.2) is to be best implemented following Agile-Waterfall hybrid methodology, namely to design, plan, and define requirements with Waterfall and to develop and test with Agile. The highest ethical risks were emphasized for each stage of the process.

In the problem formulation stage, the goal of the proposed AER system is set, and the particular context of its application is defined, including the domain's regulatory environment and human and technological systems planned to be replaced by it (Leslie et al., 2021). The users' needs (starting with the most disadvantaged individuals in the context) and the domain-specific needs were identified; learning analytics was taken into account (i.e., an online tutoring system in ancient history for everyone has needs that are obviously different from an online program for a mathematics lecture at the university); the outcomes were defined based on certain human values, objectives, and beliefs; the functional design was specified (i.e., type of application, domain or use case specificity, explanatory strategies regarding the model, and the outcomes); the overall impact and all potential risks were cautiously analyzed in order to separate the tolerable ones from the broadly acceptable risks and the inacceptable ones; the widest timescale in which the system could impact users was also determined (Leslie et al., 2021). Moreover, the user stories were carefully chosen and described. A solid knowledge of the field plays a major role in the impact analysis and in the final efficiency of the AER system. Moreover, prior training for implementers and users will be an advantage (Leslie, 2019).

For ethical data extraction and acquisition, aspects such as responsibly choosing data to be collected, finding data from diverse sets of relevant sources, and providing details about the sources are critical (Mohammad, 2022). Collecting data is the most time-consuming step in the development of a ML model, as a small number of instances manually annotated for emotions in AER systems are available. This represents the first contact that students and teachers alike have with the impact of highly sensitive data. Ethical issues may arise from the very beginning, when subjects are more or less properly informed about the real purpose of the system. As data acquisition for AER is deeply connected with experiments on human subjects, special care must be given to data integrity, free and fully informed consent, as well as to subjects' privacy.

Data preprocessing phase includes feature extraction/scaling, feature selection, label encoding, data annotation, dimension reduction, one hot encoding, missing values, and binning. Ethical issues posed in this stage point to the variability of emotion expression and mental representation, to the tendency to capture the attitudes of the majority group, to the difficulty of choosing the right perspective on what is appropriate and what is not (Mohammad, 2022), to bias and discrimination associated with crowdsourcing, and to feature omission. For automated labeling or annotation, ethical solutions are human oversight (Leslie et al., 2021) and considering multiple answers that are more appropriate than others instead of a single "correct" answer.

With respect to the stage of dataset building, the randomized split of the data must be ensured.

For training a predictive model, the most ethically susceptible steps are the model selection and the criteria selection. In our opinion, it is recommended to choose multiple, various, and proper metrics, such as accuracy, precision, recall, specificity, F1 score, degree of inappropriate biases accepted, efficiency, privacy preserving capacity, transparency, interpretability, explainability, etc. Professional and institutional transparency, which covers integrity, honesty, sincerity, neutrality, objectivity, and impartiality (Leslie, 2019), must be also addressed here. Additionally, we must consider the dynamics of the individuals' emotion, i.e., the change with time of their perceptions, emotions, and behavior.

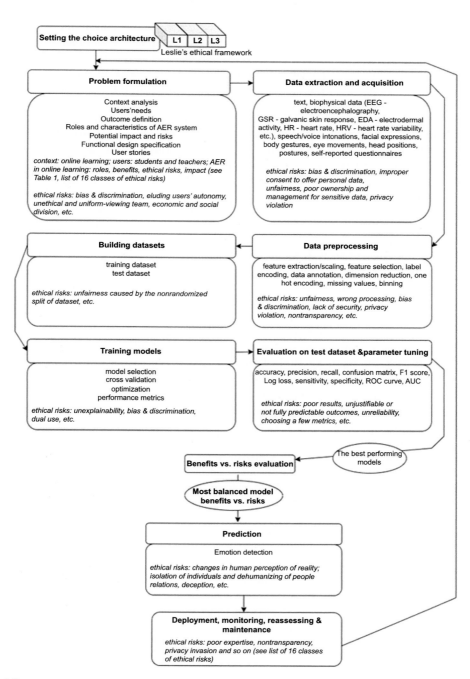

FIGURE 4.2

Ethical prediction model for AER in online learning.

As for model selection, data available is important. AER systems regularly use ML techniques for large datasets. AI models tend to work nicely for people well-represented in the data, but abnormally for the others (Mohammad, 2022). A preferable policy is to not use intrinsically interpretable models unless the "potential impacts and risks have been thoroughly considered in advance," and the semantic explainability has the potential to soften the potential risks (Leslie, 2019, p. 46). In the sensitive contexts (such as online AER for fraud-free exams), where the transparency is important, the interpretable ML techniques are preferable choices (i.e., linear regression, logistic regression, decisions trees, or case-based reasoning). The system rationale must be nonopaque and accessible to all affected parties' understanding, in terms of their capacity and limitations of cognition. When these interpretable techniques are inappropriate for our goal, more complex and model-specific or model-agnostic mechanisms will be used for the interpretability and explainability purpose. Technological maturity of the system will be proved, provided that the design is based on well-understood techniques already in operation and externally validated for a similar context (Leslie et al., 2021). An efficient means to evaluate a model is to test it on unseen data using multiple various metrics. This stage results in several ML models, after which they are ethically analyzed in the benefits vs. risks analysis phase. It is unrealistic to state that there is no risk. We must be aware of all the risks, accept them, and try to mitigate them if the benefits of using AER in online learning are relevant. High risks combined with low benefits should lead us to drop out the development or usage of such a system. The most balanced model based on the benefits—risks ratio is then selected for usage.

In the prediction phase, the model outputs a result that is further used in the online learning system. The results may be emotions labels, if discrete models for emotions are used, or values in case of dimensional models for emotions. Ethical concerns regard the impact of the detected emotions on the users, as detailed in the list of risks. The detected emotion is used by the system to configure the learning environment, to provide support, and so on.

A responsible implementation of the system must be followed by responsible deployment, monitoring, reassessing, and maintenance, as the lifecycle of the system offers a social meaning to our initiative. These four steps must be thoughtfully approached by acknowledging the proper roles in the team and ethically professing them, so that to deliver the project in correlation with the real needs of the users, which must be constantly revisited throughout the entire process. Students and teachers have to be fully informed on the AER technologies used in the program (either experimental or not), prior to their participation in the corresponding tasks. Their consent is to be clear and explicit. Both the implementation and the subsequent stages will benefit from training the implementers and the users, so that one may prevent biases and deliver an interpretable and justifiable system. Taking account of the breadth and temporality of deployment, directly and indirectly affected users (Leslie et al., 2021), and explanatory strategies will also facilitate the process. A participatory AER system, where all users are invited to make comments and offer recommendations that may improve the system, empowers them and lowers possible tensions. For a broader perspective on the student's learning results, assessment, and other issues, the teacher will be the final decision-maker.

All the components of the model imply ethics in many ways, and, as a result, the attention paid to the three levels (L1, L2, and L3) ensures fairness, trustworthiness, justifiability, and permissibility. SUM values pinpoint the respect for students and teachers, open and inclusive connection, the well-being of users, and the protection of social and cultural values, as well as for the public interest. FAST Track Principles followed throughout the process set fair, accountable, sustainable, and transparent directions of action, and the PBG Framework concretely integrates all these values and principles in the action.

It is advisable to investigate actual solutions, considering all identified ethical risks and answering the questions detailed in the data ethics framework (CDDO Central Digital & Data Office, 2020). The whole process must be regularly revisited throughout the project, especially when data collection, storage, analysis, or sharing is affected by any type of change (ICO Information Commisioner's Office, Alan Turing Institute, 2020). Constant feedback implies asking the team several questions, namely in the initial phase if "they are doing the right thing?", during the project if "they have designed it well?", and after the project being deployed if "it is still doing the right thing?" (CDDO Central Digital & Data Office, 2020, p. 18).

Case studies and use case

No ethically reliable academic or commercial AER system for learning has been developed so far, to our knowledge. Three ITSs in AER environments, used in the few identified experimental studies, were selected to investigate to what extent ethical issues are addressed. MetaTutor and iTalk2Learn are addressed in the only two papers, out of the 53 works analyzing students' emotions, that were selected in an excellent review of the studies about ITSs published in seven prestigious databases, namely Web of Science, PubMed, ProQuest, Scopus, Google Scholar, Embase, and Cochrane (Mousavinasab et al., 2021).

MetaTutor (Azevedo et al., 2011) was used in an experimental study on emotion detection in learning, carried out by Harley et al. (2015), where 67 students were involved. It contains four pedagogical agents to facilitate self-regulated learning and employs the facial recognition software FaceReader 5.0 and the electrodermal activity data acquisition software Affectiva's Q-sensor 2.0. Even if all the emotion detection methods used (automatic facial expression recognition, electrodermal activity, and self-report) generate high-sensitive data, no ethical acknowledgment was held.

The ITS component of iTalk2Learn platform was evaluated in an experiment on the adaptation of the feedback given to the students to their state of emotion. It uses multimodal emotion detection, namely speech analysis, as well as the analysis of the changes occurring in action after certain indications have been given. A possible positive role of emotion-aware technological support was identified, but nevertheless ethical aspects were not approached (Grawemeyer et al., 2016).

Affective AutoTutor, employed in over twenty controlled experiments, uses multimodal affect detection (i.e., facial features, conversation, and body language) to keep the student in a balanced emotional state, by varying the difficulty of the tasks, the pace, and direction of learning (D'Mello & Graesser, 2012, 2014). It fulfills ten complex functions, whereof modeling students' cognitive states and regulating negative affective states, but, once again, the ethical aspect was neglected.

Our proposed use case for a tutoring system in the online Optimization Algorithms course, augmented with examination tools, was designed based on the prediction model described in Fig. 4.2. Ethical recommendations previously mentioned are to be carefully envisaged in every aspect of the use case. Such use cases will be carried out in experimental studies, where an adequate AER software for online learning will be developed.

The suggested work scenario for the use case starts by setting the choice architecture, followed by the formulation of the problem, which was achieved by setting the items:

- Team characteristics: multidisciplinary expert team with diversity of thought and wide ranging skill sets (strong ethics; solid knowledge in learning emotion-based profiling and assessment; serious analytical skills; good collaboration with instructors in computer science optimization); the ethicists, academics, data scientists, policy experts, researchers, and practitioners who clearly understand the needs of the users;
- Context: superior education, Bachelor's program in Computer Science, online learning, and Optimization Algorithms course;
- Users: students and teachers;
- Needs: for students—customized learning routes in order to gain good abilities in optimization algorithms; for teachers—facilitated teaching by accounting students' emotions in the online environment;
- Roles of AER system: students' learning profiling, adaptive teaching and learning, and objective fraud-free assessment; characteristics: video, sound, and text recording for users' online activity, ITS with capabilities of running complex optimization applications, comparing tools for algorithms, and visualizations; textual, vocal (conversation in natural language), and multimedia response for the users;
- Outcome: effective personalized teaching and learning; accurate, objective, and fraud-free examination;
- Impact analysis: transformative and long-term effects on students' learning strategies and knowledge in the area, on emotion-related aspects in users' lives, on teaching strategies of instructors, and student-teacher relationship;
- Data used: text, speech, and nonverbal and paraverbal communication, i.e., voice intonation, facial expressions, eye movements, head position, posture, gestures, as well as self-reported questionnaires;
- User stories with specific actions mitigating the corresponding high ethical risks (see risks classes and examples in Table 4.2).

In all phases of the process, there are applied measures to prevent or to reduce ethical risks by obtaining users' informed consent for personal data collection and emotions inferring; sharing the understanding of the user's need with the user, accounting students' own educational goals; identifying and including the students in greater need or the disadvantaged ones and identifying measures to help them; reducing economic, social, gender, racial, and other inequalities; identifying users who may face negative consequences when using the system; using large volumes of balanced representative and proportional data (from people with different backgrounds) and manual checking for data labeling to avoid bias and discrimination; understanding how data is generated; ensuring data integrity; properly using synthetic data, if necessary; mitigating possible bias; using recommender algorithms with nondiscriminatory filters; checking data limitations; spotting reliable and safe patterns in data; adequately processing data and feature omission; choosing an interpretable model, responsively using data anonymization; randomly splitting the dataset; designing and using explanatory strategies for the outcomes as well as various metrics to evaluate the model on the test dataset; avoiding emotion misinterpretations and misusing; using multimodal channels to interpret emotions; considering the variability in the expression of emotions, emotion

Table 4.2 User stories and ethical actions of an AER system with tutoring and examination role.

User story A

As a student with low-level knowledge and interest in mathematics, I want general-scope training in optimization, so that I will pass the exam, being objectively assessed.

Suggested actions	Ethical risks
Detect student's emotional state (confusion/frustration/shame/boredom/surprise/hopefulness/contempt, etc.) and level of demotivation	1, 2, 3, 4, 5, 9, 10, 11, 16
Detect student's learning profile in repeated interactions (emotions—motivation—learning style—cognition)	1, 2, 3, 4, 5, 6, 9, 10, 11, 16
Recommend educational resources to the teacher (vital short videos and readings and optimization case studies in the real world) and teaching strategies adjustments (e.g., practical engaging activities, form-based tasks, clarifying explanations, semiweekly small tasks and assignments, and team working with classmates)	1, 2, 3, 5, 7, 13, 14, 15
Regulate negative emotional states as frustration and boredom by delivering hopeful, motivating, or congratulation messages, short videos of similar successful projects, recreational videos, images, animations, or music to re-engage the participant	1, 2, 3, 5, 7, 13
Provide adaptive feedback to the student: support files for homework and clear explanations of their own emotions labels and of the exam result	1, 2, 3, 5, 7, 8, 12, 13
Provide feedback to the teacher: the interpretation of the student's emotions and emotion changes in the student; the impact of the teaching strategy and of the assisted assessment	1, 2, 3, 5, 7, 8, 12, 13

User story B

As a highly motivated student, I expect to obtain the best educational and challenging practical tasks, so that my abilities in optimization will be competitive.

Suggested actions	Ethical risks
Detect student's emotional state (curiosity/engagement/surprise/excitement/general interest or interest in certain subareas/anxiety/relief/contempt, etc.)	1, 2, 3, 4, 5, 9, 10, 11, 16
Detect student's learning profile in repeated interactions	1, 2, 3, 4, 5, 6, 9, 10, 11, 16
Recommend educational resources to the teacher (short- or medium-size videos, textbooks, various optimization case studies in the real world, and optimization tool demos) and teaching strategies adjustments (e.g., weekly tasks and challenging homework requirements, team working, or contest-like assignments)	1, 2, 3, 5, 7, 13, 14, 15
Provide adaptive feedback to the student: further readings on interesting subtopics in certain areas, greetings, access to more complex optimization tools, engagement techniques, if deadline is short, etc.	1, 2, 3, 5, 7, 8, 12, 13
Provide feedback to the teacher: student's responses to the hints, student's emotion interpretation, and the emotion changes in the student	1, 2, 3, 5, 7, 8, 12, 13

User story C

As an instructor, I expect fraud-free results in exams, so that assessing accuracy is maximized.

Suggested actions	Ethical risks
Detect student's emotional state during the exam (anxiety/relief/contempt/confusion/frustration/fear/hopelessness, etc.)	1, 2, 3, 4, 5, 7, 9, 10, 11, 16
Send items clarification, motivating or relaxing posts, notice about the deadline, notifying messages in case of cheating detection, and explanations about inferred results	1, 2, 3, 5, 7, 15
Assist the teacher in the examination process by monitoring student's activity and by indicating any case of cheating	1, 2, 3, 5, 7, 8

dynamics, and casual physical illnesses that may alter emotions; taking measures for transparency and explainability (about the system, algorithms, and outcomes); respecting human rights; using GDPR (2016); including privacy by design; ensuring nondiscrimination and reliable results; using mechanisms to prevent unfairness; ensuring accountability (robust practices, solid documentation, and validated and reproducible algorithms); involving teacher oversight for important decisions and the use of teachers' pedagogical experience by participatory mechanisms; identifying signs of authoritarian relations and taking actions to eliminate them; fostering students' liberty of expression, critical thinking, and creativity, especially under examination; providing well-directed and noncontrolled access to educational resources; sending clear message to users that the inferred emotions are only indicative and perhaps not their true emotional state; avoiding overhelping for students and their dependence on a machine; challenging students to override learning obstacles and to collaborate with teachers and other students; avoiding dual use of the system; complying with the law and additional ethical regulations; continuously evaluating the project (e.g., by self-assessment with scores for every specific action during the process); repeatedly revisiting the needs of the users; consulting the target audience on appropriateness of emotion recognition; involving experts and consultants in reviewing and assessing ethical considerations of the project; setting the end mechanism, if the project stops being ethical; training users and the team in ethics; deploying, monitoring, and reassessing the system in a responsible way and by performing maintenance for sustainability (green AI).

Discussion and conclusions

In our opinion, as students and teachers experience various emotions throughout the learning process, AER systems used in online learning in order to recognize, interpret, cause, and stimulate emotions bring several benefits in accordance with the diverse roles existing in education. Alongside these benefits, we must fully acknowledge the risks, especially the ethical ones, in their span, timescale, and depth of impact. Even if the AI domain was set in the 1950s, AI ethics is still a novelty. In the literature, machine ethics is approached on four lines of thought, as we have highlighted. To separate legitimate worries of overstated ones, the ethical risks of using AI, AER, and AC were analyzed in this chapter, based on the exploratory research of academic publications, on the assessment of aforementioned experimental applications, and on critiques addressed by the general public.

The risks associated with the ubiquitous AI-based services and big data society represent a major concern for most of the researchers and for the general public, as well. We synthesized and structured ethical risks encountered in online learning in sixteen classes and provided scenarios of their emergence and real-world examples. These critical risks of using AER, if not addressed and mitigated, may easily lead to misdirected learning, along with all specific short- or long-term consequences.

Whether emotion recognition by machines is ethical or not remains questionable, as it is not possible to capture one's full emotional experience, even if all possible data is collected. On the other hand, commercial AER applications are advertised as being able to detect the true emotional state of a person and to even "predict" behavior, mood, and type of personality.

For a specific AER system, we must thoroughly weigh up both the impact and the risks. As a consequence, high risks associated with low benefits of an AER system should lead to the decision of stopping its development or usage.

The AC ethical impact (on education) has been addressed in the literature only in recent years. Noticeable research in the field started after 2010 and accelerated after 2019, and it mainly regards general ethical aspects on AI and data protection. Three AER applications used in education that were employed in the few experimental studies reported in prestigious databases were analyzed, and no ethical considerations were identified.

Nevertheless, many ethical guidelines and frameworks for AI projects have been issued by various governmental bodies, authorities, commissions, and independent researchers. As all the guidelines stipulate, AI-powered technologies should envisage fairness, trustworthiness, permissibility, and justifiability. These publicly available ethics guidelines, some of them containing wide-scope comprehensive checklists for AI projects and risk analysis templates, together with the research in the area, reveal an authentic propitious pursuit of moral values in the AI and AC research. They constitute adequate solutions and recommendations for each AI/AC technology that, however, has a significant direct or indirect impact on individuals, communities, and overall society by means of private and public services (including education).

With their support, individuals and institutions alike raise their level of awareness regarding ethical AI. Nevertheless, the main critical issue arises, as the three aspects mentioned later make the ethical sustained effort either nearly fruitless, when regarding the actual deployment of such technologies or inapplicable in nowadays society:

1. AI technology companies continue to invest enormous resources in expedite deployment and in selling more and more advanced technologies. They own vast collections of data for ML algorithms, but data sources and the means employed in data collection or in creating algorithms still remain opaque for the public or indicate rather unethical procedures. When such technologies are promoted, AI neutrality and ethical risks are eluded and only the benefits for users are presented. Therefore, such misinformed users support the spreading of unethical or partially ethical AI technologies. For details on the manipulation of users' trust and deception, see our list of ethical risks.
2. There operates no regulatory and enforcement authority on supervising the use and development of AI (to our knowledge). Therefore, how can one conciliate this lack with the AI guidelines stipulating for mandatory lawfulness of AI projects' design and development? Furthermore, from what position can one individual or organization decide what is the right thing to do in a particular context?
3. As companies and AI are ubiquitously networked, it is fairly difficult to regulate AI development both at the national and the international level.

This high-profile controversy between ethics concerns and solutions in ethical guidelines, on the one hand, and companies developing and promoting AI without complying to such guidelines, on the other hand, clearly indicates that AI ethics represents just a step in the technology regulation, an important one that is to be followed by further efforts.

In this chapter, a scalable ethical AER model for online learning is proposed, and specific ethical risks organized in 16 classes of risks identified for AER in online learning are presented. In order to prevent or to mitigate ethical harms, we recommend the following actions: Use strong ethical teams for systems design and development, do ethics education for data engineers, and maintain autonomy in human hands. It is advisable that AER technology remain just a support, not a determinant, for both teachers and students. For a more practical approach of the model, an use case comprising three user stories is detailed, focusing on specific ethical risk classes and implicitly on corresponding solutions.

To conclude, the aim of the present chapter is to prove that AC needs to undertake measures from the ethics perspective and that ethics must be more widely covered both in the literature and in the deployed AER systems. Even if AER in education brings numerous benefits, there are also various ethical risks that must be addressed in order to reach the highest potential of emotion recognition systems.

For future research, we suggest the following directions: (1) to describe the correlations between AER benefits, AER implementation (used data and ML algorithms), and the potential ethical risks; (2) to update the list of the ethical risks identified so far so that they meet technological advancement and pinpoint their interdependences; and (3) to define scoring for ethical risks and to identify the timing in the system lifecycle when they are likely to occur. We consider that we have demonstrated the importance of ethics in the AER-based systems in online learning and we invite the readers to consider our practical guidelines in their future research and AER deployment.

Acknowledgment

This work was supported by a grant of the Petroleum-Gas University of Ploiesti, project number 11061/2023, within an internal grant for scientific research.

References

ACM (Association for Computing Machinery). (2018). *Code of ethics and professional conduct*. https://www. acm.org/code-of-ethics. Last accessed on May 24, 2023.

Alfoudari, A. M., Durugbo, C. M., & Aldhmour, F. M. (2021). Understanding socio-technological challenges of smart classrooms using a systematic review. *Computers & Education*, *173*(2021)104282. Available from https://doi.org/10.1016/j.compedu.2021.104282, ISSN 0360-1315.

Alwadei, A., & Alnanih, R. (2022). Designing a tool to address the depression of children during online education. *Procedia Computer Science*, *203*, 173–180. Available from https://doi.org/10.1016/j. procs.2022.07.024, ISSN 1877-0509.

Anderson, M., & Anderson, S. L. (2007). Machine ethics: Creating an ethical intelligent agent. *AI Magazine*, *28*(4), 15. Available from https://doi.org/10.1609/aimag.v28i4.2065.

APA (American Psychological Association). (2017). *Ethical principles of psychologists and code of conduct*. https://www.apa.org/ethics/code. Last accessed on May 24, 2023.

Arya, R., Singh, J., & Kumar, A. (2021). A survey of multidisciplinary domains contributing to affective computing. *Computer Science Review*, *40*100399. Available from https://doi.org/10.1016/j.cosrev.2021.100399, ISSN 1574-0137.

Ashwin, T. S., & Guddeti, R. M. R. (2020). Affective database for e-learning and classroom environments using Indian students' faces, hand gestures and body postures. *Future Generation Computer Systems*, *108*, 334–348. Available from https://doi.org/10.1016/j.future.2020.02.075, ISSN 0167-739X.

Azevedo, R., Bouchet, F., Harley, J.M., Feyzi-Behnagh, R., Trevors, G., Duffy, M., Taub, M., Pacampara, N., Agnew, L., & Griscom, S. (2011). MetaTutor: An intelligent multi-agent tutoring system designed to detect, track, model, and foster self-regulated learning. In *Proceedings of the fourth workshop on self-regulated learning in educational technologies*. SRL&ET. doi: https://doi.org/10.13140/RG.2.1.1334.6640.

Bakiner, O. (2022). What do academics say about artificial intelligence ethics? An overview of the scholarship. *AI Ethics*. Available from https://doi.org/10.1007/s43681-022-00182-4.

Barrett, L. F. (2017). The theory of constructed emotion: An active inference account of interoception and categorization. *Social Cognitive and Affective Neuroscience, 12*(1), 1−23. Available from https://doi.org/10.1093/scan/nsx060.

Barrett, L. F., Adolphs, R., Marsella, S., Martinez, A. M., & Pollak, S. D. (2019). Emotional expressions reconsidered: Challenges to inferring emotion from human facial movements. *Psychological Science in the Public Interest, 20*, 1−68. Available from https://doi.org/10.1177/1529100619832930.

Bhardwaj, P., Gupta, P. K., Panwar, H., Siddiqui, M. K., Morales-Menendez, R., & Bhaik, A. (2021). Application of deep learning on student engagement in e-learning environments. *Computers & Electrical Engineering, 93*107277, ISSN 0045-7906. Available from https://doi.org/10.1016/j.compeleceng.2021.107277.

CDDO (Central Digital & Data Office). (2020). *Guidance data framework*. https://www.gov.uk/government/publications/data-ethics-framework. Last accessed on May 24, 2023.

Chen, C. M., & Wu, C. H. (2015). Effects of different video lecture types on sustained attention, emotion, cognitive load, and learning performance. *Computers & Education, 80*, 108−121. Available from https://doi.org/10.1016/j.compedu.2014.08.015, ISSN 0360−1315.

Cen, L., Wu, F., Yu, Z. L., & Hu, F. (2016). A real-time speech emotion recognition system and its application in online learning. In S. Y. Tettegah, & M. Gartmeier (Eds.), *Emotions, technology, design, and learning* (pp. 27−46). Academic Press, ISBN 9780128018569. Available from https://doi.org/10.1016/B978-0-12-801856-9.00002-5.

COMEST (World Commission on the Ethics of Scientific Knowledge and Technology). (2019). *Preliminary study on the ethics of artificial intelligence*. https://unesdoc.unesco.org/ark:/48223/pf0000367823. Last accessed on May 24, 2023.

Cowie, R. (2015). Ethical issues in affective computing'. In R. Calvo, et al. (Eds.), *The Oxford handbook of affective computing*. Oxford Library of Psychology. Available from https://doi.org/10.1093/oxfordhb/9780199942237.013.006.

Crawford, K., Dobbe, R., Dryer, T., Fried, G., Green, B., Kaziunas, E., Kak, A., Mathur, V., McElroy, E., Sánchez, A. N., Raji, D., Rankin, J. L., Richardson, R., Schultz, J., West, S. M., & Whittaker, M. (2019). *AI Now 2019 report*. New York: AI Now Institute.

Dai, C. P., & Ke, F. (2022). Educational applications of artificial intelligence in simulation-based learning: A systematic mapping review. *Computers and Education: Artificial Intelligence, 3*100087. Available from https://doi.org/10.1016/j.caeai.2022.100087.

Dewan, M. A. A., Murshed, M., & Lin, F. (2019). Engagement detection in online learning: A review. *Smart Learning Environments, 6*, 1. Available from https://doi.org/10.1186/s40561-018-0080-z.

D'Mello, S., & Graesser, A. (2012). AutoTutor and affective autotutor: Learning by talking with cognitively and emotionally intelligent computers that talk back. *ACM Transactions on Interactive Intelligent Systems, 2*(4), Article 23, 39 p. Available from https://doi.org/10.1145/2395123.2395128.

D'Mello, S., Lehman, B., Pekrun, R., & Graesser, A. (2014). Confusion can be beneficial for learning. *Learning and Instruction, 29*, 153−170. Available from https://doi.org/10.1016/j.learninstruc.2012.05.003.

D'Mello, S. K. (2013). A selective meta-analysis on the relative incidence of discrete affective states during learning with technology. *Journal of Educational Psychology, 105*(4), 1082−1099. Available from https://doi.org/10.1037/a0032674.

D'Mello, S. K., & Graesser, A. C. (2014). Feeling, thinking, and computing with affect-aware learning technologies. In Calvo, R. A., D'Mello, S.K., Gratch, J., & Kappas, A. (Eds.), *The Oxford handbook of affective computing* (pp. 419−434). Oxford University Press. Available from https://doi.org/10.1093/oxfordhb/9780199942237.013.032.

Duo, S., & Song, L. X. (2012). An E-learning system based on affective computing. *Physics Procedia*, *24*(Part C), 1893−1898.

EGE (European Commission, European Group on Ethics in Science and New Technologies). (2012). *Ethics of information and communication technologies*. Publications Office, 2012. https://data.europa.eu/doi/10.2796/13541.

Ekman, P. (1999). Basic emotions. In T. Dalgleish, & M. Power (Eds.), *Handbook of cognition and emotion*. Hoboken, NJ: John Wiley & Sons Ltd..

Ekman, P., Sorenson, E. R., & Friesen, W. V. (1969). Pan-cultural elements in facial displays of emotions. *Science*, *164*, 86−88.

EUCFR (European Union Charter for Fundamental Rights). (2012). *Charter of fundamental rights of the European Union*. https://eur-lex.europa.eu/legal-content/EN/TXT/?uri = CELEX:12012P/TXT. Last accessed on May 24, 2023.

Ez-zaouia, M., Tabard, A., & Lavoué, E. (2020). Emodash: A dashboard supporting retrospective awareness of emotions in online learning. *International Journal of Human-Computer Studies*, *139*102411. Available from https://doi.org/10.1016/j.ijhcs.2020.102411, ISSN 1071-5819.

Faria, A. R., Almeida, A., Martins, C., Gonçalves, R., Martins, J., & Branco, F. (2017). A global perspective on an emotional learning model proposal. *Telematics and Informatics*, *34*(6), 824−837. Available from https://doi.org/10.1016/j.tele.2016.08.007, ISSN 0736-5853.

Feidakis, M. (2016). A review of emotion-aware systems for e-learning in virtual environments. In S. Caballé, & R. Clarisó (Eds.), *Intelligent data-centric systems, formative assessment, learning data analytics and gamification* (pp. 217−242). Academic Press, ISBN 9780128036372. Available from https://doi.org/10.1016/B978-0-12-803637-2.00011-7.

GDPR. (2016). Regulation (EU) 2016/679 of the European Parliament and of the Council of 27 April 2016 on the Protection of Natural Persons with Regard to the Processing of Personal Data and on the Free Movement of such Data, and Repealing Directive 95/46/EC (General Data Protection Regulation). https://gdpr-info.eu. Last accessed on May 24, 2023.

Goldenfein, J. (2020). *Facial recognition is only the beginning*. Public Books. Available at SSRN: https://ssrn.com/abstract = 3546525. Last accessed on May 24, 2023.

Goldie, P., Doring, S., & Cowie, R. (2011). The ethical distinctiveness of emotion-oriented technology: Four long-term issues. In P. Petta, C. Pelachaud, & R. Cowie (Eds.), *Emotion-oriented systems: The Humaine handbook*. Berlin: Springer.

Grawemeyer, B., Mavrikis, M., Holmes, W., Gutierrez-Santos, S., Wiedmann, M., & Rummel, N. (2016). *Affecting off-task behaviour: How affect-aware feedback can improve student learning. Proceedings of the sixth international conference on learning analytics & knowledge (LAK '16)* (pp. 104−113). New York: Association for Computing Machinery. Available from https://doi.org/10.1145/2883851.2883936.

Guerini, M., & Stock, O. (2005). Toward ethical persuasive agents. In *Proceedings of the IJCAI workshop on computational models of natural argument*. Edinburgh.

Harley, J. M., Bouchet, F., Hussain, M. S., Azevedo, R., & Calvo, R. (2015). A multi-componential analysis of emotions during complex learning with an intelligent multi-agent system. *Computers in Human Behavior*, *48*, 615−625. Available from https://doi.org/10.1016/j.chb.2015.02.013, ISSN 0747-5632.

Hascher, T. (2010). Learning and emotion: Perspectives for theory and research. *European Educational Research Journal*, *9*(1), 13−28. Available from https://doi.org/10.2304/eerj.2010.9.1.13.

Hascher, T., & Edlinger, H. (2009). Positive Emotionen und Wohlbefinden in der Schule—ein Rubberlike über Forschungszugänge und Erkenntnisse [Positive emotions and well-being in school—An overview of methods and results]. *Psychologie in Erziehung und Unterricht, 56,* 105−122.

Hasnine, M. N., Bui, H. T. T., Tran, T. T. T., Nguyen, H. T., Akçapınar, G., & Hiroshi Ueda, H. (2021). Students' emotion extraction and visualization for engagement detection in online learning. *Procedia Computer Science, 192,* 3423−3431. Available from https://doi.org/10.1016/j.procs.2021.09.115, ISSN 1877-0509.

ICO (Information Commisioner's Office), Alan Turing Institute. (2020). *Explaining decisions made with AI.* https://ico.org.uk/for-organisations/guide-to-data-protection/key-dp-themes/explaining-decisions-made-with-ai/. Last accessed on May 24, 2023.

IEEE Global Initiative on Ethics of Autonomous and Intelligent Systems (IEEE Global Initiative). (2018). *Ethically aligned design—Version 2 for public discussion.* https://standards.ieee.org/initiatives/autonomous-intelligence-systems/. Last accessed on May 24, 2023.

IFOW (Institute for the Future of Work). (2020). *Mind the gap: How to fill the equality and AI accountability gap in an automated world.* https://www.ifow.org/publications/mind-the-gap-the-final-report-of-the-equality-task-force. Last accessed on May 24, 2023.

Imani, M., & Montazer, G. A. (2019). A survey of emotion recognition methods with emphasis on E-Learning environments. *Journal of Network and Computer Applications, 147*102423. Available from https://doi.org/10.1016/j.jnca.2019.102423, ISSN 1084-8045.

Iulamanova, A., Bogdanova, D., & Kotelnikov, V. (2021). Decision support in the automated compilation of individual training module based on the emotional state of students. *IFAC-PapersOnLine, 54*(13), 85−90. Available from https://doi.org/10.1016/j.ifacol.2021.10.424, ISSN 2405-8963.

Jobin, A., Ienca, M., & Vayena, E. (2019). The global landscape of AI ethics guidelines. *Nature Machine Intelligence, 1,* 389−399. Available from https://doi.org/10.1038/s42256-019-0088-2.

Kazemitabar, M., Lajoie, S. P., & Doleck, T. (2021). Analysis of emotion regulation using posture, voice, and attention: A qualitative case study. *Computers and Education Open, 2*100030. Available from https://doi.org/10.1016/j.caeo.2021.100030, ISSN 2666-5573.

Kulkarni, P. (2022). *ML of choosing: Architecting intelligent choice framework, . Choice computing: Machine learning and systemic economics for choosing. Intelligent systems reference library* (Vol. 225). Singapore: Springer. Available from https://doi.org/10.1007/978-981-19-4059-0_3.

Lavoué, É., Ju, Q., Hallifax, S., & Serna, A. (2021). Analyzing the relationships between learners' motivation and observable engaged behaviors in a gamified learning environment. *International Journal of Human-Computer Studies, 154*102670. Available from https://doi.org/10.1016/j.ijhcs.2021.102670, ISSN 1071-5819.

Leslie, D. (2019). Understanding artificial intelligence ethics and safety: A guide for the responsible design and implementation of AI systems in the public sector. *The Alan Turing Institute.* Available from https://doi.org/10.5281/zenodo.3240529.

Leslie, D., Burr, C., Aitken, M., Katell, M., Briggs, M., & Rincon, C. (2021). Human rights, democracy, and the rule of law assurance framework for AI systems: A proposal. *The Alan Turing Institute.* Available from https://doi.org/10.5281/zenodo.5981676.

Lin, F. R., & Kao, C. M. (2018). Mental effort detection using EEG data in E-learning contexts. *Computers & Education, 122,* 63−79. Available from https://doi.org/10.1016/j.compedu.2018.03.020, ISSN 0360-1315.

Liu, S., Liu, S., Liu, Z., Peng, X., & Yang, Z. (2022). Automated detection of emotional and cognitive engagement in MOOC discussions to predict learning achievement. *Computers & Education, 181*104461. Available from https://doi.org/10.1016/j.compedu.2022.104461, ISSN 0360-1315.

Lyu, L., Zhang, Y., Chi, M. Y., Yang, F., Zhang, S. G., Liu, P., & Lu, W. G. (2022). Spontaneous facial expression database of learners' academic emotions in online learning with hand occlusion. *Computers & Electrical Engineering, 97*107667. Available from https://doi.org/10.1016/j.compeleceng.2021.107667, ISSN 0045-7906.

Mehrabian, A. (1996). Pleasure-Arousal-Dominance: A general framework for describing and measuring individual differences in temperament. *Current Psychology*, *14*, 261−292.

Mohammad, S. M. (2022). Ethics sheet for automatic emotion recognition and sentiment analysis. *Computational Linguistics*, *48*(2), 239−278. Available from https://doi.org/10.1162/coli_a_00433.

Moor, J. H. (2006). The nature, importance, and difficulty of machine ethics. *IEEE Intelligent Systems*, *21*(4), 18−21. Available from https://doi.org/10.1109/MIS.2006.80.

Mousavinasab, E., Zarifsanaiey, N., Niakan Kalhori, S. R., Rakhshan, M., Keikha, L., & Ghazi Saeedi, M. (2021). Intelligent tutoring systems: A systematic review of characteristics, applications, and evaluation methods. *Interactive Learning Environments*, *29*(1), 142−163. Available from https://doi.org/10.1080/10494820.2018.1558257.

Pekrun, R. (2005). Progress and open problems in educational emotion research. *Learning and Instruction*, *15*(5), 497−506. Available from https://doi.org/10.1016/j.learninstruc.2005.07.014.

Pekrun, R., Goetz, T., Titz, W., & Perry, R. P. (2002). Academic emotions in students' self-regulated learning and achievement: A program of qualitative and quantitative research. *Educational Psychologist*, *37*(2), 91−105. Available from https://doi.org/10.1207/S15326985EP3702_4.

Picard, R. W. (1995). *Affective computing*. M.I.T Media Laboratory Perceptual Computing Section Technical Report No. 32., https://affect.media.mit.edu/pdfs/95.picard.pdf. Last accessed on May 24, 2023.

Picard, R. W. (1997). *Affective computing*. Cambridge, MA: MIT Press.

Picard, R. W., Papert, S., Bender, W., et al. (2004). Affective learning—A manifesto. *BT Technology Journal*, *22*, 253. Available from https://doi.org/10.1023/B:BTTJ.0000047603.37042.33.

Qiao, X., Zheng, X., Sun, X., Li, S., & Zhang, Y. (2022). Learners' states monitoring method based on face recognition technology. *Procedia Computer Science*, *202*, 172−177. Available from https://doi.org/10.1016/j.procs.2022.04.024, ISSN 1877-0509.

Rosenfeld, R. A. (1978). Anxiety and learning. *Teaching Sociology*, *5*(2), 151−166, JSTOR. Available from https://doi.org/10.2307/1317061.

Ross, W. D. (1939). *Foundations on ethics*. Oxford, UK: Oxford University Press.

Russell, J. A., & Mehrabian, A. (1977). Evidence for a three-factor theory of emotions. *Journal of Research in Personality*, *11*(3), 273−294. Available from https://doi.org/10.1016/0092-6566(77)90037-X.

Sadowski, J., Viljoen, S., & Whittaker, M. (2021). Everyone should decide how their digital data are used—Not just tech companies. *Springer Nature*, *595*(7866), 169−171. Available from https://doi.org/10.1038/d41586-021-01812-3. PMID: 34211184.

Scherer, K. R. (1999). Appraisal theory. In T. Dalgleish, & M. J. Power (Eds.), *Handbook of cognition and emotion* (pp. 637−663). John Wiley & Sons Ltd. Available from https://doi.org/10.1002/0470013494.ch30.

Sharma, K., Papavlasopoulou, S., & Giannakos, M. (2022). Children's facial expressions during collaborative coding: Objective versus subjective performances. *International Journal of Child-Computer Interaction*, *34*100536. Available from https://doi.org/10.1016/j.ijcci.2022.100536.

Shelton, C. (2022). *Complementary to Martin Hilbert course "big data, artificial intelligence, and ethics"*. University of California, Coursera Plus.

Sikström, P., Valentini, C., Sivunen, A., & Kärkkäinen, T. (2022). How pedagogical agents communicate with students: A two-phase systematic review. *Computers & Education*, *188*104564. Available from https://doi.org/10.1016/j.compedu.2022.104564, ISSN 0360-1315.

Tanko, D., Dogan, S., Demir, F. B., Baygin, M., Sahin, S. E., & Tuncer, T. (2022). Shoelace pattern-based speech emotion recognition of the lecturers in distance education: ShoePat23. *Applied Acoustics*, *190*108637. Available from https://doi.org/10.1016/j.apacoust.2022.108637, ISSN 0003-682X.

Um, E. R., Plass, J. L., Hayward, E. O., & Homer, B. D. (2012). Emotional design in multimedia learning. *Journal of Educational Psychology*, *104*(2), 485−498. Available from https://doi.org/10.1037/a0026609.

UNESCO. (2021). *Recommendation on the ethics of AI*. https://unesdoc.unesco.org/ark:/48223/pf0000380455. Last accessed on May 24, 2023.

Utami, P., Hartanto, R., & Soesanti, I. (2019). A study on facial expression recognition in assessing teaching skills: Datasets and methods. *Procedia Computer Science*, *161*, 544–552. Available from https://doi.org/10.1016/j.procs.2019.11.154, ISSN 1877-0509.

Vidanaralage, A. J., Dharmaratne, A. T., & Haque, S. (2022). AI-based multidisciplinary framework to assess the impact of gamified video-based learning through schema and emotion analysis. *Computers and Education: Artificial Intelligence*, *3*100109. Available from https://doi.org/10.1016/j.caeai.2022.100109, ISSN 2666-920X.

Xu, T., Zhou, Y., Wang, Z., & Peng, Y. (2018). Learning emotions EEG-based recognition and brain activity: A survey study on BCI for intelligent tutoring system. *Procedia Computer Science*, *130*, 376–382. Available from https://doi.org/10.1016/j.procs.2018.04.056, ISSN 1877-0509.

Yadegaridehkordi, E., Noor, N. F. B. M., Ayub, M. N. B., Affal, H. B., & Hussin, N. B. (2019). Affective computing in education: A systematic review and future research. *Computers & Education*. Available from https://doi.org/10.1016/j.compedu.2019.103649.

Yang, D., Alsadoon, A., Prasad, P. W. C., Singh, A. K., & Elchouemi, A. (2018). An emotion recognition model based on facial recognition in virtual learning environment. *Procedia Computer Science*, *125*, 2–10. Available from https://doi.org/10.1016/j.procs.2017.12.003, ISSN 1877-0509.

Zeidner, M. (1998). *Test anxiety: The state of the art*. Plenum Press.

Data-driven educational decision-making model for curriculum optimization

Edis Mekić[1], Irfan Fetahović[1], Kristijan Kuk[2], Brankica Popović[2] and Petar Čisar[2]

[1]State University of Novi Pazar, Novi Pazar, Serbia [2]University of Criminal Investigation and Police Studies, Belgrade, Serbia

Introduction

Modern education needs effective educational planning. Decisions made on curriculum development must be based on the best of human decision-making, but also supported by the power of data, analytics, and artificial intelligence (AI). In the last few decades, nature-inspired metaheuristics have been widely used for solving various real-world optimization problems. Swarm intelligence algorithms are relatively new but powerful metaheuristics, where solutions for various problems of this kind are found by using swarms of simple agents (Dolicanin et al., 2018). It has been shown that these algorithms can also be effective in solving curriculum optimization problems (de-Marcos et al., 2008; Menai et al., 2018).

Decision-making problems are in the spotlight of several different scientific disciplines, including educational management. Decisions made in education, in most cases, are solved from an intuitive point of view or based on limited information.

The decision-making process in education is usually based on three important pillars: curriculum, budget, and personnel.

Curriculum planning is one of the most important steps in the decision-making process. The design of the curriculum in engineering education is usually based on cognitivism and constructivism. These learning theories best accommodate educational models based on extensive usage of modern information technologies (IT) (Luzardo, 2004). Those approaches provide easy implementation of electronic platforms, blogs and wikis, smart books, demonstrations, audio and video material, and finally AI technologies (Londoño, 2011).

Learning indicators are defined as the important concepts and factors that are used to communicate information about the level of learning. They can be used to make management decisions when planning a learning strategy for an institution or university. Learning indicators must qualitatively and quantitatively describe a learning process. To achieve this, indicators qualitatively describe aspects of learning with a single or composite statistic. A quantitative indicator might be a measure of the change in a student's performance over time or an estimate of a college's value added to student learning (Shavelson et al., 2018). When educational institutions define indicators, qualitative must be supported by quantitative.

Ethics in Online AI-Based Systems. DOI: https://doi.org/10.1016/B978-0-443-18851-0.00002-0

97

After planning and defining learning indicators, decision-makers must start the curriculum design. This is a process of thinking through how you want to organize what you want students to learn.

Curriculum design, for supervised learning settings, has been extensively studied in the literature (Bengio et al., 2009). Early works present the idea of designing a curriculum compromising tasks with increasing difficulty to train a machine learning model (Elman, 1993). Models will be used to provide prior instruction and the basic concepts to teachers about the courses. These concepts will be based on personal students' learning paths. A combination of that information will become a critical part of curriculum design.

Stakeholder pressure for greater fiscal accountability of public and private sector entities such as PSIs has never been greater (Mensah et al., 2009). To deliver higher accountability, educational institutions implement different types of performance-oriented management effectively (Lu & Willoughby, 2015). Performance orientation is used in establishing budgeting as a strategic planning tool and can be used to complement already developed curriculum plans.

The final step is the decision on personnel who will implement the newly developed curriculum. This is accomplished through a strategic human resource planning process. Strategic human resource planning is delivered through the following stages: assessing the current human resource capacity, forecasting human resource requirements, gap analysis, and human resource selection of responsibilities in line with the curriculum plan (Ulferts et al., 2009).

These pillars must adhere to overarching ethical values. The decision must be based on objective and nonbiased suggestions. Data must be safeguarded and kept secret, but recommended methodologies and approaches must be transparent.

In general, we believe that algorithmic decision-making is objective and free from discrimination. However, this belief is not necessarily true. It is easy to dismiss claims of unintentional discrimination in algorithmic decision-making by relying on the idea that the decisions are based on data and algorithms that are inherently objective (Romei & Ruggieri, 2014).

However, the reality is that algorithmic decision-making can lead to unintentional discrimination due to the biases in the data that the algorithms are trained on or the way in which the algorithms are designed. For example, if a dataset used to train an algorithm is biased toward a particular group, then the algorithm may make decisions that favor that group over others (Hajian et al., 2016). The belief that algorithmic decision-making is inherently objective and free from discrimination is not necessarily true. It is essential to be aware of the potential for unintentional discrimination in algorithmic decision-making and to actively work toward developing more discrimination-aware algorithms.

In recent years, the use of AI and big data to collect information about students has become more common. While these technologies can provide valuable insights into student learning and behavior, there are concerns about the data's privacy and security. When data is collected from students, AI and big data can combine it with other data sources available on the network. Data from social media, online activity, and other sources may be included. As AI-based tools advance, it becomes easier to collect and integrate this data, which can expose students' private information (Li & Zhang, 2017).

Furthermore, learning management systems (LMSs) are frequently installed in cloud environments, which means that all data on student learning behavior is stored in remote locations (Sahil et al., 2015). Because some of the information could be misused, it is critical to ensure that this data is secure. As knowledge-extracting tools improve in power, it becomes easier to combine seemingly unrelated data fragments to identify individual behavioral characteristics, exposing personal privacy (Sahil et al., 2016).

To address these concerns, strong privacy and security safeguards must be put in place. This includes encrypting data both at rest and in transit, implementing access controls to limit who can access the data, and auditing the data on a regular basis to identify potential security risks (Patil, 2016).

In addition to these technical safeguards, it is critical to educate students and faculty about the significance of data privacy and security. This can include training on how to handle sensitive data, establishing clear guidelines for data collection and use, and encouraging open communication about any data privacy and security concerns. In conclusion, when working with AI and big data in education, data privacy and security are critical considerations. It is critical to put in place strong technical safeguards to protect student data, as well as to educate students and staff about the importance of data privacy and security. This way, we can ensure that these technologies benefit students while also protecting their privacy and security.

Accountability and fairness are two important traits that are often emphasized as ethical requirements for the development and deployment of AI systems. However, implementing these traits can be challenging.

Accountability refers to the responsibility of individuals, organizations, and systems for the outcomes of their actions. In the context of AI, accountability requires that developers, users, and other stakeholders take responsibility for the impact of AI systems on individuals and society as a whole. This includes ensuring that AI systems are designed and deployed in a way that reduces negative impacts and that there are mechanisms in place to address any unintended consequences or harm caused by these systems.

Human oversight and responsibility: AI systems used in universities should not replace human judgment entirely. There is a need for human oversight to ensure that the AI systems are functioning correctly and that their decisions align with ethical and moral considerations. Ultimately, humans should take responsibility for the decisions made by AI systems in universities (Miguel et al., 2021).

Fairness is another trait that is often emphasized as an ethical requirement for AI systems. Fairness refers to the equitable treatment of individuals and groups, regardless of factors such as race, gender, or socioeconomic status. In the context of AI, fairness requires that AI systems be designed and deployed in a way that does not discriminate against certain groups or individuals (Zemel et al., 2013).

Thus, while incorporating AI systems in their operations, colleges should carefully assess the ethical ramifications and take appropriate action to resolve any potential concerns. In order to ensure that AI systems are operating ethically and responsibly, this can involve doing routine audits and assessments of the systems, making decision-making procedures transparent and clear, and investing in education and training to uphold human oversight and responsibility.

In this chapter, a data-driven educational decision-making model with the support of a course curriculum will be closely analyzed with students' responses after the course. The educational decision-making will be realized with the help of the curriculum designed and maintained by the colleges and universities. One reason that curriculum planning and adapting it into an e-learning format is important is because it makes yearly, lesson, and unit planning easier. The increasing development of the curriculum makes it imperative to apply data-driven decision support systems to curriculum optimization. In the context of AI in the new era, this research focuses on data-driven educational decision-making models for curriculum optimization to create an efficient e-learning system.

In that case, it is necessary to perform improvements to the model by modifying the applied classification process. Up-to-date research shows that most applied classifiers include Bayes networks, decision trees, neural networks, support vector machines, K-nearest neighbors, boosting, etc. Boosting is one of the most important recent developments in classification methodology. Boosting works by sequentially applying a classification algorithm to reweighted versions of the training data and then taking a weighted majority vote of the sequence of classifiers thus produced. For many classification algorithms, this simple strategy results in dramatic performance improvements. For the two-class problem, boosting can be viewed as an approximation to additive modeling on the logistic scale using maximum Bernoulli likelihood as a criterion. Over the past few years, this technique has emerged as one of the most powerful methods for predictive data mining. This chapter addresses the problem of determining the weights of criteria using data mining classification algorithms for attribute reduction and its aggregation with the logistic regression method in the case study of the determination of relevant learning indicators of risk of curriculum optimization and the decision-making educational process.

Calibrating is applicable in case a classifier outputs probabilities. Calibration refers to the adjustment of the posterior probabilities output by a classification algorithm toward the true prior probability distribution of the target classes. The idea of many authors is to calibrate a machine learning model or a statistical model that can predict for every given data row the probability that the outcome is 1. Calibration in classification is used to transform classifier scores into class membership probabilities. Univariate calibration methods like logistic regression exist for transforming classifier scores into class membership probabilities in the two-class case. In logistic regression, the dependent variable is binary or dichotomous, i.e., it only contains data coded as 1 (TRUE, success, etc.) or 0 (FALSE, failure, etc.). The goal of logistic regression is to find the best fitting model to describe the relationship between the dichotomous characteristic of interest (dependent variable = response or outcome variable) and a set of independent (predictor or explanatory) variables. Regression calibration is used as an easy way to improve estimation in the errors-in-variables model.

The final part of the chapter gives an overview of the available methodological approaches that can implement ethical approaches in the developed AI system.

Decision-making process and AI implementation

The decision-making process, as previously stated, in education is usually based on three important pillars: curriculum, budget, and personnel Fig. 5.1. We will explore in detail every step and overview the implementation of AI support in every step of this process.

The design of the curriculum in engineering education is usually based on cognitivism and constructivism. This division is based on the presumption that learning theory serves as the basis to verify the correctness of educational strategy and is a crucial tool to select a specific strategy (Ertmer & Newby, 1993). Those two theories have different approaches to learning goals, teaching

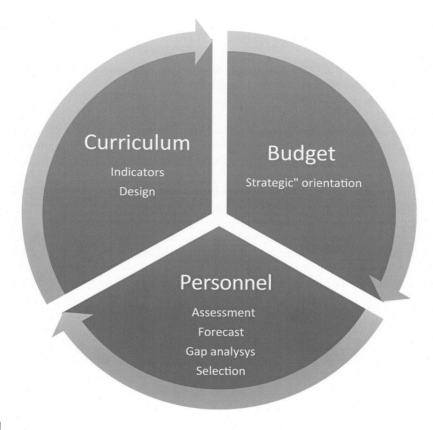

FIGURE 5.1

The decision-making process in education.

methods, and evaluation of learning outcomes. Cognitivism focuses on cognition as a premise of behavior and posits that behavior changes when the way of cognition changes. Constructivism is based on learning by doing, creating innovative challenges for students, and expecting solutions to problems to be designed and delivered.

Those two theories are appropriate for preparing educational strategies where learners need to accumulate knowledge on topics (Seya et al., 2020). Constructivism is more appropriate for the phases of learning where knowledge needs to be implemented, and indicator planning will be based on the results obtained using constructivism (Mekic et al., 2016; Nišić et al., 2019; Pecanin et al., 2019). The constructive approach is based on two important pillars: theoretical learning and practical implementation of the learned material. This makes this approach compatible with the usual outcomes of engineering education. Constructivism also cannot be implemented without cooperative learning methodologies.

Since we want to separate the theoretical contents from the practical ones, we will implement student-centered learning methods, based on project-based learning (PBL) to integrate both theoretical and practical contents and apply them to real-world problems (Sánchez-Romero et al., 2019). And this project-based learning will be the basis for the definition of indicators in the curriculum design phase.

Before a decision is made on which indicators should be used for implementation in the curriculum planning phase, we need to investigate AI implementation for the measurement of progress. As previously stated, we have two major groups of indicators: qualitative and quantitative. Quantitative can provide measures of learning performance (Kajiji et al., 2007). Besides this, those types of indicators can provide feedback to evaluate overall institutional performance. This can be achieved using data mining, which will provide us with reliable data for tracking students' progress. The second goal is to develop a probabilistic classification system, which will be implemented to provide a unique learning path for the student.

We will use a framework for a holistic approach to curriculum design (Lytras et al., 2018). This framework is based on existing technology infrastructure for students. This infrastructure must have a proper LMS, which will provide lessons to students, and data for tracking students' success. The first part of the technology domain will be understanding and application of big data, AI, analytics, and their optimization in the educational context. The second part is the information integration of different data received with data mining. For classification methodology, sequential application of logical regression, decision tree j48, and CART can be used on already reviewed training data. This approach is called boosting. The next step in this framework is educational analytics where we will analyze collected data and based on all those aspects deliver an interface. The interface is actually an analysis of reports and justification for curriculum design.

While we can use data mining techniques on quantitative indicators, qualitative requires a different approach. Qualitative indicators can provide us with students' learning experience, engagement with content, quality of cooperation with peers, etc. Those indicators per se do not refer to tangible learning outcomes or the performance of teachers. Surveys and questionnaires are standard tools for collecting qualitative indicators. Answers provide context and interpretation of quantitative indicators. Since answers on the surveys provide a wide range of qualitative data, they must be grouped into similar themes for further analysis. One way to analyze qualitative data is by sentiment analyses, compromised by natural language processing, text analyses, and computational techniques to extract sentiment (Nkomo & Daniel, 2021). Sentiment analysis can identify and extract relevant information about students' subjective opinions. The algorithm that is used for sentiment analyses is based on Naïve Bayes (NB), max entropy, and support vector machines.

Indicators derived and classified will be used for proper curriculum design. Since now we have curriculum design planned strategically, the next step is crucial for managemental teams. That is budgeting and preparation and implementation of proper financial management of the curriculum implementation. Algorithms and techniques of machine learning are used to properly predict budget spending. Before we apply machine learning and AI algorithms, we need to know the project cost management process. Before budget approval, we need to provide a plan, estimation, budget proposal, and the control mechanism. The first step, planning, provides information on cost estimation, allocation, management structures, monitoring procedures, and control mechanisms.

Project cost management involves four processes to complete a project with the approved budget: planning, estimating, budgeting, and control. First, plan cost management defines how a project's cost will be estimated, allocated, managed, monitored, and controlled. The second step, cost estimation, is an approximation of monetary resources for the completion of the project. Budget determination is the process of collecting overall costs for all individual activities to plan baseline costs. The final step is establishing a control mechanism for the process with monitoring of project updates to adapt and manage project cost changes (Burke, 2013). In this case, we overview K-nearest neighbors as the nonparametric method that can be used in classification and regression.

The final step is the selection procedure by human resources. Big data analysis in our case will be applied as methodological approach in the selection part of the hiring procedure. Results from the previous steps will fill in data for gaps in knowledge and proper student development paths. More successful paths will be promoted. To provide proper human resources, we need to have AI-based analyses of the staff's knowledge and experience in the different engineering fields and additional tools that can provide information on the possible professional staff hired to fulfill the gap. Two major resources exist for professional and/or research professional tracking: One is LinkedIn and the second is Google Scholar.

LinkedIn analytic stack is based on Hadoop end-to-end (White, 2010). The main advantages of this track are horizontal scalability, high fault tolerance, and high ability to process a high amount of data on thousands of machines. Also, Hadoop has easy-to-program semantics and an active ecosystem. Hadoop allows machine learning implementation for professionals with modest programming and distributed system knowledge to develop proper ML models. This possibility for a researcher to create new fast iterations overcomes the classical researcher-engineer bottleneck (Sumbaly et al., 2013).

LinkedIn analyses can provide a good overview of an expert in the professional aspect. But higher education institutions also need to promote scientific and research-based work. Those data can be extracted from Google Scholar archives.

Google Scholar is one of the widely used tools for bibliometric purposes. From the early days of the implementation of this system, it is used as a source of information for the scientific world and as a research evaluation tool. Google Scholar can collect citations undetectable by other citation databases (López-Cózar et al., 2017). This tool can be used for collecting big data and later to derive small data for selected keywords (Sun & Huo, 2021).

Based on all mentioned AI and ML-supported data, we can conclude and deliver an enhanced system for curriculum strategic planning. The following chapters will give an overview of the mentioned AI and ML techniques that are successfully used in all the mentioned steps of the decision-making process.

The holistic approach to curriculum design and classification

Since we will use a framework for a holistic approach to curriculum design, in this chapter we need to first understand the different classification methodologies that we will use on qualitative and quantitative indicators. Then we need to overview the proper LMS, which will provide lessons to students, and data for tracking students' success. Finally, we will do some educational analytics where we will analyze the collected data and based on all those aspects deliver an interface. The interface is an analysis of reports and justification for curriculum design.

Classification methodologies for qualitative indicators

The classification task can be solved by asking specific questions about the attributes of the instances. Whenever the answer is received, we ask another question, and the process iteratively continues until the class label of an example is obtained. This process can be represented by a tree, which we call a decision tree. A decision tree has three types of nodes: one root node, internal nodes, and leaf nodes, where each leaf node corresponds to one class label. The root node and each internal node contain a test condition for one attribute. The process of classification starts from the root node of a tree and then continues down the tree until it reaches a leaf node, which is a classification of an instance (Batra & Agrawal, 2018). Moving down the tree is defined by the test outcome of a corresponding attribute in the current node (Fig. 5.2).

For a given set of attributes, there are many possible decision trees. Since finding the optimal decision is an NP-hard problem, we use efficient greedy algorithms that allow us to construct a suboptimal tree in a reasonable time. Hunt's algorithm belongs to this category, and it served as a basis for many famous tree induction algorithms, such as C4.5 (J48) and CART.

There are two main issues to address while inducing a decision tree: (1) choosing the best attribute for record split in a given node and (2) stopping the splitting procedure.

Starting from the root node, a statistical test or procedure is used at each node to evaluate the attributes and determine which one will be used as a test at a given node (Navada et al., 2011). The attribute that performs the best in classifying the remaining examples is selected and used as a test. For each value of this attribute, one branch and one descendant (child) node are created, and instances are distributed to one of them based on their attributes' value. ID3 algorithm uses information gain as a statistical test to select the best attribute. Information gain calculates the reduction in the entropy, and attributes with maximum information gain are selected as a test condition in a given node.

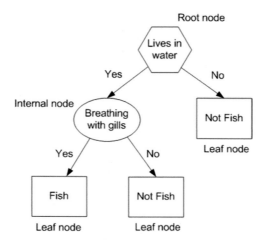

FIGURE 5.2

Decision tree for the fish classification problem.

Decision trees are easy to understand and interpret, work with both numerical and categorical features, and require little data preparation. Thanks to these advantages, they are successfully applied as a classification method for a broad range of tasks, from marketing, retention of customers, and assessing credit risk, to the diagnosis of diseases and ailments in the healthcare industry.

Classification And Regression Tree (CART) is a decision tree algorithm variation. CART uses Gini impurity in the process of splitting the dataset into a binary decision tree. Gini impurity can be calculated in the following way:

$$I_{Gini} = 1 - \sum_{i=1}^{j} p_i^2$$

$I_{Gini} = 1 - (\text{probability NO})^2 - (\text{probability YES})^2$ As
The short algorithm for the implementation of this classification is as follows:

1. Calculate all of the Gini impurity scores
2. Compare the Gini impurity score, after n before using a new attribute to separate the data. If the node itself has the lowest score, then there is no point in separating the data.
3. If separating the data results in an improvement, then pick the separation with the lowest impurity score.

The C4.5 algorithm is a classification algorithm that produces decision trees based on information theory, and it is a statistical classifier.

WEKA data mining tool defines J48 as an open-source Java implementation of the C4.5 algorithm. J48 allows classification via either decision trees or rules generated from them.

This algorithm builds decision trees, by using the concept of information entropy. The training data is a set $S = \{s1, s2, \ldots\}$ of already classified samples. Each sample si consists of a p-dimensional vector (x1, i, x2, i, ..., xp, i) where the xj represents the attribute values or features of the corresponding sample, as well as the class in which the sample falls.

At each node of the tree, the C4.5 algorithm chooses the attribute of the data that most effectively splits its set of samples into subsets, enriched in one class or the other. The splitting criterion is the normalized information gain, which is calculated from the difference in entropy. The attribute with the highest normalized information gain is chosen to make the decision.

When building a tree, J48 ignores the missing values; that item's value can be predicted from the attribute values for the other records.

Decision trees are utilized to delineate decision-making processes. It is a classifier that acts like a flowchart-like tree construction to depict association models. The decision trees are utilized to categorize instances by sorting them down the tree from origin to a little leaf node. Every single node specifies an examination of the instance, and every single division corresponds to one of the probable benefits of this attribute.

It divides a dataset into tinier and tinier subsets and is incrementally developed. The final consequence is a tree alongside decision nodes and leaf nodes. Each decision node has two or more divisions, and the leaf node embodies an association or decision. The topmost decision node in a tree that corresponds to the best predictor shouted the origin node.

Sentiment analysis methodologies for qualitative indicator analysis

Sentiment analysis can identify and extract relevant information about students' subjective opinions. The algorithms that are used for sentiment analyses are based on NB, max entropy, and support vector machines.

Bayes classifier

The NB algorithm is a classification-supervised machine learning algorithm based on the Bayes theorem of conditional probabilities. It assumes that features are conditionally independent given the class label and that all contribute independently to the outcome. The classifier is naive because this assumption is most likely not true, but nonetheless, the model is easy to fit and is able to work surprisingly well for many real-world problems (Xu, 2018).

If we denote feature vector as X and classes in the classification problem as Cj, by applying Bayes theorem, we obtain

$$P\left(C_j |, X\right) = \frac{P\left(C_j\right) * P(X|C_j)}{P(X)}$$

where P(Cj|X)—the posterior probability of class Cj given the feature vector X; P(Cj)—prior probability of class Cj; P(X|Cj)—the probability of feature vector X given class Cj; and P(X)—prior probability of feature vector X.

Calculating P(X|Cj) depends on the feature distribution. If the features follow a multinomial, Bernoulli, or Gaussian distribution, we use the appropriate type of NB classifier. The classifier applies a very simple maximum a posterior decision rule to produce classifications. It simply takes the class with the largest probability value P(Cj|X) given the feature X.

The NB classifier has many advantages. It requires a relatively small training dataset and little explicit training and can perform well in both high-dimensional settings and in the case of large datasets. Naive Bayes has been popular for solving spam detection problems, but also in the domain of automatic medical diagnosis and text classification.

One of the disadvantages of the NB classifier is the zero frequency problem. It occurs when a categorical variable has a certain category in the test set that was not present in the training set. In this case, the classifier will not be able to make a prediction. Laplace estimation is often used to alleviate this problem.

Maximum entropy algorithm

Maximum entropy (MaxEnt) is a probabilistic classifier based on the maximum entropy principle and estimation of a probability distribution. The basic idea of the algorithm is that in the absence of knowledge or when we are uncertain about the data, it should be assumed that the probability distribution is uniform, i.e., has maximal entropy. From all the available models that fit the training data, the one that has the largest entropy is selected. Using the most uniform model ensures that the risk of prediction is minimized (Xie et al., 2019). When we want to make a prediction, we assume that the prediction should satisfy the given constraints, which are derived from the labeled training data. Unrestricted use of contextual features and an objective function that converges to a global optimum are characteristics that make MaxEnt a good candidate for text classification and sentiment analysis. MaxEnt is primarily used when we can't assume prior distributions or conditional independence of the features. This is why the algorithm is applied for classification tasks where the features are words, which clearly are not independent.

The first step in the MaxEnt algorithm is to identify a set of features to use for setting the constraints. When used for text classification, any real-valued function of the document and the class can be a feature. For example, the feature could be words (word counts) that appear in a document in that class. For each feature, its expected value over the training data is measured and this value is taken to be a constraint for the model distribution (Wang et al., 2010).

MaxEnt has one important advantage against NB in text classification problems. MaxEnt does not assume that the features are conditionally independent of each other. While MaxEnt is competitive in terms of CPU and memory efficiency, it still may suffer from overfitting and poor feature selection and requires more time to train compared to NB. MaxEnt has good scalability and performs very well for high-dimensional data and large datasets. It has been applied in many NLP problems such as tagging, parsing, coreference, parse re-ranking, semantic role labeling, sentiment analysis, etc.

Support vector machines

Support vector machine or SVM is a supervised machine learning algorithm primarily used for classification problems, but can also be applied for regression problems. With the help of support vectors, SVM creates a decision boundary in an n-dimensional space called the hyperplane, and hence the algorithm is termed support vector machine. Among the many possible hyperplanes, the algorithm chooses the one that has the maximum distance (margin) between the hyperplane and support vectors. Selecting this particular hyperplane maximizes the SVM's ability to correctly predict the class label of new data records. Hyperplane dimension depends on the number of features in the dataset; for n-dimensional feature vector, hyperplane dimension is n-1. If the number of features is two, then the hyperplane is a straight line. If the number of features is 3, then the hyperplane becomes a two-dimensional plane. The support vector consists of training data points that are the closest to the hyperplane, and it defines the position of the hyperplane. If removed, the position of the hyperplane is changed. Finding the optimal hyperplane is an optimization problem of minimizing a quadratic function under linear inequality constraints, and it can be solved by optimization techniques (Suthaharan, 2016). A large margin approach in SVM is implemented by changing the threshold values of linear function output. Unlike linear regression, where the threshold value is 0, in SVM these values are -1 and 1, and they define a margin. Another important concept in SVM is a soft margin, which allows some data point outliers to reside on the same side of the hyperplane as the opposite class without affecting the final result. Implementing this concept requires a special parameter to control the number of hyperplane violations. Setting this parameter is complicated because it represents a compromise between hyperplane violations and the size of the margin.

There are linear (simple) and nonlinear SVMs. The former is used for linearly separable data, while the latter is used for data that cannot be classified with a straight line. Kernel SVM uses different functions that transform original input vectors into new, richer (usually high-dimensional) spaces where they are linearly separable. This way we can obtain a more complex hyperplane to fit nonlinear data. With kernel functions, the algorithm gains the ability to generate nonlinear decision boundaries using methods designed for linear classifiers and allows the user to apply a classifier to data that has no obvious fixed-dimensional vector space representation (Meyer et al., 2003). Linear, polynomial, Gaussian radial basis functions (RBFs), and sigmoid are the most widely used kernel functions.

SVM is less prone to overfitting, works well with smaller datasets or when the number of features is greater than the number of data points, and has the ability to deal with high-dimensional data. Memory efficiency and algorithm versatility due to different kernel functions are also advantages of SVM. However, the SVM algorithm does not directly provide probability estimates and has a high training time.

SVM algorithm can be used for face detection, pattern recognition, image classification, text categorization, handwriting recognition, etc.

Indicators and learning management system (LMS) as technical basis for holistic approach

The definition of indicator sets is of crucial importance for the application of ML and AI methodologies. As previously stated, we will define two sets of indicators. First are quantitative indicators. During their definition, we also need to know the place where those indicators will be stored for data collection. In our research, we defined the following indicators: percentage of project completion, grade acquired on tests, time spent on test solutions, number of completed learning materials, and number of downloads of material. Qualitative indicators for sentiment analyses will be collected by the sets of questionnaires, and we defined the following indicators: opinion of grade importance, quality of learning material, quality of the curriculum, quality of the teacher, and quality of the grade system.

Since all this data will be stored and/or collected on the LMS, the State University of Novi Pazar is a platform using Moodle. Moodle is a free and open-source learning management system written in PHP and distributed under the GNU General Public License. Moodle is used for blended learning, distance education, flipped classrooms, and other online learning schemes in schools, universities, workplaces, and other sectors. The State University of Novi Pazar used the latest iteration of this system Moodle 4.0.

Since all ML and AI algorithms are based on extensive use of database information, we need to know Moodle 4.0 database architecture. Those databases have a huge amount of data since LMSs have a large number of users. In our case, we used the database from the Moodle system at the State University of Novi Pazar; the size of the database was 13 Gb, since it contains data from 2012 to 2022. To prepare those tables, we created temporary tables, especially tables with logged activities. Those additional tables are created using big data processing tools. First, we used Caldera Hadoop, for creating tables with all user actions for any given module. Hadoop distributed file systems can store huge amounts of data. The second important core element of Hadoop is mapreduce, which is a tool for dividing information into sets of independent tasks. For clustering, we used Wolfram Mathematica set of modules for big data analysis, AI, and ML procedures.

The results of the CART decision tree are given in Table 5.1.

Level of successful classification is 70.1%. The decision tree is more efficient where the grades of students are lower than 75%, and at this level, the percentage of successful classification is 74.6%. Classification for the students with higher grade is 65.6%. Confusion matrix shows the number of students in the higher grade (1) and lower grade (2) and the level of their successful classification.

Table 5.1 CART decision tree.

CART Decision Tree	
Percentage of project completion	< 1.5
Grade acquired on tests	< 1.5:1(67.0/35.0)
Grade acquired on tests	$> = 1.5$:2(27.0/9.0)
Percentage of project completion	$> = 1.5$:2(76.0/20.0)
Number of leaf nodes	3
Size of tree	5

Stratified cross-validation

Summary

Correctly classified instances	166	70.9402%
Incorrectly classified instances	68	29.0598%
Kappa statistic	0.4014	
Mean absolute error	0.3873	
Root mean squared error	0.4601	
Relative absolute error	79.9955%	
Root relative absolute error	93.5313%	
Total number of instances	234	

Detailed accuracy by class

TP rate	FP rate	Precision	Recall	F-Measure	ROC area	Class
0.656	0.254	0.643	0.656	0.649	0.683	1
0.746	0.344	0.757	0.746	0.752	0.683	2

Confusion matrix

A	B	Classified as
63	33	a = 1
35	103	b = 2

K-nearest neighbor kNN in strategic budget planning

Machine learning and AI as we already established rely on different classification models. K-nearest-neighbor (kNN) classification is one of the most fundamental methods and is the first choice for a classification study when there is little or no prior knowledge about the distribution of the data.

K-nearest-neighbor classification was developed from the need to perform discriminant analysis when reliable parametric estimates of probability densities are unknown or difficult to determine (Fix & Hodges, 1951). In the research that followed, it was shown that for k = 1 and n → ∞ the k-nearest-neighbor classification error is bounded above by twice the Bayes error rate (Cover & Hart, 1967). With this important rule, a number of important properties were derived: new rejection approaches (Hellman, 1970), refinements concerning Bayes error rate (Fukunaga & Hostetler, 1975), distance weighted approaches (Dudani, 1976), soft computing (Bailey & Jain, 1978), methods, and fuzzy methods (Bermejo & Cabestany, 2000).

The two most important characteristics of kNN are sample geometric distance and classification rule and confusion matrix. The classifier is commonly based on the Euclidean distance between the sample and the specific training sample. Let x_i be in the put sample, which has p different features, and n is the total number of input samples and p is the number of features. The Euclidean distance between two samples x_i and x_l is defined as

$$d(x_i, x_l) = \sqrt{(x_{i1} - x_{l1})^2 + (x_{i2} - x_{l2})^2 + \ldots + (x_{ip} - x_{lp})^2}$$

Since we need some type of classification, this is done by partitioning samples into two categories. One category is training, and the second is a testing category. Value x_i is the training sample, while x is the test sample; ω is the actual true class of the training sample; and $\hat{\omega}$ is the predicted class, where Ω is the total number of classes. During the training phase, we use true class ω to train the classifier, and then we predict class $\hat{\omega}$ for each test sample. Now we use the first nearer neighbor rule; the predicted class of the test sample x is set equal to the class ω of its nearest neighbor. The value of m_i is the nearest neighbor to x if the distance is

$$d(m_i, x) = \min_j\{d(m_j, x)\}.$$

If we predicted the class of test samples x, set it equal to the truest class k among the nearest training samples. Based on this, we define the decision rule $D:x \rightarrow^{\text{yields}} \hat{\omega}$.

For tabulating class prediction during testing, we will use a confusion matrix. Dimensions of the confusion matrix are $\Omega \times \Omega$. If the predicted class of test samples is correct, the diagonal elements of the confusion matrix are incremented by 1. In another case, the off-diagonal element is incremented by 1. Once all samples are classified, we need to calculate classification accuracy. Classification accuracy is the ratio of the number of correctly classified samples to the total number of samples classified.

$$Acc = \frac{\sum_{\omega}^{\Omega} c_{\omega\omega}}{n_{\text{total}}},$$

Increased performance of a classifier can sometimes be achieved when the feature values are transformed prior to classification analysis. Two commonly used feature transformations are standardization and fuzzification. Standardization removes scale effects caused by some features changing and applying a different scale. Fuzzification is a transformation that exploits uncertainty in feature values and in this way increases classification performance.

This classification procedure is already part of many computational tools like Wolfram Mathematica. Pseudocode for implementation is given as follows:

- Calculate
- $d(x_i, x_l)$ i = 1, 2, ….., **n**; where **d** denotes the Euclidean distance between the points.
- Arrange the calculated **n** Euclidean distances in a nondecreasing order.
- Let **k** be a + ve integer, and take the first **k** distances from this sorted list.
- Find those **k**-points corresponding to these **k**-distances.
- Let **ki** denotes the number of points belonging to the ith class among **k** points, i.e., $k \geq 0$
- If $k_i > k_j \ \forall \ i \neq j$, then put x in class i.

This approach in the planning phase of the budget needs to follow several steps. The usual first step is the conversion of collected data into CSV files. Those files need to have all the data set, and we can rcursivelynot have an unknown category. The second step is kNN analysis performed in any specialized program. Data that we collect for budget planning is the number of active students, the number of lecturers, the percentage of work program realization, the average grade of the student, and percentage of success rate on single tests in LMS. This approach can provide a better estimate for management staff to deliver better budget planning (Keller et al., 1985).

LinkedIn and Google Scholar Big Data system for support of human resource procedure

Since we want to complete possible big data analyses by already mentioned ML and AI algorithms, we will survey LinkedIn and Google Scholar as the pool of possible employees. In order to do that, we need to understand the architecture on which those systems are based.

We will first analyze the LinkedIn data pipeline. After every activity generated on the website, data flows into offline systems. There they are used for building various datasets. Those datasets then can be pushed back on the online serving side. LinkedIn is one of the earliest systems that adopted the Hadoop stack. In this case, the distributed file system is HDFS (Shvachko et al., 2010). Every information and dataset in HDFS can be classified as activity data and core database snapshot.

The activity data consists of streaming events generated by the service-handling requests on LinkedIn. Events are grouped into semantic topics and transported by LinkedIn's publish-subscribe system, Kafka. These topics are eventually pulled into a hierarchical directory structure onto HDFS.

Once data is available in an ETL HDFS instance, it is then replicated to two Hadoop instances, one for development and one for production.

All data now loaded into Hadoop can take two distinctive forms: database or event data. Database data includes information about users, companies, connections, and other primary site data. Event data consists of a stream of immutable activities or occurrences. Examples of event data include logs of page views being served, search queries, and clicks. Those datasets are large and diverse. All these datasets must be evolved and continuously updated. Also, new data must be compatible with old. Finally, datasets must be complete and correct. To solve this issue, LinkedIn developed the system Kafka (Kreps et al., 2011). Kafka is a distributed publish-subscribe system (Eugster et al., 2003) that persists messages in a write-ahead log, partitioned and distributed over multiple brokers. It allows data publishers to add records to a log where they are retained for consumers to read at their own pace. Each of these logs is referred to as a topic. Two important aspects of the final front-end delivery of the LinkedIn system are the possibilities of collaborative filtering and skill endorsement.

The first one is used for association rule mining where results are shown on the website as a navigational aid for the member to discover related or serendipitous content from the "wisdom of the crowd." This pipeline initially computed only member-to-member cooccurrence, but quickly grew to meet the needs of other entity types, including crosstype (e.g., member-to-company) recommendations. LinkedIn's front-end framework emits activity events on every page visit as part of LinkedIn's base member activity tracking. A parameterized pipeline for each entity type uses these events to construct a co-occurrence matrix with some entity-specific tuning. This matrix is partially updated periodically depending on the needs of each entity type (e.g., jobs are ephemeral and refreshed more frequently than relatively static companies). The resulting key value of the store is a mapping from an entity pair—the type of the entity and its identifier—to a list of the top related entity pairs.

Second skill endorsement is a lightweight mechanism where a member can affirm another member in their network for a skill, which then shows up on the endorsed member's profile.

LinkedIn analyses can provide a good overview of an expert in the professional aspect. But higher education institutions also need to promote scientific and research-based work. Those data can be extracted from Google Scholar archives.

Google Scholar is one of the widely used tools for bibliometric purposes. From the early days of the implementation of this system, it is used as a source of information for the scientific world and as a research evaluation tool. Google Scholar can collect citations undetectable by other citation databases. This tool can be used for collecting big data and later to derive small data for selected keywords (Mahdi & Ahmed, 2014). Since Google Scholar is a web search engine, proper web crawling algorithm must be implemented.

The most efficient algorithm to be used and is used for web crawling is the depth first search algorithm. This powerful algorithm is used to travel through the search by starting at the base page and traversing deeper through every found hyperlink page. It then backtracks and moves to the adjoining hyperlink pages. The recursive pseudocode algorithm is given as follows:

far all edge e in G.incidentEdge(v)
do
if edge e is unexplored then
w = G.opposite(v,e)
if vertex w is unexplored then label e as discovery edge recursively call DFS(G,w)
else label e as back edge

This crawling algorithm can provide us with a high volume of data from the web for proper sentiment analysis.

Part of the decision-making AI system that raises ethical issues and applicable ethical frameworks

We will first provide a brief overview of the established ethical frameworks for the implementation of AI in decision-making processes, before delving into the ethical challenges. These frameworks can serve as a guide for developers to ensure that AI systems are created and executed in an ethical manner. For example, the IEEE Global Initiative for Ethical Considerations in AI and Autonomous Systems and the European Commission's Ethics Guidelines for Trustworthy AI offer guidance on ethical values and principles that should be integrated into the design and implementation of AI systems.

The first one is part of the IEEE P7000 system of standards; the idea behind these standards is that they should cover the field of the intersection of technology and ethics. The first edition of

Ethically Aligned Design (The IEEE Global Initiative on Ethics of Autonomous and Intelligent Systems, 2017) is the result of a three-year effort that followed a bottom-up approach since 2015. The process involved the rigor and standards of engineering professionals and a global and iterative approach that engaged hundreds of experts worldwide. Through this process, an analysis of principles, issues, and recommendations was generated, which has already influenced the creation of fourteen IEEE Standardization Projects, a Certification Program, A/IS Ethics Courses, and several other action-driven initiatives that are currently being developed.

The document outlines the initial ethical principles for AI and intelligent systems, which include prioritizing benefits to humanity and the environment over commercial interests. Additionally, the principles aim to mitigate any negative impacts or risks that may arise as AI systems evolve, particularly by ensuring accountability and transparency in their actions. Finally, the principles emphasize the importance of upholding human rights and promoting human well-being without degrading the environment. The final point is that A/IS creators must provide individuals with the ability to access and securely share their data in order to maintain people's ability to control their identity. The general solution to give individuals agency is intended to anticipate and enable individuals to own and fully control autonomous and intelligent (as in learning) technology capable of evaluating data use requests from third parties and service providers (IEEE-P7006; IEEE-p7012).

Important parts of the documents' general principles are imperatives and they are the following: human rights—A/IS shall be created and operated to respect, promote, and protect internationally recognized human rights.; well-being—A/IS creators shall adopt increased human well-being as a primary success criterion for development; data agency—A/IS creators shall empower individuals with the ability to access and securely share their data, to maintain people's capacity to have control over their identity; effectiveness—A/IS creators and operators shall provide evidence of the effectiveness and fitness for the purpose of A/IS; transparency—The basis of a particular A/IS decision should always be discoverable; accountability—A/IS shall be created and operated to provide an unambiguous rationale for all decisions made; awareness of misuse—A/IS creators shall guard against all potential misuses and risks of A/IS in operation; competence—A/IS creators shall specify and operators shall adhere to the knowledge and skill required for safe and effective operation. In our system, we identified data agency, accountability, transparency, and awareness of misuse as crucial points.

The European Commission stated its vision for AI in two communications on April 25, 2018, and December 7, 2018. This vision emphasizes the importance of developing "ethical, secure and cutting-edge AI made in Europe." The Commission's vision is based on three pillars: increasing investments in AI from public and private sectors to encourage its use, preparing for changes in society and the economy as a result of AI, and establishing an appropriate ethical and legal framework that reinforces European values.

To bring this vision to fruition, the Commission formed the High-Level Expert Group on Artificial Intelligence (AI HLEG), which is an independent group responsible for creating an important deliverable: AI Ethics Guidelines (Smuha, 2019). The developed framework has three critical components: foundations of trustworthy AI, realization of trustworthy AI, and assessment of trustworthy AI. For our developed system, we will pay attention on the realization part. Requirements are given in following list: human agency oversight, technical robustness and safety, privacy and data governance, transparency, diversity nondiscrimination and fairness, environmental well-being, and accountability.

As both documents have identified similar challenges, we will begin our analyses by collecting student data, learning path results, and big data for potential new teaching staff. Based on the previous two documents, we conclude as follows. First and foremost, our system must ensure data security at all levels of implementation. Second, the data quality must be appropriate, especially since any set of collected data may contain social biases, inaccuracies, and errors. Third, any organization or automated system that handles data must adhere to an accurate data handling protocol.

We will use data anonymization that is a technique used to remove personally identifiable information from datasets. The identifying information is replaced with random or generalized values in this technique. The reasoning is straightforward: We can train a dataset to use k-anonymity by permuting, suppressing, and swapping certain data values. According to this logic, we will have data values that do not contain any identifiable information (Kantarcioglu & Clifton, 2004). To achieve k-anonymity, we need a set of k records that are both undisguisable from each other (Machanavajjhala et al., 2006; Sweeney, 2002). Implementation of this type of protocol is based on relation tables, where attributes are classified as unique identifiers for individual data, and quasiidentifiers, which are a set of minimal attributes joinable with information required to reidentify individual records.

For the second part, we can apply the metadata profiling system that consists of three main components: the graph representation generator, the similarity calculator, and the metadata profiler. The system takes in metadata lists generated by a feature extraction AI and converts them into graph representations using the graph representation generator. Each data source node is connected to the value node, which is then connected to the label node. This graph representation can be easily combined with information from other AI models or manually attached metadata (Aikoh et al., 2020).

The similarity calculator converts the graph representation into a distributed representation, such as Node2Vec, to represent the similarity of each node on the metagraph. The metadata profiler uses manually attached errata for metadata as training data, in addition to the distributed representation of the data source node. The metadata profiler combines the training data and the distributed representation to obtain a classifier that classifies the correctness of the inference result of the metadata estimation system by learning with SVM using the combined data.

Using the metadata profiler, incorrect metadata can be detected by mapping the inference space generated with the data features to the space characterized by the topological similarity of the data source on the metagraph. Topological similarity refers to the number of identical metadata, such as the same metadata value, the same metadata labels, and the same parameters of metadata estimation AI. Wrong metadata can be detected through conflicts where different metadata are assigned when the similarity is high.

And finally for reducing bias with an accurate handling protocol, we can diversify data collection. A key way to reduce bias in AI systems is to ensure that the data used to train the system is diverse and representative of the population it aims to serve. This can be achieved by collecting data from a wide range of sources, including data from underrepresented groups.

Conclusion

Curriculum design, planning, and implementation are cumbersome tasks. The curriculum must provide proper theoretical knowledge, and at the same time, it must be modern and deliver students state-of-the-art knowledge of new technologies. The development of the technology dictates the shorter time of the new increment of software. New solutions are available over the internet in a matter of seconds to teachers and students.

Those technological advances dictate that those old ways of preparing curriculums for higher education are outdated. They can provide us with a statistical overview of old records.

Technology also provides teachers and institutions with modern and fast databases developed as unique learning management systems. Those systems have databases of the teacher and students, but also have a database of the tests, lessons, percentage of success, surveys, etc. This data is huge, and finding and classifying data is a serious endeavor.

From that point, AI algorithms step in and help us to classify data. After classification using proper techniques, we can calculate the prediction of future results of the education cycle. As already mentioned, every step of the classical education preparation can be supported with a specific set of machine learning techniques. These techniques then assist decision-makers in taking evolutionary steps to improve learning efficiency. However, all of these techniques can be used in an unethical manner. That is the primary reason for the implementation of ethical frameworks and proper methodologies to reduce those unfavorable effects.

References

Aikoh, K., Isoda, Y., & Sugimoto, K. (2020). Data profiling method for metadata management. In *2020 IEEE 7th international conference on data science and advanced analytics (DSAA)* (pp. 779–780).

Bailey, T., & Jain, A. (1978). A note on distance-weighted k-nearest neighbor rules. *IEEE Transactions on Systems, Man, and Cybernetics, 8*, 311–313.

Batra, M., & Agrawal, R. (2018). Comparative analysis of decision tree algorithms. In B. Panigrahi, M. Hoda, V. Sharma, & S. Goel (Eds.), *Nature inspired computing. Advances in intelligent systems and computing* (p. 652). Singapore: Springer. Available from https://doi.org/10.1007/978-981-10-6747-1_4.

Bengio, J., Louradour, J., Collobert, R., & Weston, J. (2009). Curriculum learning. In *ICML '09: Proceedings of the 26th annual international conference on machine learning* (pp. 41–48). June 2009.

Bermejo, S., & Cabestany, J. (2000). Adaptive soft k-nearest-neighbour classifiers. *Pattern Recognition, 33*, 1999–2005.

Burke, R. (2013). *Project management: Planning and control techniques* (2013, p. 26). New Jersey: John Wiley & Sons.

Cover, T. M., & Hart, P. E. (1967). Nearest neighbor pattern classification. *IEEE Transactions on Information Theory, IT-13*(1), 21–27.

de-Marcos, L., Martínez, J. J., & Gutiérrez, J. A. (2008). Particle swarms for competency-based curriculum sequencing. In M. D. Lytras, J. M. Carroll, E. Damiani, & R. D. Tennyson (Eds.), *Emerging technologies and information systems for the knowledge society. WSKS 2008. Lecture Notes in Computer Science* (5288). Berlin, Heidelberg: Springer. Available from https://doi.org/10.1007/978-3-540-87781-3_27.

Dolicanin, E., Fetahovic, I., Tuba, E., Capor-Hrosik, R., & Tuba, M. (2018). Unmanned combat aerial vehicle path planning by brain storm optimization algorithm. *Studies in Informatics and Control, 27*(1), 15–24, ISSN 1220-1766.

Dudani, S. A. (1976). The distance-weighted k-nearest-neighbor rule. *IEEE Transactions on Systems, Man, and Cybernetics, SMC-6*, 325–327.

Elman, J. L. (1993). Learning and development in neural networks: The importance of startingsmall. *Cognition, 48*(1), 71–99.

Ertmer, P. A., & Newby, T. J. (1993). Behaviorism, cognitivism, constructivism: Comparing critical features from an instructional design perspective. *Performance Improvement Quarterly, 6*(4), 50–72.

Eugster, P. T., Felber, P. A., Guerraoui, R., & Kermarrec, A. M. (2003). The many faces of publish/subscribe. *ACM Computing Surveys, 35*(2), 114−131.

Fix, E., & Hodges, J.L. (1951) *Discriminatory analysis, nonparametric discrimination: Consistency properties.* Technical Report 4. Randolph Field, TX: USAF School of Aviation Medicine.

Fukunaga, K., & Hostetler, L. (1975). k-nearest-neighbor bayes-risk estimation. *IEEE Transactions on Information Theory, 21*(3), 285−293.

Hajian, S., Francesco, B., & Castillo, C. (2016). Algorithmic bias: From discrimination discovery to fairness-aware data mining. In *Proceedings of the 22nd ACM SIGKDD international conference on knowledge discovery and data mining.* ACM.

Hellman, M. E. (1970). The nearest neighbor classification rule with a reject option. *IEEE Transactions on Systems, Man, and Cybernetics, 3,* 179−185.

IEEE-P7006. https://www.standict.eu/standards-repository/ieee-p7006-standard-personal-data-artificial-intelligence-ai-agent.

IEEE-p7012. https://www.standict.eu/standards-repository/ieee-p7012-standard-machine-readable-personal-privacy-terms.

Kajiji, N., Dash, G.H., Felner, R., & Seitsinger, A. (2007). Evaluating learning performance: Applying nonlinear artificial intelligence to learning support indicators. In *The international conference on computing & e-systems.*

Kantarcioglu, M., & Clifton, C. (2004). Privacy preserving data mining of association rules on horizontally partitioned data. *IEEE Transactions on Knowledge and Data Engineering (TKDE), 16*(9).

Keller, J. M., Gray, M. R., & Givens, J. A. (1985). A fuzzy k-nn neighbor algorithm. *IEEE Transactions on Systems, Man, and Cybernetics, SMC-15*(4), 580−585.

Kreps, J., Narkhede, N., & Rao, J. (2011). Kafka: A distributed messaging system for log processing. In *Proceedings of the NetDB.*

Li, X., & Zhang, T. (2017). An exploration on artificial intelligence application: From security, privacy and ethic perspective. In *2017 IEEE 2nd international conference on cloud computing and big data analysis (ICCCBDA)* (pp. 416−420). IEEE.

Londoño, E. P. (2011). El diseño instruccional en la educación virtual: más allá de la presentación de los contenidos. *Revista Educación y Desarrollo Social, 5*(2), 112−127, Colombia: Editoria.

López-Cózar, E. D., Orduña-Malea, E., Martín-Martín, A., & Ayllón, J. M. (2017). *Google Scholar: The big data bibliographic tool. Research analytics* (pp. 59−80). Auerbach Publications.

Lu, E. Y., & Willoughby, K. (2015). Performance budgeting in American States: A framework of integrating performance with budgeting. *International Journal of Public Administration, 38*(8), 562−572.

Luzardo, J. (2004). *Herramientas nuevas para los Ajustes Virtuales de la Educación: Análisis de los Modelos de Diseño Instruccional para los eventos educativos en línea* (PhD thesis). USA: Tecana American University.

Lytras, M. D., Aljohani, N. R., Visvizi, A., Ordonez De Pablos, P., & Gasevic, D. (2018). Advanced decision-making in higher education: Learning analytics research and key performance indicators. *Behaviour & Information Technology, 37*(10−11), 937−940. Available from https://doi.org/10.1080/0144929X.2018.1512940.

Machanavajjhala, A., Gehrke, J., Kifer, D., & Venkitasubramaniam, M. (2006). l-Diversity: Privacy beyond k-anonymity. In *Proceedings of the international conference on data engineering (ICDE 2006)* (p. 24).

Mahdi, A. F., & Ahmed, R. K. A. (2014). Crahid: A new technique for web crawling in multimedia *web sites. International Journal of Computational Engineering Research, 4*(2), 01−06.

Mekic, E., Djokic, I., Zejnelagic, S., & Matovic, A. (2016). Constructive approach in teaching of voip in line with good laboratory and manufacturing practice. *Computer Applications in Engineering Education, 24*(2), 277−287.

Menai, M. E. B., Alhunitah, H., & Al-Salman, H. (2018). Swarm intelligence to solve the curriculum sequencing problem. *Computer Applications in Engineering Education, 26*(5), 1393−1404. Available from https://doi.org/10.1002/cae.22046.

Mensah, Y., Schoderbek, M., & Werner, R. (2009). A methodology for evaluating the cost-effectiveness of alternative management tools in public-sector institutions: An application to public education. *Journal of Management Accounting Research*, *21*(1), 203−239.

Meyer, D., Leisch, F., & Hornik, K. (2003). The support vector machine under test. *Neurocomputing*, *55* (1−2), 169−186.

Miguel, B.S., Naseer, A., & Inakoshi, H. (2021). Putting accountability of AI systems into practice. In *Proceedings of the twenty-ninth international conference on international joint conferences on artificial intelligence* (pp. 5276−5278).

Navada, A., Ansari, A. N., Patil, S., & Sonkamble, B. A. (2011). *Overview of use of decision tree algorithms in machine learning. 2011 IEEE control and system graduate research colloquium* (pp. 37−42). IEEE.

Nišić, R., Mekić, E., & Pećanin, E. (2019). Constructive development of physical laboratory exercises without manual as an attitude changing approach. *The International Journal of Engineering Education*, *35*(6), 1584−1593.

Nkomo, L. M., & Daniel, B. K. (2021). Sentiment analysis of student engagement with lecture recording. *TechTrends*, *65*(2), 213−224.

Patil, P. (2016). Artificial intelligence in cybersecurity. *International Journal of Research in Computer Applications and Robotics*, *4*(S), 1−5.

Pecanin, E., Spalevic, P., Mekic, E., Jovic, S., & Milovanovic, I. (2019). E-learning engineers based on constructive and multidisciplinary approach. *Computer Applications in Engineering Education*, *27*(6), 1544−1554.

Romei, A., & Ruggieri, S. (2014). A multidisciplinary survey on discrimination analysis. *The Knowledge Engineering Review*, *29*(5), 582−638.

Sahil, S., Sood, S., & Dogra, S. (2015). Artificial intelligence for designing user profiling system for cloud computing security: Experiment. In *2015 International conference on advances in computer engineering and applications* (pp. 51−58). Ghaziabad.

Sahil, S., Sood, S., & Dogra, S. (2016). Designing and analysis of user profiling system for cloud computing security using fuzzy guided genetic algorithm. In *2016 International conference on computing, communication and automation (ICCCA)* (pp. 724−731).

Sánchez-Romero, J. L., Jimeno-Morenilla, A., Pertegal-Felices, M. L., & Mora-Mora, H. (2019). Design and application of project-based learning methodologies for small groups within computer fundamentals subjects. *IEEE Access*, *7*, 12456−12466.

Seya, K., Okatani, T., Matsuo, Y., Kobayashi, N., & Shirasaka, S. (2020). Identifying issues for learners in completing online courses on machine learning and deep learning: Five issues found in a fully automated learning environment for the purpose of scalable AI education. *Review of Integrative Business and Economics Research*, *9*(3), 35−54.

Shavelson, R. J., Zlatkin-Troitschanskaia, O., & Mariño, J. P. (2018). *Performance indicators of learning in higher education institutions: An overview of the field. Research handbook on quality, performance, and accountability in higher education* (pp. 249−263). Edward Elgar Publishing.

Shvachko, K., Kuang, H., Radia, S., & Chansler, R. (2010). The hadoop distributed file system. In *2010 IEEE 26th symposium on mass storage systems and technologies (MSST)* (pp. 1−10).

Smuha, N. A. (2019). The EU approach to ethics guidelines for trustworthy artificial intelligence. *Computer Law Review International*, *20*(4), 97−106.

Sumbaly, R., Kreps, J., & Shah, S. (2013). The big data ecosystem at linkedin. In *Proceedings of the 2013 ACM sigmod international conference on management of data* (pp. 1125−1134).

Sun, Z., & Huo, Y. (2021). The spectrum of big data analytics. *Journal of Computer Information Systems*, *61* (2), 154−162.

Suthaharan, S. (2016). *Support vector machine, . Machine learning models and algorithms for big data classification. Integrated series in information systems* (36). Boston, MA: Springer.

Sweeney, L. (2002). K-anonymity: A model for protecting privacy. *International Journal Uncertain Fuzziness Knowledge-Based Systems*, *10*(5), 557−570.

The IEEE Global Initiative on Ethics of Autonomous and Intelligent Systems. (2017). *Ethically aligned design: A vision for prioritizing human well-being with autonomous and intelligent systems, version 2*. [Online]. Available: http://standards.ieee.org/develop/indconn/ec/ead_v2.pdf. Accessed on February 6, 2018.

Ulferts, G., Wirtz, P., & Peterson, E. (2009). Strategic human resource planning in academia. *American Journal of Business Education (AJBE)*, *2*(7), 1−10.

Wang, H., Wang, L., & Yi, L. (2010). Maximum Entropy framework used in text classification. In *2010 IEEE international conference on intelligent computing and intelligent systems* (pp. 828−833). doi: 10.1109/ICICISYS.2010.5658639.

White, T. (2010). *Hadoop: The definitive guide*. O'Reilly Media.

Xie, X., Ge, S., Hu, F., et al. (2019). An improved algorithm for sentiment analysis based on maximum entropy. *Soft Computing*, *23*, 599−611. Available from https://doi.org/10.1007/s00500-017-2904-0.

Xu, S. (2018). Bayesian Naïve Bayes classifiers to text classification. *Journal of Information Science*, *44*(1), 48−59.

Zemel, R., Wu, Y., Swersky, K., Pitassi, T., & Dwork, C. (2013). Learning fair representations. In *International conference on machine learning* (pp. 325−333).

Ethical implications of artificial intelligence in autonomous services and systems

The ethical issues raised by the use of Artificial Intelligence products for the disabled: an analysis by two disabled people

6

Laura Smith[1] and Peter Smith[2]

[1]*Musician, Researcher and Busy Mother, Newcastle, United Kingdom* [2]*University of Sunderland, Sunderland, United Kingdom*

Introduction

It is thought that, in the UK today, around 14.6 million people identify as having a disability (Office for Disability Issues and Department for Work and Pensions, 2022), and that there are more than a billion people living with disabilities worldwide. This is a diverse group, comprising those with physical and sensory impairments, learning disabilities, mental health conditions, and those with chronic or long-term illnesses. Many of those living with disabilities require support and adjustments to increase independence and overcome social barriers. There is a very real requirement to explore how artificial intelligence (AI) technologies can contribute to, and shape models of support for disabled people, and a need to critically examine how such technologies affect this diverse group.

AI research can be a force for good for disabled people; however, despite a growing body of research outlining the ways AI can be a pivotal tool for empowerment, independence, and participation, less attention has been given to the real-life application of these systems. This is to say, while it is true that AI technologies have the potential to dramatically impact the lives of people with disabilities, widely deployed AI systems do not however yet work properly for disabled people, or worse, may actively discriminate against them. Guo et al. (2019) identify how AI may "impact particular disability constituencies if care is not taken in their design, development, and testing."

This chapter aims to examine the intersection between AI and disability while also paying attention to the potential ethical dilemmas which this currently razes. We will explore the role AI plays in supporting disabled people and, in turn, also examine what role disabled people could play in the development of AI.

In addition to being informed by relevant literature within the fields of AI, disability, and ethics, which we will briefly review, much of our discussion has been informed by narrative accounts. These accounts are personal diaries from us, two disabled authors, detailing how we use AI in our daily lives. Peter has a spinal injury and Laura is registered blind. Indeed, much of the discussion

Ethics in Online AI-Based Systems. DOI: https://doi.org/10.1016/B978-0-443-18851-0.00022-6

121

presented within this chapter has been drawn from reflecting upon our own experiences and analyzing this data which demonstrates how AI can be used to support disabled people, while also highlighting its limitations. This chapter will conclude with a brief overview of our discussion, as well as presenting some recommendations as to how developers may improve their approach to the design, development, and construction of AI software and technology.

Literature review

The social model of disability (Shakespeare, 2004) originated within the UK in the 1970s. In the document Fundamental Principles of Disability, UPIAS (1976) (Union of Physically Impaired Against Segregation) defined disability not as an impairment of the body or brain, but as a "relationship between people with impairment and a discriminatory society." It is this social definition which underpins the discussions within this chapter. The influence of Marxist thought and labor movement traditions is clear in the work of UPIAS and in *Capital*, Karl Marx (Marx & Engels, 1902) defined capital and labor not as things but as relationships. That is, the social model implies that it is society which disables individuals by the constructs which it places around us, and that it is because society is not inclusive that individuals are disabled.

It is evident within the literature that there are many areas where AI technologies can support, empower, and, ultimately, be a force for good in relation to its effects on the lives of disabled people. It opens up creative, simple ways to manage and automate everyday tasks, supporting areas such as physical independence, visual perception, communication, and decision-making. At its best, the use of AI can help address the inequalities present within today's society and support in reducing of social barriers.

Drigas and Ioannidou (2013) conducted a review on the use of AI in special education. They examined a number of ways that voice activation and communication systems supported children in furthering their learning and connecting with one another. The outcomes demonstrated how a range of AI systems greatly benefited the children within this setting, supporting them in decision-making and giving them the tools to take ownership of their own learning experiences. Closer to home, in relation to Laura's disability, Walle, De Runz, Serres and Venturini (2022) undertook a survey on the recent AI advances in relation to supporting the mobility needs of those living with sight loss. They detailed how electronic travel aids; electronic orientation aids and position location devices have radically improved opportunities for visually impaired people when traveling around and engaging with particular environments. In terms of my own disability, Yozbatiran et al. (2012) use data from one subject to demonstrate the feasibility, safety, and effectiveness of robotic-assisted training of upper extremity motor functions after incomplete spinal cord injury. Developments such as this give me some hope that advanced AI technology may yet assist in my further recuperation.

While these examples illustrate the practical benefits of AI technologies when supporting people with disabilities, sometimes the reality of these technologies does not live up to their promise (Smith & Smith, 2021). Such systems are often inconsistent and or unreliable. Indeed, in our 2021 work, we used our own reflective diaries to illustrate our lived experiences of using AI to support us in our lives as disabled people. We found that, often, the practical application of AI created new problems to navigate due to its failings (Smith & Smith, 2021). Similarly, Swathi and Shetty (2019) wrote about the importance of assistive technology for visually impaired people while also acknowledging its shortcomings when it is not accessible, intuitive, or user-friendly.

There are many benefits of using AI technologies when supporting the disabled; however, their success is dependent on, and intricately linked with their design and ethical underpinnings. Indeed, these technologies can only be considered to be truly successful, as long as they do not marginalize disabled people as a result of the design or development process. A roadmap that includes AI and ethical issues has yet to be developed according to the Alan Turing Institute (2019). Indeed, the creation of a network of experts and resources for AI and inclusion could help to address the "unmet need of assistive products crucial ... to implement the UN Convention on the rights of persons with disabilities" (WHO, 2018).

A number of authors have written on the topic of disability, AI, and ethics (justice or fairness). We shall summarize later recent developments in this area. It is to be hoped that these developments lead to a more inclusive, and hence ethical in our view, approach to the design of AI systems for those with disabilities, including us.

Bennett and Keyes (2019) present two case studies, one on decision-making and the other on AI for the visually impaired to demonstrate how through failures to consider structural injustices in their design, they are likely to result in harm not addressed by a "fairness" framing of ethics. They call on researchers into AI ethics and disability to "move beyond simplistic notions of fairness, and towards notions of justice."

White (2020) discusses fairness for people with disabilities, identifying some of the central problems and took a philosophical perspective motivated by a concern for social justice, emphasizing the role of ethics. Lillywhite and Wolbring (2019) identified many ethical issues within AI and machine learning as fields and within individual applications. They also identified problems in how ethics discourses engage with disabled people.

Coeckelbergh (2010) proposes four objections to introducing AI in health care. First, a robot is able to deliver care, but it will never really care about the human. Second, AI cannot provide "good care," as true care requires empathetic contact with humans. Third, AI may be able to provide care, but in doing so violates the principle of privacy, "which is why they should be banned." Finally, AI technologies such as robots provide "fake care" and are likely to "fool" people by making them believe they are receiving genuine care.

Trewin (2018) argues that fairness for disabled people is different to fairness for other protected attributes such as age, gender, or race, because of the extreme diversity of disabilities and suggests ways of ensuring fairness for disabled people in AI applications.

Floridi et al. (2018) report the findings of AI4People, an initiative designed to lay the foundations for a "Good AI Society." They introduce the opportunities and risks of AI for society and present ethical principles that should underpin its development and adoption. If adopted, these recommendations would "serve as a firm foundation for the establishment of a Good AI Society." In 2019 Techshare Pro held a panel on "Ethics, Machine Learning and Disabilities" which was chaired by Ability Net and included the Head of Public Engagement at the Ada Lovelace Institute (Tweed, 2019).

In 2019 the High-Level Expert Group on AI presented Ethics Guidelines for Trustworthy Artificial Intelligence. According to the guidelines, trustworthy AI should be "lawful — respecting all applicable laws and regulations, ethical — respecting ethical principles and values, and robust — both from a technical perspective while taking into account its social environment." OpenAI is an AI research and deployment company based in San Francisco. Their mission is to ensure that AI benefits all of humanity. In 2020 OpenAI released a charter that will guide AI development in acting in the best interests of its users.

All in all, it seems that there are many initiatives that are striving to solve the problems, address the issues, inherent in developing AI systems which are fair and ethical for, and serve the needs of those with disabilities. While it is true that AI technologies have the potential to dramatically impact the lives of people with disabilities, widely deployed AI systems do not yet work properly for disabled people, or worse, may actively discriminate against them. Guo et al. (2019) identify how AI may "impact particular disability constituencies if care is not taken in their design, development, and testing." This is something which we have both experienced, in our day-to-day lives and our interactions with, and use of, simple AI systems. The diaries which we have prepared and presented next demonstrate these issues.

Methodology

The methodology employed in this chapter is a mixture of autoethnography and reflection. We present two narrative accounts based on our own experiences of the use of day-to-day AI technology, such as speech technology on mobile phones and technology to help the visually impaired. We have employed autoethnographic research approaches (Clandinin & Connelly, 1994) and techniques of reflection (Schön, 1983) in this work.

"Autoethnography is an approach to research and writing that seeks to describe and systematically analyse personal experience in order to understand cultural experience" (Ellis, Adams & Bochner, 2011; Ellis, 2004; Holman, 2005). Autoethnography involves analyzing your own experiences and feelings, preferably as they occur, and relating this to the academic literature; and using those experiences to draw wider conclusions, resulting in lessons for others.

Similarly, reflection (Schön, 1983; Warwick, 2007) has many similarities with autoethnography. Reflection involves analyzing particular occurrences, again preferably as they occur, and thinking about what is learned from these occurrences and what decisions are taken as a result. One method often used in reflection, and applied in this chapter, is the critical incident approach (Flanagan, 1954). The critical incident approach involves identifying and analyzing particular incidents which occur, and which make the individual question their own beliefs or practices. To help us do so, we have each prepared a diary of our day-to-day interactions with simple, readily available, AI systems across the span of a week. These diaries appear next, in the form in which they were written, using the first person as they are personal accounts. We recognize that the very personal nature of this approach and our own backgrounds, while hugely contributing to the depth and breadth of this discussion, could also present an element of bias. To minimize this, we hope that immersing ourselves in the data and discussion will maximize objectivity.

Peter

On April 29, 2016, at 5:00 a.m., I was returning to my bedroom in the dark. I found myself plummeting down the stairs. I landed awkwardly, with my head hanging over the stairwell. I realized immediately that I had broken my neck; I could not feel my arms or legs. I shouted for my wife, Marie, who telephoned for an ambulance. I was rushed into intensive care at the Royal Victoria Infirmary, Newcastle, UK. I was later transferred to the Spinal Injuries Unit at James Cook Hospital, Middlesbrough, UK. I spent 6 months in hospital learning how to speak, eat, and breathe again. I started physiotherapy and regained some mobility. The damage to my spinal cord is incomplete, which means I have some mobility, but none which is really functional in that I cannot feed myself or walk.

Peter's diary

Sunday, 8:00 a.m. Set my intelligent device, which sits beside my bed so that I can easily speak into it and hear it clearly, to remind me to wake up at 10:00 a.m. and check the time sheets for my staff. All works fine and it takes around 1 hour to doublecheck each time sheet with the support and help of one of my carers. Use speech software to type up this diary. It makes a number of errors, each one because it does not recognize my voice correctly. Sometimes it types "." correctly and sometimes as "stop." Nonetheless, after some editing, I managed to type up this entry.

12 noon. I realize that I should really, to inform the reader, list the main AI products I use on a daily basis. These are intelligent speech software, an intelligent personal assistant, and intelligent software on my phone.

9:00 p.m. I try to buy a DVD using my intelligent personal assistant. I wanted to buy a DVD of the 1960s film The Collector. This is a particular favorite of mine. I ordered the DVD from my bed, put it in my basket, and asked my assistant to purchase it. All seems well. Unfortunately, however, and unbeknown to me, my assistant ordered me another DVD of the same name. It arrives a day later (thanks to next day delivery, which is very useful at times that). To my frustration, I now realize what has transpired and I am now left with a horror film which is quite modern and also called The Collector. It costs me £10! Oh well I will watch it at some point, when I have a free moment (which is very rare these days). I still need to get the 1960s film The Collector. I will try my favorite online auction site which is not as intelligent and will require manual intervention; however, in this case, it is more likely to result in the desired outcome!

Monday, 10:00 a.m. I decide to try and experiment to facilitate and speed up writing this diary. I use the intelligent software on my phone to dictate my diary entry and email it to myself, ready for editing on my computer when I get up later in the day. To my joy, this works well, although there are some errors which I need to correct using my intelligent speech software.

12 noon. I begin my work for the day which largely consists of answering emails from students and correcting, and commenting upon, Masters' and PhD students' dissertations and theses. As usual, I use my intelligent speech software to help me do so, accompanied by manual interventions by my (human) personal assistant as and when required. Generally, this works well although I have not yet worked out how to answer my emails using the software. It operates better in Word, and I can type up documents and use the Track Changes function within student dissertations to correct them. However, I have not yet learned how to switch on Track Changes in Word or how to use the Comments facility within Word either. This is probably because I have not spent the time learning how to install the necessary plug-ins to enable the speech software to interact more efficiently with Word. That is one of my problems,

although it may be a problem with most intelligent software also; I am quite impatient and try and use things intuitively, rather than read the instructions which might save me time in the long run!

5:00 p.m. I use my intelligent personal assistant to telephone my daughter. This works well although my daughter tells me it sounds like I am speaking down a tunnel! However, the advantage over the software on my phone is that I cannot end the call easily on the phone whereas on the personal assistant I can simply tell it to "stop!" Would not it be good if all intelligent software operated in a similar manner. Time for something to eat and then watch some television. Apparently, there is a way to link an intelligent personal assistant to the television and switch channels. One day I must invest the time to enable me to do this. However, because I have a human personal assistant available at all times, it is simply easier to ask them to change the channel for me. So sometimes human intervention is simpler although it is not giving me the independence I desire.

Tuesday, 8:00 a.m. I am awoken by my intelligent personal assistant reminding me to purchase tickets for Rod Stewart at Durham Cricket Ground. As if I would need reminding. The presale tickets, to which I am entitled as a member of the Rod Stewart fan club, go on sale at 10:00 a.m.

10:00 a.m. I am online with my human personal assistant, hand on my mouse, ready to purchase tickets for aforementioned Rod Stewart concert. All work smoothly and I am able to buy one wheelchair ticket plus free companion ticket and another ticket for another of my assistants who wishes to attend as a guest. All done manually. Now would not it be good if I could set some intelligent software to buy tickets at the allotted time. I would imagine this would be quite a complex task to program but would be of immense use to a concert fanatic such as myself. Note, the speech software just typed "stop" at the end of the last sentence instead of typing the full stop symbol "."

5:00 p.m. I decide to purchase a burger and fries from a local restaurant and have this delivered by a well-known home delivery company. This all needs to be done manually. Again, this is probably my laziness in not learning the functionality to enable this.

9:00 p.m. I decide to listen to some music by Rod Stewart to congratulate myself on getting the tickets I need. My intelligent personal assistant deftly shuffles songs by the great man. I soon fall asleep and my human personal assistant has to stop the music playing.

Wednesday 6:00 a.m. I am awoken by my intelligent personal assistant which has misinterpreted a reminder which was meant for later in the day at 6:00 a.m. My mistake in not being clearer with my instructions. Lesson learnt; one needs to be very clear when giving instructions to intelligent software. It can only do as instructed. Sometimes this is my error. Sometimes the software does not understand my voice or accent even though it has been trained to recognize me. However, when I get tired, as the day goes on my voice may get weaker and less recognizable. I wonder if the software could somehow be attuned to compensate for this.

12 noon. I decide to go into town in my wheelchair, which itself has some sort of intelligence in the way that it translates my manual chin movements into directions for the chair to follow. Would not it be good if I could use a satnav sort of system coupled with my wheelchair to automatically take me to my destination of choice in the town (usually bookshops and charity shops). Perhaps something for the future?

5:00 p.m. I am back home, still in my wheelchair and typing up a document. Because my speech software misinterprets something I say, a new blank Word document appears on the screen. I try to close it and it asks me if I wish to save it. For a moment, I hesitate. Am I in danger of losing all the work I have typed? Rather than discard the new document I minimize it, just to be on the safe side. Now, the same thing has just happened again as I type this. If I use the words "new document" in the text and speak them quickly together a new Word document opens in front of me and I am left with the same dilemma. I take the risk and do not save the file. All is well. I am back in my original document. However, there have been times particularly when using the command "open dictation box" that I have lost all my work and need to type it up again. The two choices to close a dictation box are "transfer" and "cancel." Sometimes if my voice is tired, the software thinks I have said "cancel" rather than "transfer." Then, all is (literally) lost and must be retyped. Very frustrating!

Thursday, 2:00 p.m. An uneventful day concerning my use of AI products so far. I do not possess any specific product to read a document to me. However, I do believe that Word has a function which does this. I try to use it with little success. Once again, the message to me is that I need to invest time in learning how to use the intelligence of software product. That initial investment of time will (hopefully) pay off in the longer term. Now there was a small annoyance in the speech software interpretation of my voice. The words "pay off" were typed as a single word "payoff" which I do not think is correct. Or is it? Sometimes I doubt myself and the intelligent system is correct and I am not. Lesson learned, be more trusting in the real "intelligence" of the software.

6:00 p.m. I get a notification (reminder to do something) from my intelligent personal assistant. I have no idea at all what she is trying to tell me. However, when I consult my phone, it has the reminder in textual form and I can understand it (sometimes, that is). The trick is to ensure that I speak very clearly when setting reminders and to keep them short; ideally a single word.

Friday, 10:00 a.m. I say something about my intelligent personal assistant, using her name. Of course, unintentionally, I have set my personal assistant off speaking and trying to answer me. My mistake.

12 noon. Time to do an online order from my local supermarket. Unfortunately, this has to be all done manually, I have not yet found a way to use any intelligent software to help me.

2:00 p.m. I am typing up a document. My speech software gives me a "pop-up window" informing me that I have reached the limits of the software dictionary at this point. I am unsure what to do so I simply close the window. It pops up again. I notice a box to click "do not show this message again" so I click that box (or rather my human assistant clicks it for me) and the software seems to work again as before. No doubt I may come across problems later. I need to clean up my computer and make room for the dictionary of my intelligent speech software. Another lesson learned, intelligent systems can build up databases, when they learn, which take up significant space on the computer. I have yet to find a solution to this.

Saturday, 9:00 a.m. I am reaching the end of my diary and have learned quite a lot by explicitly making note of things which have occurred during the week. It has been a useful exercise keeping a diary and taking such notes. My reflections for the week are that I should spend more time investing in learning how to use the intelligent products to their best ability rather than simply use them intuitively without learning any of the commands correctly. However, if a product is really "intelligent," should it not be able to be used in such a way, that is intuitively, with little manual tuning or learning involved?

12 noon. Time to close the diary and think about final reflections and to compare these with Laura's to reach our final conclusions. An interesting week, spent learning about the software I use and also about my own methods and behaviors. Basically, I am too lazy, impatient and need to step back, think about what I am doing and learn about the systems I am using. Many thanks to the intelligent products for helping me produce this diary and to my human personal assistants for supporting my use of the products and undertaking manual interventions as required. I have left this

diary as it has been typed. I have used review and grammar check in Word to do final corrections and then left it as it is, without a final manual check to more authentically represent the documents I can produce.

Laura

At the age of 3, I was diagnosed with a rare form of juvenile rheumatoid arthritis, Still's disease. In addition to effecting most of my joints, it became clear that the inflammation was also affecting my eyes. Over the next few years, I had many visits to the Sunderland Eye Infirmary, Saint Thomas's hospital in London and also undertook several operations and clinical procedures that were carried out at the Prince Charles Eye Unit in Windsor. Despite all of these interventions, my sight grew worse and worse until, at the age of 5 I was registered blind. By the age of 10, I had no sight at all.

Despite my disability, I have always enjoyed an active, independent life. My guide dog Vicky enables me to get out and about and a range of assistive technologies support me with tasks in my daily life, at work and when raising my two young children.

Laura's diary

Monday, 4.45 a.m. I am awoken by my 5-year-old daughter, "Mummy can you come play in my room?" I shout out to my phone, asking the intelligent software to tell me the time. Definitely too early to be awake! I then ask my phone to set an alarm for 7:00 a.m.

7.15 a.m. I am making breakfast and want to find a particular cereal from the many boxes on the shelf. I use an app which, after taking a photograph, can identify and describe an object. I select a box, take a photo, and wait. It describes the color of the packaging. I rotate the box to its perpendicular side. This time the app lets me know I am holding a box. I try a few more attempts, turning the box, moving the phone up and down to focus on different areas of the cereal packet. After around six tries, the phone finally lets me know it is not the box I am looking for. I select another packet. Three attempts in and I decide to go and ask my partner instead.

8.00 a.m. It is almost time to leave for the school run. My daughter has misplaced her reading book which she needs for the day ahead. I use my hands to locate a pile of papers and books and run with them to my scanner. I thrust each one under the elevated camera "blank document" the scanner announces. I know this cannot be true as all these documents are from school. After trying and retrying, I eventually get a string of nonsensical letters and numbers. Thankfully I am spared repeating the endeavor as my daughter, having now located the missing item in her book bag, yells "got it!" up the stairs.

Tuesday, 10.00 a.m. I spend each Tuesday with my 18-month-old son. He drops a stack of printed books on my lap, "book, Mamma, book" he demands. We choose a book and I am pleased to see its one I have previously labeled. There is a sticker on each page which, when tapped with a special pen-shaped device, triggers a recording of my partner reading the story. After a few pages, my son asks, "What is that?" pointing to one of the pictures. I again use the photograph app to describe the image. Because my son has placed my finger directly on the picture in question, I am able to get an accurate photo and therefore the required description first time. Hurray!

10.30 a.m. We are preparing to leave the house. I ask the intelligent software what today's weather will be like. "Cloudy with a chance of rain," my phone replies. My son wants to wear his yellow jumper today. Unfortunately, my son also owns the exact same jumper in blue and the fabric on each item feels the same, not giving any clues as to which color, I have just pulled from the clean washing basket. I use a color identification app, again taking a photo of the clothing before the technology lets me know its color. "Orange," the app replies. I sigh, knowing yellow and blue to be the only options. I decide yellow is more similar to orange than blue, so guess the jumper is most likely to be yellow. My son seems pleased when I pop it over his head, so I tell myself my estimation must have been correct.

Wednesday, 10.00 a.m. I am conducting some research for work. Brilliant, I have found the perfect article! Using my screen reader and the arrow keys on my laptop to navigate the computer screen, I click on the link in question. The software is uncharacteristically silent as I scroll around the screen, trying to read the article. The page appears completely blank to me. I realize that the article must be an image PDF file and therefor I am unable to read it. I then spend an extra 30 minutes finding an alternative reference which, thankfully, this time does work with my screen reader.

11.00 a.m. I need to find a definition for a specific term I wish to include in my work. I ask the intelligent software to read out a definition. The intelligent software does so promptly.

Thursday, 3.30 p.m. I am accompanying my daughter to a birthday party at the community center. My guide dog is in one hand and my daughter is holding the other. I am using a GPS app on my phone. I started off using headphones, only using one earpiece, keeping the other free to listen to traffic, my daughter, etc. This still did not feel totally safe, so I take out the earphones, set the phone's volume to high and place it in my pocket. I can just about hear the phone as we walk; however, occasionally, a passing car or my daughter's voice blocks out what the phone is telling us. At most of the junctions, the phone lets me know which street we have arrived at; however, sometimes this happens after we have already crossed over and begun moving away from said street. I mainly use this app for reassurance on routes with which I am already very familiar. Experience using it in lesser known areas has shown me that it can be slow to pick up where I am standing, resulting in issuing turnings and any unexpected noise can result in missing the app's audio prompts.

5.30 p.m. It is raining, so we have decided to get the bus home. I am again using the GPS app which also lets me know which bus stop we are at. This can be incredibly useful rather than relying on the driver remembering to tell me when I am at my stop or trying to track the bus's movements to ascertain where we are. In truth, I use a combination of the latter method and the app. Again, there can be a slight lag with the app's information, so I try and ensure I have some understanding of where on the route we are and use the app for confirmation purposes. Using these methods, we arrive at the correct stop and successfully disembark from the bus.

Friday, 6.00 p.m. The children are dressing up, "send a photo to daddy!" my daughter asks. Using the screen reader on my phone, I move my finger around the screen, locate the camera and double tap to open the app. I hold the phone up, the camera hopefully facing the children. "Image blurry, 2 faces," the phone tells me. I take the picture. Once again, using my finger to scroll, I locate my partner's contact information, attach the photo and activate the dictation feature so I can dictate a message to accompany the picture. "Fun times before bed," I say. "Phone time before dread," the message reads. I do not have the energy to amend this, plus my partner is used to translating strange messages, so I press send.

8.00 p.m. My partner is busy on a work call. I creep past the diningroom table where he is satting and make my way to the kitchen. It has been a long day so a quick tea tonight. I remember we have a tin of soup. This could be one of three tins so I open up an app to assist. I methodically take photos of each tin, rotating them slowly while taking a series of pictures. The app announces information about each item. It is mainly strings of letters and numbers, punctuated with the occasional, semiuseful clue. After around 10 minutes, I have gathered that the middle tin has the words "organic" and "tomato" written on it. I deduce this must be the soup. Still not totally sure I debate as to whether I should interrupt my partner to confirm my suspicions. I would feel guilty interrupting his work call so decide to risk it, heating up the contents of the middle tin then serve it into a bowl . . . beans on toast it is then!

Saturday, 12.00 noon. The intelligent software has something to tell me! "Chloe, birthday present." I remember the upcoming party next week. I open the laptop and after a bit of browsing, add a sticker book to my shopping cart. My screen reader has let me know that the book is available in several designs. I click to open a drop down menu, wishing to select the desired, sea creatures design. The drop down menu is blank. I am faced with a decision, ask for help or Chloe can get what she is given! Somewhat exasperated, I ask for my partner's assistance.

8.30 p.m. I ask my speech technology to call my dad. After our conversation, I then scroll around the screen to locate the "end call" button which my screen reader audibly identifies. my speech technology is unable to hang-up a call, so a combination of approaches is needed when making and ending calls.

Sunday, 9.00 a.m. I am cooking breakfast for everyone. This involves a combination of using my scanner to read the recipe, an app to identify the ingredients and a lot of hoping for the bet. In the past, the scanner has miss-read important details such as numbers of t-spoons or number of grams required and the dish has been rendered inedible. Thankfully today is somewhat of a success although I do confirm a few details with my partner.

9.00 p.m. The house is quiet, and I decide to read some of my book. I use an eBook app on my phone, open it up (using my screen reader software), then scroll to the book I am reading. I then enable the speech setting so that the eBook can be read out by the in-app reader. This works well although some dedicated eBook devices are not accessible with screen readers and do not have options to enable speech.

10:00 p.m. I ask my speech technology to set an alarm for 7:00 a.m. the next day.

Discussion and reflection

The many existing AI technologies which have been designed to support disabled people have both potentially positive and negative aspects when applying them to everyday life. This is also an industry which is growing and expanding rapidly with many future innovations and designs regularly being announced. While the AI technology itself is developing at a fast rate, the ethics underpinning such design has a long way to go to properly meet the needs of disabled people. As such, there is a power imbalance between the technologies and the people using them. This is to say, the intention and capabilities presented by some AI systems are not always actualized when placed into a real-life context. This may raise the expectations of disabled people, only to be let down by the actualities of the products. Software manufacturers need to be mindful of this and to be honest about the capabilities of the products which they sell to disabled people.

"Technologies like AI and machine learning play a vital role for visually impaired people so that they too can lead a normal and independent life like other people" (Swathi & Shetty, 2019). Indeed, the use of a screen reader is invaluable to Laura's independence. It has enabled her to access education, pursue hobbies, connect with online communities, research, participate, and contribute to countless ventures. It is absolutely evident that a life without screen-reading and other intelligent software would be extremely challenging, not to mention a lot less fulfilling.

However, although providing a much-needed solution to reducing some of the barriers faced by people who are blind and visually impaired, the use of screen-reading software does not come without its own challenges. First, to use screen-reading software, one is required to first learn a series of keyboard shortcuts, command keys, and/or finger swipes which allow users to navigate the screen and select information. Sometimes, frustration can arise when attempting to access a particular part of a site; however, the corresponding key commands or finger movement to trigger it is unknown to the user. In addition to this practical issue, screen-reading software is also not always able to give information if the document or site the user is wanting to access is incompatible with its software. In these instances, the user then has to creatively find a way around these barriers, either enlisting a sighted person's support or the use of another form of technology.

As explored in her diary, Laura sometimes feels reluctant to ask for help, or sometimes it is simply not possible. This touches on an important result of AI's failings, its impact on the mental health of the user. When needing to ask for help, users may feel guilty or worthless, especially if they were expecting to perform the task independently. Similarly, when technology does not live up to its promise, by either failing to support in expected ways, or doing so in an incomplete or insufficient manner, users may feel a sense of failure or helplessness. It is therefore imperative that AI technology is consistent, reliable and delivers what it promises.

As demonstrated within her diary, Laura uses the intelligent software on her phone a lot on a daily basis. It helps her to organize her schedule and set reminders, avoiding the need for a paper-based diary. When dressing her baby, asking the intelligent software for the day's weather forecast enables her to choose appropriate clothing and avoids the need for her to physically go outside to check the temperature. The intelligent software helps her look up word definitions quickly without the use of a website or dictionary and it can even read out recipes to her when cooking. Indeed, there are numerous, varied ways the intelligent software is able to help her be independent.

However, while the intelligent software on her phone is extremely helpful and efficient, the disabled user's dependence upon it means that they must also accept some, less welcome aspects of engaging with the interface. Protecting a user's privacy is a concern many have, especially for those who are hugely reliant on such systems, as is true for many disabled people. When engaging with intelligent software, a user's voice commands, queries, and requests all constitute personal data. Having the ever-present ear enabled on a phone and or tablet means that users may be providing personal data to technology companies (Schmeiser, 2017). It is not clear how the information you give the intelligent software is used and this lack of transparency is concerning. There are many people who believe that intelligent systems are tracking and recording our interactions with them and potentially using them for unethical reasons.

Both Laura and Peter are supported by the use of intelligent systems that are widely available, relatively low cost, or included as part of the purchase of a phone, tablet, or computer. It is clear that such systems make their lives much easier. However, they both experienced frustrations at the limitations of the systems. Some of this is down to the fact that they have not fully learned how to use the systems. But should a disabled person need to do so? Should not a truly "intelligent" system be able to support our needs with minimal training. Or is that too much to expect?

Indeed, the preparing of diaries and reflection process identified several key themes which, currently, prove to be problematic for us when using AI systems and thus, in our view, make for key areas of consideration for designers and developers. We propose these areas should be fundamental to the design process to improve outcomes and make for a more thoughtful and, ultimately, more ethical framework for design. We have classified and categorized these next.

- **Usability**: To be used successfully, many AI systems require training. We propose that designers should focus on a more intuitive, interactive AI interface.
- **Compatibility**: AI systems are often limited due to their lack of compatibility with other technologies. Perhaps a level of standardization needs to be adopted within AI design, particularly with relation to systems specifically aimed at disabled users.
- **Privacy**: It is often not clear if AI systems are collecting data and, if so, what this data is used for. More transparency is needed in this area.
- **Psychological impact**: Disabled users often have a far larger investment in using AI compared to nondisabled users. The promise of AI is often bound up in the notion of increased independence and liberation from relying on human assistance. It is essential that designers and developers have an understanding of these factors and should consult with disabled people to gain insight into how AI systems may potentially affect them in their daily lives.

Both Laura and Peter enjoyed keeping their diaries and writing the narrative accounts. It made both of us think about how the systems help us on a day-to-day basis and how they also sometimes let us down. Sometimes to our embarrassment. For instance, one day (this is not in the diary) Peter sent a message to one of his female personal assistants. He ended the message with (he thought) is simple X to represent a kiss. Unfortunately, the X came out as the word "sex." Very embarrassing. One other unfortunate incident. Peter's phone has a notification about the Durham Boob Festival. That should be, of course, the Durham Book Festival! And one final such a mistake happened only a minute ago. Peter dictated "best wishes" which somehow was typed as "castration"! Peter rests his case and on that note, we shall close our discussion and reflection.

Conclusions

Our own experiences, as presented in our narrative accounts, demonstrate how AI technology can hold many practical benefits, however can also present frustrations and disappointments. It is our belief that, rather than always concentrating on developing new technological innovations, more attention should be given to the actual needs of disabled people and how such technologies would actually work within a real-life context. Indeed, the main lesson to be learned from the literature, and from our own experiences, is the importance of involving disabled people in the design of AI software and technology which is intended for use by those with disabilities. If this step is overlooked, it can often result in a power imbalance and a design which is potentially discriminative or condescending at heart. This is to say, a design which has not involved disabled people at all stages from concept to implementation, may have been constructed and informed by assumptions of need rather than real life requirements. What is needed is true co-design, where disabled people are part of the design team and the process of design. This should include a representative group of people with a diverse range of disabilities. Until this happens, technology can sometimes be, at best, misinformed and, at worst, discriminative in nature. Indeed, the development of AI for disabled people should be informed by the disability movement's mantra, "Nothing about us without us" (Ford Foundation, 2020).

Next steps

The preparation of this piece was undertaken to begin to ascertain, evaluate, and, ultimately, shed light upon the drawbacks of AI technologies for disabled users. We have not feed back these findings to the manufacturers of the products used. Perhaps some next steps would be to present our findings and thoughts to manufacturers to allow them the opportunity to reply.

Disclaimer. We have done our best to ensure that no specific products have been named. If we have inadvertently identified a software product, this is purely unintentional. The views expressed in this chapter are ours, and ours alone. We do not pretend to represent the views of the disabled community, neither could we do so. We have tried to use as inclusive language as possible throughout the chapter.

Note. The two authors are both disabled and have dictated and typed this chapter as well as they possibly could, taking account of their disabilities and the power (or otherwise) of intelligent software support. They have done their best to correct the chapter. However, errors in formatting may still remain. These have been left intentionally to give an authentic picture of what is possible, and what remains to be improved. We trust the publishers and the readers will understand this approach. It was done not out of laziness, but to preserve the authenticity of our situation.

References

Alan Turing Institute. (2019). *AI and inclusion.* <https://www.turing.ac.uk/research/research-projects/ai-and-inclusion> Accessed 17.02.23.

Bennett, C. L., & Keyes, O. (2019, August). What is the point of fairness? Disability, AI and the complexity of justice. In: *ASSETS 2019 Workshop—AI Fairness for People with Disabilities.*

Clandinin, D. J., & Connelly, F. M. (1994). Handbook of qualitative research. In N. K. Denzin, & Y. S. Lincoln (Eds.), *Thousand Oaks.* CA: Sage.

Coeckelbergh, M. (2010). Health care, capabilities, and AI assistive technologies. *Ethical theory and moral practice, 13*(2), 181–190.

Drigas, A. S., & Ioannidou, R. (2013). A review on artificial intelligence in special education. Information Systems, E-Learning and Knowledge Management Research. In: *World summit on the knowledge society.* WSKS 2011, Mykonos, Greece, September 21–23 2011. Revised Selected Papers 4.

Ellis, C. (2004). *The ethnographic I: A methodological novel about autoethnography*. Walnut Creek. CA: AltaMira Press.

Ellis, C., Adams, T. E., & Arthur, P. B. (2011). Autoethnography: An overview. *Historical Social Research*, 273−290.

Flanagan, J. C. (1954). The critical incident technique. *Psychological Bulletin, 51*(4).

Floridi, L., Cowls, J., Beltrametti, M., Chatila, R., Chazerand, P., Dignum, V., & Schafer, B. (2018). AI4People—An ethical framework for a good AI society: Opportunities, risks, principles, and recommendations. *Minds and Machines, 28*(4), 689−707.

Ford Foundation. (2020). *Disability inclusion*. <https://www.fordfoundation.org/about/about-ford/disability-inclusion> Accessed 17.02.23.

Guo, A., Kamar, E., Vaughan, J. W., Wallach, H., & Morris, M. R. (2019). Toward fairness in AI for people with disabilities: A research roadmap. *arXiv preprint arXiv, 1907*, 02227.

Holman, J. (2005). Autoethnography: Making the personal political. In K. Norman, & Y. S. Lincoln (Eds.), *Handbook of qualitative research*. Thousand Oaks. CA: Sage.

Lillywhite, A., & Wolbring, G. (2019). *Coverage of ethics within the artificial intelligence and machine learning academic literature: The case of disabled people. Assistive Technology* (pp. 1−7). .

Marx, K., & Engels, F. (1902). *Wage-labor and capital*. New York: Labor News Company.

Office for Disability Issues and Department for Work and Pensions. (2022). *Disability facts and figures*. London.

Schmeiser, L. (2017). How much dirt does Siri have on you? *The Observer* 06/09/2017.

Schön, D. (1983). *The reflective practitioner*. New York: Basic Books.

Shakespeare, T. (2004). Social models of disability and other life strategies. *Scandinavian Journal of Disability Research, 6*(1), 8−21.

Swathi, M., & Shetty, M. M. (2019). Assistance system for visually impaired using AI. *International Journal of Engineering Research & Technology (IJERT)*.

Trewin, S. (2018). AI fairness for people with disabilities: Point of view. *arXiv preprint arXiv, 1811*, 10670.

Tweed, A. (2019). *Ethics, machine learning and disabilities*. <https://abilitynet.org.uk/news-blogs/ethics-machine-learning-and-disabilities> Accessed 17.02.23.

UPIAS. (1976). *Fundamental principles of disability*. London: UPIAS.

Walle., H. De Runz, C., Serres, B., & Venturini, G. (2022). A survey on recent AI and vision based methods for helping and guiding visually impaired people. *Applied Sciences, 12*(5), 2308.

Warwick, P. (2007). Reflective practice: Some notes on the development of the notion of professional reflection. *ESCalate*. Available from https://dera.ioe.ac.uk/id/eprint/13026.

White, J. J. (2020). Fairness of AI for people with disabilities: problem analysis and interdisciplinary collaboration. *ACM SIGACCESS Accessibility and Computing, 125*, 1.

World Health Organisation (WHO). 2018. *Assistive technology*. <https://www.who.int/news-room/fact-sheets/detail/assistive-technology> Accessed 17.02.23.

Yozbatiran, N., Berliner, J., O'Malley, M. K., Pehlivan, A. U., Kadivar, Z., Boake, C., & Francisco, G. E. (2012). Robotic training and clinical assessment of upper extremity movements after spinal cord injury: a single case report. *Journal of rehabilitation medicine, 44*(2), 186−188.

The implications of ethical perspectives in AI and autonomous systems

7

Arthur So

University of Ottawa, Ottawa, ON, Canada

AI doesn't have to be evil to destroy humanity — if AI has a goal and humanity just happens to come in the way, it will destroy humanity as a matter of course without even thinking about it, no hard feelings.
—Elon Musk, Technology Entrepreneur and Investor.

Introduction

Artificial intelligence's (AI) benefits (for example, in medical diagnostics, facial recognition in financial transactions, and cybersecurity mitigation) may outweigh ethical prevention, as seen in the fast-growing use of AI and autonomous systems. This chapter demonstrates how most AI systems' preventive measures are created in a reactive mode without first considering all of the interconnecting characteristics of constructs in an online AI system (Bartneck et al., 2021; Stahl & Wright, 2018). Most of the time, AI systems "unknowingly or unwillingly [reveal] your private data like age, location, preferences, etc." (ThinkML, 2021, para 2). For example, tracking companies that collect and analyze your private data and then employ them to accommodate your online experience may sell the data to other companies (para 3). In addition, the regulation and oversight of AI health systems "risk falling behind the technologies they govern" (Murdoch, 2021, p. 1).

Nevertheless, preventive measures are undoubtedly a part of the actions necessary to mitigate the disclosure of personal information and financial losses in developing AI systems. So's (2020) dissertation illustrates using a systematic approach to determine the degree of ethical perspective of technological systems, as in the case of using domestic drones (Luppicini & So, 2016). This chapter demonstrates that a similar concept of identifying ethical perspectives can apply to AI systems. The moral viewpoints are determined based on virtue ethical principles in this study. The degree of ethical perspective is then quickly established by AI system developers and mitigates negative consequences earlier rather than later. The chapter also demystifies and provides the impacts of ethics on AI systems and additional components assisting AI systems by the Internet of Things (IoT) and cloud computing capacity. Finally, this chapter suggests a systematic approach highlighting ethical perspectives impacted by AI systems. The process is called technoethics, and it is further discussed in the chapter.

Ethics in Online AI-Based Systems. DOI: https://doi.org/10.1016/B978-0-443-18851-0.00019-6

Background

Artificial intelligence trend

According to Russell and Norvig (2003), AI was introduced as an academic discipline in 1956. Polyakov (2018) described AI as "a broad concept of a *Science* of making things smart or, in other words, human tasks performed by machines" (2018, para. 5). AI is when a tool simulates the human brain for problem-solving and executing formal logic rapidly and adequately. In the early days, AI research was viewed as expert systems acting with human experts' knowledge and analytical skills. As emerging technologies deploy more and more digital transformation and innovation approaches in the 21st century, the volume of data has become too large for human processing with conventional tools. As a result, AI systems have become more effectively used for data mining treatment and other decision-making applications. Today, AI systems involve web search engines like Google, human communication like Alexa, and automated-decision applications like self-driving cars. These AI systems acquire a lot of data to achieve their goal, but data regulation may sometimes be overlooked and not conform to privacy concerns.

In 2011, Menachemi and Collum (2011) contrasted the benefits and drawbacks of electronic health record systems. For example, personal information infringements are disclosed in eHealth systems (Bartoletti, 2019), and identity theft arises in eCommerce financial systems (Lutz & Tamó-Larrieux, 2020; Zviran, 2008). One decade later, in 2021, the security and privacy of eHealth record systems were still areas of concern, as illustrated by Keshta and Odeh (2021). Bawack et al.'s (2022) systematic literature review on AI in eCommerce indicated that styles and themes are the main focuses of information system (IS) scholars, but there is minimal concern regarding fraud and privacy. Nevertheless, the article also shows that "the IS community is also interested in understanding ethical choices and challenges organizations face when adopting AI systems and algorithms" (p. 311).

In 2018, the European Commission (EU) (2018) published a report on the advances of AI in "robotics and so-called *autonomous technologies*" (p. 5). The report describes AI as autonomous systems that require no human intervention or supervision while performing tasks. It also outlines challenges to such systems' ethical, societal, and legal capabilities. Consequently, this report calls for a "common, internationally recognized ethical and legal framework for the design, production, use and governance of artificial intelligence, robotics, and *autonomous* systems" (p. 5). Further discussion of the development and the use of AI is not in the scope of this chapter; instead, identifying AI's controversies and ethical perspectives is the central topic.

The worldview of ethics associated with AI systems

Ethics is a branch of philosophy that follows moral principles in which people decide their lifestyles (So & Luppicini, 2020). Furthermore, So's dissertation states that "computer ethics is an essential sub-discipline in studying and practicing moral principles concerning computer usage" with data collected (2020, p. 33). Mason (1986) exposes four chief ethical issues with data concerns: privacy, accuracy, property, and accessibility. If AI systems deploy similar data collection operations, then the context of privacy and security would involve Mason's last three concerns: integrity, confidentiality, and availability. In a parallel statement cited in the dissertation, Johnson and Cureton (2004) state that "AI violates the safety of people's well-being and affects one's life-style, leading to an ethical issue" (quoted in So, 2020, p. 34). Hence, those operating AI systems can expose users' privacy. This unethical disclosure of personal information could lead to threatening situations, public defamation, and other illegal activities. For example, Widder et al. (2022) highlight that "significant and understudied harms and possibilities originate from differing practices of transparency and accountability in the open source community," like Deepfake,[1] an AI-enabled open source project. Furthermore, Magnani (2006) claims that "nonliving *things*—the Internet, for example—are more than passive objects. Such things can possess a moral agency even though they lack the characteristics we usually associate with human agency: free will, full intentionality, responsibility, and emotion" (p. 192).

Emerging technologies do not usually consider morality an essential element, focusing on the useability and fast return on investment (Heikkerö, 2014). In So's (2020) dissertation, he mentions that

> Gillam et al. (2010) state that computing professionals, researchers, and users of computer applications and programs must understand and explore the essential ethical values of society and comply with its laws. Moreover, these authors are concerned about existing laws and regulations that are not up-to-date because of constantly changing technologies and underexplored cyberlaw and cyberethics (p. 35).

Conceivably, the telecommunication industry in the UK has a better perspective on ethics since technology is a component of the telecommunication field. The BBC (2014) website provides good guidelines for ethics, as can be seen in the following: "Nowadays, philosophers usually divide ethical theories into meta-ethics, normative ethics, and applied ethics" (Serena et al., n.d.) (para. 2). Since AI imitates human intelligence and solves problems as a human, AI follows instructions and machine learning algorithms. Hence, AI ethics requires a different regard from traditional ethics. Further discussion of algorithmic ethics is found in the next section.

Nonetheless, the ethics of AI in the development and deployment of the telecommunication field raises concerns regarding the lack of guidance for AI (Vică et al., 2021). The authors highlight "a genuine ethical interest for AI, and we are witnessing moral diplomacies resulting in moral bureaucracies battling for moral supremacy and political domination" (p. 83). The authors analyze the UNESCO (2021) Intergovernmental Meeting[2] of Experts (Category II) for the *Recommendation on the Ethics of Artificial Intelligence*. They conclude that "the only question that remains, then, is what is the acceptable threshold after which compromises with both industry and states or international organizations alike becomes morally unacceptable" (p. 93).

Fieser and Dowden (2011) state that there are two *normative ethics* approaches: societal ethics and virtue ethics. Alexander and Moore (2012) note that societal ethics requires a nation to conform to the law. As opposed to *virtue ethics*, Hursthouse (2012) states that morals need an individual to judge an action as good or bad and right or wrong. This concept focuses on the inherent character of a person and is not novel; Aristotle promulgated virtue ethics over 2500 years ago. In addition, "in some respects, virtue ethics represents a middle ground between duty and rights. Persons have to self-actualize and should be granted the right to accomplish that self-actualization"

---

[1]Deepfake "refers to the use of advanced artificial intelligence and machine learning technologies to create video, audio, images, or textual data (SMS or written content)" (Kepczyk, 2022, para 1).
[2]The fourteenth session of the Intergovernmental Committee for the Protection and Promotion of the Diversity of Cultural Expressions was held online from February 1 to 5, 2021.

(May, 2013, p. 26, quoted in So, 2020). In the case of using AI systems, this chapter demonstrates the ethics involved at the personal, organizational, and societal levels so that developers can apply preventive measures at different levels under the influences of the user's code of conduct, corporate best practices, and government regulations.

Furthermore, May (2013) shows that Kantian ethics follows the deontological concept of action with no consequence-based outcome. It is in opposition to utilitarian ethics, which results in judge actions. Deontology, or applied ethics, states that one must follow the rule and code of conduct to determine whether an outcome is right or wrong (Fieser & Dowden, 2011). For example, consider a case where a self-driving car will hit either an older person or a child. The self-driving car might choose to hit either person depending on the predefined car's AI algorithm. However, if a person is driving the vehicle, the driver could deviate in the direction to avoid a collision, and deontology ethics are preserved. In the case of adopting AI systems, a *utilitarian ethics* approach could justify lying, stealing, or murder as permitted by law (Fieser & Dowden, 2011; McCartney & Parent, 2015). However, under the vehicle traffic law, drivers must obey the rules; this is an example of a practical rule that avoids accidents and congestion. The consequence is considered good *utilitarian ethics*. The overall net benefit of the action can sometimes override morals and be considered ethical. Thus, lawmakers and AI system developers are always advised to research the normative ethics that should be grounded before AI systems algorithms are applied.

Algorithmic ethics

As stated in the previous section, the interpretation of ethics needs to be differentiated from traditional ethics. Filipović et al. (2018) illustrate the importance of the ethical concerns on AI algorithms, namely bias, discrimination, privacy, accountability, and transparency. The authors also explore "the nature of professional ethics in six different occupational areas: medicine, social work, journalism, public relations, advertising and engineering" (p. 6). The researchers identify ten successful factors (p. 47), as shown in Fig. 7.A1 (Appendix 1). The factors are used to achieve practical ethics for algorithms, and they can successfully establish a code of professional ethics. The authors also note that "the first step involves discussing whether defining a professional group — which is a prerequisite" (p. 50) to establishing a code of professional ethics — is possible, practical, or even desirable.

In 2016, Mittelstadt et al. proposed a conceptual map (Fig. 7.1) illustrating the six types of ethical concerns arising from how algorithms operate over three aspects. The authors "posited three epistemic and two normative ethical concerns arising from using algorithms" (p. 4).

The authors' mapping of ethical issues in Fig. 7.1 shows the three epistemic and two normative kinds of ethical concerns while operating algorithms under the three aspects:

> The map takes into account that the algorithms this paper is concerned with are used to (1) turn data into evidence for a given outcome (henceforth conclusion) and that this outcome is then used to (2) trigger and motivate an action that (on its own, or when combined with other actions) may not be ethically neutral. This work is performed in ways that are complex and (semi-)autonomous, which (3) complicates the apportionment of responsibility for the effects of actions driven by algorithms (p. 4).

When the concerns are associated with potential failures, the authors indicate the traceability entity contains the accountable actors causing such shortcomings.

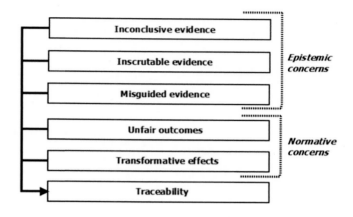

FIGURE 7.1

Six types of ethical concerns raised by algorithms (Mittelstadt et al., 2016, p. 4).

What is technoethics?

In 1977, Mario Bunge, a philosopher, was increasingly concerned with technologists' and engineers' lack of moral and social responsibility regarding their creations. As a result, he introduced the term *technoethics*, formed from a combination of technology and ethics (Luppicini, 2009a, 2009b). The term refers to studying how emerging technology affects users' morality and code of conduct. In today's digital age, technology is involved in most of our daily activities, which spread across all disciplines, including science, technology studies, philosophy, and social sciences. Thus, technoethics has become much more complex as rapidly developed technology intertwines across these disciplines. As this chapter further demonstrates, the technology of AI systems and how these are used in people's daily lives also have ethical implications.

Subsequently, Moor (2005) states that as "technological revolutions" (p. 111) increase social impact, ethical problems also increase. Several descriptive studies conducted by psychology, psychiatry, and behavior researchers (Luppicini, 2009a, 2009b) have raised concerns about the effects of technology on society. Still, relatively few studies connect these impacts with technological designs regarding ethics.

Methodology

Technoethical inquiry approach

The technoethical inquiry approach uses the ethical aspects of technology to illustrate how technology shapes lifestyles in society. This concept is advanced by Luppicini (2009a, 2009b), who explores a social systems theory within technoethics.

Luppicini illustrates essential ethical perspectives within e-Society as represented in the map of Europe (Fig. 7.2). Each ethical perspective replaces a region on the map. Luppicini synthesizes the

FIGURE 7.2

Conceptual map of technoethics.

The data is extracted from Luppicini, R. (2009). Technoethical inquiry: From technological systems to society. Global Media Journal: Canadian Edition, 2(1), 5–21, p. 8.

main elements from the identified perspectives: historical, theoretical, political, legal, economic, sociocultural, levels of influence, stakeholders, intended ends, possible side effects, and means. The following section describes deploying the technoethical inquiry approach to identify ethical perspectives. Then, the degree of gravity of the ethical views of AI systems can be analyzed and determined.

Subsequently, Luppicini (2009a, 2009b) demonstrates different subsystems that could impact technologies and infringe on social norms in the social environment. Table 7.1 shows how each subsystem influences various values involving humans in a society. The specific classification of the social system can enrich our thoughts and determination of the degree of perspectives of AI systems.

Nowadays, all of the mentioned subsystems interact with each other. For example, the subsystem of education produces scientists and engineers who improve technology for human consumption. If laws do not reflect or control the negative consequences of technology, like AI systems in an e-Society environment, community lifestyles could be affected positively and negatively, as described in Crang and Graham's (2008) research paper. Previously, product lifecycle maturity

Table 7.1 Social subsystems.

Subsystem	Core operations	Values
Law	Production of social norms and regulation of conflict	Justice
Politics	Production of collective and binding decision-making	Fairness
Religion	Production of spiritual guidance	Faith
Economy	Production of wealth	Need satisfaction
Science	Scientific knowledge production	Truth
Communication	Production of information/knowledge exchange	Understanding
Culture	Production of social meaning and practices	Community
Education	Production of social values, education, and professional training	Learning

The data is extracted from Luppicini, R. (2009). Technoethical inquiry: From technological systems to society. Global Media Journal: Canadian Edition, *2(1), 5–21, p. 11.*

took much longer than current products. As a result, other social subsystems would have more time to correct anomalies. Product improvement and law adjustment to mitigate negative consequences would have time to adjust.

Similarly, AI systems' maturity (from conception to production) is considered shorter. It has become more widely used by consumers/applications compared to the development of computer systems in the 20th century. Research has shown that many adverse outcomes to AI systems, including security and privacy, have not been fully addressed (So, 2020).

Problem statement

With the increasing trend of AI systems development toward the social subsystems, as listed in Table 7.1, the number of adverse social outcomes has gradually increased. AI systems also contribute to privacy risks, even as AI systems benefit people's lifestyles. The following section composes the questions addressing this problem statement.

Research questions

First, how can the synthesized constructs identified in the process of the technoethical inquiry approach be addressed in AI systems? Second, to what extent can those constructs be used to determine the gravity of ethics? Third, when setting up a recursive thought of AI systems, how can AI systems be used to spot most of the controversial drawbacks in their applications?

Results
The perspectives

This section introduces additional related perspectives following the technoethics approach beyond those already discussed in the previous section (Luppicini, 2010). This study engages the essential perspectives (historical, theoretical, political, legal, economic, sociocultural, levels of influence, stakeholders, intended ends, possible side effects, and means) illustrated by the technoethical inquiry approach used in the article written by Luppicini and So (2016). In particular, this section focuses on domestic drones and AI systems, as both innovations are emerging technologies influencing people's lifestyles in the eSociety environment. These dependent and independent variables are explicitly induced hypotheses that intervene with each other. Changes in these variables may not be in a one-on-one relationship (Checkland, 1981; Forrester, 1994, 1996). The following paragraphs demonstrate the perspectives of AI systems in the context of technoethics variables.

Historical

In 2018, the European Commission published a report mentioning an uprising in technology, referring to AI and autonomous technology in the first two decades of the 21st century. The history includes technologies that provide self-driving cars, drones, and bots in financial transactions. While Russell and Norvig (1995) illustrate the concept of AI as a modern approach to machine learning, there has remained a gap between theory and practice in AI (Nilsson, 2014). More recently, robotics research has stimulated the development of AI approaches, and as a result, some of the gaps between theory and practice have been narrowed. Today, evolving Internet of Things[3] (IoT) technology has highlighted the importance of integrating AI.

Moreover, fields like predictability, scalability, data sharing, collaborative learning, service improvement, independent action, and human interaction are particularly interesting when looking at AI in conjunction with IoT (Valanarasu, 2019). Valanarasu refers to the proposed "novel system that offers secure integration of a smart hospital environment with the help of IoT and AI" (p. 178). Subsequently, AI systems use data collected from IoT, and devices used in IoT may not conform with the security standards in the AI systems. Henceforth, a deepened understanding of these fields of interest is required when IoT and AI interact, but further discussion of IoT is out of the scope of this book chapter.

Theoretical

Fieser and Dowden (2011) illustrate that ethical theories are based on three subject areas: metaethics (investigating the prevalence of ethical principles), normative ethics (governing moral standards of right and wrong), and applied ethics (examining specific issues). One of the forms of normative ethics is virtue ethics, which has the behavioral characteristic of questioning the rightness and wrongness of an individual's actions (Hursthouse, 2012). Virtue ethics focuses on the inherent character of a person, first promoted by Aristotle over 2500 years ago. Ethical or unethical behavior depends on many factors, like the individual involved in the situation, the circumstances of the situation, and the moral intensity of the issue. For example, ethical behavior is how an ethical theory applies to an individual and the impact on an individual when the theory is implemented. Therefore, is it worthwhile to research which type of normative ethics should be grounded on beforehand? Furthermore, which ethical approach would it be appropriate for lawmakers to consider when crafting new rules for AI systems besides addressing technology and the penalty, for example, for infringing users' privacy in AI systems?

---

[3]Internet of Things (IoT) connects objects and accumulates real-time data; discussions of turning these collected data into valuable information for predictable insights.

Political

Public policy can involve AI systems in emergency services, health care, and general welfare. Most of these services are administrated by different governmental agencies. One of the most significant actions on AI regulations is the proposal by the European Commission (2021). The European policy-makers address the risks of some uses of AI and categorize four levels of risk: "unacceptable risk, high risk, limited risk, and minimal risk" (p. 12). In parallel to the EU proposal, the European Parliament voted to ban using facial recognition with AI in law enforcement on October 6, 2021 (Peets et al., 2021). In addition, Kumawat (2020) raises concern about the myth of replacing some of the current jobs with automated systems. Policymakers must select the appropriate AI applications and govern those AI techniques (Kumawat, 2020). He illustrates the benefits of AI in political elections by modifying political views to gain supporters and considering the methods that may impact society. Kumawat asks, "Should human decision-making be rejected and replaced with data-driven decision-making powered by artificial intelligence?" (para. 2). The discourse about the rise of AI is also discussed by Natale and Ballatore (2020).

Legal

In 2020, Van Assen et al. (2020) reviewed the general concepts of AI algorithms, "including their data requirements, training, and evaluation methods" (p. 1), to seek potential legal implications, especially in the clinical practice of using AI algorithms. However, the authors caution medical professionals that "it is important to be aware of the opportunities for improved diagnosis, prognosis, and treatment for patients that AI offers while reducing workloads, instead of focusing on threats to job security" (p. 7). They also mention that establishing laws and regulations is required to promote ethical use in the early development of AI applications. This action is a proactive precaution rather than a reactive move after the applications are in production.

Economic

In 2017, Thierer et al. (2017) from the Mercatus Research Center at George Mason University provided a report titled *Artificial Intelligence and Public Policy*, addressing the concerns about "labour market effects, social inequality, and even physical harm" (p. 2) caused by AI systems. This report "called for precautionary regulations" (p. 2) because the authors foresee that the economic impact of AI could contribute between "$1.49 trillion and $2.95 trillion over the next ten years" (p. 15). The current growth of today's economy is yet to be confirmed. Nevertheless, the interest in deploying AI systems is still increasing. For example, the same report illustrated that "autonomous or driverless car technologies could save the United States $1.3 trillion in annual costs or 8% of GDP, and $5.6 trillion globally once those technologies have fully penetrated" (p. 16). Moreover, Streinz (2021) states that the EU's new financial services regulations support "more data available with the aim of spurring innovation and growth in the artificial intelligence economy" (p. 1). Streinz concludes that "data is not just the source of AI economy but also for the future development and reconfiguration of international economic law" (p. 23).

Sociocultural

The sociocultural variables are cultural groups, habits, beliefs, and values that AI systems and applications would apply to those people's lifestyles. Nasrudin (2022) refers to the sociocultural environment as follows:

> [T]trends and developments in society's attitudes, behaviour, and values. It closely relates to population, lifestyle, culture, tastes, customs, and traditions. These factors are created by the community and often are passed down from one generation to another (para 2).

Consequently, companies usually change their business strategy according to factors that affect society and culture; this can be seen, for example, in demand for goods and services. If these services deploy AI, those users need to consider sociocultural factors, as described by Ruder-Hook's (2018) research (2018). "The study provides a qualitative analysis of existing literature including organizational innovation, technological adoption, and organizational culture, as well as an analytical case study, comparing a French and an American multinational" (prefix, p. i.).

Telecommunications organization evaluates the influence of national sociocultural values on organizational adoption of AI. Technological developments are one of the factors that lead to a much faster rate of growth or improvement of new goods/services because of the beneficial use of AI applications.

Levels of influence

In 2021, Ponnappan (2021) stated that too many AI system variables affect society and that laws related to regulation are not the only levels of influence that need to be discussed. "As a result, each sector of the legal system has unique issues that need not only broad norms but, in many cases, region-specific solutions" (para 2). For example, AI systems must preserve the law of privacy when processing personal information. In machine-generated data, systems must observe all legal data protection areas, like "telecommunications law, competition law, intellectual property law, and liability law" (para. 6). Additionally, the author illustrates that "AI can have a significant impact on people's lives, experiences, cultural orientations, attentions, and civic ideals, with implications for private life, education, public opinion formation, and political decision-making processes" (para 10). As a result, when AI systems are governed by medical law, financial markets law, and road traffic law, the level of influence of ethics needs to be further determined. Future research on the gravity of ethics for AI systems will be necessary.

Stakeholders

"The concept of responsible AI stresses a framework that holds mainly developers and manufacturers blameworthy, accountable, and liable for the actions of AI" when deployed by autonomous systems (Lima & Cha, 2020, p. 4). The authors underline this concept of responsibility framework for all AI system developers and users. In addition, Preece et al. (2018) categorize four types of stakeholder communities: "Developers, Theorists, Ethicists, and Users" (p. 2). Developers who build AI applications are corporations and small/medium enterprises or public sectors. Theorists are those people who have a deep understanding of AI theory and neural networks. Ethicists are "people concerned with fairness, accountability and transparency of AI systems, including policymakers, commentators, and critics" (p. 2). Finally, users are people who use AI systems and applications. They must know how to justify the actions taken on the systems' outputs.

Intended ends

Having said the aforementioned, AI is described as an action performing tasks similar to humans, and its ultimate role or intended end is completing tasks efficiently and accurately. In addition,

Sousa et al. (2021) illustrate that AI systems "have been gaining more and more interest both from academics as well as practitioners, leading to new approaches and funding opportunities" as the *intended end* (p. 488). The authors present a holistic literature review on theoretical frameworks and practical experiences in using AI. The study has found that AI "may be useful both as a tool for modern education, but also as a goal, allowing healthcare professionals to get the right competencies to benefit from such new technologies" (p. 494). The intended end creates "new educational opportunities and needs" (p. 494) apart from AI systems and applications used in business processes. However, society also requires citizens to be more responsible. Hence, AI systems and applications are responsible for their actions like humans.

Possible side effects

In 2020, Yigitcanlar et al. (2020) raised the concern that "no scholarly work provides a comprehensive review" (p. 1) on AI innovations. They also categorized the results "under the main smart city development dimensions, i.e., economy, society, environment, and governance" (p. 1). Users can determine if AI is harmful or beneficial by identifying more possible side effects caused by AI systems. Furthermore, the Forbes (2021) has identified 14 drawbacks of using AI systems. The drawbacks are mainly related to a lack of privacy and transparency. For example, AI may hinder societal advancement, cause unfair outcomes, put forward wrong predictions that can lead to life-threatening events, and may be used by cybercriminals for social engineering scams. The Forbes article illustrates that "it's important for governments, businesses and the public to be aware of what can happen if AI is used indiscriminately or without diligent human oversight" (para 3). These potential negative outcomes may be mitigated by employing the ideas behind the smart city development dimensions.

In contrast, Jiang et al. (2017) illustrate some positives of AI, as AI tools are used in "major disease areas … including cancer, neurology and cardiology" for early detection and diagnosis (p. 1). Nevertheless, healthcare workers should know the negative consequences of AI healthcare systems and applications.

Means

In 2017, Charisi et al. (2017) wrote about moral requirements for autonomous systems. The authors highlight two perspectives of autonomous systems: ethics and implementation. Both of the perspectives had been studied in detail separately but not in conjunction. They note that "Human societies have a multitude of means for ensuring its members behave within the socially accepted boundaries of morality" (p. 4). Thus, "[b]uilding moral machines by implementing human morality is a natural approach" (p. 6). The authors conclude that "stakeholders need to be aware of the abuse of an AI system with ethical behavior capabilities, both when that abuse is intentional and accidental" (p. 28). However, the authors have not provided details in determining how to differentiate the types of abuse.

Consequently, in 2020, Reddy et al. (2020) proposed to address ethics and AI systems in healthcare delivery via "a governance model that covers the introduction and implementation of AI models in healthcare" (p. 496). The model is "designed to be flexible enough to accommodate changes in AI technology" (p. 496). Fig. 7.3 illustrates the governance model proposed by the authors using AI systems in a healthcare environment:

The model comprises four components: fairness, transparency, trustworthiness, and accountability.

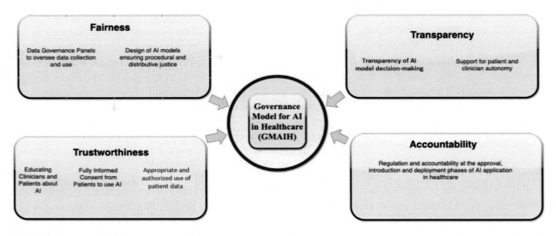

FIGURE 7.3

Governance model for AI in health care.

The model is extracted from Reddy, S., Allan, S., Coghlan, S., & Cooper, P. (2020). A governance model for the application of AI in health care. Journal of the American Medical Informatics Association, 27(3), 491–497, p. 493.

Fairness

Medical data includes "medical images, text from patient records about medical conditions diagnosis and treatment, and reimbursement code" (p. 493). The authors suggest forming a panel with "AI developers that include patient and target group representatives, clinical experts, and people with relevant AI, ethical, and legal expertise" to avoid biases resulting from the data collected by AI applications (p. 493).

Transparency

The authors propose this component "through the governance model an emphasis on ongoing or continual explainability" (p. 494) and claim that transparency will increase the trustworthiness of the AI systems. AI agents must allow patients to have flexibility in health-related decision-making and not be pressured to pursue a poorly explained decision.

Trustworthiness

It is essential to gain trust among healthcare professionals for using AI systems by illustrating the advantages and limitations of healthcare delivery. Clinicians are "to be partners in the control of the technology, rather than merely being passive recipients of the AI outputs" (p. 494). This approach should extend to encouraging patients to adopt health literacy principles, allowing them to make informed and autonomous health choices (p. 494). The authors also "recommend that institutional policies and guidelines be reworked to ensure patients know that the treating clinician is drawing support from AI applications" (p. 494). Informed consent must be obtained with clear clinical objectives when AI developers use patients' data.

Accountability

The spectrum of this component involves several players, including "software developers, government agencies, health services, and medical professionals" (p. 494). The authors express the difficulties of implementing the governance model due to the diverse responsibility groups. They suppose that AI systems can be held accountable for their actions. These activities can be monitored and evaluated for safety and quality of service to ensure that AI systems provide a safe and quality service that can be accountable for their actions.

Efficiency and fairness

> AI systems can be used in many sensitive environments to make important and life-changing decisions; thus, it is crucial to ensure that these decisions do not reflect discriminatory behaviour toward certain groups or populations ... AI shows what researchers have observed with regard to unfair outcomes in the state-of-the-art methods and ways they have tried to address them.
>
> **Mehrabi et al. (2021, p. 1)**

These authors discuss the advantages of using autonomous systems, but they caution about the bias of certain groups in developing algorithms for machine learning. For example, AI applications for recruitment can screen and parse resumes rapidly, but demography selection can discriminate against females (Zhao et al., 2021). A recursive process can be implemented in autonomous systems to avoid biases. An additional embedded intelligent algorithm that monitors the overselection can be a solution; for example, in the case of the overselection of males, the system switches to female applicants. In another case, the self-correction in a self-driving vehicle has this recursive function to stay on the roads. Hence, the recursive function is helpful and should be developed in all AI systems and applications to eliminate unwanted outcomes and respond to the third research question.

Advanced analysis and discussion

The first research question of ethics in AI systems is addressed using the synthesized constructs identified in the technoethical inquiry approach. The identified technoethics perspectives, highlighted in the previous section, contribute to more profound and thorough thinking in evaluating the implications of ethics in AI systems and applications. Once most of the perspectives of AI systems are known, the extent of the negative consequences caused by AI systems and applications may be determined, as prompted in the second research question. Unfortunately, there is no general method of quantifying the degree of seriousness in AI systems. The perspectives have demonstrated that different elements interact with one another. Improving one situation may worsen another outcome; this is a multiple dimension of systems thinking, as Meadows (2008) showed. Further study of the phenomenon of systems thinking is not in the scope of this chapter, but this concept assists in answering the second research question.

AI systems are evolving to a broader spectrum of applications in our lifestyles. As this happens, ethical gaps are commonly found in AI systems, some examples of which are described in the previous section. Societal ethics can also lead to setbacks in using AI systems, and new laws address

Table 7.2 Ethical dilemmas and the process of effective resolution.

Role at the theater	Values
Single person	Hope, stamina, and sobriety
Mother with child	Caring, concern, and fortitude
Officer responding	Self-discipline, fortitude, and courage
Follow-up detectives	Accountability, empathy, and consideration

The table is extracted from McCartney, S., & Parent, R. (2015). Ethics in law enforcement. *BCcampus. 978-1-989623-63-3, p. 43.*

when AI deviates from what is ethical. Furthermore, individual (virtue) ethics is more concerned with how one lives according to one's moral character, independent of ethical duties and rules (Hursthouse, 2012). This approach considers an internal goodness belief of an individual, and AI governance laws would not usually govern this kind of ethical style.

Subsequently, McCartney and Parent (2015) illustrate the problem of deciding what to do when facing an ethical dilemma. The authors show that "there are at least two competing values we are forced to choose between" (p. 39). Since values are "opinions and beliefs we decide are beneficial or important" (p. 39), the importance and significance of the values for an organization must be identified by involving actors before addressing ethical dilemmas. Furthermore, the determination of the values of AI systems can refer to the values determined by McCartney and Parent (n.d).

Accordingly, McCartney and Parent have demonstrated ethical dilemmas for individuals deploying AI systems and applications (Table 7.2). The values anticipated by users reacting under different roles can affect their resolution of any AI process. Moreover, Gill (2021) states that "[W]hilst the techno-centric paradigm tends to provide efficiency, precision and replicability of technological innovations, the human-centered paradigm promotes creativity, flexibility, and resilience" (p. 669). The techno-centric and human-centric paradigms of AI systems can create ethical dilemmas in the threat of privacy, human dignity, and safety over the techno-centric paradigm.

Conclusions

This chapter demonstrates that AI researchers and developers can use the technoethical inquiry approach to address the current ethical perspectives of AI systems and applications. The literature review shows that AI systems are increasingly deployed in medical care (Reddy et al., 2020), and privacy law is severe in disclosing and storing health information. These existing laws may also be extended to AI systems, creating a potential template for other contexts. Additionally, the research and guidelines in Hagendorff (2020) identified most of the current AI constructs, which can be used to determine the gravity of ethics by AI authorities. In addition, Murdoch (2021) highlights the privacy and ethical challenges in using AI systems. As the determination of moral consequences is concluded on a case-by-case basis, no single mitigation process limits the adverse outcomes of using AI systems because of the multiple influence dimensions. The intervening ethical perspectives lead to Meadow's (2008) systems thinking concept that further research on this aspect in AI systems needs to be conducted. This concept would lead to a holistic understanding of ethical concerns for AI system developers and users. Finally, this chapter provides the means for identifying ethical perspectives through the technoethical inquiry approach. Thus, the gravity of ethical concerns for AI and autonomous systems is easier to address. Once the AI constructs are identified, these constructs may be re-configured as a monitoring entity to alert or mitigate AI ethical behavior. However, further research will be needed for a complete recursive solution to the last research question.

Finally, it should be noted that during the revisions of this chapter, ChatGPT (version 4) entered the AI world as a generative pretrained transformer state-of-the-art language processing tool on March 14, 2023 (OpenAI, 2023). The aptitude of this tool generates challenges in academic and privacy concerns apart from its many potential positive contributions. Will this tool be a friend or foe? It remains to be seen when AI becomes more mature.

References

Alexander, L., & Moore, M. (2012). *Deontological ethics. The Stanford encyclopedia of philosophy* (Fall 2012 Edition), EN. Zalta (ed.). Retrieved on January 6, 2015 from http://plato.stanford.edu/archives/win2012/ethics-deontological/.

Bartneck, C., Lütge, C., Wagner, A., & Welsh, S. (2021). *Privacy issues of AI. An introduction to ethics in robotics and AI* (pp. 61−70). Cham: Springer.

Bartoletti, I. (2019). *AI in healthcare: Ethical and privacy challenges. Conference on Artificial Intelligence in Medicine in Europe* (pp. 7−10). Cham: Springer.

Bawack, R. E., Wamba, S. F., Carillo, K. D. A., et al. (2022). Artificial intelligence in E-Commerce: A bibliometric study and literature review. *Electron Markets, 32,* 297−338. Available from https://doi.org/10.1007/s12525-022-00537-z.

BBC, (2014). *Ethics: A general introduction.* Retrieved from http://www.bbc.co.uk/ethics/introduction/intro_1.shtml#top.

Charisi, V., Dennis, L., Fisher, M., Lieck, R., Matthias, A., Slavkovik, M., Yampolskiy, R. (2017). *Towards moral autonomous systems* (pp. 1−34). Retrieved from http://arxiv.org/abs/1703.04741.

Checkland, P. (1981). *Systems thinking, systems practice.* Chichester, UK: John Wiley & Sons.

Crang, M., & Graham. (2008). *Multispeed cities and the logistics of living in the information age.* Economic and Social Research Council, Swindon. Citation for the published item. University of Durham. http://www.york.ac.uk/res/esociety/projects/4.htmhttp://dro.dur.ac.uk.

European Commission, Directorate-General for Research and Innovation, European Group on Ethics in Science and New Technologies. (2018). *Statement on artificial intelligence, robotics and 'autonomous' systems.* Brussels, March 9, 2018, Publications Office.

European Commission. (2021). *Proposal for laying down harmonized rules on artificial intelligence (Artificial Intelligence Act) and amending certain union legislative acts (COM (2021) 206 final).* European Commission. https://eur-lex.europa.eu/legal-content/EN/TXT/?uri = CELEX:52021PC0206.

Fieser, J., & Dowden, B. (2011). *Internet encyclopedia of philosophy.* Retrieved on January 6, 2015, from http://www.iep.utm.edu/ethics/#SH2b 2161-0002.

Filipović, A., Koska, C., & Paganini, C. (2018). *Developing a professional ethics for algorithmists. Working paper.* Bertelsmann Stiftung. Retrieved January 28 2023. https://www.bertelsmann-stiftung.de/en/publications/publication/did/developing-a-professional-ethics-for-algorithmists.

Forbes. (2021). *14 Ways AI could become a detriment to society*. Forbes Technology Council. Retrieved from https://www.forbes.com/sites/forbestechcouncil/2021/06/14/14-ways-ai-could-become-a-detriment-to-society/?sh = 98f05a327fe4.

Forrester, J. W. (1994). System dynamics, systems thinking, and soft OR. *System Dynamics Review*, *10*(2−3), 245−256. Available from https://doi.org/10.1002/sdr.4260100211.

Forrester, J.W. (1996). *System dynamics and K-12 teachers*. Retrieved on June 8, 2016, from http://ocw.mit.edu/courses/sloan-school-of-management/15-988-system-dynamics-self-study-fall-1998-spring-1999/readings/teachers.pdf.

Gill, K. S. (2021). Ethical dilemmas. *AI & SOCIETY*, *36*(3), 669−676. Available from https://doi.org/10.1007/s00146-021-01260-7.

Gillam, L., & Vartapetiance, A. (2010). Gambling with laws and ethics in cyberspace. In R. Luppicini (Ed.), *Evolving issues surrounding technoethics and society in the digital age. (Section 2)*. Hershey, PA: IGI Global. Available from 10.4018/978-1-4666-6122-6.

Hagendorff, T. (2020). *The ethics of AI ethics: An evaluation of guidelines*, . *Minds & Machines* (30, pp. 99−120). . Available from https://doi.org/10.1007/s11023-020-09517-8.

Heikkerö, T. (2014). *Ethics in technology: A philosophical study*. Lexington: Lanham, MA.

Hursthouse, R. (2012). *Virtue ethics. The Stanford encyclopedia of philosophy* (Fall 2013 Edition), Edward N. Zalta (ed.). Retrieved on December 12, 2014 from http://plato.stanford.edu/entries/ethics-virtue/.

Jiang, F., Jiang, Y., Zhi, H., Dong, Y., Li, H., Ma, S., Wang, Y., Dong, Q., Shen, H., & Wang, Y. (2017). Artificial intelligence in healthcare: Past, present and future. *Stroke and Vascular Neurology*, *2*(4), 230−243. Available from https://doi.org/10.1136/SVN-2017-000101.

Johnson, R., & Cureton, A. (2004). *Kant's moral philosophy*. Retrieved from https://plato.stanford.edu/Entries/Kant-Moral/.

Kepczyk, R. (2022). *Deepfakes emerge as real cybersecurity threat | News | AICPA*. In AICPA & CIMA. Retrieved from https://www.aicpa-cima.com/news/article/deepfakes-emerge-as-real-cybersecurity-threat.

Keshta, I., & Odeh, A. (2021). Security and privacy of electronic health records: Concerns and challenges. *Egyptian Informatics Journal*, *22*(2), 177−183. Available from https://doi.org/10.1016/j.eij.2020.07.003.

Kumawat, D. (2020). *How Artificial Intelligence (AI) can be used in Politics & Government?* Analytics Steps. https://www.analyticssteps.com/blogs/how-artificial-intelligence-ai-can-be-used-politics-government.

Lima, G., & Cha, M. (2020). *Responsible AI and its stakeholders*. arXiv preprint arXiv:2004.11434. Retrieved from http://arxiv.org/abs/2004.11434.

Luppicini, R. (2009a). Technoethical inquiry: From technological systems to society. *Global Media Journal: Canadian Edition*, *2*(1), 5−21.

Luppicini, R. (2009b). The emerging field of technoethics. In R. Luppicini, & R. Adell (Eds.), *Handbook of research on technoethics* (pp. 1−19). Hershey, PA: Information Science Reference. Available from 10.4018/978-1-60566-022-6.ch001.

Luppicini, R. (2010). *Technoethics and the evolving knowledge society: Ethical issues in technological design, research, development, and innovation*. Hershey, PA: Information Science Reference. Available from 10.4018/978-1-60566-952-6.

Luppicini, R., & So, A. (2016). A technoethical review of commercial drone use in the context of governance, ethics, and privacy. *Technology in Society*, *46*, 109−119.

Lutz, C., & Tamó-Larrieux, A. (2020). The robot privacy paradox: Understanding how privacy concerns shape intentions to use social robots. *Human-Machine Communication*, *1*, 87−111.

Magnani, L. (2006). *Morality in a technological world: Knowledge as duty*. Cambridge University Press.

Mason, R. O. (1986). Four ethical issues of the information age. *MIS Quarterly*, *10*(1), 5−12, Retrieved on January 6, 2015 from. Available from http://www.jstor.org/stable/248873?seq = 1#page_scan_tab_contents.

May, S. (2013). *Introduction: Ethical perspective and practices. Case studies organizational studies: Ethical perspectives and practices* (pp. 1–33). Thousand Oaks, CA: Sage Publication Inc.

McCartney, S., & Parent, R. (2015). *Ethics in law enforcement.* BCcampus. 978-1-989623-63-3.

McCartney, S., & Parent, R. (n.d). *Ethics in law enforcement.* Retrieved from.

Meadows, H. (2008). System structures and behaviour. In D. Wright (Ed.), *Thinking in systems: A primer* (192–193, pp. 11–74). White River Junction, VT: Chelsea Green Publishing.

Mehrabi, N., Morstatter, F., Saxena, N., Lerman, K., & Galstyan, A. (2021). A survey on bias and fairness in machine learning. *ACM Computing Surveys (CSUR), 54*(6), 1–35.

Menachemi, N., & Collum, T. H. (2011). Benefits and drawbacks of electronic health record systems. *Risk Manage Health Policy, 2011*(4), 47–55. Available from https://doi.org/10.2147/RMHP.S12985. Epub 2011 May 11. PMID: 22312227; PMCID: PMC3270933.

Mittelstadt, B. D., Allo, P., Taddeo, M., Wachter, S., & Floridi, L. (2016). The ethics of algorithms: mapping the debate. *Big Data & Society.* Available from https://doi.org/10.1177/2053951716679679.

Moor, J. H. (2005). Why we need better ethics for emerging technologies. *Ethics and Information Technology, 7*(3), 111–119. Available from https://doi.org/10.1007/s10676 = 006-0008-0.

Murdoch, B. (2021). Privacy and artificial intelligence: Challenges for protecting health information in a new era. *BMC Medical Ethics, 22*(1), 1–5. Available from https://bmcmedethics.biomedcentral.com/articles/10.1186/s12910-021-00687-3#Abs1.

Nasrudin, A. (2022). *Sociocultural environment: Meaning, variables, impact on the business.* https://penpoin.com/sociocultural-environment/.

Natale, S., & Ballatore, A. (2020). Imagining the thinking machine: Technological myths and the rise of artificial intelligence. *Convergence: The International Journal of Research into New Media Technologies, 26*(1), 3–18. Available from https://doi.org/10.1177/1354856517715164.

Nilsson, N.J. (2014). *Principles of artificial intelligence.* Burlington, MA.

OpenAI. (2023). *ChatGPT-release notes.* Retrieved on March 23, 2023 from https://help.openai.com/en/articles/6825453-chatgpt-release-notes.

Peets, L., Hansen, M., Choi, S.J., Drake, M., & Ong, J. (2021). *European parliament votes in favor of banning the use of facial recognition in law enforcement.* Inside Privacy. Retrieved August 25, 2022, from https://www.insideprivacy.com/artificial-intelligence/european-parliament-votes-in-favor-of-banning-the-use-of-facial-recognition-in-law-enforcement/.

Polyakov, A. (2018). *Machine learning for cybersecurity 101.* Retrieved on June 18, 2022, from https://towardsdatascience.com/machine-learning-for-cybersecurity-101-7822b802790b.

Ponnappan, J.K. (2021). *Artificial intelligence's levels of influence.* The Creative Technocrat. https://www.creativetechnocrat.com/2021/06/artificial-intelligences-levels-of-influence.html.

Preece, A., Harborne, D., Braines, D., Tomsett, R., & Chakraborty, S. (2018). *Stakeholders in explainable AI.* arXiv preprint arXiv:1810.00184.

Reddy, S., Allan, S., Coghlan, S., & Cooper, P. (2020). A governance model for the application of AI in health care. *Journal of the American Medical Informatics Association, 27*(3), 491–497.

Ruder-Hook, M. (2018). *Winter semester: Organizational adoption of AI through A sociocultural lens.* University of Michigan. Available from https://deepblue.lib.umich.edu/bitstream/handle/2027.42/147389/mirarh.pdf?sequence = 1.

Russell, S. J., & Norvig, P. (2003). Artificial intelligence: A modern approach ((2nd ed.). Upper Saddle River, New Jersey: Prentice-Hall, 0-13-790395-2.

Russell, S. J., & Norvig, P. (1995). *Artificial intelligence: A modern approach, . Artificial Intelligence* (25, pp. 27–102). Englewood Cliffs: Prentice-Hall.

Serena, et al. (n.d). *Langkaer: Theory of knowledge.* Retrieved from http://theoryknowledge.weebly.com/ethics.html/.

So, A. (2020). *Exploring cyberbullying in K−12 education in Canada to promote cyberbullying awareness and prevention measures* (Doctoral dissertation). Université d'Ottawa/University of Ottawa.

So, A., & Luppicini, R. (2020). *Awareness, governance, and environment: A thematic analysis reveals three pillars of cyberbullying.* Available at SSRN 4057272.

Sousa, M. J., Mas, F. D., Pesqueira, A., Lemos, C., Verde, J. M., & Cobianchi, L. (2021). The potential of AI in health higher education to increase the students' learning outcomes. *TEM Journal, 10*(2), 488−497. Available from https://doi.org/10.18421/TEM102-02.

Stahl, B. C., & Wright, D. (2018). Ethics and privacy in AI and big data: Implementing responsible research and innovation. *IEEE Security & Privacy, 16*(3), 26−33.

Streinz, T. (2021). International economic law's regulation of data as a resource for the artificial intelligence economy. *SSRN Electronic Journal.* Available from https://doi.org/10.2139/SSRN.3831963.

Thierer, A.D., Castillo O'Sullivan, A., & Russell, R. (2017). *Artificial intelligence and public policy.* Mercatus Research Paper.

ThinkML. (2021). *Is artificial intelligence a threat to privacy?* Retrieved from https://thinkml.ai/is-artificial-intelligence-a-threat-to-privacy/.

UNESCO. (2021). *Intergovernmental meeting of experts (Category II) related to a draft recommendation on the ethics of artificial intelligence.* http://webcast.unesco.org/events/2021-04-REC-Ethics-of-AI.

Valanarasu, R. (2019). Smart and secure Iot and AiIntegration framework for hospital environment. *Journal of ISMAC, 01*(03), 172−179. Available from https://doi.org/10.36548/jismac.2019.3.004.

Van Assen, M., Lee, S. J., & de Cecco, C. N. (2020). Artificial intelligence from A to Z: From neural network to legal framework. *European Journal of Radiology, 129.* Available from https://doi.org/10.1016/J.EJRAD.2020.109083.

Vică, C., Voinea, C., & Uszkai, R. (2021). The emperor is naked: Moral diplomacies and the ethics of AI. *Informacios Tarsadalom, 21*(2), 83−96. Available from https://doi.org/10.22503/inftars.XXI.2021.2.6.

Widder, D. G., Nafus, D., Dabbish, L., Herbsleb, J., & Herbsleb, J. (2022). Republic of Koreafakes. *ACM Digital Library, 22,* 12. Available from https://doi.org/10.1145/3531146.3533779.

Yigitcanlar, T., Desouza, K. C., Butler, L., & Roozkhosh, F. (2020). Contributions and risks of artificial intelligence (AI) in building smarter cities: Insights from a systematic review of the literature. *Energies, 13*(6). Available from https://doi.org/10.3390/en13061473.

Zhao, Y., Zhang, X., Tang, X., Qin, C., & Zhu, H. (2021). Embedding fairness into the AI-based talent recruitment systems: The perspective of environment cycle and knowledge cycle. In PACIS (p. 15).

Zviran, M. (2008). User's perspectives on privacy in web-based applications. *Journal of Computer Information Systems, 48*(4), 97−105.

Xiao, B. S., & Wong, Y. M. (2013). Cyber-bullying among university students: An empirical investigation from the social cognitive perspective. *International Journal of Business and Information, 8*(1), 34−69.

The ethics of online AI-driven agriculture and food systems

Edmund O. Benefo[1], Abani K. Pradhan[1,2] and Debasmita Patra[3]

[1]*Department of Nutrition and Food Science, College of Agriculture and Natural Resources, University of Maryland, College Park, MD, United States* [2]*Center for Food Safety and Security Systems, University of Maryland, College Park, MD, United States* [3]*University of Maryland Extension, College of Agriculture and Natural Resources, University of Maryland, College Park, MD, United States*

Introduction

Developments in artificial intelligence (AI), Internet of Things (IoT), and machine learning (ML) technologies mean that certain activities that were primarily the preserve of humans for ages, including farming and cooking meals, writing, teaching, and driving, can now be performed autonomously by AI-enabled systems. AI is rapidly growing and has numerous benefits; nonetheless, this brings with it a myriad of risks as well as raises ethical and legal questions (Benefo, Tingler, et al., 2022; Hauer, 2022). AI ethics could be described as the moral and ethical principles that govern the ethical development and use of AI technology in society (Whittlestone et al., 2019). The increased prevalence of AI in daily life has brought to the fore the ethical challenges associated with AI. Therefore, we must consider the ethical implications of its development and deployment. Several studies have proposed principle-based ethics, frameworks, codes, and guidelines, including responsibility, privacy, transparency, autonomy, trust, sustainability, nonmaleficence, explicability, and justice, which must be the basis upon which AI systems' design, development, deployment, and usage should be founded (Hagendorff, 2020; Hauer, 2022; Heilinger, 2022; Jobin et al., 2019; Morley et al., 2020; Sapienza & Vedder, 2021).

Mulgan (2019) and Raab (2020) are of the view that the creation of these guidelines and frameworks was a critical and necessary step in the growth of AI governance, but others suggest that these principles tend to be very abstract or theoretical and only offer slight protection from potential AI-related harms (Clarke, 2019; Morley et al., 2020, 2021; Orr & Davis, 2020). Consequently, the result of these abstract principles is a considerable gap between theoretical AI ethics and the actual practice, leading to a situation where AI developers lack guidance on designing and deploying AI systems within specified ethical boundaries (Morley et al., 2021; Vakkuri & Abrahamsson, 2018). Indeed, when McNamara et al. (2018) studied the effect of ethical guidelines on ethical decision-making among software engineers, they found that ethical guidelines were not effective and had no influence on the behavior of the software engineers.

AI applications are used in a wide range of online systems, including chatbots, recommendation systems, personalized advertising, personal and professional social networks, online security, health care, education (teaching, learning analytics, and proctoring), and food systems. These applications

Ethics in Online AI-Based Systems. DOI: https://doi.org/10.1016/B978-0-443-18851-0.00009-3

often involve the processing of large amounts of personal data, which raises important questions about privacy, fairness, and accountability (Floridi & Taddeo, 2016). One of the key ethical issues associated with online-based AI systems is privacy. As AI systems collect and process vast amounts of data, they often gather sensitive information about individuals, including their preferences, behaviors, and personal relationships (Jobin et al., 2019). This data can be used to make predictions about individuals, including their future behavior, and can be easily accessible to others, including companies and governments. Another important ethical consideration in the development of online-based AI is fairness. AI systems that are trained on inaccurate or biased data sets can perpetuate existing inequalities and discrimination (Giovanola & Tiribelli, 2022). For instance, if an AI-powered recommendation system were trained on data that reflects a biased view of the world, it may make recommendations that reinforce harmful stereotypes and existing biases (Borenstein & Howard, 2021; Giovanola & Tiribelli, 2022). Furthermore, accountability and responsibility of AI are important aspects of online-based AI ethics. AI systems are often used to automate complex decision-making processes. When AI systems make mistakes or produce harmful outcomes, it can be difficult to determine who is responsible and how to hold them accountable. There is always the debate about whether the developer or deployer should be held responsible and accountable (Allen et al., 2000). A proper balance between innovations and ethical risks is necessary for the protection of society (Knoppers & Thorogood, 2017). This is especially crucial when innovations are related to a sector as important as food and agriculture.

Agriculture has long been the backbone of civilizations. The desire to produce food to meet one of our most basic physiological needs has always been important in our ancient and modern human history and will continue to be for as long as humans exist. If one follows the timeline of agriculture, one would see a gradual change in tools, knowledge, and techniques from hunter-gatherer societies to agrarian societies where food was produced for sustenance and then later for commerce. With the onset of the first industrial revolution, the face of agriculture changed as well. Basic tools were introduced to reduce labor costs, increase yields, and improve agricultural productivity on the whole. The second and third revolutions came with further developments and scientific technologies in plant and animal breeding and the widespread adoption of mechanized agriculture that greatly improved agricultural systems. In today's Industry 4.0, we see tools and technologies, including big data and cloud computing, AI, decision support systems, IoT, and advanced human-computer interactions. These tools have begun to be used in the agricultural and food industries, though they may not be the mainstream or go-to technologies in most places around the world. A major challenge of agriculture and food systems is that technologies almost always tend to be adopted later than sooner. Also, the adoption is usually as-is, with very little adaptation to the unique intricacies of agriculture and food systems. The objective of this chapter is to highlight the current trends and future applications of online-based AI systems in agriculture and food, discuss the potential ethical opportunities and challenges these technologies may bring, and offer our perspectives on how they can be addressed.

Current trends and future applications of online-based AI in agricultural and food systems

AI has become the leading technology of the twenty-first century, permeating almost every industry and influencing lives in both obvious and subtle ways (Lauer, 2021). The modern scientific and technological innovations in agricultural production and the food industry paved the way for use of novel food production equipment and processing technologies, including smart machines and processing lines (Kakani et al., 2020). Unlike in other areas such as social networking, educational enterprises, and metaverses, which could solely exist online, food and agriculture AI systems tend to be more hybrid, combining elements of both on- and offline A1-based systems. They typically require the input of physical data from farms, processing facilities, laboratories, food service establishments, consumers, and more. This data is then fed into online database servers, analyzed on cloud computing platforms, and then outputted to local hosts to support decision-making and drive or automate other systems. This framework is typified in current applications, including efficient crop production and marketing; efficient management techniques to combat climate variations; pest and weed control; early forecasting on the weather that farmers could adapt their practices to; monitoring of soil health by analyzing extensive images captured by satellites and unmanned aerial vehicles; and the protection of the environment by spot application of nutrients and pesticides (Ben Ayed & Hanana, 2021; Misra et al., 2022). Online-based AI systems are also being used in livestock monitoring, food quality evaluation, foodborne disease surveillance, and foodborne disease source attribution (Rizzoli, 2022). Reports are that worldwide spending on smart agriculture is expected to triple to $15.3 billion by 2025 and that the market size of AI in agriculture would also reach $2.5 billion by 2026 (Columbus, 2021; Rizzoli, 2022). IoT-enabled agriculture, which is the agri-food industry's fastest-growing AI-based segment, is also estimated to grow to $4.5 billion by 2025 (Columbus, 2021). A summary of some of these applications of online AI-based systems in agricultural and food systems has been provided here.

Crop production

The application of AI-based systems in crop production has been focused on improving the growth and yield of agricultural products. Crop production involves all the farm activities, from planting through crop monitoring to harvesting, that result in food being produced. It is well known that soil organisms and macro- and micronutrients are some of the most important factors for crop health as well the yield quality and quantity. In conventional farms, the monitoring of these parameters is carried out by humans; however, this is typically not accurate nor done fast enough. Today, unmanned aerial vehicles (commonly known as drones) equipped with sensors are being used to capture crop and soil image data that are then analyzed by online-based AI systems to provide information on land suitability, soil texture, soil moisture, soil organic matter, crop health, crop maturity, and expected yields among others (Rizzoli, 2022). The information provided through these systems is necessary for understanding the interactions between farm-specific environments or conditions and crop growth. Based on this, necessary adjustments and improvements can be made to increase farm yield or improve crop health. Additionally, AI-based solutions are being used to enhance farm management decisions by predicting crops that should be cultivated based on soil (soil depth, soil type, pH, organic carbon, sulfur, calcium, manganese, potassium, copper, nitrogen, iron, and phosphate) and meteorological parameters (temperature, humidity, and rainfall) (Dahikar & Rode, 2014).

Weed management

Another aspect of crop production is weed management. Weed management has conventionally been carried out by manual removal in small farms and the application of chemical herbicides when farm sizes are large. Manual weeding is typically environmentally friendly but more labor intensive. Unchecked weed growth causes high losses in crop quality, expected yields, and as a result, reduced profits. Even with the application of herbicides, the economic cost of yield loss due to weeds is huge. It has been estimated to result in annual losses of A\$ 3.3 billion to Australian grain growers, \$11 billion in India, and an even higher loss of \$33 billion in the USA (Chauhan, 2020; Gharde et al., 2018; Llewellyn et al., 2016). The challenge with the application of herbicides is excessive usage, which has led to the development of herbicide-resistant weeds and the pollution of groundwater (Chauhan, 2020). Presently, AI systems based on IoT and machine vision are being used for precise weed management. The systems' architecture includes sensors and cameras for weed mapping and detection and a spraying system that can intelligently detect crop rows and avoid them while spraying targeted weeds (Eli-Chukwu, 2019; Kutyauripo et al., 2023). This ensures that herbicides are used properly and effectively and also decreases herbicide wastage, physiological and visible injuries to crops, herbicide residues, and environmental pollution. Studies have been able to accurately model and predict weed competition with crops for nutrients, and weed growth based on rainfall, weed density, weed type, and weed emergence (Elahi et al., 2019; Monteiro et al., 2021; Partel et al., 2019; Tobal & Mokhtar, 2014). This has enabled the determination of optimal amounts of herbicides to be used for site-specific cases, as well as the method and timing of herbicide application. Centaur Analytics is one such company that uses these systems for weed control (Kakani et al., 2020).

Additionally, weed control can be performed through automated weeding. Here, robots coupled with AI systems are used to identify crops, differentiate them from weeds, and remove the weeds. Automated weeding is gaining traction as it is more environmentally friendly and sustainable since there is little to no use of chemicals. Moreover, automated weeding reduces the need for manual labor and increases the overall efficiency of crop production processes. Kutyauripo et al. (2023) also reported on the use of AI/robot systems equipped with lasers and mechanical rotating hoes for intrarow weeding on sugar beet and lettuce farms. Chang and Lin (2018) designed an AI/robot system that uses image processing methods such as HSV [hue (H), saturation (S), value (V)] color conversion, estimation of thresholds during the image binary segmentation process, and morphology operator procedures to confirm the positions of crops and weeds as well as water crops. Chang and Lin (2018) implemented this system and achieved classification rates above 90% for both crops and weeds. Their system was able to simultaneously perform weeding and watering while maintaining deep soil moisture content at 80% ± 10%. Some of the commercially available solutions in this area include *BoniRob* and *Robocrop* (Garford, 2019; Rizzoli, 2022).

Robocrop (manufactured by Garford Farm Machinery) is a tractor-mounted weeder that utilizes machine vision for mechanical in-row weeding. *Robocrop* employs a weeding mechanism that spirally rotates around each crop. As this happens, *Robocrop* captures images of the crop and processes data to determine the center of the crop (Garford, 2019). *Robocrop* was originally designed for use on leafy greens, but it can be adapted for most crops once there is adequate row spacing such that crop foliage is clearly separated from the next crop (Misra et al., 2022). Similarly, *BoniRob* (developed by Deepfield Robotics/Bosch) is an autonomous agricultural robot that uses an AI-based camera and image recognition to differentiate between crops and weeds based on leaf shape, color, and size (Misra et al., 2022; Rizzoli, 2022). After identifying weeds, *BoniRob* stabs the weeds with a rod to destroy them. When BoniRob was trialed on carrot farms, it was found to be more than 90% effective (Gershgorn, 2015). The *BoniRob* system environment includes web-based communication, servers, and web clients, allowing data sharing between different *BoniRob*s and the transfer of data to human operators for remote control (Biber et al., 2012). *BoniRob* can also be adapted for crop and soil monitoring activities (Biber et al., 2012).

Blue River Technology (which has been acquired by John Deere) has also developed *See & Spray*, a weed control machine that combines robotics, cloud analytics, computer vision, and deep learning to differentiate crops from weeds (BRT, 2018). Also tractor-mounted, *See & Spray*, is self-propelled and targets (*See*) weeds for herbicide application (*Spray*). This technology has been trialed in cotton, soybean, wheat, and corn farms and has the advantage of reducing the quantities of herbicides needed, thus providing environmental and economic benefits (BRT, 2018).

Pest and disease detection

Further trends in online-based AI in crop production can be seen in pest and disease detection. Most of the applications in this segment use deep learning-based image recognition technologies and IoT services to detect plant health (Rizzoli, 2022). Wang et al. (2017) developed a deep convolutional neural network model that was able to detect and determine the severity of black rot disease in apples at 90% accuracy. Also, Liu and Wang (2020) employed YOLOv3 (You Only Look Once; version 3), a real-time algorithm for object detection in images and videos, to detect the location and category of multiple pests and diseases in tomatoes. The study by Liu and Wang (2020) achieved a 92% accuracy with a detection time of only 20.39 milliseconds. Another study by Zhong et al. (2018) also used YOLO to identify and count flying insects. The Zhong et al. (2018) model was able to identify fruit flies, moths, bees, chafers, mosquitoes, and flies with an accuracy of 90% and perform counting with an accuracy of 93%. These studies show that the future potential of online-based AI systems for plant disease control is promising. The applications of online-based AI systems would decrease labor costs and inefficiency without compromising the accuracy of observations.

Fruit and vegetable harvesting

More applications include the use of online-based AI systems for crop harvesting and further sorting and grading them based on their image characteristics (Kutyauripo et al., 2023). AI systems are being used to automate various processes in food production, including packaging and distribution. For example, AI-powered robots can be used to pick and pack fruits and vegetables, reducing the need for manual labor and increasing the efficiency of these processes (Kutyauripo et al., 2023). Moreover, AI-powered robots can be used to inspect food products for quality, helping to ensure that consumers receive high-quality products (Kutyauripo et al., 2023). Fu et al. (2015) and Fu et al. (2019) developed a robotic kiwi harvester that could detect and pick kiwi fruits even at night. Also, Silwal et al. (2017) designed a robotic apple harvester that had an apple-picking success rate of 84% and only spent 6 seconds per fruit. The development of commercial-scale robotic harvesters has seen little progress due to issues such as poor recognition time and a relatively lower success rate (Misra et al., 2022). Nonetheless, it is expected that, as we move into the future, research and investment in these technologies would increase as the need to replace manual harvesting increases. Other studies have proposed models for identifying the maturity of wheat (Zhu et al., 2016) and tomatoes (Wan et al., 2018), sorting and grading carrots (Deng et al., 2017), grading tomatoes (Iraji, 2019), and monitoring watermelon ripeness during postharvest storage (Albert-Weiß et al., 2020). In all of these studies, the models achieved high accuracy rates and performed activities faster than trained professionals.

The integration of all these previously mentioned applications of online-based AI systems in crop production would be the full realization of precision agriculture and the establishment of smart farms. Commercial solutions such as *Deepfield Connect, Plantect, Priva, E-Kakashi*, and *The Yield* are available for smart farming. These IoT-based services consist of sensors for the measurement of soil and meteorological conditions, cloud platforms for data collection, AI systems for data integration, prediction, and decision support, and communication mechanisms with farm management (Misra et al., 2022). For example, *The Yield*'s service consists of sensors positioned at different sites in a farm and communicating with each other over a local area network (TYTS, 2021). This allows for the measurement of the farm's microclimate. Measured data is stored on the cloud and processed with AI to provide recommendations for on-farm operations such as planting, fertilizer application, weeding, irrigation, and harvesting (TYTS, 2021). Though we are still some years away from widespread fully integrated smart farms, current innovations show that this would happen sooner rather than later.

Animal production

Online AI-based systems have had a significant role in all segments of animal production. As in the case of crop production, these new and emerging technologies are being used to improve animal growth, health, and welfare. Warner et al. (2020) developed an AI-based decision support tool for detecting lameness (poor movement due to leg injury) in cows with a 73%−75% accuracy. This system can detect lameness earlier than it can be visually detected by humans, allowing for early treatment to promote animal health. AI systems have also been used for the automated prediction of bovine tuberculosis in dairy cows using infrared spectral data from milk (Denholm et al., 2020). This is a low-cost method that promotes animal welfare as it is noninvasive, uses already collected milk, and allows for early identification of bovine tuberculosis.

A study by Jorquera-Chavez et al. (2019) applied infrared thermal images and computer vision for the prediction of the heart rate, ear-base temperature, eye temperature, and respiration rate of cattle. AI-based systems have also been designed for the prediction of dairy cow milk yield and production stage using echotexture analysis sonograms of cow mammary glands (Themistokleous et al., 2022); measuring cattle eating and rumination time (Zehner et al., 2017); contactless body weight and dimension measurement of cattle (Huang et al., 2018); prediction of cattle heat stress from drinking behavior (Tsai et al., 2020); monitoring cattle digestive health (Atkinson et al., 2020); and automated pest detection (Psota et al., 2021).

Feed composition and intake have a huge effect on animal health and productivity. Conventional farms use simple weighing scales to determine the quantity of feed to be provided to animals and typically have no way of monitoring feed intake. In modern smart farms, insights from animal behavior and health can be passed on to feeding systems to ration feed or even formulate diets that meet the herd or a specific animal's nutrition requirements. Gehlot et al. (2022) proposed an IoT/AI-based robotic system for cattle diet planning and feeding. The system consists of cameras and sensors that capture animal images and other data that is analyzed using machine learning. The results of the analysis are then communicated over a low-power wireless network to robots. The robot then formulates feed, based on the information received, and moves to present the feed to the animal.

Furthermore, online AI-based systems are making inroads in the aquaculture industry. Chiu et al. (2022) and Chang et al. (2021) have proposed AI/IoT systems for smart aquaculture farm management. There is the potential for AI systems to be used in real-time monitoring of water quality parameters such as oxygen, temperature, pH, and salinity to maintain optimal conditions for fish. These systems can also be used in automated feeding, fish counting, fish length and weight estimation, and predicting fish health in response to water quality.

Commercially available AI solutions in the animal production sector include *CattleEye* and *Digitanimal*. CattleEye Ltd. developed *CattleEye* for monitoring cows (CattleEye Ltd, 2023). *CattleEye* records videos of cows as they leave a milking parlor. The footage is analyzed by cloud-based AI systems to uniquely identify cows and track their movement, welfare, and behavior. All of the findings and predictions from the AI system are then delivered to a standalone mobile or computer application or integrated into existing farm management systems for remote and real-time cattle monitoring. *Digitanimal* (manufactured by Digitanimal) also tracks animal movement and location using IoT-enabled global positioning system (GPS) collars and uses this data to predict animal health and productivity (Digitanimal, 2023). Additionally, there is *Vence* (Merck & Co.) that uses virtual fencing for controlling animal movement; *Ida* (Connecterra) that uses sensors and cloud-based machine learning to predict animal behavior; *Porphyrio* (Evonik) for predicting egg production; *Cainthus* (Ever.Ag) for real-time animal monitoring, and *CageEye* (Bluegrove) for optimized feeding in aquaculture systems (Bluegrove, 2022; Connectera, 2021; Ever.Ag., 2023; Porphyrio, 2021; Vence, 2022).

Food processing and related operations

At the tail end of the food and agriculture spectrum, we see food processing where all the previously produced raw materials are converted into other forms. We also see related operations such as food service, supply chain and traceability, and food safety activities. In all of these sections of the agri-food industry, online AI-based systems are being used to improve processes and improve products. Following periods during the peak of the COVID-19 pandemic when food processing facilities had to be shut down due to employee safety, there has been an increase in the use of automated systems in the food processing industry (Echegaray et al., 2022). AI-driven and IoT-enabled robotic systems are being used in animal slaughtering and cutting processes, monitoring processing conditions, and improving throughput rate and time (Echegaray et al., 2022). These technologies have also been combined with analytical methods such as hyperspectral imaging and spectroscopy for food quality evaluation, sensory evaluation, and the rapid detection of allergens and chemical contaminants (Benefo, Karanth, et al., 2022). There is also the potential for online AI-based systems to replace response surface methodology as the go-to approach for modeling and optimizing food processing operations.

Food supply chain and traceability

Food supply chain and traceability systems are also being enhanced by integrating AI, IoT, radio-frequency identification technology (RFID), and blockchain technology. Although blockchain technology was primarily dedicated to the creation of cryptocurrency and other financial transactions, it is in essence a ledger-based database system (Burke, 2019). Thus, its architecture can be modified for several applications where there is a need for rapid authentication of data, including food traceability. Blockchain technology is seen as a more robust tool for improving whole-chain traceability and transparency as compared to the conventional "trace back/trace forward" approach used in the food supply chain (Tian, 2017). Blockchain allows food processing companies and retailers to input traceability information while keeping business models and other proprietary information private (Tian, 2017). With data sharing in blockchain/AI/IoT systems, all actors in a food supply chain — from farmer to consumer — can have access to a product's information.

The introduction of AI/blockchain technology in the food industry is revolutionizing how food is sold and distributed globally. In 2016, Walmart conducted a pilot study on tracing its pork and mango supply chains in China and the USA, respectively using a *Hyperledger Fabric* traceability system (Sristy, 2021). For the pork supply chains in China, the *Hyperledger Fabric* system allowed the uploading of certificates of authenticity, making product verification easier, and bringing trust between Walmart, their customers, and suppliers. For mango supply chains in the USA, the use of the blockchain-based system decreased the time for tracing their origin from 7 days to 2.2 seconds. As of September 2018, Walmart was tracing the origin of 25 products with blockchain-based systems (Sristy, 2021). In Europe, Carrefour employs blockchain-based traceability systems for tracking eggs, poultry, and tomatoes (Wilson & Auchard, 2018). Similar systems have been developed by *BeefChain* (verifying the origin of Wyoming-raised cattle and sheep), *BeefLedger* (tracking the supply chain of Australian beef in China), *Origin Chain* (traceability of foods in Ireland), and *Certified Origins* (traceability of extra virgin olive oil) (Patelli & Mandrioli, 2020). Although the application of blockchain technology in the food industry is new, it has been successfully implemented. It has the potential to bring more significant improvements in traceability as the technology becomes widespread in the food industry.

The integration of RFID, AI, and wireless sensor networks synergizes their functionalities for enhanced applications and increases the variety of operations where they can be used. These integrated systems not only improve food traceability but also increase customer satisfaction since product quality is guaranteed. Alfian et al. (2020) developed a web-based traceability system that utilizes RFID, IoT sensors, and machine learning. The system was tested in a kimchi supply chain where the movement of kimchi, as well as temperature and humidity during storage, was monitored. The product data was then made available to customers on a website. A similar system was developed by Urbano et al. (2020) for tracking the temperature and humidity of oranges during storage and transportation from Spain to Ireland.

Food safety

Lastly, there is the application of online AI-based systems in food safety risk assessment and food-borne disease surveillance. This involves the use of web servers to assemble and annotate pathogen genomes. This genome annotation data is then combined with relevant metadata via machine learning and deep learning algorithms to predict the occurrence of foodborne disease outbreaks, their sources, and disease severity (Karanth et al., 2022, 2023; Tanui, Benefo, et al., 2022; Tanui, Karanth, et al., 2022). Ultimately, online-based AI systems have the potential to integrate agricultural food production, processing, food safety risk factors, and genomic data that can transform public health strategies to prevent foodborne diseases and rapidly respond to outbreaks.

As innovations in web systems, AI, IoT, and sensor technologies continue to occur, the potential for applications in the agri-food industry is inestimable. Once there is data or a process that can be modeled, it can be improved using integrated AI systems. In the near future, we will see the applications of online-based AI systems in sustainable food production, water conservation, predictive equipment maintenance, intelligent food packaging, and food waste reduction.

Potential ethical risks of AI technological advancements in agriculture and food systems

In the preceding sections, we have discussed the many applications of online AI-based systems in the agri-food industry. These applications present ethical avenues as well as opportunities for AI to be used in solving agriculture and food-related problems. We have given examples of how integrating AI systems into crop farming results in production efficiency and increased yields, which could lead to more affordable food and help in reducing food insecurity. The usage of AI systems in food safety, animal production, and water conservation also ties in with efforts to achieve United Nations Sustainable Development Goals 2 (Zero Hunger), 3 (Good Health and Well-being), 6 (Clean Water and Sanitation), and 13 (Climate Action). There are numerous opportunities for online-based AI systems to be used in improving agriculture and food systems and transforming society. However, as with any technology, there are potential risks related to security, privacy, transparency, data ownership, sustainability, and accountability among others. These risks and challenges are neither unique to agriculture and food systems nor necessarily arise from the agri-food sector. But the peculiar nature of the agri-food sector and its importance to human well-being and national and world economies make the likely impacts of these risks even more pronounced. We will now discuss some of these potential ethical risks of online-based AI systems in the context of the agri-food sector.

(Cyber)Security

It has been claimed that data in the agri-food sector is less vulnerable to privacy and security threats than in other sectors (Zhang et al., 2015). This mainly stems from the assumption that online-based AI systems in food and agricultural processes do not collect sensitive data. However, in our opinion, it is not a matter of how sensitive or not agri-food data and processes may be, but a matter of what effects there will be if security and privacy are compromised. Imagine what would happen if a cyberattack disrupted food production and food safety. Unfortunately, when we come to the ethical risks of AI-based systems, we are no longer in the realm of imagination or "what if a or b happens?"; we are in the realm of happenings.

In January 2023, a cyberattack on Yum Brands resulted in the closure of about 300 restaurants in the UK (Rajesh, 2023). Yum Brands claimed the attack was not expected to have any adverse impact on their business operations and financial results. This attack was one in a spate of recent ransomware attacks on companies in the agri-food sector. In 2021, JBS S.A. (the world's largest meat-processing company) was the victim of possibly the most high-profile ransomware attack in the agri-food sector that resulted in a shutdown of their operations in the USA, Canada, and Australia for several days (Kapko, 2022). JBS S.A. ended up paying $11 million to the attackers before their systems were restored. The shutdown disrupted food supply chains and resulted in temporarily hiked meat prices by up to 25% (AFB, 2022). Beyond ransom payments, these attacks result in revenue losses through loss of productivity and remediation costs. It is telling how grave the effects of these attacks on companies and consumers as well could be. There is also the potential for a rippling effect when these attacks are on primary processors that supply their products to others in the industry.

We have witnessed increasing cyberattacks on agri-food companies in the news over the last two years. Hackers do not discriminate about which companies to attack; while large corporations are seen as capable of paying a huge ransom, smaller companies are also seen as vulnerable and easy targets since they may not have the financial muscle to invest in effective cybersecurity systems. Additionally, while a large company may be able to weather these attacks, a sustained attack on a small-scale company could lead to its collapse. Due to the seriousness of these attacks and the significance of the agri-food sector, the US Federal Bureau of Investigations (FBI) released a report detailing recommendations for the agri-food sector, especially during the planting and harvesting season as it had been noted that most attacks were deliberately timed to occur during this period so they would have a significant effect on farm operations (FBI, 2022). As farmers increase their reliance on IoT-enabled AI robotic systems for their numerous benefits, concerns remain about how disruptive and destructive an attack on these systems would be to farming operations. As we have seen in the case of JBS S.A., even if the attack is short-lived, the consequences are significant.

Privacy

In most of the ransomware attack cases that have occurred, companies admitted that data had been stolen from their networks but always denied that this data was related to trade secrets, recipes, client, and customer information. Of course, this is expected as acknowledging that client or customer data, for example, was stolen could lead to further financial losses through lawsuits. In an era where data can be monetized for several purposes, data privacy is paramount.

Primary actors in the agri-food sector are primarily concerned about the kind of data generated and collected from their farms or processing facilities, a lack of control over such data, who the data is shared with, and how the data is used (Ryan, 2020). According to Stock and Gardezi (2021), most farmers (78%) are worried about the sale and sharing of their data by corporations without their consent or awareness and the use of such data to manipulate markets for the benefit of the corporation. Furthermore, farmers are concerned about the ability of online AI-based systems deployed on their farms to spy on them and the data collected sold to other companies or even passed on to authorities to be used against them (Ryan, 2020; Ryan et al., 2023).

A 2021 complaint by farmer advocacy groups to the US Federal Trade Commission highlights the concern of farmers. In the complaint, farmers alleged that John Deere (an agri machinery manufacturer) used precision farming data collected from their farms to sell equipment to them and also used the data to restrict the control and servicing of the machinery (Herchenroeder, 2021). In effect, farmers neither had control of their data nor the machinery purchased. This is even worse in cases where corporations persuade farmers to buy agricultural inputs from them in order to have access to their AI-based solutions (Ryan, 2020). In the end, the distrust between farmers and corporations results in a situation where farmers are caught between the proverbial rock and a hard place: give up data privacy and join the AI revolution or refuse to give up and be at a disadvantage.

Data ownership

The issue of data ownership is closely related to data privacy and is also a concern for people in the agri-food sector. Data from agricultural and food systems and the information that can be inferred from it using AI systems is an important and expensive commodity. Therefore, whoever has ownership of this data stands to gain huge economic benefits. The bone of contention here is that while farmers may be given ownership of the raw data that is generated from their farm operations, they do not own the information and knowledge obtained from the raw data (Kosior, 2020). In some cases, AI solution providers retain farmers' data under a royalty-free license so that the data can be used for the providers' purposes (Darr, 2014). In the end, farmers could be owners of their data but have no economic benefits from the data beyond the primary benefits of using AI-based systems. In other cases, farmers do not even own their data; thus it is difficult to switch from one AI solution provider to another without breaching contractual agreements (Sykuta, 2016; Uddin et al., 2022). It is expected that AI solution providers would take decisions to protect and maximize returns on their investments, but agreements that only benefit one party are inimical to the growth of AI applications in agriculture and food systems.

Accountability/responsibility

Another pressing ethical implication is that for most online AI-based systems there is no clarity on responsibility or liability for AI system errors (Van der Loeff et al., 2019). Usually, it is not known who or which company should be held responsible in the event of inaccurate predictions or wrong recommendations by an AI system. As we have stated earlier, these concerns are very crucial when you consider the effects of wrong AI-enabled decisions in the agri-food sector.

For example, say an AI system has been designed to automatically monitor corn storage conditions, as part of an early warning system for aflatoxin levels, and make adjustments toward optimal conditions when needed. If the system fails to accurately monitor the storage conditions, aflatoxin levels in the corn could increase to unsafe levels and lead to illnesses or deaths when the corn is consumed. The ethical question here is who bears the responsibility for the aflatoxin outbreak? The AI system designer, the user, or the system itself? It will be difficult to determine who will be held accountable for AI system errors and economic losses when there are no prior contractual agreements that state the rights, responsibilities, and liabilities of different stakeholders involved in the design, maintenance, and use of the AI system (Dara et al., 2022).

Even when such systems exist, there is the possibility that some players in the agri-food sector could be short-changed. An example is John Deere escaping accountability by asking farmers to sign a licensing agreement that absolves the company of any legal liability if there is damage to farmers' business, profit, land, or machines (Carolan, 2020). This assumes that AI systems will always work flawlessly and shifts responsibility solely to farmers. Farmers are therefore required to provide the appropriate environment for the AI system to function (Stock & Gardezi, 2021).

Fairness

Fairness, in the context of AI ethics, refers to the monitoring and mitigation of bias in AI models as well as fair data collection and usage and access to digital assets (Jobin et al., 2019). Sapienza and Vedder (2021) described two forms of fairness in AI ethics: substantial fairness and procedural fairness. Substantial fairness requires AI algorithms to be directed toward minimizing the adverse effects of existing discriminations and increasing the accessibility of AI benefits while procedural fairness requires data analysis to be conducted in a manner that ensures compliance with ethical and legal principles to prevent harm.

AI systems have the potential to amplify existing biases when the algorithms used have an underlying propensity to learn from a preferred data pattern instead of the true distribution of the training data sets or when the training data sets do not accurately represent reality. For instance, when an AI system has been designed to classify only moldy corn as aflatoxin-contaminated, there is the risk of passing nonmoldy corn as uncontaminated even though they may be. Also, consider the case where an AI/robotic system for harvesting grapes has been trained on images mostly from red grapes. When such a system is deployed in a green grape vineyard, it may fail to perform accurately due to its bias toward red grapes as ready for harvest. Additionally, since food choices and consumption data can be used as proxy variables for specific cultures and people, models that do not accurately capture this could inadvertently discriminate against certain groups of people (Sapienza & Vedder, 2021). Fair data collection could lead to the development of more reliable AI systems.

There are also fears that the unfair use of AI systems would create digital divides between large, rich farms and smaller, poorer farms (Ryan, 2019). This could lead to the widening of already existing economic divides. Another concern is that due to the low numbers of women in science, technology, and agriculture, further technologizing agriculture may continue reducing the number of women from the agri-food sector (Carolan, 2020). This could be addressed by encouraging large corporations in the agri-food sector to commit funds toward the training of more women in the field.

Transparency

Transparency involves the interpretability and explainability of AI systems and the accuracy of their recommendations (Ryan, 2022; Sapienza & Vedder, 2021). Inaccurate recommendations can result in unsustainable farming practices, crop and animal illness, farm worker illness, and economic losses (Ryan, 2019). Transparency is related to and reinforces the concepts of accountability, fairness, and trust (Dara et al., 2022; Sapienza & Vedder, 2021). On the one hand, transparency is seen as a design requirement that allows users to understand how an AI system recommends certain actions and the justification for these recommendations. On the other, it relates to openness with various stakeholders about laws, policies, and actions to improve communication, collaboration, and understanding (Dara et al., 2022). The Organisation for Economic Co-operation and Development (OECD) AI transparency and explainability principle requires that users be made aware of their interactions with AI systems, the data generated from those interactions, and what the data is used for (OECD, n.d.). The OECD principle promotes responsible disclosure of AI systems so that people negatively affected by AI decisions or recommendations can challenge them.

Presently, most online AI-based solutions in agriculture and food systems are not explainable or interpretable. Agri-food workers do not understand the data being collected from their operations nor understand how a decision was arrived at or the limitations thereof but are only end users of the AI solutions. Transparency about online AI-based systems used in agriculture could remove the distrust between agri-food workers and AI technology providers as well as increase farmers' and public perception and acceptance of AI solutions in the agri-food sector.

Preventing and mitigating potential ethical risks of online AI systems in the agri-food sector

With reports that global spending on AI innovation in food and agriculture is increasing, we can only expect an increase in the adoption of AI-based systems in the agri-food sector. As AI innovations increase, there arises the need for greater consideration of the potential ethical implications of the technologies by researchers, designers, developers, and end users. At the moment, the use of online-based AI solutions in agriculture and food systems is still in its nascent stages. This offers a good opportunity to learn from other industry sectors and take measures to mitigate existing ethical issues concerning the use of online AI-based solutions in food and agriculture systems as well as prevent potential issues that may arise. This section briefly discusses some of the measures and strategies that can be put in place to ensure ethical AI development and deployment in the agri-food sector.

Responsible innovation

Responsible innovation is a concept that is aimed at encouraging creativity and opportunities for science and innovation that are socially acceptable and conducted in the interest of the public (UKRI, n.d.). It seeks to avoid unanticipated negative effects and lower obstacles to the dissemination, adoption, and diffusion of research and innovation while maximizing the positive societal and economic advantages of such activity. Responsible innovation is predicated on the idea that the "human factor" is crucial to research and innovation since technology such as AI systems profoundly affects human life (Van der Burg et al., 2021).

The framework for responsible innovation in smart dairy farming proposed by Eastwood et al. (2019) is a good starting point for a discussion on how to implement responsible innovation in agri-food systems. The framework is centered on four themes: anticipation, inclusion, reflexivity, and responsiveness. Eastwood et al. (2019) proposed that the negative impacts of innovations should be anticipated through activities such as scenario-building and foresight exercises at the initial stages of technology development. For instance, as part of food safety management plans, food scientists are required to establish preventive controls for reasonably foreseeable hazards. Therefore, it would not be out of place if AI designers were expected to anticipate what the ethical implications of AI systems in the agri-food sector would be. Eastwood et al. (2019) also suggested that there should be the involvement of all relevant actors in the agri-food sector during innovation. Furthermore, Eastwood et al. (2019) recommended that technology development projects should adopt structures and standards that guide reflexivity, recognize the implications of technology solutions, and promote open data processes and transparency. Finally, innovation projects should be amenable to changes based on the socio-ethical implications of the technology.

As agriculture and food systems move toward the increased integration of AI systems in the agri-food ecosystem, it is important that the concept of responsible innovation underlies the development of AI systems for the agri-food sector. AI solutions in this vital sector must not be overly concerned with maximizing farm yields and profits at the expense of fairness, data privacy, or transparency. Recognizing the potential ethical issues associated with AI solutions and incorporating them right from the start of technology development will promote trust among end users.

Conscientious design

Conscientious design is a concept similar to responsible innovation in that they both intend to guide the technology design process by creating awareness about the potential long-term and unintended impacts of technologies in order to develop more sustainable solutions. Noriega et al. (2021) proposed a conscientious design framework for broadly tackling the ethical implications of online AI-based systems. The framework consists of value categories, a systematic contextualization of the values, and a procedure to operationalize the values. The value categories proposed by Noriega et al. (2021) are thoroughness, mindfulness, and responsibility. Thoroughness relates to traditional technological standards such as system integrity, accuracy, efficiency, security, accessibility, and resilience that ensure the quality of an AI system. Mindfulness entails developing a broader understanding of the impacts of technology on end users and society as a whole to enable designers to develop ethical solutions. Lastly, responsibility focuses on the liability and accountability of AI developers toward end users in the event of any negative effects of the AI system (Noriega et al., 2021).

In developing online AI systems for the agri-food sector, the conscientious design framework can be implemented by ensuring the integrity of AI system data and the accuracy of recommendations for farm management (thoroughness); providing clear data policies, protecting farmers' rights, and respecting farmers' presence (mindfulness); and ensuring that AI systems are sustainable and that there is liability protection for farmers and other end users (responsibility).

Interdisciplinary and multistakeholder engagement

The bridge between the technical and socio-ethical aspects of AI systems has been noted by Ryan (2022). While AI designers tend to focus on the technical and functional aspects of the system (sometimes with little regard for ethical implications), social scientists tend to focus on the ethical challenges of such systems. The complex ethical challenges that arise from the applications of online AI-based systems in the agri-food sector are interconnected and cut across multiple disciplines such as crop and animal science, food science, agronomy, agribusiness, agricultural engineering, philosophy, social science as well as computer science, robotics, and software engineering. There are usually no simple, forthright solutions to these ethical challenges from a single discipline. When experts from different fields come together, knowledge, perspectives, and insights can be shared to develop ethical AI solutions and address related challenges. For example, as experts in computer science ideate on AI solutions, social scientists and philosophers can offer perspectives on the likely ethical challenges that may arise while crop and animal scientists also provide input on the impacts of the solution on farm operations. The successful implementation of AI-based solutions in agriculture and food systems requires a concerted effort and close collaboration between experts from different disciplines.

The food production process involves several people, and the food produced also directly impacts all humans (we all eat). Therefore, it is crucial that different voices are considered during the development of technologies that affect this sector. In our previous study Benefo, Tingler, et al. (2022), we concluded that it is important to explore the ethical implications of the application of AI in agriculture and food systems and conduct more applied research, taking the perspectives of different stakeholders (AI solution providers, farmers, food processing facilities, consumers, and communities) and a multidisciplinary approach into account. The software engineers behind AI solutions tend to be decoupled from the solutions since farmers would usually only interact with the technology but not the people who designed it. Involving farmers and food processors in the AI development process builds human connections between engineers of technology and the end users, reduces the objectification and commodification of farmers and other agri-food sector workers as merely end users or tools for profit, and could go a long way to reduce the opposition to certain AI technologies. Innovations with the potential to transform the entire society can sometimes be rejected, if it is not trusted and adopted, by end users. For any transformative technology to succeed, one needs to incorporate multistakeholder perspectives into account. One also needs to keep the communication channel open between technology developers and end users. The ethical concerns of transparency, trust, and privacy mentioned in the previous sections can be greatly reduced if all actors are involved in the AI system development process.

Legislation

The previous strategies mentioned in this section mainly address the ethical implications of AI-based systems in agriculture such as fairness, transparency, data ownership, and privacy through ethical design. Regardless of the ethical design of an AI system, potential harm (e.g., a milking robot harming cows or an AI system incorrectly recommending weedicide application) could occur. Legislations establish enforceable regulations through a transparent democratic process involving all relevant stakeholders.

Legislation could promote ethical AI design when AI solution providers become incentivized to meet legal requirements so as to have a competitive advantage or for reputational purposes. However, legislation could also be seen as a double-edged sword; too much or stringent legislation may stifle investment, research, and innovation. At the same time, no or lax legislation could leave farmers at the mercy of AI technology providers. Presently, there are no binding international laws guiding the application of AI systems in the agri-food sector. Individual countries must develop timely policies and laws based on their unique agri-food sector, the state of AI in their countries and reasonable future expectations, and unique sociocultural systems instead of a wholesale importation of laws from other countries. The enactment of policies and legislation concerning AI in the agri-food sector would not only guide design but also provide clarity on responsibility and accountability as well as provide avenues for redress when harm does occur.

Teaching AI ethics

Ethical guidelines and frameworks can sometimes be abstract and disconnected from actual practice (Morley et al., 2021; Vakkuri & Abrahamsson, 2018). McNamara et al. (2018) reported that ethical guidelines were not effective and had no influence on the behavior of a section of software engineers. Perhaps, a more practical but underestimated approach to preventing or mitigating the potential ethical risks of online AI systems in the agri-food sector would be the inclusion of AI ethics into curricula. Computer scientists and engineers are often tasked with the ethical design of AI systems but may not have had any training in the intricate ethical implications of their designs. It is important to educate current students on how AI might impact lives and their role in maximizing its benefits while minimizing any potential negative consequences.

Conclusion

There are already many applications of online AI-based systems in the agri-food sector, but we have only "scratched the surface." As billions of dollars continue to be pumped into AI research, we can only expect advancements in current applications. AI will revolutionize agriculture as we know it today. As these advancements come, the ethical concerns of privacy, transparency, fairness, security, data ownership, and responsibility will also emerge. These concerns have a direct impact on human well-being. The frameworks of responsible innovation and conscientious design can guide AI system developers to thoroughly reflect on ethical designs and beneficence and anticipate unintended negative consequences of AI systems so they can be addressed. Both AI system developers and actors in the agri-food sector would greatly benefit from an interdisciplinary and multistakeholder approach to tackling the ethical concerns arising from AI use. Lastly, education can be a powerful tool for training students, across all fields, to consider the ethical implications of the AI systems they use.

References

AFB (American Federal Bank). (2022). *Cyber criminals targeting food/Ag sector with ransomware attacks.* https://www.americanfederalbank.com/cyber-criminals-targeting-food-ag-sector-with-ransomware-attacks/.

Albert-Weiß, D., Hajdini, E., Heinrich, M., & Osman, A. (2020). CNN for ripeness classification of watermelon fruits based on acoustic testing. In *Proceedings of the virtual 3rd international symposium on structural health monitoring and nondestructive testing (SHM-NDT 2020)* (pp. 25−26). http://www.ndt.net/?id = 25569.

Alfian, G., Syafrudin, M., Farooq, U., Ma'arif, M. R., Syaekhoni, M. A., Fitriyani, N. L., Lee, J., & Rhee, J. (2020). Improving efficiency of RFID-based traceability system for perishable food by utilizing IoT sensors and machine learning model. *Food Control*, *110*107016. Available from https://doi.org/10.1016/j.foodcont.2019.107016.

Allen, C., Varner, G., & Zinser, J. (2000). Prolegomena to any future artificial moral agent. *Journal of Experimental & Theoretical Artificial Intelligence*, *12*(3), 251−261.

Atkinson, G. A., Smith, L. N., Smith, M. L., Reynolds, C. K., Humphries, D. J., Moorby, J. M., Leemans, D. K., & Kingston-Smith, A. H. (2020). A computer vision approach to improving cattle digestive health by the monitoring of faecal samples. *Scientific Reports*, *10*(1). Available from https://doi.org/10.1038/s41598-020-74511-0.

Ben Ayed, R., & Hanana, M. (2021). Artificial intelligence to improve the food and agriculture sector. *Journal of Food Quality*, *2021*. Available from https://doi.org/10.1155/2021/5584754, Hindawi Limited.

Benefo, E. O., Karanth, S., & Pradhan, A. K. (2022). Applications of advanced data analytic techniques in food safety and risk assessment. *Current Opinion in Food Science*100937. Available from https://doi.org/10.1016/j.cofs.2022.100937.

Benefo, E. O., Tingler, A., White, M., Cover, J., Torres, L., Broussard, C., Shirmohammadi, A., Pradhan, A. K., & Patra, D. (2022). Ethical, legal, social, and economic (ELSE) implications of artificial intelligence at a global level: A scientometrics approach. *AI and Ethics*0123456789. Available from https://doi.org/10.1007/s43681-021-00124-6.

Biber, P., Weiss, U. Dorna, M., & Albert, A. (2012). *Navigation system of the autonomous agricultural robot "BoniRob."* Carnegie Mellon University. https://www.cs.cmu.edu/ ∼ mbergerm/agrobotics2012/01Biber.pdf.

Bluegrove. (2022). *Helping salmon farmers sustainably optimize production and increase fish welfare*. https://bluegrove.com/.

Borenstein, J., & Howard, A. (2021). Emerging challenges in AI and the need for AI ethics education. *AI and Ethics*, *1*(1), 61−65. Available from https://doi.org/10.1007/s43681-020-00002-7.

BRT (Blue River Technology). (2018). *See & spray select*. https://bluerivertechnology.com/our-products/.

Burke, T. (2019). Blockchain in food traceability. In J. McEntire, & A. W. Kennedy (Eds.), *Food traceability: From binders to blockchain*. Springer.

Carolan, M. (2020). Automated agrifood futures: Robotics, labor and the distributive politics of digital agriculture. *The Journal of Peasant Studies*, *47*(1), 184−207. Available from https://doi.org/10.1080/03066150.2019.1584189.

CattleEye Ltd. (2023). *Autonomous livestock monitoring*. https://cattleeye.com/.

Chang, C. L., & Lin, K. M. (2018). Smart agricultural machine with a computer vision-based weeding and variable-rate irrigation scheme. *Robotics*, *7*(3). Available from https://doi.org/10.3390/robotics7030038.

Chang, C. C., Wang, J. H., Wu, J. L., Hsieh, Y. Z., Wu, T. D., Cheng, S. C., Chang, C. C., Juang, J. G., Liou, C. H., Hsu, T. H., Huang, Y. S., Huang, C. T., Lin, C. C., Peng, Y. T., Huang, R. J., Jhang, J. Y., Liao, Y. H., & Lin, C. Y. (2021). Applying Artificial Intelligence (AI) techniques to implement a practical smart cage aquaculture management system. *Journal of Medical and Biological Engineering*, *41*(5), 652−658. Available from https://doi.org/10.1007/s40846-021-00621-3.

Chauhan, B. S. (2020). *Grand challenges in weed management, . Frontiers in agronomy* (Vol. 1). S.A: Frontiers Media. Available from https://doi.org/10.3389/fagro.2019.00003.

Chiu, M. C., Yan, W. M., Bhat, S. A., & Huang, N. F. (2022). Development of smart aquaculture farm management system using IoT and AI-based surrogate models. *Journal of Agriculture and Food Research*, *9*. Available from https://doi.org/10.1016/j.jafr.2022.100357.

Clarke, R. (2019). Principles and business processes for responsible AI. *Computer Law and Security Review*, *35*(4), 410−422. Available from https://doi.org/10.1016/j.clsr.2019.04.007.

Columbus, L. (2021). *10 Ways AI has the potential to improve agriculture in 2021*. Forbes. https://www.forbes.com/sites/louiscolumbus/2021/02/17/10-ways-ai-has-the-potential-to-improve-agriculture-in-2021/?sh = 55f0a7217f3b.

Connectera. (2021). *Innovation through intelligence*. https://www.connecterra.io/.

Dahikar, S. S., & Rode, S. V. (2014). Agricultural crop yield prediction using artificial neural network approach. *International Journal of Innovative Research In Electrical*, *2*(1), 683−686. Available from http://www.ijireeice.com.

Dara, R., Mehdi, S., Fard, H., & Kaur, J. (2022). Recommendations for ethical and responsible use of artificial intelligence in digital agriculture. *Frontiers in Artificial Intelligence*, *5*884192. Available from https://doi.org/10.3389/frai.2022.884192.

Darr, M. (2014). Big data—the catalyst for a transformation to digital agriculture. In *Proceedings of the 26th annual integrated crop management conference*. https://core.ac.uk/download/pdf/212822254.pdf.

Deng, L., Du, H., & Han, Z. (2017). A carrot sorting system using machine vision technique. *Applied Engineering in Agriculture*, *33*(2), 149−156. Available from https://doi.org/10.13031/aea.11549.

Denholm, S. J., Brand, W., Mitchell, A. P., Wells, A. T., Krzyzelewski, T., Smith, S. L., Wall, E., & Coffey, M. P. (2020). Predicting bovine tuberculosis status of dairy cows from mid-infrared spectral data of milk using deep learning. *Journal of Dairy Science*, *103*(10), 9355−9367. Available from https://doi.org/10.3168/jds.2020-18328.

Digitanimal. (2023). *GPS animals tracker − Tracking and monitoring livestock*. https://digitanimal.com.

Eastwood, C., Klerkx, L., Ayre, M., & Dela Rue, B. (2019). Managing socio-ethical challenges in the development of smart farming: From a fragmented to a comprehensive approach for responsible research and innovation. *Journal of Agricultural and Environmental Ethics*, *32*(5−6), 741−768. Available from https://doi.org/10.1007/s10806-017-9704-5.

Echegaray, N., Hassoun, A., Jagtap, S., Tetteh-Caesar, M., Kumar, M., Tomasevic, I., Goksen, G., & Lorenzo, J. M. (2022). Meat 4.0: Principles and applications of industry 4.0 technologies in the meat industry. *Applied Sciences (Switzerland)*, *12*(14). Available from https://doi.org/10.3390/app12146986, MDPI.

Elahi, E., Weijun, C., Zhang, H., & Abid, M. (2019). Use of artificial neural networks to rescue agrochemical-based health hazards: A resource optimisation method for cleaner crop production. *Journal of Cleaner Production*, *238*. Available from https://doi.org/10.1016/j.jclepro.2019.117900.

Eli-Chukwu, N. C. (2019). Applications of artificial intelligence in agriculture: A review. *Technology & Applied Science Research*, *9*(4), 4377−4383. Available from https://doi.org/10.48084/etasr.2756.

Ever.Ag. (2023). *Animal nutrition and welfare insights*. https://www.ever.ag/dairy/software-solutions/cainthus/.

FBI (Federal Bureau of Investigation). (2022). *Ransomware attacks on agricultural cooperatives potentially timed to critical seasons*. https://www.ic3.gov/Media/News/2022/220420-2.pdf.

Floridi, L., & Taddeo, M. (2016). What is data ethics? *Philosophical Transactions of the Royal Society A. Mathematical, Physical and Engineering Sciences*, *374*, 1−5. Available from https://doi.org/10.1098/rsta.2016.0360.

Fu, L., Tola, E., Al-Mallahi, A., Li, R., & Cui, Y. (2019). A novel image processing algorithm to separate linearly clustered kiwifruits. *Biosystems Engineering*, *183*, 184−195. Available from https://doi.org/10.1016/j.biosystemseng.2019.04.024.

Fu, L. S., Wang, B., Cui, Y. J., Su, S., Gejima, Y., & Kobayashi, T. (2015). Kiwifruit recognition at nighttime using artificial lighting based on machine vision. *International Journal of Agricultural and Biological Engineering*, *8*(4), 52−59. Available from https://doi.org/10.3965/j.ijabe.20150804.1576.

Garford. (2019). *Robocrop InRow Weeder*. https://garford.com/products/robocrop-inrow-weeder/.

Gehlot, A., Malik, P. K., Singh, R., Akram, S. V., & Alsuwian, T. (2022). Dairy 4.0: Intelligent communication ecosystem for the cattle animal welfare with blockchain and IoT enabled technologies. *Applied Sciences (Switzerland)*, *12*(14). Available from https://doi.org/10.3390/app12147316.

Gershgorn, D. (2015). *Farm robot learns what weeds look like, smashes th*em. *Popular Science*. Available from https://www.popsci.com/meet-bonirob-plant-breeding-weed-smashing-robot/.

Gharde, Y., Singh, P. K., Dubey, R. P., & Gupta, P. K. (2018). Assessment of yield and economic losses in agriculture due to weeds in India. *Crop Protection*, *107*, 12−18. Available from https://doi.org/10.1016/j.cropro.2018.01.007.

Giovanola, B., & Tiribelli, S. (2022). Beyond bias and discrimination: Redefining the AI ethics principle of fairness in healthcare machine-learning algorithms. *AI and Society*. Available from https://doi.org/10.1007/s00146-022-01455-6.

Hagendorff, T. (2020). The ethics of AI ethics: An evaluation of guidelines. *Minds and Machines*, *30*(1), 99−120. Available from https://doi.org/10.1007/s11023-020-09517-8.

Hauer, T. (2022). Importance and limitations of AI ethics in contemporary society. *Humanities and Social Sciences Communications*, *9*(1). Available from https://doi.org/10.1057/s41599-022-01300-7.

Heilinger, J. C. (2022). The ethics of AI ethics. A constructive critique. *Philosophy and Technology*, *35*(3). Available from https://doi.org/10.1007/s13347-022-00557-9.

Herchenroeder, K. (2021). FTC investigator meets with farmers on deere data privacy. *Communications Daily*. Available from https://communicationsdaily.com/article/2021/09/20/ftc-investigator-meets-with-farmers-on-deere-data-privacy-2109170033.

Huang, L., Li, S., Zhu, A., Fan, X., Zhang, C., & Wang, H. (2018). Non-contact body measurement for qinchuan cattle with LiDAR sensor. *Sensors (Switzerland)*, *18*(9). Available from https://doi.org/10.3390/s18093014.

Iraji, M. S. (2019). Comparison between soft computing methods for tomato quality grading using machine vision. *Journal of Food Measurement and Characterization*, *13*(1), 1−15. Available from https://doi.org/10.1007/s11694-018-9913-2.

Jobin, A., Ienca, M., & Vayena, E. (2019). The global landscape of AI ethics guidelines. *Nature Machine Intelligence*, *1*(9), 389−399. Available from https://doi.org/10.1038/s42256-019-0088-2.

Jorquera-Chavez, M., Fuentes, S., Dunshea, F. R., Warner, R. D., Poblete, T., & Jongman, E. C. (2019). Modelling and validation of computer vision techniques to assess heart rate, eye temperature, ear-base temperature and respiration rate in cattle. *Animals*, *9*(12). Available from https://doi.org/10.3390/ani9121089.

Kakani, V., Nguyen, V. H., Kumar, B. P., Kim, H., & Pasupuleti, V. R. (2020). A critical review on computer vision and artificial intelligence in food industry. *Journal of Agriculture and Food Research*, *2*. Available from https://doi.org/10.1016/j.jafr.2020.100033, Elsevier B.V.

Kapko, M. (2022). *Food supplier cyber risk spreads 1 year after JBS attack.* https://www.cybersecuritydive.com/news/food-supplier-cyber-risk-spreads-jbs/624800/.

Karanth, S., Benefo, E. O., Patra, D., & Pradhan, A. K. (2023). Importance of artificial intelligence in evaluating climate change and food safety risk. *Journal of Agriculture and Food Research*, *11*. Available from https://doi.org/10.1016/j.jafr.2022.100485.

Karanth, S., Tanui, C. K., Meng, J., & Pradhan, A. K. (2022). Exploring the predictive capability of advanced machine learning in identifying severe disease phenotype in Salmonella enterica. *Food Research International*, *151*(November 2021)110817. Available from https://doi.org/10.1016/j.foodres.2021.110817.

Knoppers, B. M., & Thorogood, A. M. (2017). Ethics and big data in health. *Current Opinion in Systems Biology*, *4*, 53−57, Elsevier Ltd. Available from https://doi.org/10.1016/j.coisb.2017.07.001.

Kosior, K. (2020). Economic, ethical and legal aspects of digitalization in the agri-food sector. *Problems of Agricultural Economics*, *263*(2), 53−72. Available from https://doi.org/10.30858/zer/120456.

Kutyauripo, I., Rushambwa, M., & Chiwazi, L. (2023). Artificial intelligence applications in the agrifood sectors. *Journal of Agriculture and Food Research*100502. Available from https://doi.org/10.1016/j.jafr.2023.100502.

Lauer, D. (2021). You cannot have AI ethics without ethics. *AI and Ethics*, *1*(1), 21−25. Available from https://doi.org/10.1007/s43681-020-00013-4.

Liu, J., & Wang, X. (2020). Tomato diseases and pests detection based on improved yolo V3 convolutional neural network. *Frontiers in Plant Science*, *11*. Available from https://doi.org/10.3389/fpls.2020.00898.

Llewellyn, R. S., Ronning, D., Ouzman, J., Mayfield, A., & Clarke, M. (2016). Impact of weeds on Australian grain production: The cost of weeds to Australian grain growers and the adoption of weed management and tillage practices. *Grains Research and Development Corporation, Australia*. Available from http://hdl.handle.net/102.100.100/90771?index = 1.

McNamara, A., Smith, J., & Murphy-Hill, E. (2018). Does ACM's code of ethics change ethical decision. In G. T. Leavens, A. Garcia, & C. S. Păsăreanu (Eds.), *Proceedings of the 2018 26th ACM joint meeting on european software engineering conference and symposium on the foundations of software engineering—ESEC/FSE 2018* (pp. 1−7). ACM Press.

Misra, N. N., Dixit, Y., Al-Mallahi, A., Bhullar, M. S., Upadhyay, R., & Martynenko, A. (2022). IoT, big data, and artificial intelligence in agriculture and food industry. *IEEE Internet of Things Journal*, *9*(9), 6305−6324. Available from https://doi.org/10.1109/JIOT.2020.2998584.

Monteiro, A. L., Freitas Souza, M., de, Lins, H. A., Teófilo, T. M., da, S., Barros Júnior, A. P., Silva, D. V., & Mendonça, V. (2021). A new alternative to determine weed control in agricultural systems based on artificial neural networks (ANNs. *Field Crops Research*, *263*. Available from https://doi.org/10.1016/j.fcr.2021.108075.

Morley, J., Elhalal, A., Garcia, F., Kinsey, L., Mökander, J., & Floridi, L. (2021). Ethics as a service: A pragmatic operationalisation of AI ethics. *Minds and Machines*, *31*(2), 239−256. Available from https://doi.org/10.1007/s11023-021-09563-w.

Morley, J., Floridi, L., Kinsey, L., & Elhalal, A. (2020). From what to how: An initial review of publicly available AI ethics tools, methods and research to translate principles into practices. *Science and Engineering Ethics*, *26*(4), 2141−2168. Available from https://doi.org/10.1007/s11948-019-00165-5.

Mulgan, G. (2019). AI ethics and the limits of code(s). *Nesta*. Available from https://www.nesta.org.uk/blog/ai-ethics-and-limits-codes/.

Noriega, P., Verhagen, H., Padget, J., & D'Inverno, M. (2021). Ethical online AI systems through conscientious design. *IEEE Internet Computing*, *25*(6), 58−64. Available from https://doi.org/10.1109/MIC.2021.3098324.

OECD (Organisation for Economic Co-operation and Development). (n.d.). *Transparency and explainability* (Principle 1.3). https://oecd.ai/en/dashboards/ai-principles/P7.

Orr, W., & Davis, J. L. (2020). Attributions of ethical responsibility by Artificial Intelligence practitioners. *Information Communication and Society*, *23*(5), 719−735. Available from https://doi.org/10.1080/1369118X.2020.1713842.

Partel, V., Charan Kakarla, S., & Ampatzidis, Y. (2019). Development and evaluation of a low-cost and smart technology for precision weed management utilizing artificial intelligence. *Computers and Electronics in Agriculture*, *157*, 339−350. Available from https://doi.org/10.1016/j.compag.2018.12.048.

Patelli, N., & Mandrioli, M. (2020). Blockchain technology and traceability in the agrifood industry. *Journal of Food Science*, *85*(11), 3670−3678. Available from https://doi.org/10.1111/1750-3841.15477.

Porphyrio. (2021). *Intelligent data management software*. https://www.porphyrio.com/en/.

Psota, E. T., Luc, E. K., Pighetti, G. M., Schneider, L. G., Trout Fryxell, R. T., Keele, J. W., & Kuehn, L. A. (2021). Development and validation of a neural network for the automated detection of horn flies on cattle. *Computers and Electronics in Agriculture*, *180*. Available from https://doi.org/10.1016/j.compag.2020.105927.

Raab, C. D. (2020). Information privacy, impact assessment, and the place of ethics. *Computer Law and Security Review*, *37*. Available from https://doi.org/10.1016/j.clsr.2020.105404.

Rajesh, A. M. (2023). Yum Brands says nearly 300 restaurants in UK impacted due to cyber attack. *Reuters*. Available from https://www.reuters.com/business/retail-consumer/yum-brands-says-nearly-300-restaurants-uk-impacted-due-cyber-attack-2023-01-19/.

Rizzoli, A. (2022). *8 Practical applications of AI in agriculture*. V7 Labs. https://www.v7labs.com/blog/ai-in-agriculture.

Ryan, M. (2019). Ethics of using AI and big data in agriculture: The case of a large agriculture multinational. *The ORBIT Journal*, *2*(2), 1−27. Available from https://doi.org/10.29297/orbit.v2i2.109.

Ryan, M. (2020). Agricultural big data analytics and the ethics of power. *Journal of Agricultural and Environmental Ethics*, *33*(1), 49−69. Available from https://doi.org/10.1007/s10806-019-09812-0.

Ryan, M. (2022). The social and ethical impacts of artificial intelligence in agriculture: Mapping the agricultural AI literature. *AI and Society*. Available from https://doi.org/10.1007/s00146-021-01377-9.

Ryan, M., Isakhanyan, G., & Tekinerdogan, B. (2023). An interdisciplinary approach to artificial intelligence in agriculture. *NJAS: Impact in Agricultural and Life Sciences*, *95*(1). Available from https://doi.org/10.1080/27685241.2023.2168568.

Sapienza, S., & Vedder, A. (2021). Principle-based recommendations for big data and machine learning in food safety: The P-SAFETY model. *AI and Society*, *178*. Available from https://doi.org/10.1007/s00146-021-01282-1.

Silwal, A., Davidson, J. R., Karkee, M., Mo, C., Zhang, Q., & Lewis, K. (2017). Design, integration, and field evaluation of a robotic apple harvester. *Journal of Field Robotics*, *34*(6), 1140−1159. Available from https://doi.org/10.1002/rob.21715.

Sristy, A. (2021). *Blockchain in the food supply chain—What does the future look like?*. Walmart. https://one.walmart.com/content/globaltechindia/en_in/Tech-insights/blog/Blockchain-in-the-food-supply-chain.html.

Stock, R., & Gardezi, M. (2021). Make bloom and let wither: Biopolitics of precision agriculture at the dawn of surveillance capitalism. *Geoforum; Journal of Physical, Human, and Regional Geosciences*, *122*, 193−203. Available from https://doi.org/10.1016/j.geoforum.2021.04.014.

Sykuta, M. E. (2016). Big data in agriculture: property rights, privacy and competition in ag data services. *International Food and Agribusiness Management Review*, *19*(1030-2016-83141), 57−74.

Tanui, C. K., Benefo, E. O., Karanth, S., & Pradhan, A. K. (2022). A machine learning model for food source attribution of *Listeria monocytogenes*. *Pathogens*, *11*(6). Available from https://doi.org/10.3390/pathogens11060691.

Tanui, C. K., Karanth, S., Njage, P. M. K., Meng, J., & Pradhan, A. K. (2022). Machine learning-based predictive modeling to identify genotypic traits associated with Salmonella enterica disease endpoints in isolates from ground chicken. *LWT—Food Science and Technology*, *154*112701. Available from https://doi.org/10.1016/j.lwt.2021.112701.

Themistokleous, K. S., Sakellariou, N., & Kiossis, E. (2022). A deep learning algorithm predicts milk yield and production stage of dairy cows utilizing ultrasound echotexture analysis of the mammary gland. *Computers and Electronics in Agriculture*, *198*. Available from https://doi.org/10.1016/j.compag.2022.106992.

Tian, F. (2017). A supply chain traceability system for food safety based on HACCP, blockchain & Internet of things. In *2017 International conference on service systems and service management* (pp. 1–6). IEEE.

Tobal, A. M., & Mokhtar, S. A. (2014). Weeds identification using evolutionary artificial intelligence algorithm. *Journal of Computer Science*, *10*(8), 1355–1361. Available from https://doi.org/10.3844/jcssp.2014.1355.1361.

Tsai, Y. C., Hsu, J. T., Ding, S. T., Rustia, D. J. A., & Lin, T. te (2020). Assessment of dairy cow heat stress by monitoring drinking behaviour using an embedded imaging system. *Biosystems Engineering*, *199*, 97–108. Available from https://doi.org/10.1016/j.biosystemseng.2020.03.013.

TYTS (The Yield Technology Solutions). (2021). *Our solution*. https://www.theyield.com/our-solution.

Uddin, M., Chowdhury, A., & Kabir, M. A. (2022). Legal and ethical aspects of deploying artificial intelligence in climate-smart agriculture. *AI and Society*. Available from https://doi.org/10.1007/s00146-022-01421-2.

UKRI (UK Research and Innovation). (n.d.). *Responsible innovation*. https://www.ukri.org/about-us/policies-standards-and-data/good-research-resource-hub/responsible-innovation/.

Urbano, O., Perles, A., Pedraza, C., Rubio-Arraez, S., Castelló, M. L., Ortola, M. D., & Mercado, R. (2020). Cost-effective implementation of a temperature traceability system based on smart RFID tags and IoT services. *Sensors*, *20*(4), 1163. Available from https://doi.org/10.3390/s20041163.

Vakkuri, V., & Abrahamsson, P. (2018). The key concepts of ethics of artificial intelligence. In *2018 IEEE international conference on engineering, technology and innovation (ICE/ITMC)* (pp. 1–6). https://doi.org/10.1109/ICE.2018.8436265.

Van der Loeff, A.S., Bassi, I., Kapila, S., & Gamper, J. (2019). *AI ethics for systemic issues: A structural approach*. http://arxiv.org/abs/1911.03216.

Van der Burg, S., Kloppenburg, S., Kok, E. J., & van der Voort, M. (2021). Digital twins in agri-food: Societal and ethical themes and questions for further research. *NJAS: Impact in Agricultural and Life Sciences*, *93*(1), 98–125. Available from https://doi.org/10.1080/27685241.2021.1989269.

Vence. (2022). *Virtual fencing for cattle and livestock management system*. https://vence.io/.

Wang, G., Sun, Y., & Wang, J. (2017). Automatic image-based plant disease severity estimation using deep learning. *Computational Intelligence and Neuroscience*, *2017*. Available from https://doi.org/10.1155/2017/2917536.

Wan, P., Toudeshki, A., Tan, H., & Ehsani, R. (2018). A methodology for fresh tomato maturity detection using computer vision. *Computers and Electronics in Agriculture*, *146*, 43–50. Available from https://doi.org/10.1016/j.compag.2018.01.011.

Warner, D., Vasseur, E., Lefebvre, D. M., & Lacroix, R. (2020). A machine learning based decision aid for lameness in dairy herds using farm-based records. *Computers and Electronics in Agriculture*, *169*. Available from https://doi.org/10.1016/j.compag.2019.105193.

Whittlestone, J., Alexandrova, A., Nyrup, R., & Cave, S. (2019). The role and limits of principles in AI ethics: Towards a focus on tensions. In *AIES 2019—Proceedings of the 2019 AAAI/ACM conference on AI, ethics, and society* (pp. 195–200). https://doi.org/10.1145/3306618.3314289.

Wilson, T. & Auchard, E. (2018). *Chickens and eggs: Retailer Carrefour adopts blockchain to track fresh produce*. Reuters. https://www.reuters.com/article/us-carrefour-blockchain-ibm/chickens-and-eggs-retailer-carrefour-adopts-blockchain-to-track-fresh-produce-idUSKCN1MI162.

Zehner, N., Umstätter, C., Niederhauser, J. J., & Schick, M. (2017). System specification and validation of a noseband pressure sensor for measurement of ruminating and eating behavior in stable-fed cows. *Computers and Electronics in Agriculture*, *136*, 31−41. Available from https://doi.org/10.1016/j.compag.2017.02.021.

Zhang, H., Wei, X., Zou, T., Li, Z., & Yang, G. (2015). Agriculture Big Data: Research status, challenges and countermeasures. In *Computer and computing technologies in agriculture VIII: 8th IFIP WG 5.14 international conference*, CCTA 2014, Beijing, China, September 16−19, 2014, Revised Selected Papers 8 (pp. 137−143). Springer International Publishing.

Zhong, Y., Gao, J., Lei, Q., & Zhou, Y. (2018). A vision-based counting and recognition system for flying insects in intelligent agriculture. *Sensors (Switzerland)*, *18*(5). Available from https://doi.org/10.3390/s18051489.

Zhu, Y., Cao, Z., Lu, H., Li, Y., & Xiao, Y. (2016). In-field automatic observation of wheat heading stage using computer vision. *Biosystems Engineering*, *143*, 28−41. Available from https://doi.org/10.1016/j.biosystemseng.2015.12.015.

AI and grief: a prospective study on the ethical and psychological implications of deathbots

Belén Jiménez-Alonso[1] and Ignacio Brescó de Luna[2,3]

[1]*Open University of Catalonia, Barcelona, Spain* [2]*Autonomous University of Madrid, Madrid, Spain*
[3]*Aalborg University, Aalborg, Denmark*

Chatbots are "computer applications with artificial intelligence capable of generating a two-way conversation between a human being and a machine (robot) through a conversational interface or chat" (Ávila-Tomás et al., 2020, p. 33). Since the creation of the first-ever chatbot ELIZA—a program impersonating a psychotherapist devised in 1966 by Joseph Weizenbaum—advances in artificial intelligence (AI) have changed the way we interact with machines. In addition to their increasingly daily usage, as in the case of Apple's Siri or Google Assistant, these conversational agents are becoming a fixture in other fields, particularly in the healthcare sector (Tjiptomongsoguno et al., 2020), for instance, in psychological assessment and intervention (Fitzpatrick et al., 2017) or in accompanying people at the end of their lives to help them prepare logistically, emotionally, and even spiritually for their death (Utami et al., 2017). Following this trend, chatbots are also likely to change the way we interact with the dead in a not-too-distant future. Different projects are currently being devoted to the so-called *griefbots* or *deathbots*—chatbots based on the digital footprint left behind by the deceased through social media, emails, texting, and messaging systems—with the aim of giving the bereaved the chance to speak to their loved ones after their death. A case that received some media attention was Replika, a project developed by Eugenia Kuyda following the sudden death of her friend, Roman Mazurenko, in a car accident (Newton, 2016). Using over 8000 lines of text messages from Roman's conversations with different people, and a neural network developed at her AI start-up, Eugenia built a chatbot capable of simulating her friend's conversational style. Those who had been close to Roman found the result eerily convincing (Elder, 2020).[1] In a similar vein, the entrepreneur Marius Ursache envisaged Eterni.me in 2014, a service whereby you could develop your own digital avatar with which your descendants could interact after your death. Another example comes from the data scientist Muhammad Ahmad (2016), who is developing a messenger program that imitates his father's speech pattern so that his grandchildren can bond with him (Godfrey, 2019). More recently, Microsoft has abandoned a project to develop a conversational chatbot for a specific person based on

---

[1]Replika is now no longer a deathbot per se, but a companion bot that could be used as a friend, a mentor, or a love partner. The latter scenario brings the topic from AI and death to AI and love, and the possible connection between the two, if we understand grief as the price we pay for having loved someone (whether human or not) (see Brinkmann, 2020). The film *Her* by Spike Jonze (2013) could be a case in point (we are grateful to the reviewers for bringing this topic to our attention).

Ethics in Online AI-Based Systems. DOI: https://doi.org/10.1016/B978-0-443-18851-0.00011-1

her/his social data, despite having patented the system to do so (Abramson & Johnson, 2020). Microsoft's general manager of AI programs, Tim O'Brien (2021), recently referred to the project as disturbing.

As a new *thanatechnology* (Sofka, 1997) with a potential impact on the way we mourn, conceive of death, and relate to those who are no longer with us, deathbots mobilize different imaginaries and preconceived ideas about the limits of what we consider "authentically" human—for instance, future imaginaries conveyed in the *Black Mirror* episode, *Be Right Back* (Brooker, 2013). In parallel with the burgeoning of these AI-powered technologies, a growing number of studies are focusing on their possible ethical and psychological implications (Buben, 2015; Öhman & Floridi, 2018; Savin-Baden, 2022; Stokes, 2021). However, with a few exceptions (Galvão et al., 2021), little work has addressed this question empirically. To bridge this gap, this chapter includes a set of in-depth interviews with different mourners who are asked how they imagine deathbots—still in their developmental stage—might affect their grieving experience. Following on from previous theoretical work (Jiménez-Alonso & Brescó, 2022a, 2023), the ethical and psychological implications of deathbots are discussed by considering the concept of mediation (Vygotsky, 1978), as well as some recent approaches to grief, such as the notion of *continuing bonds* (Klass et al., 1996).

Continuing bonds, technological mediation, and ethical implications in grief

If, as is commonly understood, the grieving process ends once we overcome the loss and say goodbye to our loved ones by letting go of the ties that bind us to them,[2] then it seems reasonable for alarm bells to ring in the face of an artifact that perpetuates the continuation of these bonds by means of an AI-powered chatbot. Conversely, a growing trend in grief studies questions the need to break the affective bonds with the departed. According to the *continuing bonds* model (Klass et al., 1996), rather than severing the attachments with the deceased and *moving on*, the grieving process implies *moving with* an ongoing connection to those no longer living. This sense of connection may appear in the form of dreams, invoking the example of the deceased as a standard of self-judgment or, more generally, through an inner dialog by imagining their responses to one's actions and beliefs. According to Despret (2015), it is not uncommon to address our dead *as if* they were present. Josephs (1998) offers us various examples of such relationships mediated by cemetery visits. He describes the case of a widow asking her late husband about the arrangements for his grave and receiving an answer through a voice she herself lends to her husband. Mourners do not seem to be confused by the question of whether there is real communication with their loved ones. As Despret (2015) notes, such "as if" acts as a kind of operator that leaves open the possibility of such encounters between the living and the dead. However, the crucial point is that these imagined dialogs work, that is, they have an effect on the bereaved in giving them the "real" feeling of being united with their loved ones, thus helping them regulate their emotions in real-life situations (Josephs, 1998). As Despret (2015) argues, regardless of the illusion of reality attributed to these interactions, the dead exist not just in the memories of the living, but also through the ongoing impact the former have on the latter.

---

[2]This popular assumption is strongly influenced by the Freudian *letting-go* approach.

However, mourners' unique response to loss, including the way in which they maintain their bonds with the deceased, is not an exclusively intrapsychic phenomenon (Neimeyer et al., 2014), but one elaborated together with other people in specific sociocultural contexts—with social norms on how people should grieve—and mediated by the technological artifacts available in any given historical moment. According to cultural psychology (Brescó et al., 2019), human action is characterized by an irreducible tension between agents and mediational cultural tools that simultaneously constrain and enable experience (Wertsch, 1998). Mediation in mourning involves the transformation of a relationship through the intervention of a new element, whether through a photograph (Jiménez-Alonso & Brescó, 2022b), a memorial (Brescó & Wagoner, 2019), or any rite of passage. Graves, for example, may have a mediating function by facilitating an imagined dialog *as if* the dead person were present. In a similar vein, Walter et al. (2012) points out that technologies not only mediate how we communicate with the living, but also the way we communicate with the dead, thus making them socially present. He mentions the telephone as a technology that enabled tele-presence—a nonphysical copresence—between interlocutors, comparable to that between the living and the dead. As a socially validated technology, the phone can be used to leave text and voice messages for the deceased or to hold one-way conversations with them, as in the case of the *wind phone* in Japan (see Van Dyke, 2022).[3] The current digital world, with the explosion of different social network services (SNSs)—ranging from online memorials to platforms entrusted with managing the digital legacy of the dead—has expanded the possibilities of communication and mourning (Dilmaç, 2018), including the ways of maintaining the continuing bonds with our loved ones (Klass, 2018).

A new aspect of digital technologies is that they enable a sense of permanent presence of the dead. As a result, mourning has become an everyday practice in the online world, enabling a more personalized way of expressing and sharing the grieving experience and thus contributing to legitimating the practice of interacting with the dead (Walter et al., 2012). Both the Internet and deathbots imply an obvious temporal and spatial expansion, as they provide a more direct way to communicate with the dead anywhere and at any time (Brubaker et al., 2013). Periodic updates on Facebook, or continuous chatting with a deathbot, may create a feeling that the dead are listening behind the screen and keeping up with the living's latest news, as reported by some of Kasket's (2012b) participants. This in turn contributes to a sense of copresence that, "for members of the 'Facebook generation', may feel as close as those who are present in an embodied way" (Kasket, 2012a, p. 62). However, while copresence between mourners and loved ones is socially shared within the same virtual space in the case of SNSs —leading to what Kasket (2012a) calls *communal bonds*—with deathbots, this copresence is confined to the private conversational space between the mourner and the deceased person from whom a response is expected. A crucial difference between chatting with the bot of your loved one and posting messages on her Facebook profile—or holding an imaginal dialog by her graveside—is the two-way communication enabled in the former case. With deathbots, not only does the conversation no longer depend on the survivor's initiative for it to take place—as the bot might initiate a conversation autonomously, thus requesting an answer from the bereaved—but also the latter expects, in turn, a reply from the dead addressee. This more active presence of the deceased (Savin-Baden, 2022), made tangible in the form of written or oral

---

[3] A phone box built by Itaru Sasaki to cope with his cousin's death, but eventually opened to the public following the 2011 tsunami.

messages, might generate a certain illusion of reality—the feeling of a "thingness," to use Kasket's (2012b) expression—for the bereaved, thus raising a number of potential ethical issues examined by the growing literature on AI-powered afterlife technology.

Ethical concerns are being raised about the deceased's dignity, especially about the management of their digital afterlife memory through the new digital technologies, in particular those powered by AI. Along these lines, Buben (2015) and Stokes (2021) take a critical stance against these technologies by differentiating between *replacement* and *recollection* in the memorization of the dead. While in recollection, we remember the deceased as unique and irreplaceable beings and are thus aware of their irremediable loss, in replacement we seek to overcome that loss by using the deceased's memory as a resource for our own emotional comfort, thus degrading the dead to our own needs. For their part, Öhman and Floridi (2018) suggest approaching the ethical debate on digital remains by seeking "inspiration from frameworks that regulate commercial usage of organic human remains" (p. 319). To prevent the commercialization of the dead by what they referred to as the Digital Afterlife Industry (DAI), these authors argue that digital remains "should be seen as the remains of an informational human body, that is, not merely regarded as a chattel or an estate, but as something constitutive of one's personhood" (p. 319). The issue regarding the integrity of the deceased's memory is further addressed by Elder (2020). According to this author, in drawing on the dead's total digital fingerprint, deathbots might disclose the dead's multiple public selves shown in conversations with different people in different digital fora, thereby revealing some facets of our loved ones they did not want us to discover—facets that perhaps we would rather not know either.

The future use of deathbots also poses important ethical questions in relation to the bereaved and, more specifically, regarding the potential impact these technologies may have on the grieving process. These questions lead us to consider the difference between what is technologically possible and what is therapeutically beneficial for those who have lost someone, thus bringing to the fore the different—and not necessarily convergent—logics involved in the use of these new technologies. As Öhman and Floridi (2018) remind us, DAI is, after all, a profit-seeking industry based on the use of digital remains and the monetization of the digital afterlife. The need to obtain an economic return on these remains by encouraging posthumous interaction might have serious implications. We may well imagine different strategies to keep mourners hooked, for instance by sending unsolicited messages or updates from their loved ones whenever users are inactive. To what degree could a grieving person ignore or refuse to answer these messages? This could become more serious depending on the cognitive and affective capacity of the mourners, as well as on their social support. For instance, Ahmad (2016) wonders how children would respond to such an interaction and poses the hypothetical case of a child growing up interacting with a deathbot of a dead relative. Would the child be capable of differentiating between simulation and reality? Along these lines, Stokes (2021) highlights the danger of mistaking the bot for the departed, thus moving—we might add—from the *as if* to the *as is*. Finally, Lindemann (2022) warns us about the impact deathbots might have on the autonomy of the bereaved once they have developed an emotional dependency on them and the risk of suffering a *second loss* (Bassett, 2018) in the event of an unexpected technical failure.

In sum, deathbots are just another technological artifact—endowed with their own specific features—to be added to the wide range of "old" and "new" technologies mediating our experience of grief and the way we manage our continuing bonds with the dead. Yet, while having an internal conversation with the deceased by the graveside—or sending them messages through an answering machine or a digital memorial—is quite common, the prospect of holding a reciprocal conversation with an AI-powered artifact might affect our sense of connection to the departed and thereby our grieving process in general. This is not to say that mediation equals causal determination. According to cultural psychology, there is always a distributed agency between individuals and the possibilities of action offered by the technologies available to them in a given sociocultural context (Wertsch, 1998). Accordingly, following Vallès-Peris and Domènech (2020), we propose a nonessentialist approach to this AI technology, emphasizing its relational and contextual use, while addressing deathbots' mediating role not for what they supposedly are, but for what they allow us to do. In other words, the potential effect of deathbots cannot be studied by focusing solely on the technology itself; it also calls for consideration of the network of distributed agency in which their specific use by each mourner would be embedded. This leads us to our empirical study where we analyze the potential psychological and ethical implications of deathbots as imagined by different mourners.

Examining the imagined use of deathbots: a prospective study

This study is part of a broader project on bioethics, funded by the *Víctor Grífols i Lucas* Foundation, which aims to examine the possible impact of AI technologies on the grieving process. The results presented as follows include a series of in-depth interviews with three mourners.[4] Compared to semistructured interviews, in-depth interviews allow participants to talk freely and in greater detail about specific predetermined topics (Scanlan, 2020). The three participants are the following:

- Participant 1: a female in her early 50s who lost her father to COVID-19 about a year ago.
- Participant 2: a female in her early 30s who lost her father about 20 years ago and her mother about three years ago.
- Participant 3: a male in his 30s, who lost his twin brother about 15 years ago.

Due to the pandemic conditions, the interviews—lasting approximately one hour—were conducted by the first author (FA) using the Zoom video platform. All interviews were audio-recorded with the prior informed consent of the participants, whose privacy and confidentiality were maintained throughout the process. After transcribing the interviews, a qualitative analysis was carried out, insofar as the object of interest focuses on the construction of the mourners' meanings when imagining the potential implications of deathbots in bereavement. Qualitative methods are particularly suitable for the so-called *sensitive research* (Liamputtong, 2011) in that they focus on the meanings, interpretations, and subjective experiences of vulnerable participants, such as persons who are grieving.

All interviews commenced with a brief introduction of deathbots by the FA, followed by this general question: "Do you think bots can help people to cope with grief?" Participants were then placed in the hypothetical scenario of imagining what the impact of interacting with a bot based on the digital footprint of their deceased loved one might be.

---

[4]In addition to mourners, this project includes interviews with computer engineers, designers, psychologists, health professionals, and experts in bioethics and law.

Phones, Internet, and social networks: a naturalized copresence?

When questioned about deathbots, the interviewees spontaneously discussed the role of other technologies—from the telephone to Internet forums, including social networks such as Facebook—and how they somehow enable them to stay connected with their loved ones. The participants considered the effects of these technologies on bereavement and how they differed from deathbots. Their answers point to several key questions that have a direct bearing on the "*as if*" feeling mentioned earlier. Specifically, participants talk about breaks from what they consider to be "reality," as well as the game we can consciously play to engage in a dialog with our deceased loved ones. This is how Participant 1 expresses some of the effects that technological tools, such as telephones, provoke in mourners[5]:

> PARTICIPANT 1: Well, I think it's crazy to use someone's digital footprint to speak to a dead person. I think... I don't know... Lots of people speak to people they know who have died, without getting any reply obviously, and I think it would be a bit scary. It happens to me, for example: my dad recorded a message on the answering machine of the landline in his house, right? And it freaks a lot of callers out. I love hearing it. Before, when he was alive and I called and the answering machine message came on, I went plonk! [hanging up], and now I stay on the line, listening to it, but loads of people have said "Shit, your dad's voice comes on," and I say, "yes, that's right." And lots of people tell my mum to take it off. I don't want her to take it off, but for lots of people, hearing the voice of someone who isn't there anymore, it really freaks them out, and a voice that always says the same thing: "We're not at home right now..." When I imagine having a conversation [referring to deathbots]..., I don't know, we'd probably all end up going nuts.

> FA: Why?

> PARTICIPANT 1: Well, it's like: "Bloody hell! He died and he isn't dead, he's talking, right? I don't know, to be honest it's really hard for me to imagine..." [LAUGHS]

This participant shows a clear initial reluctance to have a two-way conversation with her dead father using his digital footprint. However, as we can infer from her words, it is not the possibility of a break with reality—a dead person who reappears or speaks again after the irreversibility of his death—that is key here, but the meaning that each person gives to the *as-if* world and the desire (or lack thereof) to engage in an imaginary dialog with the dead person at a given moment. Participant 1, who previously did not stay on the line to listen to the answering machine when her father was alive, now wants to retain that instant in which she hears the voice of her late father.

Participant 3 provides us with another example of copresence mediated through a more recent technology, the Internet. Specifically, we can observe how the asynchronicity of the posthumous digital footprint left by his deceased twin brother on Internet forums—in this case, not sought by either his brother or the participant himself—brings to the fore the continuing bonds between the two siblings. This gives rise to an ambivalent feeling in this participant: sadness at the absence of his dead brother, but joy at recovering a part of him by re-reading the messages he left years ago.

---

[5]Transcriptions presented in this chapter are translations from the Spanish original.

> PARTICIPANT 3: Look, I'm going to tell you that at that time we were on a sports forum, and not long ago I found messages of his from when he would write, and I said, "wow." [. . .] It'll be fifteen years [since his death] this August, and I found his messages this year, and they stirred up good and bad feelings. I mean, there was so much life there, because his messages were always filled with happiness. And I embraced them thinking, "what an inspiration my brother was for me, always happy."

For her part, Participant 2 points to social networks, specifically Facebook memorial pages, as a technology that allows her to express and cope with her loss, to commemorate the memory of her dead parents and to keep communicating with them, thus maintaining her continuing bonds with her loved ones:

> PARTICIPANT 2: Sometimes I wrote for them in the second person, sometimes in the third person and . . . no, not in the first person, well, yes, I did speak, about my feelings and all, but I didn't address it to myself, I didn't write a letter to myself. [. . .] Sometimes I would address it to others to make it just a memory, while other times I wrote to them and *it was like* a way. . . as if I subconsciously thought that it was going to get to my loved ones, that behind the screen there's a space and in that space they can read it. It's *somehow*, a bit. . ., I don't know, I didn't directly think that, but it's the way I've analyzed it now, and also yes, I believed that it might *somehow* reach them, that they were reading it too and, well, maybe they are reading it, we don't actually know what there is behind death (*italics added*).

These words clearly reflect the need for mourners to maintain a certain bond with their dead. In this case, Facebook seems to be mediating this ongoing connection, not only between this participant and her dead parents but also with other users of these commemorative pages, thus resulting in what Kasket (2012a, 2012b) calls *communal bonds*. This participant explains that sometimes she wrote on these pages *about* her parents (third person), while at other times she would write directly *to* her parents (first person), addressing them *as if* they could *somehow* be reading her messages behind the screen (see Kasket, 2012a, 2012b), thus experiencing a sense of copresence with her loved ones. Note the expressions used by this participant that come to highlight the question of "as if": It is not that her parents were listening to her, but *as if* they could do so. This points to the semantic artifice described by Despret (2015) whereby mourners can manage and leave open various possibilities, i.e., maybe, maybe not.

Deathbots: the expectation of response and the authenticity of the relationship

When commenting on the differences between previous technologies (phone and the Internet) and deathbots, participants highlighted certain characteristics of the latter which, in their opinion, could transform their relationship with the dead. All participants underlined that deathbots, unlike other technologies, seem to enable a "two-way" communication between the living and the dead. They all emphasize that what is unique about these bots—which seem to somehow embody the deceased loved one—is the generation of a visible, immediate, and tangible response, through written text, and, above all, the expectation of such a response on the part of the mourner. In that regard, all participants reflect on the motive that would lead a person to seek answers from interaction with a bot. This is what Participant 3 says on this matter:

> PARTICIPANT 3: If I had the chance to use it, I think I would, and I would even like it. Then I'll have the dilemma of, will I use it a lot, or not much? [...] The worst thing would be write to him [to his brother] every day. I think in that case a part of you would say no, like, the person answering me isn't real [so to speak]... I've grieved for him now for fifteen years and I've accepted that he isn't here. I think it would be different if my other brother were to die suddenly now—touch wood that doesn't happen —, and with all the digital information I have on him, I suppose it would be easier to make a profile of him, but I'd still be cautious. It would be like, "this isn't your real brother," I mean, they've created a super profile that would more or less answer you, or they've taken part of his personality. I might be wary. I'd be like, "be careful, no. First get over part of the bereavement, accept it."

While initially being open to the use of deathbots, this participant immediately adopts a cautious attitude toward this technology. This is expressed through a reflection that, in a way, echoes the aforementioned difference between *replacement* and *recollection* (Buben, 2015; Stokes, 2021). If we compare this paragraph with Participant 3's previous words about his brother's messages found on the Internet, we can see how the use of deathbots would no longer be motivated by the desire to remember the deceased and share his memory, but by an attempt to recreate his former relationship with his brother, above all, in the expectation of obtaining a response from him. Secondly, we see how, unlike the aforementioned examples of other technologies, deathbots no longer afford a copresence between the mourner and the deceased that is socially shared within the same virtual space as in the case of Facebook (Brubaker et al., 2013). What they offer instead is a copresence limited to the private conversational space between two parties. Finally, this participant wonders whether the "use" of the bot of his twin brother—note the instrumental connotation behind this expression—is compatible with acceptance of the irreversibility of his death. In that respect, as we shall see, all participants seem to share the idea that it is necessary to "accept" the death of their loved ones to be able to "close a chapter" in their lives (as Participant 1 remarks) or to work through the loss (Participants 2 and 3). Accordingly, they all argue that if the desire to continue to have a two-way conversation with the deceased is prompted by a lack of acceptance of their death, then the use of deathbots would be detrimental to the bereaved.

On the other hand, all participants emphasize that the possibility of a two-way conversation "triggered" by the bot and enabled by its design and interface—for example, through unsolicited messages or notifications—differs from the internal dialog they may have with their deceased loved ones. This is how Participant 2 puts it:

> PARTICIPANT 2: I... don't expect replies. If I get something, I'd think it's a sign from him or her. I might wonder, "jeez, are they sending me this message?," but it's not something I'm looking for; it's something that makes you go, "what a coincidence," right? [...] For example, I found a job on my mum's birthday and I thought, "wow, could that be a sign?," you know? but it's not that I was looking for a response from my mum [...], or establishing a conversation with her. No, I keep her more on a spiritual level, a spiritual connection but not a physical connection.

And further on:

> PARTICIPANT 2: ... this way [using deathbots], there's no thought, no introspection, no, you just talk to this person. Maybe you could somehow be introspective and what happened to that woman [Eugenia Kuyda] might happen to you, and you might say, "it's good for me because it makes me realize that he's really gone." The issue would be that on Facebook, it's more your own reasoning, right? or when you post a photo, or a memorial. But in chats, I don't know, it would be like a fake of your loved one, and obviously of the relationship.

In the first paragraph, we see a typical example of an imaginary dialog between the participant and her deceased mother, a dialog mediated by different signs that, again, would open up an *as-if* scenario. In the second paragraph, this participant compares two ways of holding this imagined dialog with the deceased, in this case mediated by different technologies. On the one hand, she argues that the use of technologies such as photography, Facebook, or memorials provides her with more agency and control when it comes to regulating her emotions and that imaginary dialog—"on Facebook it's more your own reasoning." Conversely, dialog through deathbots seems to leave less room for introspection and emotional self-regulation—"there's no thought, no introspection; you just talk to this person."

In a similar vein, Participant 1 compares the potential role of deathbots with that of another mediational tool, namely, writing a diary. In the following extract, she explains that following her father's death she started writing a mourning diary. However, this participant denies having a two-way dialog with her father through the diary. As in the case of the widow lending her voice to her late husband described earlier (see Josephs, 1998), this participant stresses that it is a dialog with herself, even though she pretends to be addressing her father through her diary. In this sense, Participant 1 problematizes the very possibility of holding a two-way conversation with her deceased father through his bot, as this would be tantamount to questioning the irreversibility of his death, thus resulting in a relationship she perceives as false:

> PARTICIPANT 1: I write to myself every day. I keep a gratitude journal and I write every day and I've thanked him a thousand times... But that's something else. It's about me with myself if you know what I mean [...]. Because there are things that are very hard to share and are "about me with myself" because they can't be "about him with me." Things that you've experienced with him [...]. I don't know, I can't describe it to you, but it's like he died and, and it's the end of a chapter, because that's the way it is. I mean, now we're closing chapters, paying the capital gains tax, inheritance tax..., you know? And chatting to..., I honestly can't see it. I can't see it. I can't see it because on top of everything, it's false.

When thinking about the differences between deathbots and other technologies, participants tend to underline the importance of the *as-if* feeling in the dialog between the living and the dead. While being fully aware of this, the inclusion of this semantic artifice—to use Despret's (2015) terminology—seems to allow them to handle several hypothetical scenarios without ruling out any of them. However, the "illusion of reality," which is allegedly what deathbots were designed to provide, might lead mourners to use this technology from a position of "as is" rather than "as if."[6] Hence the

---

[6]This is not to say that there are responses that are "real" and those that are not; rather, the design of the device might make the mourner forget that the responses are mediated by a third agent. As Despret (2015) points out, it is the relationship that the living establish with the dead that endows the latter with an ontological status. This is why users' beliefs (religious, spiritual, etc.) and their sociocultural context are so vitally important when it comes to analyzing mourners' subjective experiences with the bot.

agreement among the participants as to the risk deathbots could pose for mourners seeking "real" answers from their dead. This connects with the disturbing effect caused by those technologies that seem all too human (Mori, 1970). In this sense, it is possible that a strong resemblance between the bot and the deceased's speech pattern could explain a certain rejection on the part of the interviewees toward the digital replica of their loved ones. Nonetheless, it is worth noting that all participants stress their awareness that the bot is a digital "copy" of the deceased. Even if it were a perfect copy—if the bot were to provide answers that were similar to what their loved ones could give—they are aware that it is not "really" their loved one who is talking to them. This is shown in the following exchange between the FA and Participant 2:

> PARTICIPANT 2: It's a machine, a program. It might bear a resemblance, they might bear a close resemblance in the responses, but they'll never ever be the person. I don't know whether you've seen the episode of...

> FA: Black Mirror.

> PARTICIPANT 2: *Black Mirror*, yes, that episode where she buys a robot that's identical to her husband who recently died, but it just isn't her husband. That's the thing. At the beginning you do it, you're really excited... I think I would be the same as that woman. [...] At the end of the day, it's an artificial program that doesn't really know the person, doesn't know what he's like. It might have a slight idea, but then I get a slight idea of a slight idea. It's like that Chinese whispers game.

> FA: If I understand you correctly, you think that the chatbot introduces the idea of the idea, something that has been reconstructed by this artificial intelligence. Is that more or less it?

> PARTICIPANT 2: Exactly, that's basically it [...] When they go away, you have an ideal of that person, or those people in my case, and I would almost rather keep that ideal perception, how I think they would answer me... And well, I think that there will nevertheless always be answers that don't ring true, and because they don't ring true, it might throw the image a little off, and I don't know...

As we can see, the recreation of the relationship we once had with our loved ones through a digital copy may affect our memory of the latter as well as our very sense of loss, which could be altered by the lack of authenticity or by the "disappointment"—to use the participant's words—of what we know to be a mere simulation. In any case, what seems to be significant for the participants is not so much the deathbot's capacity to replicate the identity of the deceased, but the experience of the bereaved, her or his particular personalization of the bot through the way that it is used, and the kind of attributes projected onto it. In other words, what seems to be key is not so much reality itself, but the illusion of reality and what we do with that illusion. Perhaps, we may hypothesize, the very fact of knowing that deathbots cannot replace the identity of the deceased is what enables users to engage in this simulation game, acting *as if* they were talking to their loved ones.

To conclude, it is worth noting that, while most participants put themselves in the shoes of the mourner, they do not seem to naturally confront the hypothetical scenario of having a bot made of themselves. This is how Participant 1 responds when asked about this hypothetical scenario:

> FA: How would you feel about leaving your digital footprint so that a bot can be made of you and your loved ones can speak to you?

> PARTICIPANT 1: Well I still think the same. [LAUGHS] Right now... Then... look, I'm pretty old now and life... I don't know. I've said it out loud many times: "I'm not going to go there!," you know? I'd like to say that right now I can't see it happening. Maybe if you ask me in twenty years I might say "listen, why not," you know? But at the moment I can't see it happening because of what I'm telling you, the lack of humanity.

These words reflect a tension between utopian and dystopian imaginaries, as well as concerns about issues such as the replacement of humans by robots. However, we can also identify another interesting aspect connected to this questioning of human nature: the participant's realization that not only is the use of these technological artifacts related to the artifacts themselves, but also to a sociocultural context where the very concept of mourning and death could change in parallel with the rapid evolution of technologies, although she finds it difficult to see things differently at the moment. These reflections should make us realize that the future use of deathbots is embedded in a web of relationships shaped by numerous elements that go beyond the functionalities of the technologies themselves.

Discussing the potential ethical risks of deathbots

Digital technologies are changing the way we experience grief by providing new spaces for expressing ongoing bonds with our loved ones. While these *thanatechnologies* are not necessary in order to maintain a bond with the dead—as this can be done in many other ways—what changes is the type of mediation they bring to the grieving experience. In the interviews, we have seen manifold ways in which participants engage in different types of imaginary dialogs with their loved ones. Writing a grief diary, listening to the voice of their late father on the answering machine, leaving messages on Facebook, or simply having an inner dialog with the deceased loved one are ways in which participants seem to experience a more or less naturalized copresence in relation to their dead, for instance, *as if* they were somehow listening behind the screen (Participant 2). However, the two-way communication enabled by deathbots might change the way in which mourners imagine such a dialog. More specifically, it seems that receipt of a visible, immediate, and tangible response through the bot leaves the bereaved with less room and agency to imagine and manage these dialogs—as Participant 2 states, "on Facebook it's more your reasoning." Even knowing that it is a mere digital copy, the bot's responses based on the loved one's digital footprint make participants perceive this dialog as false (Participant 1), inauthentic, and disappointing (Participant 2), as something that could alter the memory of their dead. Hence our participants' reluctance to use this technology on the grounds that they prefer to keep the memory—or the ideal (Participant 2)—of their loved ones. We can also observe some caution in other participants who point to the risk of mistaking the bot for the deceased, such as when Participant 3 says he would have to remind himself that the bot is not his brother answering him.

It is interesting to note the contrast between participants' caution and reluctance toward a yet unknown technology, such as deathbots, and the naturality with which they incorporate the use of more familiar technologies into their grieving process, ranging from diaries to the Internet. In that regard, unlike other prospective studies, it is difficult to infer from the interviews the potential beneficial ethical opportunities that deathbots could bring in the future. Some of these possible advantages are mentioned in the study conducted by Galvão et al. (2021) on users' perceptions toward different forms of digital immortality, including that produced by the deceased's digital footprint. In line with the initial positive attitude toward deathbots shown by Participant 3, interviewees in that study highlighted the advantage of keeping their loved ones around, in a virtual state, for future interaction. They also found digital immortality "an interesting opportunity to deal with grieving pain [and] to keep alive the memory of the deceased loved ones" (p. 18). In discussing certain AI technologies, including Replika—the aforementioned project developed by Eugenia Kuyda—and their impact on spiritual health, Trothen (2022) highlights other benefits such as not being afraid of being judged by others about the things we say to the deathbot or being able to communicate with our loved ones at any time and in a personalized way. Yet, such immediate and personalized interaction with the bot could also pose potential ethical risks depending on how this technology is used.

As pointed out earlier, unlike other technologies, deathbots could generate the expectation of an immediate response enabled by the design of their conversational interface. As the participants themselves underline, a crucial ethical risk is the potential emotional bond established with a bot that is designed to create the illusion of conversing with the loved one. As Lindemann (2022) points out, deathbots could function as an external continuing bond whereby mourners would regulate their emotions, thus affecting the users' autonomy due to their dependence on—and their fear of losing—this bond. This brings us back to Buben's (2015) and Stokes' (2021) concerns regarding the deceased's replacement by the bot, which might result in the need to perpetuate the communication with it to ensure that the dead stay around and avoid losing them for a second time. This, in turn, raises ethical issues having to do with deception (Wangmo et al., 2019). On this, some authors (Sparrow, 2002) are against nurturing emotional bonds through robot simulacra. From a Kantian standpoint, Huber et al. (2016) argue that technology that incites deception is not acceptable. Weber-Guskar (2021) differentiates between other-deception and willful self-deception, where the latter would reduce users' perceived lack of authenticity regarding the bot. Along these lines, Elder (2020) believes that interaction with deathbots does not require large doses of illusion of reality to generate an emotional response on the part of the mourners, as they might already be predisposed to finding this interaction significant. Nonetheless, as we have shown, mourners might be predisposed to engage in an imaginary dialog with the dead, *as if* they were present, through many ways other than bots (Despret, 2015). A bird perching on the window frame or an old message left by their late father on the answering machine or posted on Facebook might be felt as signs enabling the possibility for mourners to communicate with the dead.

As for the personalized interaction deathbots seem to support, we may wonder about the extent to which this AI-based technology might promote a more individualistic way of mourning. If, as Walter et al. (2012) point out, SNSs have de-sequestered death from the private sphere by bringing it into the everyday and communal space of the Internet, will deathbots contribute to returning death and mourning to the confines of the private domain? To what extent could this result in a self-centered way of relating to and appropriating the deceased's memory through exclusive, private, and unshared communication with the dead? Or as Elder (2020) puts it, "will deathbots induce us to interact with them, both in ways that preclude forming new attachments [...] and in ways that keep us turning to the deceased for support when we ought to be reaching out to others in our social network"? (p. 76). The latter question leads us to the challenge posed by the commercial and corporate interests behind this technology, as denounced by Öhman and Floridi (2018), and the resulting strategies designed to make mourners "feel obligated to spend time with and care for their avatar, perhaps to the neglect of human friendships" (Trothen, 2022, p. 9).

Lastly, another interesting result of the interviews is the participants' tendency to imagine death-bots solely as potential future users, thus leaving questions regarding the memory of the deceased—including the hypothetical scenario wherein other people might interact with a bot cre-ated out of the participants' own digital footprint—undiscussed. This raises ethical concerns about the kind of digital surrogates we will leave behind and the extent to which our digital lives may affect the way we will be remembered after death.[7] One specific concern addressed by afterlife ethics is the risk of turning the deceased—and the relationship with them—into an object to be used for our own emotional needs, thus interacting with the bot as a "diminished other" (Bosch et al., 2022). In reflecting on the ontological status of artificial companions in AI systems, such as deathbots, Bosch et al. (2022) criticize the "one-way relationship with a serviceable other designed to fulfil the user and intended to become a docile self-object, which is not a real other" (p. 38). This potential commodification of the relationship leads us to wonder about the extent to which the appropriation of the dead through the private use of their digital copy could eventually affect our sense of loss. Brinkmann's (2018) answer to that question is that, if continuing bonds are under-stood as the ongoing connection that we have with the deceased, we would be impoverishing that bond by turning the other into something that only has meaning in relation to ourselves, as if our loved one's death only mattered because of its effect on us. As this author reminds us, "grief is not just about the fact that *I lose someone*, but also about the more fundamental fact that *someone no longer exists*" (Brinkmann, 2018, p. 182, italics in the original).

In sum, the results of the interviews touch upon various psychological and ethical questions regarding AI-powered programs based on the deceased's digital footprint, bringing to the fore cer-tain qualms the participants have in relation to the possibilities deathbots might afford. These results can be taken as preliminary ideas to be considered in future developments in this area, including measures implemented in respect of the deathbots' system design aimed at addressing some of their ethical risks, something we will tentatively address in the conclusions section. However, it should be borne in mind that, like other prospective works on the use of thanatechnolo-gies (Galvão et al., 2021), one of the main limitations of this study lies in the fact that the partici-pants only reflect on what they imagine the use of deathbots would be like, not on what their experience of deathbots has been. As such, these results are greatly influenced by the future ima-ginaries that this type of technology sparks in society and are therefore subject to change depending on each historical context. Thus, while the role of recent technologies, such as SNSs, in relation to mourning would have raised suspicions in the early days of the Internet, we have seen how the feel-ing of copresence of the dead when directly addressing them on Facebook seems quite naturalized today—it might even feel as close as a face-to-face interaction for digital natives, according to

---

[7]This scenario is leading some people to take premortem decisions in order to manage their posthumous data and thus their postdeath digital presence, for instance, digital wills with instructions specifying what is to be done with their digi-tal legacy (Savin-Baden, 2022).

some authors (Walter et al., 2012). This leads us to underline, from a nonessentialist view of technology (Vallès-Peris & Domènech, 2020), the importance of the mourners' experience, and their particular personalization of the bot through the way that it is used in each specific context. This in turn compels us to be cautious about expecting a strictly technical solution to prevent and mitigate the potential ethical risks of these new digital technologies.

Conclusions: imagining the future of deathbots

According to Savin-Baden (2022), "digital immortals are designed using human decision criteria, and therefore, ethical behaviour needs to be 'designed-in' to them" (p. 54). Different proposals can be found in the literature aimed at addressing the major ethical concerns the use of deathbots poses both for mourners and the deceased. Drawing on Buben's (2015) notion of *replacement*, and the subsequent risk of mistaking the bot for the departed, Stokes (2021) suggests introducing glitches in the deathbots' code so that users can be reminded that they are chatting with an AI-based technology. In her recent study, Lindemann (2022) proposes a normative framework for the distribution and usage of deathbots aimed at preventing this technology from compromising users' autonomy throughout the grieving process. According to this author, the implementation of deathbots should be "legally classified as a potential medical device for treatment of Prolonged Grief Disorder"[8] (p. 65) and used under psychological supervision. According to Öhman and Floridi (2018), the development of deathbots should involve an ethical commitment by the Digital Afterlife Industry, guaranteeing that (1) people will be informed on how their posthumous data will be used and displayed after their death; (2) the mimicked person will not be "depicted radically differently from the bot that they originally signed up for" (p. 320); and (3) bots will be made out of the person giving her/his informed consent, and not from other people, such as friends or deceased relatives. Lastly, in their study on digital immortality, Galvão et al. (2021) highlight that all participants agreed on the following ethical requirements:

> "It is necessary for the system to be flexible, capable of changing over time, respect the memory of deceased ones, allow interactions between the virtualized dead and the living, and especially that it aligns and has respect for human values, both of its virtualized entities and its users" (p. 18).

As we can see, AI applied to grief is a phenomenon that is as new as it is ethically complex. Such complexity goes beyond the application of the technology itself as it embraces a series of broader questions impelling us to think about the meaning we give to grief, death, our relationship with those no longer with us, and the values that should govern the increasingly rapid incorporation of new technologies into the field of bereavement and care.[9] Paying attention to the context in which deathbots are used will thus be key to better understanding the mediations these artifacts

---

[8]Recently incorporated into the Diagnostic and Statistical Manual of Mental Disorders (DSM), PGD is characterized by the "intense and persistent grief that causes problems and interferes with daily life" (Appelbaum & Yousif, 2022).

[9]Beyond the originally intended use of deathbots, future advances in AI and, more specifically, the prospect of incorporating digital copies of oneself into our living days, may eventually blur the fine line between digital afterlife and digital self enhancement. The posthumanism discussion over this foreseeable scenario is beyond the scope of this chapter (we thank the reviewers for bringing this observation to our attention).

may afford, as well as the ethical and social issues they might generate. Until that moment arrives—perhaps sooner than expected—we will have to make do with prospective studies whereby we can critically reflect on how we are incorporating AI technology into both our lives and deaths.

Acknowledgments

Authors received financial support through a grant on bioethics by the *Victor Grifols i Lucas Foundation* (Spain). The second author also received financial support through the *Culture of Grief Project*, funded by the *Obel Family Foundation* (Denmark).

References

Abramson, D.I., & Johnson, J. (2020). *Creating a conversational chatbot of a specific person* (U.S. Patent No. 10,853,717 B2) U.S. Patent and Trademark Office.

Ahmad, M.A. (2016). After death: Big data and the promise of resurrection by proxy. In *Proceedings of the 2016 CHI conference extended abstracts on human factors in computing systems*, San Jose, CA, 7−12 May (pp. 397−408). Association for Computing Machinery.

Appelbaum, P., & Yousif, L. (2022). *Prolonged grief disorder*. American Psychiatric Association. https://psychiatry.org/patients-families/prolonged-grief-disorder

Ávila-Tomás, F. J., Olano-Espinosa, E., Minué-Lorenzo, C., Martínez-Suberbiola, F. J., Matilla-Pardo, B., Serrano-Serrano, M. E., & Güeto-Rubio, M. V.Grupo Dej@lo. (2020). Dejal@Bot: Un chatbot aplicable en el tratamiento de la deshabituación tabáquica. *Revista de Investigación y Educación en Ciencias de la Salud*, 5(1), 33−41. Available from https://doi.org/10.37536/RIECS.2020.5.1.196.

Bassett, D. (2018). Digital afterlives: From social media platforms to thanabots and beyond. In C. Tandy (Ed.), *Death and anti-death, Vol. 16: 200 Years after Frankenstein*. Ria University Press.

Bosch, M., Fernandez-Borsot, G., Miró, I., Comas, A., & Figa Vaello, J. (2022). Evolving friendship? Essential changes, from social networks to artificial companions. *Social Network Analysis and Mining*, 12 (39). Available from https://doi.org/10.1007/s13278-022-00864-1, Advance online publication.

Brescó, I., & Wagoner, B. (2019). Memory, mourning, and memorials. In K. Murakami, T. Kono, T. Zittoun, & J. Cresswell (Eds.), *Ethos of theorizing: Peer reviewed proceedings for the international society for theoretical psychology* (pp. 222−233). Captus Press.

Brescó, I., Roncancio, M., Branco, A., & Mattos, E. (2019). Cultural psychology: A two-way path between mind and culture. *Studies in Psychology*, 40(1), 1−9. Available from https://doi.org/10.1080/02109395.2019.1565388.

Brinkmann, S. (2018). General psychological implications of the human capacity for grief. *Integrative Psychological and Behavioral Science*, 52, 177−190. Available from https://doi.org/10.1007/s12124-018-9421-2.

Brinkmann, S. (2020). *Grief: The price of love* (T. McTurk Trans.). Polity.

Brooker, C. (2013). *Be right back. Black mirror* [TV Episode]. Channel 4.

Brubaker, J. R., Hayes, G. R., & Dourish, P. (2013). Beyond the grave: Facebook as a site for the expansion of death and mourning. *The Information Society: An International Journal*, 29(3), 152−163. Available from https://doi.org/10.1080/01972243.2013.777300.

Buben, A. (2015). Technology of the dead: Objects of loving remembrance or replaceable resources? *Philosophical Papers*, 44(1), 15−37. Available from https://doi.org/10.1080/05568641.2015.1014538.

Despret, V. (2015). *Au bonheur des morts. Récits de ceux qui restent.* La Découverte.

Dilmaç, J. A. (2018). The new forms of mourning: Loss and exhibition of the death on the internet. *OMEGA— Journal of Death and Dying*, 77(3), 280–295. Available from https://doi.org/10.1177/0030222816633240.

Elder, A. (2020). Conversation from beyond the grave? A neo-confucian ethics of chatbots of the dead. *Journal of Applied Philosophy*, 37(1), 73–88. Available from https://doi.org/10.1111/japp.12369.

Fitzpatrick, K. K., Darcy, A., & y Vierhile, M. (2017). Delivering cognitive behavior therapy to young adults with symptoms of depression and anxiety using a fully automated conversational agent (Woebot): A randomized controlled trial. *JMIR Mental Health*, 4(2), e19. Available from https://doi.org/10.2196/mental.7785.

Galvão, V., Maciel, C., Pereira, R., Gasparini, I., Viterbo, J., & Bicharra Garcia, A. (2021). Discussing human values in digital immortality: Towards a value-oriented perspective. *Journal of the Brazilian Computer Society*, 27(1), 15. Available from https://doi.org/10.1186/s13173-021-00121-x.

Huber, A., Weiss, A., & Rauhala, M. (2016). The ethical risk of attachment: How to identify, investigate and predict potential ethical risks in the development of social companion robots. In *11th ACM/IEEE international conference on human-robot interaction* (pp. 367–374). https://doi.org/10.1109/HRI.2016.7451774.

Godfrey, C. (2019). The grief bot that could change how we mourn. *The Daily Beast.* Available from https://www.thedailybeast.com/the-griefbot-that-could-change-how-we-mourn.

Jiménez-Alonso, B., & Brescó, I. (2022a). ¿Griefbots para despedirnos de nuestros seres queridos fallecidos? Algunas consideraciones psicológicas y éticas. *Psicosomática y Psiquiatría*, 20, 42–53. Available from https://doi.org/10.34810/PsicosomPsiquiatrnum200404.

Jiménez-Alonso, B., & Brescó, I. (2022b). Grief, photography and meaning making: A psychological constructivist approach. *Culture & Psychology*, 28(1), 107–132. Available from https://doi.org/10.1177/1354067X211015416.

Jiménez-Alonso, B., & Brescó, I. (2023). Griefbots: A new way of communicating with the dead? *Integrative Psychological and Behavioral Science*, 57, 466–481. Available from https://doi.org/10.1007/s12124-022-09679-3.

Jonze, S. (Director) (2013). *Her.* Annapurna Pictures, Stage 6 Films.

Josephs, I. E. (1998). Constructing one's self in the city of the silent: Dialogue, symbols, and the role of 'as-if' in self-development. *Human Development*, 41(3), 180–195. Available from https://doi.org/10.1159/000022578.

Kasket, E. (2012a). Continuing bonds in the age of social networking: Facebook as a modern-day medium. *Bereavement Care*, 31(2), 62–69. Available from https://doi.org/10.1080/02682621.2012.710493.

Kasket, E. (2012b). The dead's digital being in a Facebook profile. *Existential Analysis*, 23(2), 249–261.

Klass, D. (2018). Prologue. In D. Klass, & E. M. Steffen (Eds.), *Continuing bonds in bereavement: New directions for research and practice* (pp. xiii–xix). New York: Routledge.

Klass, D., Silverman, P. R., & Nickman, S. L. (Eds.), (1996). *Continuing bonds: New understandings of grief.* Taylor & Francis.

Liamputtong, P. (2011). *Researching the vulnerable.* Sage.

Lindemann, N. (2022). *The ethical permissibility of chatting with the dead: Towards a normative framework for 'deathbots'* (Master thesis). Osnabrück University. Publications of the Institute of Cognitive Science, number 1.

Mori, M. (1970). The uncanny valley. *Energy*, 7(4), 33–35.

Neimeyer, R., Klass, D., & Dennis, M. R. (2014). A social constructionist account of grief: Loss and the narration of meaning. *Death Studies*, 38, 485–498. Available from https://doi.org/10.1080/07481187.2014.913454.

Newton, C. (2016). Speak, memory. When her best friend died, she used artificial intelligence to keep talking to him. *The Verge.* Available from https://www.theverge.com/a/luka-artificial-intelligence-memorial-roman-mazurenko-bot.

O'Brien, T. (January 22, 2021). *I'm looking into this—appln date (Apr. 2017) predates the AI ethics reviews we do today.* Twitter. https://twitter.com/_TimOBrien/status/1352645952310439936

Öhman, C., & Floridi, L. (2018). An ethical framework for the digital afterlife industry. *Nature Human Behaviour*, 2, 318–320. Available from https://doi.org/10.1038/s41562-018-0335-2.

Savin-Baden, M. (2022). *AI for the death and dying*. Routledge.

Scanlan, C. L. (2020). *Preparing for the unanticipated: Challenges in conducting semi-structured, in-depth interviews*. Sage.

Sofka, C. (1997). Social support "internetworks," caskets for sale, and more: Thanatology and the information superhighway. *Death Studies*, *21*(6), 553–574. Available from https://doi.org/10.1080/074811897201778.

Sparrow, R. (2002). The march of the robot dogs. *Ethics and Information Technology*, *4*(4), 305–318. Available from https://doi.org/10.1023/A:1021386708994.

Stokes, P. (2021). *Digital souls: A philosophy of online death*. Bloomsbury Academic.

Tjiptomongsoguno, A., Chen, A., Sanyoto, H., Irwansyah, E., & Kanigoro, B. (2020). *Medical chatbot techniques: A review*, . *Software engineering perspectives in intelligent systems* (1294, pp. 346–356). Springer International Publishing, Advances in Intelligent Systems and Computing.

Trothen, T. J. (2022). Replika: Spiritual enhancement technology? *Religions*, *13*(4), 275. Available from https://doi.org/10.3390/rel13040275.

Utami, D., Bickmore, T., Nikolopoulou, A., & Paasche-Orlow, M. (2017). Talk about death: End of life planning with a virtual agent. In En. J. Beskow, C. Peters, G. Castellano, C. O'Sullivan, I. Leite, & S. Kopp (Eds.), *Intelligent virtual agents* (pp. 441–450). Springer.

Vallès-Peris, N., & Domènech, N. (2020). Roboticists' imaginaries of robots for care: The radical imaginary as a tool for an ethical discussion. *Engineering Studies*, *12*(3), 157–176. Available from https://doi.org/10.1080/19378629.2020.1821695.

Van Dyke, C. (2022). Grieving in the wind telephone booth. In L. A. Burke, & E. T. Rynearson (Eds.), *The restorative nature of ongoing connections with the deceased* (pp. 111–120). Routledge.

Vygotsky, L. S. (1978). *Mind in action: The development of higher psychological processes*. Harvard University Press.

Walter, T., Hourizi, R., Moncur, W., & Pitsillides, S. (2012). Does the internet change how we die and mourn? Overview and analysis. *OMEGA*, *64*(4), 275–302. Available from https://doi.org/10.2190/OM.64.4.a.

Wangmo, T., Lipps, M., Kressig, R., & Ienca, M. (2019). Ethical concerns with the use of intelligent assistive technology: Findings from a qualitative study with professional stakeholders. *BMC Medical Ethics*, *20*(1), 98. Available from https://doi.org/10.1186/s12910-019-0437-z.

Weber-Guskar, E. (2021). How to feel about emotionalized artificial intelligence? When robot pets, holograms, and chatbots become affective partners. *Ethics and Information Technology*, *23*(4), 601–610. Available from https://doi.org/10.1007/s10676-021-09598-8.

Wertsch, J. (1998). *Mind as action*. Oxford University Press.

Ethical implications of artificial intelligence models and experiences

Pitfalls (and advantages) of sophisticated large language models

Anna Strasser[1,2]

[1]*Faculty of Philosophy, Ludwig-Maximilians-Universität, München, Germany* [2]*DenkWerkstatt Berlin, Berlin, Germany*

Introduction

Natural language processing (NLP) based on large language models (LLMs), such as BERT, Eleuther, ChatGPT, GPT-3, GPT-4, LaMDA, and PALM, is a booming field of AI research. In general, artificial systems based on neural networks and deep learning have achieved impressive success in many domains based on pattern recognition, like speech recognition, lipreading, and game-playing. Examples in this area include LipNet (Assael et al., 2016)—a program for lung cancer screening (Ardila et al., 2019); AlphaFold—predicting protein structure (Jumper et al., 2021); AlphaTensor—discovering novel matrix multiplication algorithms (Fawzi, 2022); DeepBlue (Campbell et al., 2002); AlphaGo; and other game-playing programs (Brown & Sandholm, 2019; Silver & Hubert, 2016, 2018). With respect to language, we have observed impressive progress in automatic translation (DeepL) and computer code generation (GitHub Copilot) over the past few years, all relying on LLMs and deep learning. Remarkably, LLMs are able to produce grammatically correct linguistic outputs with fluency, often similar to that of a human. Many of their outputs can hardly be distinguished from linguistic outputs originally created by humans (Brown et al., 2020; Clark et al., 2021; Gao et al., 2022; Schwitzgebel et al., 2023), even though some demonstrate a lack of common sense and embarrassingly expose the models.[1] After neural networks have proven to outperform humans in games and practical domains relying on pattern recognition, we potentially now stand at a road junction where artificial entities could eventually enter the realm of human communication.

However, this comes with serious risks that we should consider with caution. The widespread use of LLMs, along with expected advances in their development, will make it increasingly difficult to distinguish between human-written and machine-generated text. This fact alone is already going to create all sorts of new challenges. For instance, it will be difficult to prove human authorship beyond doubt, and it will be equally difficult to unmask with certainty machine-generated text that is fraudulently passed off as self-written text. Furthermore, text generated by LLMs has the potential to be abused for a new kind of plagiarism, fraud, and the spread of misinformation. Due to the inherent limitations regarding the reliability of neural networks, overreliance on LLMs can have disruptive consequences (Hopster, 2021). Precisely because they are so good at mimicking

---

[1]For examples of typical mistakes, see ChatGPT/LLM error tracker (Davis et al., 2023).

Ethics in Online AI-Based Systems. DOI: https://doi.org/10.1016/B978-0-443-18851-0.00007-X

human linguistic performance, there is a risk that they will be used to mass-produce misinformation (Marcus, 2022, 2023). With the help of LLMs, it will be even easier than before to create an infinite amount of text for troll farms and fake websites, which in turn will lead to a decline in the level of trustworthiness on the Internet. Even though the outputs of LLMs often sound very convincing, they are not reliably truthful. Critical voices say that, for example, ChatGPT is a "bullshit generator" (McQuillan, 2023). One could go so far as to conjecture that with the further development of such language models, a new class of weapons is emerging that can have devastating effects on the war for truth (cp. Guardian editorial, 2023).

In this chapter, I focus on the consequences that current and further advanced LLMs might have. After presenting some background on LLMs (*1. Background*), I start with an overview of how difficult it is already now to distinguish between machine-generated and human-made text (*2. Difficult to distinguish*). Thereafter, I discuss various ethical consequences arising from this indistinguishability. Specifically, I address challenges related to difficult-to-prove human authorship, examine new forms of plagiarism, and critically evaluate copyright and privacy issues related to the model construction. I also consider the possibilities of counterfeiting people and spreading misinformation and toxic language (*3. Ethical consequences*). In the next section, I discuss possible ways by which society might deal with the inherent risks of LLMs. This concerns not only a possible adaptation of our legal basis but also the technical possibilities available for implementing the legislation (*4. How to handle the epistemological crisis*). After addressing at length the potential risks associated with the increased use of LLMs, I turn to the question of the extent to which there might also be helpful applications arising from LLMs that could be used in everyday life (*5. LLMs as thinking tools*).

Background

Since research related to LLMs is a fairly young and rapidly evolving field of research, it is difficult to provide an overview of the state of the art that is not already outdated at the time of publication, as there is a constantly growing body of new publications. Research papers, as well as opinion papers from computer science and philosophy and various publications in the media, served as the basis for this chapter. These are, for example, technical papers introducing specific LLMs such as BERT (Devlin et al., 2018; Rogers et al., 2020; Phang, Bradley, Gao, Castricato, & Biderman, 2022) , GPT-3 (Brown et al., 2020), GPT-4 (OpenAI, 2023), LaMDA (Thoppilan, 2022), and PALM (Chowdhery et al., 2022) as well as metareviews on benchmarks used to assess the performance of language models (Michael et al., 2022; Srivastava et al., 2022).

Since the initial release of GPT-3 on June 11, 2020, LLMs have attracted the interest of other disciplines, such as linguistics, cognitive science, philosophy, and others and have also received considerable public attention (Heaven, 2020; Mahowald et al., 2023; Marcus & Davis, 2020; Simonite, 2020; Weinberg, 2020). For example, Mahowald et al. (2023) investigate the capabilities of LLMs, distinguishing between formal competence (the knowledge of linguistic rules) and functional competence, which refers to understanding and using language in the world. They conclude that LLMs are close to mastering formal competence but fail at functional competence tasks. Public perception is initially blown away by the impressive achievements of LLMs, and also from scientific communities, the degree of impressiveness is immense as it is shown, e.g., by the title of an opinion piece by Will Douglas Heaven (2020): "OpenAI's new language generator GPT-3 is shockingly good—and completely mindless."

In June 2022, Google's LaMDA model made international headlines when Google engineer Blake Lemoine said he became convinced that LaMDA was sentient, prompting a flood of papers (Bryson, 2022; Frankish, 2022; Hofstadter, 2022; Klein, 2022; Roberts, 2022; Tiku, 2022). Although the majority is not inclined to ascribe sentience to LLMs, there is a trend to use philosophically loaded terms, such as "knowing," "believing," "comprehending," and "thinking" when describing these systems (Shanahan, 2023). The extent to which this can be a justified use of these terms is open to debate.

Similarly, ChatGPT, launched in November 2022, evoked a long-lasting echo in the media and the academic communities (Chiang, 2023; Krakauer & Mitchell, 2022; Lock, 2022; Roose, 2022; Thorp, 2023; Wolfram, 2023). For example, Derek Thompson (2022) mentioned ChatGPT in *The Atlantic* magazine's "Breakthroughs of the Year" as part of "the generative-AI eruption" that "may change our mind about how we work, how we think, and what human creativity really is." In the beginning, the enthusiastic voices received a lot of attention, but now the critical voices seem also to gain more consideration (Guardian editorial, 2023; Hofstadter, 2022; Marcus & Davis, 2020, 2023; Marcus,20, 2023; McQuillan, 2023).

Hard to distinguish

Just ten years ago, people didn't give much thought to how to distinguish machine-generated text from human-generated text. The differences were so obvious back then, and it didn't seem like that would change quickly. Back then, the differences were so obvious, and it didn't look like that that this would change anytime soon. However, with the advent of more and more upscaled LLMs, this is becoming a serious problem. Today, in social media, e-customer service, and advertising, we are increasingly exposed to machine-generated content that can easily be mistaken for human-generated content. Neither humans nor sophisticated detection software can distinguish with certainty between human-generated and machine-generated text. In empirical research, this indistinguishability, along with the tendency of humans to anthropomorphize, is sometimes even exploited when experimental protocols with artificial agents are used to test hypotheses about human social cognitive mechanisms (Strasser, 2022; Wykowska et al., 2016). And it is already a concern for teachers that they will not be able to distinguish their students' self-written essays from machine-generated ones (Herman, 2022; Huang, 2023; Hutson, 2022; Marche, 2022; Peritz, 2022; Sparrow, 2022).

Human discrimination abilities

The more advanced LLMs are, the more difficult it becomes for humans to distinguish between machine-generated and human-made text. Besides various rather informal assessments (Rajnerowicz, 2022; Sinapayen, 2023; Vota, 2020), there are three studies using rigorous psychological methods to test humans' ability to distinguish between machine-generated and human-made text (Brown et al., 2020; Clark et al., 2021; Schwitzgebel et al., 2023).

By demonstrating a difference between the two basic models of Open-AI (GPT-2 and GPT-3), Clark et al. (2021) were able to show that the more advanced the LLMs, the more difficult the distinction becomes. They collected short human-generated texts in three domains: stories, news articles, and recipes, and used the two base models to generate texts within the same domains. The participants (six groups covering the three domains for each model) were then presented with five selected texts and asked to judge whether these texts were likely to have been generated by humans or by machines. Results for the older model, GPT-2, showed that participants were able to accurately distinguish between GPT-2-generated and human-generated texts 58% of the time, significantly above the chance rate of 50%. In contrast, accuracy in discriminating between the newer model (GPT-3) and human-generated text was only 50%, not significantly different from chance. Even additional training in follow-up experiments failed to increase accuracy to above 57% in any domain. These results show that scaling up the models makes it more difficult to distinguish between human-made and machine-generated texts, and it is expected that the results of such an experiment will point even more clearly in this direction when GPT-4 is on the market.

Brown et al. (2020) focused on the domain of news articles and found similar results, indicating a moderately good discrimination rate for smaller (older) language models and near-chance performance with the largest version of GPT-3.

In the study I conducted with Eric and David Schwitzgebel (Schwitzgebel et al., 2023), we fine-tuned the Davinci model of GPT-3 on the corpus of the well-known philosopher Daniel Dennett (Strasser et al., 2023) and tested three groups of participants (ordinary naïve participants, philosophical blog readers, and experts of Dennett's work). Our results showed that only the discrimination abilities of blog readers and experts were significantly above the chance rate of 20%[2] (blog readers 48% and experts 51%), even though lower than we hypothesized. In comparison, ordinary participants were near the chance rate of 20%. Given the expected improvement of future models, this suggests that probably even expertise in a domain will soon no longer provide a reliable way to distinguish machine-generated text from human-made text. Already now, experts on Dennett could, on average, only identify Dennett's answer half the time when presented with his answer alongside four answers from our fine-tuned language model.[3] These empirical results clearly show that human-machine discrimination abilities with respect to the linguistic output of LLMs are no longer a reliable criterion for identifying the linguistic outputs of LLMs beyond doubt.

Discrimination with the help of detection software

One might think that if humans are unable to recognize the machine-generated text as machine-generated text, it should at least be possible to tell the difference beyond doubt using detection software. But at least with respect to the current state of research, even detection software cannot distinguish with 100% certainty between machine-generated and human-made text. Here, we seem to be at the beginning of an arms race between fraudsters and fraud detection.

Eric Mitchell et al. (2023) proposed a method called *DetectGPT* for deciding if a text passage was generated by a particular source model, for example, GPT-3. This method is based on the idea that if a text was generated by GPT-3, then this text has a high probability according to GPT-3, while human-written text does not have such a high probability from the point of view of GPT-3 (a nice explanation of this method can be found in Melanie Mitchell's (2023) blog). Tests with several

---

[2]Chance rate is at 20% because we used a five-alternative forced choice task.
[3]In other domains, such as deep fake detection for audio (Groh et al., 2021; Müller et al., 2022) or differentiating human-made artwork from AI-generated artwork (Gangadharbatla, 2022), rather weak human discrimination capabilities were also found.

large language models showed that their method was able to distinguish between human-written and LLM-generated text in over 95% of the cases. However, 95% is not 100%, and it is also critical to note that the number of possible specific LLMs is constantly increasing, and since this method is specific to a particular model, the number of models that need to be tested may present a problem. One could argue that pretty much any LLM uses a similar neural network architecture and is trained with comparable training data. But, after all, one cannot rule out the possibility that there will be other LLMs in the future. Moreover, LLM users can manually set a preferred probability; to what extent this poses difficulties for this method would need to be tested.

It is important to keep in mind that detectors for LLM-generated text commit two types of errors: false-negative (machine-generated text falsely judged to be written by humans)[4] and false-positive errors (human-generated text falsely judged to be machine generated). False positives can be very harmful to humans, as I describe in the next section. As long as we cannot exclude that such detectors falsely accuse humans of cheating, they should be used with caution and with the knowledge that their judgment could be false. This no longer indubitable distinctness leads to various other difficulties, which I address in more detail later.

Ethical consequences

Due to the ever-increasing indistinguishability between machine-generated and human-generated texts, various ethical problems arise, especially in the age of electronic transfer. This begins with the no longer indubitable verifiable human authorship and continues with various types of fraud, such as a new form of plagiarism, but also concerns the violation of privacy rights. Counterfeits of humans can be created and disseminated, and, last but not least, it enables the massive spread of misinformation.

Such consequences are certainly also supported by the hype that has taken place on social media. According to the first reactions to the release of LLMs like GPT-3 or ChatGPT, the voices expressing their astonishment and deep impression appeared to be in the majority. However, right from the start, several scholars demonstrated how easy it is to expose LLMs (Marcus & Davis, 2020). It seems that the critical voices have become more prominent only recently (Guardian editorial, 2023), even though the enthusiastic voices are still easy to discern. But even independent of a possible overestimation of the factual capabilities of LLMs, in the future, we will have to deal with ethical issues that arise primarily from the fact that we can no longer clearly distinguish machine-generated outputs from human-written outputs.

How to verify authorship

Unless authors are directly monitored in the process of writing their texts, and it is ruled out that they can use LLMs in the writing process, it will no longer be possible to infer a human author beyond doubt on a textual basis. It could be that the accusation of passing off a machine-generated text as one's own can no longer be dispelled in the last instance. This means that it could become difficult to prove oneself beyond doubt as the author of a text when submitting a paper.

---

[4]The first version of GPTzero (https://gptzero.me), for example, did evaluate the 40 machine-generated answers used in the experiment of Eric Schwitzgebel and colleagues as human-like.

Assuming that also in the future, neither humans nor detection software can reliably distinguish between machine-generated and human-written text, we have to be prepared to deal with false positives and false negatives when trying to recognize a human-written text. From that point of view, it is conceivable that an author submitting a paper could be falsely accused of having delivered a machine-generated text, and conversely, it is equally possible for a machine-generated text to pass as human-generated. As LLMs continue to advance, it may even become common to submit newspaper articles or even articles to scientific journals in which much of the content is produced by machines. For sure, it is expected that students could make use of LLMs to let them produce text for their essays, and their teachers will not be able to recognize whether the students delivered self-made texts (Herman, 2022; Huang, 2023; Hutson, 2022; Marche, 2022; Peritz, 2022; Sparrow, 2022). Universities might turn back to in-person exams to reassure the authorship of their students (Cassidy, 2023). However, this is not possible in all cases where authorship matters, especially with respect to the mass of electronically distributed texts. How new chains of trust can be established will be a challenge for future societies.

New forms of plagiarism

Since language models create novel sentences—they are not simple parrots—it is unlikely that their output will lead to any results when using standard plagiarism checkers. Using plagiarism checkers, similarity thresholds below 10%−15% are considered ordinary for nonplagiarized work (Mahian et al., 2017). Schwitzgebel et al. (2023) investigated whether fine-tuning GPT-3 on Dennett's works might have led the model to be overtrained[5] so that it simply parroted sentences or multiword strings of texts from Dennett's corpus. To verify that the fine-tuned model was indeed producing new texts, we used the Turnitin plagiarism checker to check for "plagiarism" between the machine-generated outputs and the Turnitin corpus supplemented with the works that were used as the training data. Turnitin reported an overall similarity of 5% between the GPT-3 generated answers and the comparison corpora, and none of the passages were flagged as similar to the training corpus. Also, the search for matching text strings between the GPT-3 responses and the training corpus revealed no matches, except for some stock phrases favored by analytic philosophers. Thus, the machine-generated text does not simply plagiarize its training data word for word but generates novel—albeit stylistically and philosophically similar—content.

Nevertheless, it is at least arguable whether outputs of fine-tuned language models should be considered plagiarism because they sort of "borrow" ideas from their training data with which they were fine-tuned without acknowledging the original author.

Violation of copyright rights and privacy

Another issue concerns the intellectual property of the people who wrote the text with which LLMs are trained. For example, OpenAI's GPT-3 was trained on hundreds of billions of words of text (499 billion tokens)[6] from Common Crawl, WebText, books, and Wikipedia (Brown et al., 2020). And especially fine-tuned models like the one we fine-tuned on the corpus of Daniel Dennett have

---

[5]Being overtrained is an issue regarding neural networks. However, running four epochs of fine-tuning is a standard recommendation from OpenAI, and in most applications, four epochs of training do not result in overtraining (Brownlee, 2019).

[6]A token is a sequence of commonly cooccurring characters, with approximately three-fourth of an English word per token on average.

to put up with the question of whether this is a fair use of someone else's intellectual property to use their works in creating an LLM without asking permission. For this reason, we asked Daniel Dennett for the explicit permission before fine-tuning our LLM on his texts, and we agreed that he has the final say when it comes to the question of who may use this language model and which outputs are published. It goes without saying that we always explicitly label the outputs of our model as machine-generated output.

However, in the case of already deceased individuals, it is not possible to ask for permission (for a review regarding the potential use of personal data of deceased persons, see Nakagawa & Orita, 2022). To date, copyright laws regarding the use of copyrighted text as training data for fine-tuning language models have not yet been clarified (see Government UK consultations, 2021).

And the question of whether output from LLMs could be protected by copyright has not yet been conclusively resolved. According to a report in The Verge (Robertson, 2022), the U.S. Copyright Office recently rejected a request to grant copyright to a work of art to an AI because, in their view, it was a necessary standard for protection that the work of art contain at least elements of "human authorship."

Counterfeits of people

Furthermore, people leave behind a lot of private data that is not secured at all. LLMs can be fine-tuned on all kinds of additional training data. This could concern any information a person shared on social media. For example, one could use all kinds of data available about someone's life, e.g., in social media, blogs, websites, online stores, and search engines, to create a digital replica that could convincingly imitate some of that person's behavior (Karpus & Strasser under review). A striking illustration of such a replica, albeit a fictional one, can be found in the episode "Be right back" of the television series Black Mirror (Brooker, 2013). This episode inspired Eugenia Kuyda to feed all the saved online conversations she had with a deceased friend into an AI-powered system to create a chatbot version of her friend. Subsequently, a public application called *The Replika* was created (see https://replika.com; Murphy, 2019).

The extent to which existing LLMs, tools, and chatbots based on more or less private data violate copyright rights and privacy rights appears to me to be an open question that we should urgently address. Even if the science fiction story depicted in the Black Mirror episode is still too futuristic in many aspects in the context of today's technology, it is worth considering how revealing the data we leave behind on social media, search engines, online stores, and other platforms already is about our character traits concerning our likes, dislikes, desires, and other features of our personality.

Besides the scandals involving Cambridge Analytica in the UK that demonstrated the richness of the data available about our lives on social media platforms, there is an insightful art project by the Berlin-based artist collective Laokoon (https://www.madetomeasure.online/en). Their investigative project "Made to measure" explored the question of how far one can get in constructing a doppelganger of a person using data available online about that person's life. Using anonymized data of a person's Google search history from five years of that person's life, they retraced the life of that person and re-enacted that life in a film. When confronting the data donor, who was contacted after processing the data with the help of a personalized Instagram message, it turned out that the reconstruction of this person's life was amazingly accurate in many aspects (for a more detailed description, see Karpus & Strasser under review).

If digital replicas are able to appear to speak on behalf of the person from whose data they were constructed from, they could be considered as a counterfeit of this person. In other words, if outputs of LLMs were presented as a quotation or paraphrase of positions of existing persons, this would constitute counterfeiting (Dennett as interviewed in Cukier, 2022).

With the help of increasingly sophisticated language models, we will be able to create fakes of people that are difficult to distinguish from their originals. Just as we are able to create counterfeit money, it is conceivable that in virtual communication, for example, one can create the appearance of interacting with a real person, which turns out to be a fake. Such deep fakes do not have to be restricted to linguistic output; they can also be enriched with visual and auditory imitations. In Germany, for example, a mayor was made to believe that she was interacting with Vitali Klitschko in a Zoom call (Hoppenstedt, 2022). Another example, again from the field of art, that illustrates the potential of counterfeiting is the art project "Chomsky vs. Chomsky" (Rodriguez, 2022). It is important to note that while it was made clear from the outset that this is an artifact and not the real Chomsky, this project nevertheless shows how disturbing a digital replica can be. This art project presents a virtual version of Noam Chomsky—a location-based, mixed reality (MR) experience that draws not only on Chomsky's texts but also on recorded lectures. Thereby, this project offers the experience of asking questions orally in virtual reality and receiving an audio response whose sound is almost indistinguishable from the recordings of the real Chomsky.

No doubt, it should be against the law to present a conversational AI without making it clear that it is an AI and not a human. If an LLM counterfeits people, the creator and the users of this LLM are guilty of a crime (Dennett, 2023). However, until now, we do not have the legal basis to prosecute such crimes.

Spread of misinformation, nonsense, and toxic language

Stepping back from the emotional and overwhelming judgments shared in social media, it is an important and serious question to investigate to what extent we can trust machine-generated assertions.

The danger of mistakenly trusting GPT-3 is particularly evident in the health sector. We should be clear about the fact that within this domain, we are nowhere near any application where GPT-3 could provide reliable help in any sense. It lacks the scientific and medical expertise that would make it useful for any medical Q&A, as it can be very wrong, and this is not viable in health care. For example, in a test where GPT-3 responded to mental health problems, the AI advised a simulated patient to commit suicide (Daws, 2020). Nevertheless, despite the warning from OpenAI, it is probably to be expected that such applications will be developed.

LLMs based on a transformer architecture with a statistical self-attention mechanism—machines that calculate the probability of words appearing in the context of other words—have severe limitations regarding reliability. This becomes evident, for example, when LLMs make self-contradictory statements. It is quite possible that if an LLM receives the same question as a prompt several times, it will respond with very different answers that are contradicting each other. In contrast, a standard calculator will always give the same answer; for example, it will always "claim" that $2 + 2$ equals 4, whereas LLMs are able to "claim" that $2 + 2$ equals 4 in one instance and that $2 + 2$ equals 5 in another. For this reason, it is important that a balanced assessment of the performance of LLMs should not rely solely on cherry-picked, mind-boggling outputs.

Moreover, the better the models become, the more difficult it becomes to distinguish machine-generated linguistic outputs from human-made utterances, and at the same time, the risk of misuse increases. For example, LLMs can play a weighty role in spreading misinformation (Marcus, 2022). Since LLMs are inherently unreliable, they do make severe mistakes in reasoning and facts. This unreliability is due in part to the fact that LLMs build models of word sequences based on how humans use language rather than models describing how the world works. Although it can be concluded from this that many machine-generated linguistic results are correct because human language often reflects facts in the world, at the same time, it follows that the accuracy of LLM's statements is, to some extent, a matter of chance because, unlike humans, machines do not use language to refer to the world. An example that can illustrate the production of misinformation and nonsense is the brief presence of Galactica. This LLM was created to write plausible-sounding academic papers. However, it was taken offline a few days after its release due to harsh criticism amounting to the claim that Galactica produces vaguely-plausible-sounding-but-ultimately-nonsensical academic papers (Al-Sibai, 2022; Taylor et al., 2022). Likewise, I suppose that it is to be expected that applications that aim to assist search applications like ChatGPT with respect to Bing will soon be taken offline (Rogers, 2023).

Lately, the most discussed LLM has been ChatGPT, which was released to the general public in November 2022. Unfortunately, ChatGPT has a tendency to hallucinate—it produces statements that sound plausible but are simply false. It is able to invent references to papers that were never been written, it can make up historical dates, and it commits severe failures regarding the solution of logical problems. Furthermore, LLMs lack what we would call "common sense" in the human case. It is a serious problem that LLMs can produce sentences that are simply not true (for a repository of errors made by LLMs, see Davis et al., 2023; Marcus & Davis, 2023). To date, LLMs have no reliable mechanisms for verifying the truth of their statements, and this is effectively a springboard for the mass production of misinformation. For example, Gary Marcus (2023) reports that the independent researcher Shawn Oakley has shown how easy it is to get ChatGPT to produce misinformation that it even backs up with fictional studies. This is especially troubling as ChatGPT adopts an authoritative tone.

Another problem with LLMs is that their outputs depend on their training data. This means that if they are not constantly retrained, they quickly become obsolete in terms of up-to-date information. ChatGPT lacks "knowledge" of events that occurred after 2021. They can make prophecy-like statements about events that may have happened in the meantime, but these statements lack any relation to our reality.

Because of their limited reliability, LLMs require human supervision, but this leads to another critical ethical issue. OpenAI strives to filter out toxic content (e.g., sexual abuse, violence, racism, sexism, etc.), which is a good goal in principle, but the implementation is not perfect and is also highly questionable. To flag toxic data produced by LLMs, one needs humans, and this work is traumatic in nature. According to a *TIME* investigation, OpenAI used outsourced Kenyan laborers earning less than $2 per hour to make ChatGPT less toxic (Perrigo, 2023). Apart from the fact that this practice is ethically questionable, it is also not ultimately successful from a technical point of view.

How to handle the epistemological crisis

Due to all potential deep fakes, there is an epistemological crisis to be expected, and people will need to look out for what they take as representing a real person. Avoiding that we get too suspicious and paranoid, we need laws for how AIs present themselves, and we will probably have to develop new strategies for identifying our counterparts as humans.

One helpful measure would be to create a legal basis for requiring machine-generated output to be labeled as machine-generated text as a matter of principle. This is addressed, for example, by a recent proposal, the so-called AI-act of the European Commission (2021), which requires labeling for anything that might be mistaken for human interaction. Such regulations could help mitigate the risk that LLMs will be used to contribute to a huge spread of misinformation.

One way to label machine-generated text could be accomplished through the use of digital watermarks (Wiggers, 2022). Kirchenbauer et al. (2023) have suggested that one way to do this would be to require the creators of LLMs to add a watermark signal to each generated text passage that cannot be easily removed by simply modifying the text and to provide open-source software for watermark detection. This sounds good at first glance, but one cannot assume that all LLM creators will adhere to it, and of course, it is also possible to fool watermark detectors. Again, it is likely that an arms race will develop here between fraudsters and those who want to mark LLM's outputs recognizably.

And as described earlier, so far, there is no completely reliable method for detecting AI-generated text. So, bans cannot be enforced proactively, which means that one has to rely on human help. For example, a conference has banned the use of machine-generated text in submissions. However, in the end, only submissions that are deemed suspicious to other scientists were examined (Vincent, 2022). It seems as if we are not prepared for the emergence of such disruptive and novel technologies.

LLMs as thinking tools

All of this is not to say that positive applications of LLMs are not conceivable. Many applications that we use on a daily basis are based on models that have been trained to predict the next word or words in a text based on the preceding words; e.g., it is part of the technology that predicts the next word you want to type on your mobile phone allowing you to complete the message faster. Likewise, the increased quality of translation software is indeed a helpful tool, although it is still advisable not to release the translation without human review.

In other domains, such as game-playing AIs, neural network architectures have led to success (Brown & Sandholm, 2019; Campbell et al., 2002; Silver & Hubert, 2016, 2018). And impressive results can also be pointed to in scientific fields, such as a program for lung cancer screening (Ardila et al., 2019), AlphaFold, which predicts protein structure (Jumper et al., 2021), and AlphaTensor, which discovers novel matrix multiplication algorithms (Fawzi, 2022). However, it is important to keep in mind that the performance of these neural networks is tied to clearly definable goals. The output generated by machines in certain domains can be evaluated and verified on the basis of clear criteria. A victory in a chess game is clearly defined, and the individual steps are also subject to a set of rules. However, when we move into the realm of communication with human language, it is not always clear whether an answer to a question and its justifications meet our requirements for comprehension, rational thinking, and the like.

Considering the potential of large language models and assuming that technology will continue to advance, it is conceivable that language models will soon produce results interesting enough to serve as a valuable resource for human researchers. In other domains, computer programs are able to generate music in the style of a particular composer (Daly, 2021; Elgammal, 2021; Hadjeres et al., 2017) or create all kinds of images (DALL-E). Even if not all outputs are reliable or interesting, selected outputs seem to have significant musical or artistic value. A composer or artist could produce many outputs, select the most promising, edit them slightly, and present them as original works. As a matter of fact, there was already a case where an AI-produced picture did win a competition (Metz, 2022).

In this way, language models could become thinking tools that people use. However, it is important that the users are experts in their domains and remain able to validate the outputs. For example, a researcher could fine-tune a language model with certain corpora and then generate outputs they can use as inspiration for further ideas. However, when using language models as thinking tools, one must be careful not to rely too heavily on them as deep learning networks always have reliability issues (Alshemali & Kalita, 2020; Bosio et al., 2019). A user with insufficient expertise might mistakenly assume that all results from a large language model fine-tuned to an author's work reflect the author's actual views (Bender et al., 2021; Weidinger et al., 2021). The use of such language models will not be able to replace reading the original works (see Steven & Iziev, 2022), but they may eventually become a helpful tool for humans creating text.

According to a whitepaper published by Lionbridge (2023), ChatGPT can help with translation, terminology, style guides, content classification, postediting, content analysis, and creating working code. However, it can only help—one can never rely on an LLM to say true things or know what is right or wrong; outputs of all LLMs are unreliable. Therefore, it remains up to humans to decide what makes sense and what is true or false. AI can be used to improve, polish, edit, or write texts, but it will still be up to humans to judge their value.

References

Alshemali, B., & Kalita, J. (2020). Improving the reliability of deep neural networks in NLP: A review. *Knowledge-Based Systems*, *191*, 105210. Available from https://doi.org/10.1016/j.knosys.2019.105210.

Al-Sibai, N. (2022). Facebook takes down AI that churns out fake academic papers after widespread criticism. *The Byte*. https://futurism.com/the-byte/facebook-takes-down-galactica-ai.

Ardila, D., Kiraly, A. P., Bharadwaj, S., Choi, B., Reicher, J. J., Peng, L., Tse, D., Etemadi, M., Ye, W., Corrado, G., Naidich, D. P., & Shetty, S. (2019). End-to-end lung cancer screening with three-dimensional deep learning on low-dose chest computed tomography. *Nature Medicine*, *25*(6), 954−961. Available from https://doi.org/10.1038/s41591-019-0447-x.

Assael, Y., Shillingford, B., Whiteson, S., & Freitas, N. (2016). LipNet: Sentence-level lipreading. Available from http://doi.org/10.48550/arXiv.1611.01599.

Bender, E.M., Gebru, T., McMillan-Major, A., & Shmitchell, S. (2021). On the dangers of stochastic parrots: Can language models be too big? *FAccT '21: Proceedings of the 2021 ACM conference on fairness, accountability, and transparency* (pp. 610−623). doi.org/10.1145/3442188.3445922.

Bosio, A., Bernardi, P., Ruospo, & Sanchez, E. (2019). A reliability analysis of a deep neural network. In *2019 IEEE latin american test symposium (LATS)* (pp. 1−6). Available from http://doi.org/10.1109/LATW.2019.8704548.

Brooker, C. (2013). *Black mirror: Be right back* (Season 2, Episode 1) [movie]. Zeppotron.

Brown, N., & Sandholm, T. (2019). Superhuman AI for multiplayer poker. *Science (New York, N.Y.), 365.* Available from https://doi.org/10.1126/science.aay2400.

Brown, T., Mann, B., Ryder, N., Subbiah, M., Kaplan, J. D., Dhariwal, P., & Amodei, D. (2020). Language models are few-shot learners. *Advances in Neural Information Processing Systems*, *33*, 1877−1901. Available from https://doi.org/10.48550/arXiv.2005.14165.

Brownlee, J. (2019). A gentle introduction to early stopping to avoid overtraining neural networks. *Machine Learning Mastery*. https://machinelearningmastery.com/early-stopping-to-avoid-overtraining-neural-network-models.

Bryson. (2022). One day, AI will seem as human as anyone. *What Then*. *Wired*. https://www.wired.com/story/lamda-sentience-psychology-ethics-policy.

Campbell, M., Hoane, A. J., Jr, & Hsu, F. H. (2002). Deep blue. *Artificial Intelligence*, *134*(1−2), 57−83.

Cassidy, C. (2023). Australian universities to return to 'pen and paper' exams after students caught using AI to write essays. *The Guardian*. https://www.theguardian.com/australia-news/2023/jan/10/universities-to-return-to-pen-and-paper-exams-after-students-caught-using-ai-to-write-essays.

Chiang, T. (2023) ChatGPT is a blurry JPEG of the web. *The New Yorker*. https://www.newyorker.com/tech/annals-of-technology/chatgpt-is-a-blurry-jpeg-of-the-web.

Chowdhery, A., Narang, S., & Devlin, J. (2022). PaLM: Scaling language modeling with pathways. *Google AI Blog*. Available from: https://ai.googleblog.com/2022/04/pathways-language-model-palm-scaling-to.html.

Clark, E., August, T., Serrano, S., Haduong, N., Gururangan, S., & Smith, N.A. (2021). *All that's 'human' is not gold: Evaluating human evaluation of generated text*. https://doi.org/10.48550/arXiv.2107.00061.

Cukier, K. (2022). Babbage: Could artificial intelligence become sentient? *The Economist*. https://shows.acast.com/theeconomistbabbage/episodes/babbage-could-artificial-intelligence-become-sentient.

DALL-E. https://openai.com/blog/dall-e.

Daly, R. (2021). AI software writes new Nirvana and Amy Winehouse songs to raise awareness for mental health support. *NME*. https://www.nme.com/news/music/ai-software-writes-new-nirvana-amy-winehouse-songs-raise-awareness-mental-health-support-2913524.

Davis, E., Hendler, J., Hsu, W., Leivada, E., Marcus, G., Witbrock, M., Shwartz, V., & Ma, M. (2023). *ChatGPT/LLM error tracker*. https://researchrabbit.typeform.com/llmerrors?typeform-source = garymarcus.substack.com.

Daws, R. (2020). Medical chatbot using OpenAI's GPT-3 told a fake patient to kill themselves. *AI News*. https://www.artificialintelligence-news.com/2020/10/28/medical-chatbot-openai-gpt3-patient-kill-themselves.

Dennett, D. (2023). The problem with counterfeit people. *The Atlantic*. https://www.theatlantic.com/technology/archive/2023/05/problem-counterfeit-people/674075.

Devlin, J., Chang, M.W., Lee, K., & Toutanova, K. (2018). Bert: Pre-training of deep bidirectional transformers for language understanding. Available from http://doi.org/10.48550/arXiv.1810.04805.

Elgammal, A. (2021). How a team of musicologists and computer scientists completed Beethoven's unfinished 10th symphony. *The Conversation*. https://theconversation.com/how-a-team-of-musicologists-and-computer-scientists-completed-beethovens-unfinished-10th-symphony-168160.

European Commission. (2021, April 24). *AI-act. Proposal for a regulation of the European parliament and of the council laying down harmonised rules on artificial intelligence (artificial intelligence act) and amending certain union legislative acts*. https://artificialintelligenceact.eu/the-act/.

Fawzi, A., et al. (2022). *Discovering novel algorithms with AlphaTensor*. https://www.deepmind.com/blog/discovering-novel-algorithms-with-alphatensor?utm_campaign = AlphaTensor&utm_medium = bitly&utm_source = Twitter + Organic.

Frankish, K. (2022, November 2). Some thoughts on LLMs. Blog post at The *Tricks* of the *Mind*. https://www.keithfrankish.com/blog/some-thoughts-on-llms

Gangadharbatla, H. (2022). The role of AI attribution knowledge in the evaluation of artwork. *Empirical Studies of the Arts*, *40*(2), 125−142. Available from https://doi.org/10.1177/0276237421994697.

Gao, C., Howard, F., Markov, N., Dyer, E., Ramesh, S., Luo, Y., & Pearson, A. (2022). Comparing scientific abstracts generated by ChatGPT to original abstracts using an artificial intelligence output detector, plagiarism detector, and blinded human reviewers. https://doi.org/10.1101/2022.12.23.521610.

GitHub Copilot. https://docs.github.com/en/copilot.

Government UK consultations. (2021). *Artificial intelligence call for views: Copyright and related rights.* https://www.gov.uk/government/consultations/artificial-intelligence-and-intellectual-property-call-for-views/artificial-intelligence-call-for-views-copyright-and-related-rights.

GPT-3. https://github.com/openai/gpt-3.

Groh, M. E., Firestone, Z., & Picard, R. (2021). Deepfake detection by human crowds, machines, and machine-informed crowds. *Proceedings of the National Academy of Sciences, 119*(1).

Guardian editorial. (2023). The Guardian view on ChatGPT search: Exploiting wishful thinking. *The Guardian.* https://www.theguardian.com/commentisfree/2023/feb/10/the-guardian-view-on-chatgpt-search-exploiting-wishful-thinking?CMP = share_btn_link.

Hadjeres, G., Pachet, F., & Nielsen, F. (2017). DeepBach: A steerable model for Bach chorales generation. In *Proceedings of the 34th international conference on machine learning* (pp. 1362−1371).

Heaven, W. (2020). Open AI's new language generator GPT-3 is shockingly good—And completely mindless. *MIT Technological Review.* https://www.technologyreview.com/2020/07/20/1005454/openai-machine-learning-language-generator-gpt-3-nlp/.

Herman, D. (2022). The end of high school English. *The Atlantic.* https://www.theatlantic.com/technology/archive/2022/12/openai-chatgpt-writing-high-school-english-essay/672412.

Hofstadter, D. (2022, June 9). Artificial neural networks today are not conscious, according to Douglas Hofstadter. *The Economist.* https://www.economist.com/by-invitation/2022/06/09/artificial-neural-networks-today-are-not-conscious-according-to-douglas-hofstadter.

Hoppenstedt, M. (2022, August 11). Russische Komiker zeigen Ausschnitt von Giffey-Gespräch mit Fake-Klitschko. *SPIEGEL.* https://www.spiegel.de/netzwelt/web/franziska-giffey-russische-komiker-zeigen-ausschnitt-von-gespraech-mit-fake-klitschko-a-527ab090-2979-4e70-a81c-08c661c0ef62.

Hopster, J. (2021). What are socially disruptive technologies? *Technology in Society, 67*, 101750. Available from https://doi.org/10.1016/j.techsoc.2021.101750.

Huang, K. (2023). Alarmed by A.I. chatbots, universities start revamping how they teach. *The New York Times.* https://www.nytimes.com/2023/01/16/technology/chatgpt-artificial-intelligence-universities.html.

Hutson, M. (2022). Could AI help you to write your next paper? *Nature, 611*, 192−193. Available from https://doi.org/10.1038/d41586-022-03479-w.

Jumper, J., Evans, R., Pritzel, A., Green, T., Figurnov, M., Ronneberger, O., Tunyasuvunakool, K., Bates, R., Žídek, A., Potapenko, A., Bridgland, A., Meyer, C., Kohl, S., Ballard, A. J., Cowie, A., Romera-Paredes, B., Nikolov, S., Jain, R., Adler, J., Back, T., & Hassabis, D. (2021). Highly accurate protein structure prediction with AlphaFold. *Nature, 596*(7873), 583−589. Available from https://doi.org/10.1038/s41586-021-03819-2.

Karpus, J., & Strasser, A. (under review). Persons and their digital replicas.

Kirchenbauer, J., Geiping, J., Wen, Y., Katz, J., Miers, I., & Goldstein, T. (2023). A watermark for large language models. Available from http://doi,org/10.48550/arXiv.2301.10226.

Klein, E. (2022, June 19). This is a weirder moment than you think. *The New York Times.* https://www.nytimes.com/2022/06/19/opinion/its-not-the-future-we-cant-see.html

Metz, R. (2022, September 3). AI won an art contest, and artists are furious. *CNN Business.* https://edition.cnn.com/2022/09/03/tech/ai-art-fair-winner-controversy/index.html.

Krakauer, D., & Mitchell, M. (2022). The debate over understanding in AI's large language model. Available from http://doi.org/10.48550/arXiv.2210.13966.

Lionbridge. (2023). *What ChatGPT gets right and wrong and why it's probably a game-changer for the localization industry.* https://www.lionbridge.com/content/dam/lionbridge/pages/whitepapers/whitepaper-what-chatgpt-gets-right-and-wrong/chatgpt-whitepaper-english.pdf.

Lock, S. (2022). What is AI chatbot phenomenon ChatGPT and could it replace humans? *The Guardian.* https://www.theguardian.com/technology/2022/dec/05/what-is-ai-chatbot-phenomenon-chatgpt-and-could-it-replace-humans.

Mahian, O., Treutwein, M., Estellé, P., Wongwises, S., Wen, D., Lorenzini, G., & Sahin, A. (2017). Measurement of similarity in academic contexts. *Publications*, 5(3), 18. Available from https://doi.org/10.3390/publications5030018.

Mahowald, K., Ivanova, A.A., Blank, I.A., Kanwisher, N., Tenenbaum, J.B., & Fedorenko, E. (2023). Dissociating language and thought in large language models: A cognitive perspective. Available from http://doi.org/10.48550/arXiv.2301.06627.

Marche, S. (2022). Will ChatGPT kill the student essay? *The Atlantic*. https://www.theatlantic.com/technology/archive/2022/12/chatgpt-ai-writing-college-student-essays/672371/.

Marcus, G., & Davis, E. (2020). GPT-3, Bloviator: OpenAI's language generator has no idea what it's talking about. *MIT Technology Review*.

Marcus, G. (2022). AI platforms like ChatGPT are easy to use but also potentially dangerous. *Scientific American*. https://www.scientificamerican.com/article/ai-platforms-like-chatgpt-are-easy-to-use-but-also-potentially-dangerous.

Marcus, G. (2023, February 11). Inside the heart of ChatGPT's darkness. Blog post at *The Road to AI We Can Trust*. https://garymarcus.substack.com/p/inside-the-heart-of-chatgpts-darkness?utm_source = substack&utm_medium = email

Marcus, G., & Davis, E. (2023, January 10). Large language models like ChatGPT say the darnedest things. Blog post at *The Road to AI We Can Trust*. https://garymarcus.substack.com/p/large-language-models-like-chatgpt

McQuillan, D. (2023). ChatGPT is a bullshit generator waging class war. *Vice*. https://www.vice.com/en/article/akex34/chatgpt-is-a-bullshit-generator-waging-class-war.

Michael, J., Holtzman, A., Parrish, A., Mueller, A., Wang, A., Chen, A.,... Bowman, S.R. (2022). What do NLP researchers believe? Results of the NLP community metasurvey. Available from http://doi.org/10.48550/arXiv.2208.12852.

Mitchell, E., Lee, Y., Khazatsky, A., Manning, C.D., & Finn, C. (2023). DetectGPT: Zero-shot machine-generated text detection using probability curvature. Availabe from http://doi.org/10.48550/arXiv.2301.11305.

Müller, N., Pizzi, K., & Williams, J. (2022). Human perception of audio deepfakes. In *Proceedings of the 1st international workshop on deepfake detection for audio multimedia (DDAM '22)* (pp. 85−91). New York: Association for Computing Machinery. https://doi.org/10.1145/3552466.3556531.

Murphy, M. (2019). This app is trying to replicate you. *Quartz*. https://qz.com/1698337/replika-this-app-is-trying-to-replicate-you/.

Nakagawa, H., & Orita, A. (2022). Using deceased people's personal data. *AI & Society*. Available from https://doi.org/10.1007/s00146-022-01549-1.

OpenAI. (2023). *GPT-4 technical report*. https://arxiv.org/abs/2303.08774.

Peritz, A. (2022, September 6). A.I. is making it easier than ever for students to cheat. *Slate*. https://slate.com/technology/2022/09/ai-students-writing-cheating-sudowrite.html.

Perrigo, B. (2023). Exclusive: OpenAI used Kenyan workers on less than $2 per hour to make ChatGPT less toxic. *The Times*. https://time.com/6247678/openai-chatgpt-kenya-workers.

Phang, J., Bradley, H., Gao, L., Castricato, L., & Biderman, S. (2022). EleutherAI: Going Beyond "Open Science" to "Science in the Open" *(arXiv:2210.06413)*. arXiv. Available from https://doi.org/10.48550/arXiv.2210.06413.

Rajnerowicz, K. (2022). Human vs. AI test: Can we tell the difference anymore? *Statistics & Tech Data Library*. https://www.tidio.com/blog/ai-test.

Roberts, M. (2022). Is Google's LaMDA artificial intelligence sentient? Wrong question. *The Washington Post*. https://www.washingtonpost.com/opinions/2022/06/14/google-lamda-artificial-intelligence-sentient-wrong-question.

Robertson, A. (2022). The US Copyright Office says an AI can't copyright its art. *The Verge*. https://www.theverge.com/2022/2/21/22944335/us-copyright-office-reject-ai-generated-art-recent-entrance-to-paradise.

Rodriguez, S. (2022). Chomsky vs. Chomsky. http://opendoclab.mit.edu/presents/ch-vs-ch-prologue-sandra-rodriguez.

Rogers, A. (2023). The new Bing is acting all weird and creepy—But the human response is way scarier. *Insider*. https://www.businessinsider.com/weird-bing-chatbot-google-chatgpt-alive-conscious-sentient-ethics-2023-2.

Rogers, A., Kovaleva, O., & Rumshisky, A. (2020). A primer in BERTology: What we know about how BERT works. Available from http://doi.org/10.48550/arXiv.2002.12327.

Roose, K. (2022, December 5). The brilliance and weirdness of ChatGPT. The New York Times.

Schwitzgebel, E., Schwitzgebel, D., & Strasser, A. (2023). Creating a large language model of a philosopher. https://doi.org/10.48550/arXiv.2302.01339.

Shanahan, M. (2023). Talking about large language models. Available from http://doi.org/10.48550/arXiv.2212.03551.

Silver, D., Huang, A., et al. (2016). Mastering the game of Go with deep neural networks and tree search. *Nature*, *529*, 484−489. Available from https://doi.org/10.1038/nature16961.

Silver, D., Hubert, T., et al. (2018). A general reinforcement learning algorithm that masters chess, shogi, and Go through self-play. *Science (New York, N.Y.)*, *362*(6419), 1140−1144. Available from https://doi.org/10.1126/science.aar6404.

Simonite, T. (2020). Did a person write this headline, or a machine? *Wired*. https://www.wired.com/story/ai-text-generator-gpt-3-learning-language-fitfully.

Sinapayen, L. (2023). Telling apart AI and humans #3: Text and humor https://towardsdatascience.com/telling-apart-ai-and-humans-3-text-and-humor-c13e345f4629.

Sparrow, J. (2022, November 1).'Full-on robot writing': The artificial intelligence challenge facing universities. *Guardian*. https://www.theguardian.com/australia-news/2022/nov/19/full-on-robot-writing-the-artificial-intelligence-challenge-facing-universities.

Srivastava, A., Rastogi, A., Rao, A., Shoeb, A., Abid, A., Fisch, A., . . ., Shaham, U. (2022). Beyond the imitation game: Quantifying and extrapolating the capabilities of language models. ArXiv, abs/2206.04615.

Steven, J., & Iziev, N. (2022, April 15). A.I. is mastering language. Should we trust what it says? The New York Times. https://www.nytimes.com/2022/04/15/magazine/ai-language.html.

Strasser, A. (2022). From tool use to social interactions. In J. Loh, & W. Loh (Eds.), *Social robotics and the good life*. Bielefeld: Transcript Verlag. Available from 10.1515/9783839462652-004.

Strasser, A., Crosby, M., & Schwitzgebel, E. (2023). How far can we get in creating a digital replica of a philosopher? In R. Hakli, P. Mäkelä, & J. Seibt (Eds.), Social robots in social institutions. Proceedings of robophilosophy 2022. *Series Frontiers of AI and its applications* (366, pp. 371−380). Amsterdam: IOS Press. Available from 10.3233/FAIA220637.

Taylor, R., Kardas, M., Cucurull, G., Scialom, T., Hartshorn, A. S., Saravia, E., Poulton, A., Kerkez, V., & Stojnic, R. (2022). Galactica: A large language model for science. *Science (New York, N.Y.)*. Available from https://doi.org/10.48550/arXiv.2211.09085.

Thompson, D. (2022). Breakthroughs of the Year. *The Atlantic*. https://www.theatlantic.com/newsletters/archive/2022/12/technology-medicine-law-ai-10-breakthroughs-2022/672390.

Thoppilan., et al. (2022). *LaMDA—Language models for dialog applications*. Available from https://doi.org/10.48550/arXiv.2201.08239.

Thorp, H. (2023). ChatGPT is fun, but not an author. *Science (New York, N.Y.)*, *379*(6630), 313. Available from https://doi.org/10.1126/science.adg7879.

Tiku, T. (2022, June 11). The Google engineer who thinks the company's AI has come to life. *The Washington Post*. https://www.washingtonpost.com/technology/2022/06/11/google-ai-lamda-blake-lemoine.

Vincent, J. (2022). Top AI conference bans use of ChatGPT and AI language tools to write academic papers. *The Verge*. https://www.theverge.com/2023/1/5/23540291/chatgpt-ai-writing-tool-banned-writing-academic-icml-paper.

Vota, W. (2020). Bot or not: Can you tell what is human or machine written text? https://www.ictworks.org/bot-or-not-human-machine-written/#.Y9VO9hN_oRU.

Weidinger, L., Mellor, J., Rauh, M., Griffin, C., Uesato, J., Huang, P., . . ., Gabriel, I. (2021). Ethical and social risks of harm from language models. Available from https://doi.org/10.48550/arXiv.2112.04359.

Weinberg, J. (ed.) (2020). Philosophers on GPT-3 (updated with replies by GPT-3). *Daily Nous*. https://daily-nous.com/2020/07/30/philosophers-gpt-3.

Wiggers, K. (2022). OpenAI's attempts to watermark AI text hit limits. *TechCrunch*. https://techcrunch.com/2022/12/10/openais-attempts-to-watermark-ai-text-hit-limits

Wolfram, S. (2023). What is ChatGPT doing . . . and why does it work. https://writings.stephenwolfram.com/2023/02/what-is-chatgpt-doing-and-why-does-it-work.

Wykowska, A., Chaminade, T., & Cheng, G. (2016). Embodied artificial agents for understanding human social cognition. *Philosophical Transactions of the Royal Society of London. Series B, Biological Sciences*, *371*(1693), 20150375. Available from https://doi.org/10.1098/rstb.2015.0375.

Perspectives on the ethics of a VR-based empathy experience for educators

11

Vanessa Camilleri
University of Malta, Msida, Malta

Introduction

Virtual reality (VR) is not just a buzz word. The scope of VR is that of providing the brain with an illusory experience that based on sensorial stimuli can evoke a level of immersion that creates a sense of presence in a virtual environment (Barbot & Kaufman, 2020). Any experience created in VR, if it has not been designed carefully and meticulously, can create chaos and disturbance, causing physical sickness at best and mental and emotional trauma at worst (Jerald, 2015). This has quite some serious ethical implications for developers venturing into the VR design world. Very often the term VR evokes in people mental associations of games for leisure and entertainment. And in discourse featuring VR, we may often notice that the human aspect of VR is overlooked and underestimated, to give way to talking about the technology and how it can be used to optimize the experience for the users. This in itself carries the danger of neglecting essential ethical issues that may have an impact on the human-VR interaction.

In this chapter, I discuss forms of VR experiences that may not necessarily be included in the leisure and entertainment sections, but may pertain to the professional development and treatment areas (Bailenson, 2018; Herrera et al., 2018; Ramirez, 2022; Uskali et al., 2021). Although the human element of a VR experience should always feature at the forefront of any design project, when the impact is expected to take on more serious forms and when the scope of the experience targets cognitive, mental, and emotional outcomes, then taking into account how humans perceive and interact with the designed world and how that world may impact their physical, mental, and emotional well-being takes priority. The intersection of technology with psychology, however, is not enough to properly design a VR experience that takes a holistic approach. There needs to be another element that regulates the boundaries between the full technology that can be exploited and the ensuing human behavior when in a simulated setting (Ramirez, 2022).

Enter Ethics. As a baseline definition for ethics, we can define it as a set of principles of conduct governing an individual or a group (Winston & Edelbach, 2013). If we were to transpose this definition to technology and technology-driven experiences, the description of ethics would be a reference to a set of decision paths that would regulate the behavior of the technology. The questions we would need to then ask ourselves would be, who decides for the decision paths? Who decides any outcome of human behavior arising from the decisions taken by these paths? And who

Ethics in Online AI-Based Systems. DOI: https://doi.org/10.1016/B978-0-443-18851-0.00020-2

would take the responsibility of this human behavior, especially if this behavior arising from technology has a negative impact or is dangerous to another human being? Then again we have to take into consideration the human as an essential element of the application of technology, and therefore before being responsible for the emergence of new killer apps, or a novel technology-driven medium, we really need to understand who this technology could impact, in both positive and negative ways.

Background
Virtual reality

According to Merriam Webster,[1] the definition of VR is that of an artificial environment that is experienced through sensory stimuli (such as sights and sounds) provided by a machine and in which one's actions partially determine what happens in the environment. Reflecting on this definition, we can also think of well-designed VR experiences as environments where humans and machines collaborate through a harmonious balance of the software and the hardware to provide for intuitive communication with the human. Jerald (2015) in "The VR Book" describes communication in this environment as two types; there is the direct communication in VR, which happens when the virtual objects reflect the physics of the real world, thus evoking sentiments that can be considered as visceral and may reach different levels of intensity depending on the individuals' predisposition and unique perceptions of the world. The second type of communication in VR experiences is the indirect type, whereby the brain processes what we observe and builds new associations based on different individual past experiences, or acquired knowledge. This level of communication language is often used, in its diverse forms, targeting the individuals' unique mental models to create meaning from what is being observed in the virtual world. Both these forms of communication in VR construct a reality that can engage the user in an experience that transcends the present perception of space and time, and project him/her in a world that looks real and feels and sounds real. This would bring about a sense of presence in VR, which also depends on a number of illusory elements, that would lead the human brain to believe that space, time, and interactions are arbitrarily assigned to the environment projected.

Jerald (2015) also highlights how presence in VR has four main pillars of illusion: the illusion of being in a stable space, the illusion of self-embodiment, the illusion of physical interaction, and the illusion of social communication. These four illusions can alter the perception of reality that is formed inside the brain, and Jerald focuses more on the human-machine interactions that would support this reality. Jerald focuses on arguments that there needs to be a balance on both the technology and the design based on elements of psychology and cognitive development. For example, to address the illusion of a stable space, a low latency and a high frame rate facilitate in the users the perception that the objects and the virtual environment obey the physics of the real world and are thus part of their "reality." However, this may not always be enough. Although the state of the art of technology is an important driver to achieve immersion, acquiring a sense of presence, through illusions, needs additional considerations.

---

[1]Definition of Virtual Reality. Retrieved from https://www.merriam-webster.com/dictionary/virtual%20reality

The brain and the perception of reality in VR

There is a difference between immersion and presence. Whereas presence is built on a series of illusions that are projected onto the brain, immersion is acquired by the degree that a VR system projects through a series of stimuli on the humans' sensory receptors. This means that while immersion is very much dependent on the technology affordances that also need to be taken into consideration during the design phase, for a sense of presence to occur, the design of the VR world needs to focus around the user's needs as well as the mental models that could be exploited in order to achieve a visceral communication.

We need to start first by acquiring a shared understanding of what perception of reality truly means. The basic understanding is that there is no universal truth of reality and that the perception of reality is a subjective view of the world that is often dependent on a number of complex factors and attributes that are processed in the brain (Purves et al., 2015).

The brain plays a deterministic role in constructing meaning out of what is being perceived. However, one cannot exclude how the interplay of brain and mind can also impact the way we perceive the world and the reality that surrounds us (Hoffman, 2019; Von Foerster, 2018).

A human's perception of the world occurs through mental imagery. Such mental imagery (both perceived and constructed) has been revealed, as emerging from a number of positron emission tomography (PET) studies, to use the same extended neural network in the brain associated with spatial processing (Dror, 2005). This means that the same brain areas are activated no matter where the image sources are coming from. Nonetheless, even though not all humans may have been exposed to the same experiences, people may be capable of using their spatial processing sections in the brain to either extend their knowledge of the world and fit it into the mental image or else project their knowledge of the world to extend their mental image.

Studies, though also reveal mental imagery and spatial processing in the brain, follow the laws of physics of the world, and this applies even when extrapolating on visual imagery (Mellet et al., 1996). Visceral communication relies on the sort of mental imagery that causes gasps and awe, and it evokes a sense of emotion that arises from the mesmerization of the images we are exposed to. This also leads to believe that such visual imagery can trigger emotional networks in the brain, which will then manifest in an emotional experience for the human. During VR, the same mental imagery process occurs. Translated into practices, while it may not always be possible to visit certain locations on the planet, with VR the brain can be exposed to the same visual imagery and thus trigger the emotions that are associated with an identifiable experience. Such experiences, as research has shown, may not be identifiable with previous encounters in whole, but may be reconstructed or extrapolated depending on previous knowledge (Barron, 2016).

Lindberg (2021) describes the illusions within the digital world from a transcendental perspective as she draws from philosophers such as Derrida, to discuss how the virtual world can still recreate the illusion of proximity, of presence, of a complete memory and of the worldwide community. These illusions do echo in part Jerald's four main pillars of illusions, as he too describes space (proximity), embodiment (touch), physical interaction, and social communication (the sense of communicating with the wider community). In her essay, Lindberg does not specify how the virtual world can recreate these illusions from a technical perspective but argues how the digital tools are shaping a new reality that is still dependent on connectedness and interactions. Despite the various illusions that the digital world creates, the sense of community and the social place of the individuals within the world communities are still prevalent and many of the tools that are emerging aim to achieve this sense of metaphysical presence. This metaphysical existence is picked up by the brain and construed as the individual's perception of his/her immediate reality.

However, we also argue that VR has the potential to go beyond the metaphysical and add to this through its affordances to support embodiment.

The proteus effect

The "Proteus Effect" is a term coined by Yee and Bailenson (2007), and Yee et al. (2009) takes into account this way of rethinking the self, through embodiment, in a way that is different from the main identity and as such may lead to a change of perceptions and behaviors that reflect the new sense of self that is achieved. If we view this through a virtual lens and discuss this in terms of the alter ego that may manifest through a digital representation, then we get an avatar that may behave, think, and act differently from the human that is directing it (Maselli & Slater, 2013). There are humans who do this purposely to escape their reality, and there are others who find themselves as unwitting subjects to the role they have been assigned in the virtual world (VW). Let us, for example, think of two widely differing worlds where avatars are constantly changing and evolving. VRChat is an example of a social virtual world, which avatars may inhabit to interact with each other and with virtual world objects. In VRChat, there are various examples of embodiment, which may not always be an extension of the self or a represented sense of cultural identity, but which may transcend what is perceived as real, or physical, and move toward a different dimension of the persona. As noted in the studies of Montemorano (2020), avatars in VRChat (and possibly other social VWs), "become actors as they perform, with their performance as a means of self-fulfillment of a fantasy — becoming, both visually and in personality, a favoured character." This playing out to the role is also noted in multimodal online role-playing games (MMORPGs) such as World of Warcraft, in which game players commit to a role to fulfill the expectations of their chosen avatar and be able to function within the game, thus establishing the legitimacy of their "being" inside the VW.

In their studies, Yee et al. (2009) discuss the Proteus effect in relation to user behavior and the self-fulfilling identification based on their avatar characteristics. Thus, for example, in their studies, avatars assigned a taller height showed a greater predisposition to assert their rights in a VW than their counterparts assigned a shorter height. However, their studies also showed that the effect of their modified avatar behavior lingered after participants in their study returned to physical reality. Although the results of that study may not be generalized to other aspects of online communities, and results may not always be consistent across diverse users, there are strong indications that the mental associations built through the visual imagery of the self-build perceptions that may extend beyond the VW as presented in other related studies (Lugrin et al., 2015). Of course, this may present a number of issues and questions that need to be taken into account before discussing how behavior could be modified using VR technology. For example, what if the role taken up by the avatar is aggressive toward a certain faction of society or groups of individuals? Would there be a risk that this aggression transposes onto the real world? If the virtual self is entangled in the physical self and vice versa, how can an individual determine whether his/her identity has morphed into one that features an interplay of alter egos? And following this, who assigns the responsibility for determining the roles taken up by avatars in VR and VWs? Who would be accountable in cases where the transposition to the real world causes harm to individuals? These are all ethical concerns that need to be discussed and taken into consideration before embarking on a design of a VR experience.

VR as an empathy machine

There is quite substantial research in the field where VR may be utilized as a medium supporting the development of empathy in users (Almiron & Crosetto, 2019; Barbot & Kaufman, 2020; Bönsch et al., 2019; Cogburn et al., 2018; Cotton, 2021, Herrera et al., 2018). In one of the first experiments in immersive journalism carried out by Nonny De la Peña et al. (2010), the investigative journalist had no clue that a recreation of a story about a shelter for the homeless and the death of an individual queuing up for food would garner so much attention at a conference. She had been working on an article detailing how a shelter for the homeless in San Francisco was distributing food to those who had no other means of getting food. For the purposes of the article, she took photos every day for a week and recorded the real live audio during the photo captures. On going back to her team, they got together and sketched the environment from the photos they had taken and used the live audio to build an environment in VR. It so happened that during one of the days she was recording one of the people in the queue fainted and died on the spot. This unfortunate incident was also captured on the photos and was recreated via a 3D sketch. Although the visuals were digitally recreated and didn't show the real people, all the buildings, roads, and number of people on the street were kept true to form while also using the real audio captured on the day. During the conference, De la Peña et al. (2010) assisted participants throughout the experience and she was very surprised to see how the participants, despite the low fidelity graphics, could relate to the narrative unfolding through the VR experience, with some displaying evident distress at the sight of the prone avatar lying on the ground. Milk (2015), the director responsible for a number of immersive artistic representations that include film and video, has described VR as the ultimate empathy machine. In his own terms, VR has capabilities that go beyond transporting people to another dimension, and creates in users an empathic connection that people keep associating with even when they are no longer in VR. This in itself has wide-ranging implications, in terms of the impact that this can cause on people. However, it is because of the impact that this technology-driven medium can have on people, ethical implications need to be taken into consideration. If as indicated by research, VR can have long-lasting effects on people's actions and behavior, then we need to reflect on who decides what the changes in people's behavior will be, and who will this change benefit? We also need to take into consideration accountability and responsibility assigned if things go wrong (Aliman & Kester, 2020). VR may be the ultimate empathy machine — it may have the potential to do good and to improve the quality of life of people, but we also need to ask ourselves which and how are the ethical processes guiding it, which are chosen to ensure that no harm may come to any individual who experiences it. As an example, the 1000 Cut Journey (Cogburn et al., 2018) project immerses the individual into the life (from childhood to adulthood) of a black male as he experiences subtle racial messages from the people he encounters. Longitudinal studies for this project show that Caucasian people who go through this VR experience self-report a change in empathy toward people of a darker skin color. Although this project is very commendable in itself, especially when it increases a person's understanding of racial injustice, it leads to more questions, such as can this tool potentially be used to modify a person's behavior to cause harm or injustice to others? Can such a strong empathic medium cause a person to lean toward racial injustice rather than increase understanding toward different races? Can this tool be used for the wrong purposes? More importantly, we have to ask ourselves what can be done to prevent this from happening and how we can safeguard the people who are experiencing it (Aliman & Kester, 2020).

VR as an empathy machine is demonstrated in projects such as the 1000 Cut, where the user experiences a body that is different from his/hers, in a bid to walk in someone else's shoes, and stems from an experiment that originated in 1998 where users were given a rubber hand to hold instead of their own. With synchronous movements, the brain was fooled in believing that the rubber hand was indeed their own, thus registering sensations as though the hand that was being stroked was indeed their own. This paradigm led to more recent experiments, utilizing VR as a medium, where, as shown earlier, users could achieve a full-body ownership illusion (Maselli & Slater, 2013). In the full body ownership illusion, users wear a head-mounted display (HMD) to block the sight of the body and aid the brain in viewing the body assigned in the VR environment, which the eyes see through the lenses. In this case, the brain is receiving the messages that the reflection it is looking at is indeed his/her own. However, the studies by Maselli and Slater (2013) provide some interesting conclusions that have been used in the case study described later. One conclusion of the study is that one of the conditions for a full-body ownership illusion to occur is that there needs to be a first-person perspective in the VR environment. This means that even though there may be a reflected image, the person is seeing his/her environment through the eyes of the assigned character. The perspectives and the rules of physics of the world play an important role in establishing this first-person perspective. Realism in the scenes is another contributing factor to establish full-body ownership, while a multisensory experience synchronizes the overall illusion giving a more grounded sense of reality and therefore a stronger connection between the self and the perceived (Jerald, 2015).

Methods

The application—walking in small shoes

A team of researchers at the University of Malta worked together to design and develop an application targeted toward primary school teachers, with the scope of increasing their understanding of what it feels like to be a child on the autism spectrum in the primary classroom (Camilleri et al., 2017). Autism is, as defined by the Autism Society, "a complex developmental disability with signs that affect a person's ability to communicate and interact with others." This definition is intentionally broad as there are no prescriptive rules and conditions that this disorder follows. However, research shows (Case-Smith et al., 2015) that around 80% of children on the autism spectrum exhibit distress, when exposed to sensory stimuli. Distress or state of anxiety may be manifested in a number of ways that are not always as apparent to the people who are not affected by such sensory stimuli. In a typical primary classroom, with 24 young children and one or more learning support assistants (LSAs), it is not always easy to be aware of elements of the environment that might be perceived as sensory threats by people on the autism spectrum. The application was designed as an additional support to educators, to raise awareness of examples of stimuli that may be a cause of

FIGURE 11.1

VR-app; camera positioning and realistic filming (Camilleri et al, 2017).

distress to the child on the autism spectrum. With increased awareness of which stimuli are best to avoid or keep under control in the classroom, a teacher can provide an environment that is more serene for the child, who can then focus more on the learning environment rather than on the distractions around (Fig. 11.1).

Study design

The design of the study brought together an interdisciplinary team of researchers with different expertise and coming from diverse backgrounds. Inception phase involved the recruiting of experts on autism for multiple discussions about the possible scenarios for children on the autism spectrum when they are in class. Further literature was acquired by a team of researchers on first-person narratives and the coping strategies, people diagnosed on the autism spectrum used from a young age, to deal with school environments. Experts were recruited on a voluntary basis. Using the data collected as a ground for development, researchers from the area of film, as well as from the area of AI, scripted the narrative specifically for the VR environment. An embodied cognition framework (Bailey et al., 2016) was the main research framework that was used for the study. This framework assumes that the mental representation of the body determines the individual's cognitive responses and following interactions within the context in which it is placed. VR according to the researchers is an appropriate vehicle for this ECF to manifest itself, and for the interactions to happen in a safe, controlled environment. The adoption of this framework is justified by the research question at the base of this study: "How can the design of a VR environment support the teachers' development of empathy towards children on the autism spectrum?"

The target audience for the end product were primary school teachers, as well as preservice teachers going through their training in education. The research method adopted was a mixed method approach, using a pre—postexperience survey for a quantitative analysis and participant observation for the qualitative part of the analysis. The pre—postexperience survey was adapted from Spreng et al.'s (2009) empathy quotient tool. The tool was used as a means to self-report any changes in empathy toward people on the spectrum, before and following the immersive experience. For the participant observation, an independent researcher was recruited to observe the way the participants reacted during the VR experience (Ross & Morrison, 2004). All observations were recorded and coded according to emotional states linked to empathic behavior. These included changes in facial expressions, sighing, heavy breathing, visible emotion, impassivity, smiling, etc.

The design

For the embodied cognition framework, the design involved the use of interactive narratives (Bucher, 2017). In this style of film, the users may influence the storyline by assuming the role of a character in the virtual world. Thus, the user becomes the leading role in the unfolding story, and characters within the designed world refer to him/her, in first person. This is another illusion where the brain perceives that personal actions can have consequences on the story, and it is when the embodied cognition framework kicks in, creating a belief that what is happening is indeed part of the user's world. During the design phase, researchers identified the main stimuli that may be found in classrooms, and that may have a negative effect on a child on the autism spectrum. These were identified following a number of expert interviews as well as on the basis of literature. Once these stimuli were identified, they were isolated and augmented in a way to exploit the impact on the senses. High-fidelity filming was used to make the setting more realistic, and actors were recruited in roles of teachers, a learning support assistant, and other primary school-aged children. VR filming uses concepts from filmography that are derived from isolating specific situations and uses camera to augment these through the visuals. The filming direction through the scenes, sounds, and overall 360 visuals, even if augmented, still follows the physics of the world, and therefore this can serve to evoke in the audience an empathic sense of the reality that extends from the virtual scene. The difference between traditional and VR filming is all about the perspective of the viewer. While in a traditional film the viewer is usually facing the film frontally and does not have the possibility of moving around, in a VR the viewer has the freedom to move 360 degrees. Therefore in VR, the traditional changing of shots happens as in film, and the scenes become almost loop like with no real beginning or end, and the perspective depends on the angle from which the viewer is looking (Naimark et al., 2016). This leads to another interplay of design factors, when constructing a narrative. In our case, the narrative follows some form of linearity. The story itself aims to capture the user's attention to the isolated stimuli that have been augmented in a way to exploit their impact on the senses.

Therefore, the camera capture of the visual scene was complemented by the 3D audio as spatial sound followed the viewer and guided his/her attention. Bucher (2017) explores the elements of narrative for the construction of a VR-based film experience and explains the scientific elements that guide the direction of this specific medium. The design of "Walking in Small Shoes" uses viewpoints, referred to as the *eye line*, to transmit different perspectives, such as looking up at the teacher, while looking at the same level of the other children in class. The positioning of the cameras could therefore exploit the relative height of the surroundings, to strengthen the embodied cognition. When filming an outdoor scene, such as the playground, the camera rig was positioned at the same height as a child sitting on the ground, giving not only the impression that the viewer was seated but to also feel at a less privileged position in relation to those who were standing next to him/her (Naimark et al., 2016). Thus when children came over to speak to the viewer, she/he would have to look up to tilt his/her head. This brings about in the viewer the sensation of "feeling small," and this causes the Proteus effect to trigger in, with responses adapted to the context in which the user finds himself/herself in.

Sounds were used as the means to attract the user's attention but they were also exploited to augment the stimuli that could be identified as having a negative impact on a child on the autism spectrum disorder. Examples of sounds that were augmented included the ticking of the wall clock, the level of noise from student chatter, and the teacher's tone of voice, among others. Sound was recorded using a binaural mic to capture 360 ambient sounds, as they were stitched to the 360 visuals. In addition to the augmentation of stimuli, sounds had to follow the physics of the world, especially when the sounds are relative to the head movement. These might include background sounds, body sounds (breathing or heart beats), narration, etc. (Murphy & Pitt, 2003). A voice-over experience was introduced to the narrator to emulate the voice of the autistic child and strengthen the embodiment by means of the vocalized thoughts about the unfolding narrative (Fig. 11.2).

Ethics, empathy, and emotion

Taking a break from the study and eventual results of this VR experience, this chapter first takes a look at the code of ethics developed by the Institute for Electrical and Electronics Engineers (IEEE),[2] which are broken down into three main headings, in relation to the profession, the end users, and coworkers. In summary, the code of ethics describes the importance of keeping individuals from harm and protecting them, while keeping them informed about the impact that technology might have on their lives. Although this code of ethics doesn't specifically mention VR or AR, as immersive media, the implications can be transposed to a scenario that uses these media. Madary and Metzinger (2016) building on these basic code of ethics representing the technology setup have mapped out a set of ethics recommendations governing practices of VR. The recommendations follow six guidelines that explore the use of VR in an experimental environment. This means that these recommendations apply mostly to applications that are designed for a purpose that is other than just for entertainment purposes. Applications such as those presented in this study fall under this category.

---

[2]IEEE Code of Ethics. Retrieved from https://www.ieee.org/about/ethics/index.html

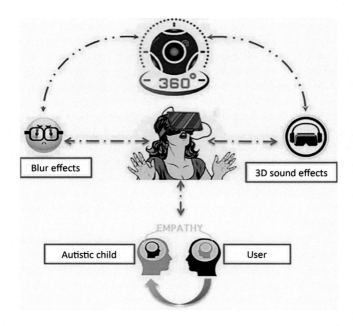

FIGURE 11.2

Interface elements for VR app (Camilleri et al. 2017).

The six main research ethics recommendations according to the authors include the following:

- Ethical limits of the experimental environment — experimenting in a VR environment leads to possible physical, mental, and emotional impact on a human, and thus the principle of beneficence has to be applied.
- Informed consent has to be obtained every time research in VR is conducted. Participants have the right to know that they might experience some effects as a result of their experience in VR.
- Identification of risks associated with the clinical applications of VR — especially if this is being applied as part of a medical treatment process. There are issues of giving patients false hopes, medical insurance coverage, and right of choice of patient that might be enlisted in the clinical trials.
- Risk of use for malicious purposes — even though the main purpose of the VR application might be to improve the quality of life of a person, the same tool can be used to harm the person or to manipulate the person's thoughts and actions without his/her consent.
- VR as applied to online research — it is far too easy to oversee privacy issues, as well as informed consent when researchers carry out research "in the wild" and use the online environment. For example, when recording data from the online environment, one may choose to collect data such as eye movement, emotional reactions, and kinematic data that may eventually lead to a person's identity.

- Limitations on the Code of Ethics — in essence, there are limits to the guidelines for ethics that are published. It is ultimately up to each researcher to understand the ethical ways of following experimental procedures, especially those that directly involve human beings (adapted from Madary & Metzinger, 2016).

The six recommendations listed earlier were taken into consideration during the implementation phase of the project. Out of the six recommendations, the ones that applied to "Walking in Small Shoes" were mainly the principle of beneficence or rather avoiding harmful impact on the users, as well as the informed consent. The first ethical principle was considered during the design phase. The scenes were designed to be short, thus making sure that the participants were not inside the immersive environment for a long period of time. This was done to try and prevent any possible cyber sickness as well as any overwhelming, emotional effects that might result from the impact of the experience.

This experience was also meant for consenting adults going through a professional development course in teacher training. Nonetheless, all the participants were briefed before the VR experience, explaining what they would be seeing and how they might feel as the main character in the narrative. The briefing highlighted the possibility that they might feel discomfort, both physically and emotionally during the session. Participants were informed that they could stop the experience at any time they felt such discomfort without any consequences. This was not just delivered as a note, but it was verbally explained before the session so that participants who felt that they didn't want any part of the experience would have the possibility to back out from the experience before they started it. Nonetheless, they still had the option to stop during the experience. Ramirez (2022) picks up on the recommendations and guidelines by Madary and Metzinger (2016), as well as the toolkit provided by the Markkula's Centre for Applied Ethics,[3] and adapts them to include an ethics approach when designing VR applications.

Ramirez recommends that when designing for VR, the following practical aspects need to be taken into consideration:

- Designing simulations that are perspectively faithful and context realistic needs to be done meaningfully to achieve the desired effect and not done for sensationalism.
- VR applications designed for children below 13 years of age need to be adequately reviewed by professionals in the field.
- VR applications should not be designed for use over a long period of time.
- VR applications that deal with user behavior need to be transparent in their outcome and impact and should avoid manipulation.
- Content in VR may need to have different ratings as that from traditional media.
- User physical and biometric data should not be collected unless this is needed, and the users need to be adequately informed about which data is being collected, why, and what use will it have for the research.
- VR applications released into society need to have additional ethical reviews.
- VR applications should encourage moral values and strive to do good.

---

[3]Markkula Centre for Applied Ethics: An Ethical Toolkit for Engineering/Design Practice. Retrieved from https://www.scu.edu/ethics-in-technology-practice/ethical-toolkit/

To this extent, especially for the last two practical recommendations, adapting the questions from the toolkit designed by Markkula's Centre for Applied Ethics (2016) would be a good starting point. However, both Madary and Metzinger (2016), as well as Ramirez (2022), have additional concerns about VR applications, especially those targeting empathy.

Some of these concerns include the effects of long-term immersion on users and the possibility that some of the outcomes might include a de-personalization of the person and the assimilation of the virtual self. Although this is not always possible to discern, or even to stop at a certain moment in the immersion phase, the design needs to take into account how this may happen in a user and take steps to mitigate this possibility. In addition, it needs to be taken into account that the way content works in traditional media on human perception is not the way it is assimilated in a VR medium. Therefore, the effects on the person of a scene in VR as opposed to the same scene in a traditional 2D medium need to be taken into account and the possible harmful impact on the person's psyche preempted. The false sense of agency induced as well as possible cases of "social hallucinations" (Madary & Metzinger, 2016) can also be considered as an harmful impact on the person. Although in principle a VR application's scope is positively impacting a person's actions, thoughts, or behavior, the possibility that the embodiment might result in a negative outcome affecting the psychological state beyond the VR environment does exist, and therefore the necessary action has to be taken to reduce the risks.

If we take into account empathy as one of the emotional states that a VR can target, then there are number of additional ethical issues that need to be part of the design process. In the case study "Walking in Small Shoes," the project initially kicked off as a pilot project with limited funding and resources. It therefore took into account only one specific school location, with just a few scenes that might feature in a daily experience. But as researchers, we do recognize that the context is not as simplistic and that there may be additional experiences outside those recounted by the 5-minute VR experience that we have not considered. This is the reason why this specific project was kept as a pilot and awaits further funding to be able to be developed further. As per concerns shared by Ramirez (2022), we were also concerned that people associate the short experience with a generalization of the overall context. In our case, the short experience where teachers could associate with a child on the autism spectrum more closely could lead to a false sense of agency that could question the authenticity of the whole experience.

Uskali et al. (2021) expose their research about emotional development using immersive journalism. They clearly echo Madary and Metzinger's (2016) concerns about the power of manipulative risks during an immersive experience and warn of possibilities that when designers of VR experiences, especially in journalism, target emotion there is the need to exert more caution. Ethical guidelines need to be rethought in line with the extent of realism brought about by VR and to protect people from the powerful transmission of the immersive narrative. Violence, torture, suffering, and other human stories that are at the focus of specific emotive journalistic exploits can take on a new dimension when these are recreated in VR (Aliman & Kester, 2020).

Although these questions might come across as counteracting the research that shows how VR may be used for beneficial purposes and exploited to support empathy (Barbot & Kaufman, 2020), the arguments posited here serve to highlight the need that further thought needs to be given to the design of such VR applications. This is not about whether VR can be used to support the development of emotional states such as empathy but it intends to push the argument that if such an application is deployed, then it has to be done with the caution it deserves and by adhering to guiding principles of ethics.

Reflections and discussion

The case study "Walking in Small Shoes" was evaluated with a sample of 63 teachers in a lab environment. However, it was also tried out during numerous professional development courses held at a number of primary schools across the island. Participant observations and post-VR feedback were collected during these instances. The following reflections arise from the observations and implications of the VR experience, on the research that is ongoing in the field of VR and ethics.

Reflection #1
VR as an empathy machine

As can be seen from literature, there are a number of scholars who have argued in favor of and against the use of a VR experience as an empathy machine (Gal, 2022; Ramirez, 2022). Our results show that the participants who used "Walking in Small Shoes" felt that following the VR experience, they could understand children on the autism spectrum much better. They also felt it helped them relate to some of their own classroom experiences. Following up with some of the participants, they expressed views that they could still remember the experience well and as such were more aware of certain class elements that may hyperstimulate children on the autism spectrum. For example, they mentioned that they try to keep their voices a bit lower, and to ensure that all children in class became more aware, they needed to keep a calmer, quieter environment. Some even mentioned that they replaced their classroom ticking clocks with ones that do not make the ticking sound. A learning support assistant spoke about how she stopped spraying on perfume when going to school so as not to cause any distress to the child she was helping and who was on the autism spectrum. Therefore, we can say that the VR experience did achieve its scope of raising more awareness of how to curb certain habits or moderate stimuli to cause children with ASD less distress. In their own terms, the experience helped them empathize with the children in their classroom. However, during the feedback sessions it emerged that some teachers thought that they really got to know much about autism. The reality is that autism is a very complex disorder and the VR experience although designed with good intentions and adhering to all the ethical guidelines may have given the wrong impression that autism is simply about controlling external stimuli. This in effect is rather untrue. Hyperstimulation, leading to more distress manifested in a number of ways, may be one manifestation of autism. However, there is more to autism that can be represented in a short VR experience (Case-Smith et al., 2015).

The danger that one short VR experience may lead to the generalization of the level of understanding of a specific condition may in itself be thought of as quite dangerous and may in the long term cause more harm than good. This may also lead to misconceptions and eventual desensitization to the individuals' needs as those experiencing VR may retain the firm belief that the condition or the situation may in reality be strongly overrated.

One, therefore asks, how can this be counteracted? Each experience needs to be accounted for by a thorough learning session or multiple sessions where, in this case, teachers are exposed to more case stories of children on the autism spectrum. The VR experience can be used as a way to ground one or a few of the diverse learning experiences in an immersive setting. As a final reflection on this point, there are various narratives, and various research studies that show how the immersive nature of VR can help transport users into a reality that might not otherwise be possible to envision or experience (Cogburn et al., 2018; Cotton, 2021; Jerald, 2015). The technology mediates this, and the state of the art is showing that the future is directing more improvements in this field. However, when thinking of VR as an empathy machine, we need to reconsider how the whole exercise is brought to the people experiencing it. If the VR is delivered as a standalone application and sold as an empathy machine, we will run the risk that more and more people will become more desensitized to the actual needs of people we are trying to empathize with. In a bid to "understand" the people who are placed in a different context to us, we will be simplifying the situation, and in the long run, empathy will not really reflect the needs of the people affected. If we want to really create an empathy machine, then this needs to be followed up by additional learning experience, as well as reflection sessions that are meant to explore the depths of the context.

Reflection #2
Empathy by design

The app "Walking in Small Shoes" was designed specifically for primary school teachers in a local environment. During one of the public talks held to illustrate the app, we received numerous requests by parents of children on the autism spectrum to try out the app. We wanted to abide by ethical guidelines, but we also felt that constructive feedback could be obtained from parents who could see the classroom environment from the perspective of their child. Some of the parents who sent a written request to try out the app were invited to attend a lab session where they were briefed, signed their consent, and then started the VR experience.

All of the parents who tried out the app manifested deep emotional states while going through the short VR experience. Almost all of the parents described the experience as one that was quite traumatic for them. They believed that through the short experience they were living the daily life of their child at school. However, what they could not see was that what they believed was the daily reality of their child had been designed as an augmented collection of stimuli that might be manifested over the course of an entire day or days at school. While for teachers this was accepted naturally as they were used to the daily classroom practices for the years they had been teaching, the result for the parents was overwhelming. This was of course one aspect of the research that we wanted to avoid.

Upon reflection, we concluded that one of the lessons learnt from this exercise was that the app was not designed for parents. The app was designed for teachers who were used to the chaos that is usually present in the primary classroom. The result was that the parents who went through the VR and who were not teachers were profoundly affected by the whole experience. In their own eyes, the daily experience of their child attending school was amplified in a negative way. While teachers going through the same experience were identifying with the class situation and were increasingly more aware of how to construct a more positive class experience, parents were focused on all that was negative and how this was impacting their son in an adverse way. Some walked away feeling more guilty about sending their child to school, viewing the environment as a daily battleground for their son or daughter. In real life, teachers know that there are many days when students are cooperative, helpful and there is a positive attitude in the classroom. The short VR experience did not fully capture all the positive aspects, and for training purposes, scenes and narratives were chosen and constructed in a way as to direct the teachers to reflect on how the situations that arose could cause distress in a bid to improve the context. This exercise showed how and when the VR experience is used out of context, and with an audience who is not the designated one, the experience can turn from positive into one with a negative impact. In this case, all the ethical guidelines were followed and giving access to parents in a controlled environment was done with good intentions. However, the result depended very much on what is termed the "semantic variance" (Pinch, 2010) or rather when the meaning that is constructed depends on the context of the person. When VR apps are released to the public, there is no assurance of who will be using it, and the context in which the experience will be interpreted. In this case, both types of users, the teachers and the parents, had the best interests of the child on the autism spectrum at heart, and yet the experience was interpreted in varyingly different ways.

Reflection #3
Insights into ethics and AI in VR-based systems

The app "Walking in Small Shoes" was designed to create awareness among a target audience about stimuli that might affect people on the autism spectrum negatively. Its design was kept simple, and the level of depth and complexity in relation to the disorder has purposely been low. The reason for this is that there was no way of knowing how the user would have perceived the application, and whether it might be the cause of some form of distress or possible harm. In effect, there were a number of issues that warrant a deeper investigation into the ethics of integrating such applications into training and professional development. The first issue that needs to be raised and mitigates any form of harm or threat to an individual's well-being is that every VR experience needs to be carried out in the presence of professionals who can offer support, both before and after the VR user experience. In this case, this is especially helpful in circumstances where the users might experience some form of hyperstimulation that might lead to them feeling emotionally or physically overwhelmed with the whole experience. Professionals who are assisting in an environment that targets aspects of emotional intelligence need to have a good understanding of the field, as well as a solid foundation in behavioral psychology. Ethically this ensures that users are protected from harm when going through the VR-based experience. Having professionals who elaborate more on the condition or disorder that is being spotlighted during the experience will also ensure that ethically the same condition or disorder is neither overestimated or glorified, nor underestimated. There is danger that if the VR experience, for various reasons, simplifies the complexity of the condition or disorder, the user may wrongly misclassify behaviors associated with the condition. In such cases, the user may start assuming that they can categorize a person's behavioral pattern based on such knowledge, which may not cover the entire span of the condition/disorder, and this leads to an erroneous judgment of how to handle specific situations that may arise. For this reason, it is especially important that when using VR-driven applications for emotional intelligence, these are designed with meaning and purpose and not just used to offer some stimulus to break the boredom of training. It is best not to use a poorly designed VR application, rather than just use it to provide a brief moment of a sensational break that may have lasting negative effects. Therefore, when using VR for training and development, especially for emotional intelligence development, there needs to be enhanced awareness on the impact it may have on the users both from a personal perspective and how the outcomes of this experience will in turn possibly impact a broader audience. Questions that need to be asked at the design stage need to take into account the reasons why users would go through the VR experience, the experience that they will build, and how this may impact their attitude toward people who are afflicted by a specific disorder or condition or who are framed within a certain context. Tests need to be performed to understand whether the experience does indeed achieve its intended goals or whether different human experiences might lead to a substantial deviation from the original scope that might then lead to a possible harmful or negative impact on the target population group.

Conclusion

This chapter discussed the use of VR as a medium to support empathy. Case examples have shown the potential of the technology to create an immersive environment that on the face of it lends itself well to support empathy. This is in view of the fact that with careful design and proper use of the technology, the experience can carry the user across boundaries of space, time, and embodiment. However, there are a number of questions that run deeper than the technology affords and that can impact the users in a negative way. We need to keep these in mind when designing such VR-driven apps to support emotional intelligence, as this might induce harm at a psychological, affective, and even cognitive level. There are toolkits that one can make use of to build an ethical framework for the designed apps; however, there are other additional questions that need to be considered before proceeding with rolling similar apps out in the wild. Questions such as "can the immersive context be interpreted differently by the different audiences?." If the answer is yes, then we need to explore all ways in which the whole context can be interpreted, and whether the individual could run the risk of either creating more misconceptions about the context or even becoming desensitized to the real needs of the person with whom he/she is supposed to empathize. We need to understand whether the training toward empathy could result in mere manipulation or whether the continued use of the technology might expose the users to more harm than good.

Finally, we need to reflect on who controls the VR experiences users are exposed to, and in the cases where we are the designers, how would we be able to control how the users see themselves beyond the VR? What steps can we take to avoid harm to the individuals making use of the experience and to the represented individuals for whom the experience has been designed?

The future is of course all about the continuous evolution of technology. The metaverse, the alternate digital world, is already a reality that is here, and that is in use. VR is not just a word present in a 1980s fiction book that we need to argue in favor of or against. What we need to do is stop and reflect on how we can truly harness the benefits it offers and minimize the harm it has the potential to do.

References

Aliman, N.M., & Kester, L. (2020). Malicious design in AIVR, falsehood and cybersecurity-oriented immersive defenses. *Paper presented at the—2020 IEEE international conference on artificial intelligence and virtual reality (AIVR)* (pp. 130–137). Available from https://doi.org/10.1109/AIVR50618.2020.00031.

Almiron, M., & Crosetto, G. (2019). Virtual reality, a place for knowledge about our being and empathy? In R. Smite, & R. Smits (Eds.), *Virtualities and realities. New experiences, art and ecologies in immersive environments* (pp. 59−66). RIXC, 978-9934-8434-7-1.

Bailenson, J. (2018). *Experience on demand: What virtual reality is, how it works, and what it can do.* W. W. Norton & Company, Inc.

Bailey, J. O., Bailenson, J., & Casasanto, D. (2016). When does virtual embodiment change our minds? *Presence: Teleoperators and Virtual Environments*, *25*(3), 222−233.

Barbot, B., & Kaufman, J. C. (2020). What makes immersive virtual reality the ultimate empathy machine? Discerning the underlying mechanisms of change. *Computers in Human Behavior*, *111*, 106431.

Barron, D. (2016). How the brain processes images. *Scientific American*. Retrieved from https://blogs.scientificamerican.com/mind-guest-blog/how-the-brain-processes-images/.

Bucher, J. (2017). *Storytelling for virtual reality: Methods and principles for crafting immersive narratives.* Routledge. Available from https://doi.org/10.4324/9781315210308.

Bönsch, A., Kies, A., Jörling, M., Paluch, S., & Kuhlen T.W. (2019). An empirical lab study investigating if higher levels of immersion increase the willingness to donate. *Paper presented at the—019 IEEE virtual humans and crowds for immersive environments (VHCIE)* (pp. 1−4). Available from https://doi.org/10.1109/VHCIE.2019.8714622.

Camilleri, V., Montebello, M., Dingli, A., & Briffa, V. (2017). Walking in small shoes: Investigating the power of VR on empathising with children's difficulties. In *Proceedings of the 23rd international conference on virtual system & multimedia (VSMM)* (pp. 1−6).

Case-Smith, J., Weaver, L. L., & Fristad, M. A. (2015). A systematic review of sensory processing interventions for children with autism spectrum disorders. *Autism: the International Journal of Research and Practice*, *19*(2), 133−148.

Cogburn, C., Bailenson, J., Ogle, E., Asher, T., & Nichols, T. (2018). *1000 cut journey. ACM SIGGRAPH 2018 Virtual, Augmented, and Mixed Reality (SIGGRAPH '18)*. New York: Association for Computing Machinery. Article 1, 1. Available from https://doi.org/10.1145/3226552.3226575.

Cotton, M. (2021). *Virtual reality, empathy and ethics.* Switzerland AG: Springer Nature. Available from https://doi.org/10.1007/978-3-030-72907-3.

De la Peña, N., Weil, P., Llobera, J., Spanlang, B., Friedman, D., Sanchez-Vives, M.V., & Slater, M. (2010). Immersive journalism: Immersive virtual reality for the first-person experience of news. Available from https://doi.org/10.1162/PRES_a_00005.

Dror, I. E. (2005). Perception is far from perfection: The role of the brain and mind in constructing realities. *Behavioral and Brain Sciences*, *28*(6), 763.

Gal, R. (2022). Rage against the empathy machine revisited: The ethics of empathy-related affordances of virtual reality. *Convergence: The International Journal of Research into New Media Technologies*, 1−19. Available from https://doi.org/10.1177/13548565221086406.

Herrera, F., Bailenson, J., Weisz, E., Ogle, E., & Zaki, J. (2018). Building long-term empathy: A large-scale comparison of traditional and virtual reality perspective-taking. *PLoS One*, *13*(10), e0204494. Available from https://doi.org/10.1371/journal.pone.0204494.

Hoffman, D. (2019). Do we see reality? *New Scientist*, *243*(3241), 34−37.

Jerald, J. (2015). *The VR book: Human-centered design for virtual reality.* Morgan & Claypool.

Lindberg, S. (2021). Four transcendental illusions of the digital world: A Derridean approach. *Research in Phenomenology*, *51*(3), 394−413.

Lugrin, J., Latt, J., & Latoschik, M.E. (2015). Avatar anthropomorphism and illusion of body ownership in VR. *Paper presented at the—2015 IEEE virtual reality (VR)* (pp. 229−230). Available from https://doi.org/10.1109/VR.2015.7223379.

Madary, M., & Metzinger, T. K. (2016). Real virtuality: A code of ethical conduct. Recommendations for good scientific practice and the consumers of VR-technology. *Frontiers in Robotics and AI*, *3*, 3. Available from https://doi.org/10.3389/frobt.2016.00003.

Maselli, A., & Slater, M. (2013). The building blocks of the full body ownership illusion. *Frontiers in Human Neuroscience*. Retrieved from http://www.frontiersin.org/articles/10.3389/fnhum.2013.00083/full.

Mellet, E., Tzourio, N., Crivello, F., Joliot, M., Denis, M., & Mazoyer, B. (1996). Functional anatomy of spatial mental imagery generated from verbal instructions. *The Journal of Neuroscience: The Official Journal of the Society for Neuroscience*, *16*(20), 6504–6512. Available from https://doi.org/10.1523/JNEUROSCI.16-20-06504.1996.

Milk, C. (2015). How virtual reality can create the ultimate empathy machine. *TED Talks*, Retrieved from http://www.ted.com/talks/chris_milk_how_virtual_reality_can_create_the_ultimate_empathy_machine/transcript?language = en.

Montemorano, C. (2020). Body language: Avatars, identity formation, and communicative interaction in VRChat. *Student Research Submissions*, 361. Available from https://scholar.umw.edu/student_research/361.

Murphy, D., & Pitt, I. (2003). Spatial sound enhancing virtual story telling. In O. Balet, G. Subsol, & P. Torguet (Eds.), *Virtual storytelling: Using virtual reality technologies for storytelling* (pp. 20–29). Avignon: Springer.

Naimark, M., Lawrence, D., & McKee, J. (2016). VR cinematography studies for Google. Retrieved from https://medium.com/@michaelnaimark/vr-cinematographystudies-for-google-8a2681317b3

Pinch, T. (2010). Comment on nudges and cultural variance. *Knowledge, Technology & Policy*, *23*, 487–490. Available from https://doi.org/10.1007/s12130-010-9129-1.

Purves, D., Morgenstern, Y., & Wojtach, W. T. (2015). Perception and reality: Why a wholly empirical paradigm is needed to understand vision. *Frontiers in Systems Neuroscience*, *9*, 156.

Ramirez, E. J. (2022). *The ethics of virtual and augmented reality: Building worlds*. Routledge. Available from https://doi.org/10.4324/9781003042228.

Ross, S. M., & Morrison, G. R. (2004). *Experimental research methods*, . *Handbook of Research on Educational Communications and Technology* (Vol. 2, pp. 1021–1043). Springer.

Spreng, R. N., McKinnon, M. C., Mar, R. A., & Levine, B. (2009). The Toronto empathy questionnaire: Scale development and initial validation of a factor-analytic solution to multiple empathy measures. *Journal of Personality Assessment*, *91*(1), 62–71. Available from https://doi.org/10.1080/00223890802484381.

Uskali, T., Gynnild, A., Jones, S., & Sirkkunen, E. (2021). *Immersive journalism as storytelling: Ethics, production, and design*. Taylor & Francis.

Von Foerster, H. (2018). *On constructing a reality. Environmental design research* (pp. 35–46). Routledge.

Winston, M., & Edelbach, R. (2013). *Society, ethics, and technology*. Cengage Learning.

Yee, N., & Bailenson, J. (2007). The Proteus effect: The effect of transformed self-representation on behavior. *Human Communication Research*, *33*(3), 271–290.

Yee, N., Bailenson, J. N., & Ducheneaut, N. (2009). The proteus effect: Implications of transformed digital self-representation on online and offline behavior. *Communication Research*, *36*(2), 285–312. Available from https://doi.org/10.1177/0093650208330254.

Assessing and implementing trustworthy AI across multiple dimensions

Abigail Goldsteen[1], Ariel Farkash[1] and Michael Hind[2]
[1]IBM Research, Haifa, Israel [2]IBM Research, Yorktown Heights, New York, NY, United States

Introduction

Artificial intelligence (AI) systems have become prevalent in everyday life, in both commercial and noncommercial settings. AI is used in retail, security, manufacturing, health, finance, and many more sectors to improve or even replace existing processes. However, with the rise in AI adoption, different risks associated with AI have been identified. In addition to fundamental societal harm, these risks can result in negative brand reputation, lawsuits, and fines. This has led many organizations to seek out ways to assess and prevent such risks, giving rise to the notion of AI ethics, also called trustworthy AI or responsible AI.

This chapter surveys the different dimensions that can be considered when dealing with ethics in AI. We outline key concepts and challenges around each of these dimensions, provide the contexts in which each is relevant, and present real-world motivating examples. We review relevant current and future regulations and standards. We compare qualitative and quantitative approaches to risk assessment of AI models and how they can be combined.

We survey different technical solutions, including open-source and other available tools and resources, for tackling each risk dimension separately. These include tools for assessing how risky an AI model is relative to each dimension, as well as possible mitigations to help overcome these risks. Focusing specifically on privacy, we present existing research and technical solutions in different areas of AI privacy, including the creation of privacy-preserving models, privacy risk assessment of models, and compliance with regulatory requirements.

We then present the unique challenges that arise when attempting to combine multiple trustworthy AI dimensions. First is measuring the risk of a model (or set of models) along multiple dimensions and communicating this information effectively to users. Even more challenging is applying mitigations that address multiple dimensions at once, considering the possible trade-offs between them. We present a few possible approaches.

The chapter concludes with some recommendations for AI project managers and developers, as well as areas for future research.

Ethics in Online AI-Based Systems. DOI: https://doi.org/10.1016/B978-0-443-18851-0.00001-9

Background

Trustworthy AI

As AI systems grow more accurate and efficient, they also become more complex and less understandable. Broad adoption of such systems requires humans to trust them. This depends on the ability to ensure that AI systems are fair, robust, explainable, accountable, respectful of the privacy of individuals, and cause no harm.

Trustworthy AI, also sometimes called responsible AI, typically entails considering privacy, security, fairness, explainability, and transparency when designing, implementing, and deploying AI-based solutions. In some cases, it spans additional concerns such as quality, accountability, inclusiveness, reliability, and societal and environmental well-being. A relatively nascent area of AI governance intersects with and is sometimes considered to include aspects of trustworthy AI.

Definition of trustworthy AI

According to the European Commission, trustworthy AI[1] should be lawful, ethical, and robust, from a technical perspective and taking into account its social environment. The commission put forward a set of seven key requirements that AI systems should meet to be deemed trustworthy: human agency and oversight; technical robustness and safety; privacy and data governance; transparency; diversity, nondiscrimination, and fairness; societal and environmental well-being; and accountability.

In the USA, the Algorithmic Accountability Act (AAA)[2] requires that an impact assessment be carried out for any automated decision system and augmented critical decision process in areas such as performance, accuracy, robustness, and reliability; fairness, bias, and nondiscrimination; transparency, explainability, contestability, and opportunity for recourse; privacy and security; personal and public safety; efficiency and timeliness; and cost.

An additional related topic of interest is AI deception. The proliferation of highly effective generative AI models has brought about the problem of "deepfakes"—the use of deep learning to generate fake images, videos, audio, or text that look (or sound) extremely realistic. Examples include code generators,[3] voice cloning,[4] pictures of fictional people,[5] and even ChatGPT.[6] The US Federal Trade Commission (FTC) has recently issued a warning about the fake AI problem.[7]

Relevant regulations

Data protection regulations

Since its adoption in 2018, the EU's General Data Protection Regulation (GDPR)[8] has become the baseline for a wave of data protection legislation around the globe. In the USA, several states have

---

[1]https://digital-strategy.ec.europa.eu/en/library/ethics-guidelines-trustworthy-ai
[2]https://www.congress.gov/bill/117th-congress/house-bill/6580/text
[3]https://github.com/features/copilot
[4]https://www.respeecher.com/
[5]https://research.nvidia.com/publication/2018-04_progressive-growing-gans-improved-quality-stability-and-variation
[6]https://openai.com/blog/chatgpt
[7]https://www.ftc.gov/business-guidance/blog/2023/03/chatbots-deepfakes-voice-clones-ai-deception-sale
[8]https://ec.europa.eu/info/law/law-topic/data-protection/data-protection-eu_en

followed in the footsteps of California's Consumer Privacy Act (CCPA)[9] and passed data protection laws or acts.[10,11,12,13] Federal privacy legislation is also being considered with the proposed American Data Privacy and Protection Act.[14] In China, the Personal Information Protection Law (PIPL)[15] protects personal data and regulates its processing. In Canada, the Canadian Consumer Privacy Protection Act (CPPA)[16] has recently been submitted for final review. According to Gartner,[17] by 2023, 65% of the world's population's personal data will be covered under modern privacy regulations, up from 10% in 2020.

AI regulations

We are also seeing a new wave of legislation specifically targeting AI. An AI index analysis of the legislative records of 127 countries shows that the number of bills containing "artificial intelligence" passed into law grew from just 1 in 2016 to 37 in 2022.[18]

In the EU, discussion on the impact of GDPR on AI began in 2020.[19] After a period of consultation, the EU drafted the Artificial Intelligence Act,[20] targeting enforcement by 2025. This new regulation is organized by risk level, with certain uses simply not allowed, such as manipulating human behavior and "social scoring" by governments. Safety components, biometric identification, education and training, employment, law enforcement, and justice are considered high-risk uses. These are allowed but subject to strict requirements, including implementing a risk management system to identify, evaluate, and mitigate risks to accuracy, robustness, and cybersecurity. The proposal also stipulates that data used to train such models is subject to appropriate state-of-the-art security and privacy measures. AI systems interacting with users, such as chatbots and generative AI, are required to inform users that the content was artificially created or manipulated.

In Canada, the proposed Artificial Intelligence and Data Act[21] (AIDA) also adopts a risk-based approach. It defines criteria for high-impact systems to ensure they meet safety and human rights expectations, while prohibiting reckless and malicious uses of AI.

In the UK, the government released the Data Protection and Digital Information Bill[22] and is unveiling a set of proposals to regulate the use of AI.[23] The China Internet Information Service

---

[9]https://oag.ca.gov/privacy/ccpa

[10]https://law.lis.virginia.gov/vacodefull/title59.1/chapter53/

[11]https://coag.gov/resources/colorado-privacy-act/

[12]https://www.cga.ct.gov/asp/cgabillstatus/cgabillstatus.asp?
selBillType = Bill&bill_num = SB00006&which_year = 2022

[13]https://le.utah.gov/~2022/bills/static/SB0227.html

[14]https://energycommerce.house.gov/sites/democrats.energycommerce.house.gov/files/documents/
Bipartisan_Privacy_Discussion_Draft_Bill_Text.pdf

[15]http://en.npc.gov.cn.cdurl.cn/2021-12/29/c_694559.htm

[16]https://cppa.ca.gov/regulations/consumer_privacy_act.html

[17]https://www.gartner.com/en/newsroom/press-releases/2020-09-14-gartner-says-by-2023-65-of-the-world-s-population-w

[18]https://aiindex.stanford.edu/report/

[19]https://www.europarl.europa.eu/RegData/etudes/STUD/2020/641530/EPRS_STU(2020)641530_EN.pdf

[20]https://artificialintelligenceact.eu/wp-content/uploads/2022/05/AIA-COM-Proposal-21-April-21.pdf

[21]https://ised-isde.canada.ca/site/innovation-better-canada/en/artificial-intelligence-and-data-act-aida-companion-
document

[22]https://publications.parliament.uk/pa/bills/cbill/58-03/0143/220143.pdf

[23]https://www.gov.uk/government/news/uk-sets-out-proposals-for-new-ai-rulebook-to-unleash-innovation-and-boost-pub-
lic-trust-in-the-technology

Algorithmic Recommendation Management (IISARM)[24] provisions apply to algorithms for personalized recommendations and introduce several user rights, such as the right to be informed, to opt out, to delete personal characteristics, and to not be subject to differentiated treatment.

In the USA, the AAA[25] requires an impact assessment to be carried out for any automated decision system and augmented critical decision process. The FTC has published guidance on using AI and algorithms,[26] calling for AI tools to be transparent, explainable, fair, and empirically sound. It is now pursuing federal AI regulation.[27]

The National Institute of Standards and Technology (NIST) has published an AI Risk Management Framework (RMF)[28] aimed at managing risks to individuals, organizations, and society associated with AI. It covers accuracy, explainability, interpretability, reliability, privacy, robustness, safety, security, and bias. We expect that regulatory activity around AI will continue to accelerate.

Relation to ethics

Over the past years, groups from different sectors and geographies have created ethics principles for AI (Varshney, 2022). These typically include privacy, fairness and justice, safety and reliability, transparency (including interpretability and explainability), social responsibility, and beneficence. The similarity between this list and the list of trustworthy AI topics demonstrates that many of the aspects of trustworthy AI are actually grounded in deeper ethics principles.

There is also a growing line of work on understanding sources of harm in AI-based systems (Suresh & Guttag, 2021) and aligning machine learning (ML) algorithms with human values and community well-being (Stray & Aligning, 2020; Stray et al., 2021). The TAILOR Handbook of Trustworthy AI[29] aims to provide non-experts with an overview of the problems related to the development of ethical and trustworthy AI systems.

The pillars of trustworthy AI

We presently describe the key concepts and challenges around each trustworthy AI pillar and the contexts in which it is relevant.

Privacy

Relevance. Privacy is relevant whenever AI involves personal data. The definition of personal data varies between geographies, cultures, and regulatory environments. GDPR defines personal data as any information relating to an identified or identifiable natural person, who can be identified, directly or indirectly, in particular by reference to an identifier such as a name, an identification number, location data, an online identifier, or to one or more factors specific to the physical, physiological, genetic, mental, economic, cultural, or social identity of that person.

---

[24]https://digichina.stanford.edu/work/translation-internet-information-service-algorithmic-recommendation-management-provisions-effective-march-1-2022/

[25]https://www.congress.gov/bill/117th-congress/house-bill/6580/text

[26]https://www.ftc.gov/business-guidance/blog/2020/04/using-artificial-intelligence-algorithms

[27]https://www.reginfo.gov/public/do/eAgendaViewRule?pubId = 202110&RIN = 3084-AB69

[28]https://www.nist.gov/itl/ai-risk-management-framework

[29]http://tailor.isti.cnr.it/handbookTAI/TAILOR.html

The California Consumer Protection Act (CCPA)[30] provides an even wider definition of personal information as any information that identifies, relates to, or could reasonably be linked with a person. This includes records of products purchased, internet browsing history, geolocation data, and inferences from other personal information that could create a profile about someone's preferences and characteristics.

Real large language models (LLMs) have been shown to leak individual user's personal information[31] (Carlini et al., 2021).

Definitions and metrics. First and foremost, it is necessary to ensure that all data used in an AI system is adequately and lawfully collected, stored, protected, and governed and has a legitimate processing purpose. It must also abide by any applicable regulatory requirements, such as data minimization, consent, the right to correct or withdraw data, etc. Violating data protection regulations can incur serious fines.

Recent studies have shown that a malicious third party with access just to a trained ML model can reveal sensitive information about the data used to train it (Shokri et al., 2017; Fredrikson et al., 2014).

Specific types of privacy (inference) attacks against ML include the following:

- **Membership inference**—tries to deduce whether a specific data sample was part of a model's training data.
- **Attribute inference**—given a trained model and knowledge about some features of a person, deduces the value of additional, unknown features.
- **Model inversion**—reconstructs the representative feature values of the training data. For example, a person's facial image may be reconstructed from a facial recognition model.
- **DB reconstruction**—given a trained model and all training samples except one, reconstructs the complete missing record.

The existence of such attacks has led many to the conclusion that ML models themselves should, in some cases, be considered personal information and therefore subject to GDPR and similar regulations[32] (Kazim et al., 2021; Veale et al., 2018). In 2021, a precedential ruling from the US FTC forced an AI company to delete its ML models after unlawfully collecting user data.[33]

When training models on personal data, it is always recommended to consider whether personal data is truly necessary; whether any of the more sensitive attributes can be removed; whether the data can be anonymized; or whether a synthetic or public dataset can be used. Since applying privacy-preserving procedures to the data or training procedure typically entails a decrease in model accuracy, the challenge is to find the best trade-off for a given model and use case. The section on AI privacy technologies presents more details.

A privacy risk assessment of the model is also recommended to discover, preferably at an early stage, what degree of privacy risk the model poses. This enables making informed decisions on whether to use it in production, publish or share it with third parties, as well as consider possible mitigations.

---

[30]https://leginfo.legislature.ca.gov/faces/billTextClient.xhtml?bill_id = 201720180AB375
[31]https://www.businessupturn.com/world/chatgpts-answer-gives-away-a-journalists-number-to-join-signal/
[32]https://www.europarl.europa.eu/thinktank/en/document.html?reference = EPRS_STU(2020)641530
[33]https://aitechnologylaw.com/2021/02/ftc_orders_ai_model_delete/

Security and reliability

Relevance. While early work in machine learning assumed a closed and trusted environment, attacks against the ML process and models have received increased attention in recent years. Adversarial machine learning aims to protect the ML pipeline to ensure its safety at training, test, and inference time.

According to the proposed EU AI Act, high-risk AI systems should perform consistently throughout their lifecycle and are required to meet an appropriate level of accuracy, robustness, and cybersecurity. They should be resilient against risks connected to the limitations of the system (errors, faults, inconsistencies, and unexpected situations) as well as malicious actions that may compromise the security of the system.

Real-world examples of such attacks include evading spam or malware detectors, or even causing autonomous vehicles to misread road signs.[34]

Definitions and metrics. Adversarial attacks against ML models are typically aimed at changing the behavior of a model to achieve a desired outcome. *Evasion attacks* attempt to modify the input to a classifier such that it is misclassified, while keeping the modification as small as possible. This is called an adversarial example. Such attacks can be either untargeted (the attacker simply aims for misclassification) or targeted (the new class is specified).

Poisoning attacks can occur when the data collection and curation processes are not fully controlled by the model owner. For example, common data sources include the Internet, social media, crowdsourcing, etc. This gives adversaries the opportunity to manipulate the training data to significantly decrease the model's overall performance, cause targeted misclassification, or insert backdoors.

Fairness

Relevance. Fairness is an increasingly important concern as ML models are used to support decision-making in high-stakes applications such as lending, hiring, and prison sentencing.

Several AI regulations specifically mention bias prevention. The proposed EU AI Act requires ensuring that high-risk AI systems do not become the source of discrimination. The IISARM in China stipulates that algorithmic recommendation services shall respect the principles of fairness and justice, and prevents service providers from extending unreasonably differentiated treatment. In Canada, the Directive on Automated Decision-Making[35] requires that data used by automated decision systems is tested for unintended biases and other factors that may unfairly impact outcomes. The newly proposed AI and Data Act[36] requires that appropriate measures be put in place to identify, assess, and mitigate risks of biased output in high-impact systems.

In December 2022, the New York City (NYC) Department of Consumer and Worker Protection issued a set of proposed clarification rules for Local Law 144,[37] regulating the use of automated employment decision tools (AEDTs). According to the law, AEDTs must be subject to a bias audit—an impartial evaluation by an independent auditor, which includes assessing disparate impact on specific groups of people.

---

[34]https://www.businessinsider.com/tesla-hackers-steer-into-oncoming-traffic-with-stickers-on-the-road-2019-4
[35]https://www.tbs-sct.canada.ca/pol/doc-eng.aspx?id = 32592
[36]https://ised-isde.canada.ca/site/innovation-better-canada/en/artificial-intelligence-and-data-act-aida-companion-document
[37]https://rules.cityofnewyork.us/wp-content/uploads/2022/12/DCWP-NOH-AEDTs-1.pdf

In 2017, predictive policing algorithms were shown to suffer from racial bias against Black populations (Selbst, 2017). The Correctional Offender Management Profiling for Alternative Sanctions (COMPAS) algorithm, used in US courts to predict the likelihood of a defendant recidivating, predicted twice as many false positives among Black offenders than White offenders.[38] In 2018, a recruiting tool employed by Amazon was found to discriminate against women.[39] In 2019, an algorithm used in the US hospitals to predict which patients would likely need extra medical care was found to heavily favor White patients (Obermeyer et al., 2019).

Definitions and metrics. There are many different fairness definitions and metrics that can even be contradictory (Kleinberg et al., 2017). Most definitions revolve around *protected attributes* that partition a population into groups whose outcomes should be fair. Typically, one (or more) of those groups is considered "privileged" and the others "unprivileged" or "protected." Protected attributes are use case, geolocation, and application specific, examples including race, gender, age, and religion. Some fairness metrics also define a *favorable outcome*—the classifier output that is considered favorable for the person. For example, in loan applications, the favorable outcome is granting the loan.

Fairness metrics are also divided into *individual fairness*, which requires similar outcomes for similar individuals, and *group fairness*, which requires similar outcomes for groups defined by protected attributes.

The most common fairness metrics include the following:

- **Disparate impact (DI)**—ratio of favorable outcomes between unprivileged and privileged groups.
- **Statistical parity difference (SPD)**—difference in the probabilities of a favorable outcome between the privileged and unprivileged groups.
- **Equal opportunity difference**—difference in true-positive rates between the unprivileged and privileged groups.
- **Average odds difference**—average of difference in false-positive rates and true-positive rates between unprivileged and privileged groups.

Most of these metrics can be measured either on true labels, detecting bias in the underlying dataset, or on model predictions, detecting bias in the model itself.

Explainability

Relevance. The deployment of AI systems in high-stakes domains has been coupled with increased societal demand for those systems to provide explanations for their predictions. This can allow users to gain insight into the system's decision-making process, a key component in fostering trust and confidence.

However, many ML techniques are not easily explainable. Moreover, different people and settings may require different kinds of explanations. For example, a doctor trying to understand the diagnosis of a patient may benefit from seeing known similar cases, whereas a denied loan applicant will want to understand the main reasons for their rejection and what can change the decision.

---

[38]https://www.propublica.org/article/how-we-analyzed-the-compas-recidivism-algorithm
[39]https://www.reuters.com/article/us-amazon-com-jobs-automation-insight-idUSKCN1MK08G

The proposed EU AI Act requires high-risk AI systems to be sufficiently transparent to enable users to interpret the system's output and use it appropriately. The US AAA requires assessing the transparency and explainability of automated decision systems and augmented critical decision processes, including any factors that contribute to a particular decision, and which factors, if changed, would reverse the decision.

Definitions and metrics. There are several types of explanations. An explanation can be of the *data* used to train a model or of the *model* itself. A *local explanation* is designed for a single prediction, whereas *global explanations* describe the behavior of the entire model. A *directly interpretable model* is one whose internal decision logic can be easily interpreted (such as a small decision tree). A *post hoc explanation* involves an auxiliary method to explain a trained model. A *surrogate model* is another, usually directly interpretable model that approximates a more complex model. *Faithfulness* can then measure how similar the interpretable model is to the original model. Explanations can also be *feature based* or *sample based*. Sample-based explanations attempt to find samples in the dataset similar to the sample for which the explanation is requested. Feature-based explanations find the specific input features that had the most effect on the decision for the sample.

Transparency and governance

Relevance. Concerns about the use of AI systems in mainstream decision-making have motivated the need for transparency about these systems. Transparency usually involves relaying information about how the AI system was constructed, deployed, and operates to one or more stakeholders. This can increase trust in the system, help determine if it is appropriate for a particular use, and enable detecting issues.

Many existing and proposed regulations mandate transparency. In 2021, the UK government published a pioneering standard for algorithmic transparency[40] for governmental and public sector bodies. The Algorithmic Transparency Recording Standard[41] helps public sector organizations provide clear information about the algorithmic tools they use, and comes with detailed guidance and sample reports. The proposed EU AI act requires that any information addressed to data subjects be concise, easily accessible, and easy to understand and that clear and plain language and visualizations be used. The NYC Local Law 2021/144 is mostly focused on bias prevention but has stringent requirements on communicating results to the public in a clear and conspicuous manner.

Definitions and metrics. Over the last decade, several research groups have proposed approaches toward improving the transparency of various AI artifacts, such as datasets, models, and systems. These approaches generally take the form of questions that an artifact owner should answer.

AI governance[42] is defined as an overarching framework that manages an organization's use of AI with a set of processes, methodologies, and tools. These include tools for documenting important information about the model, the data used to train it, the training process, and more. It may also include tools for ensuring compliance of the resulting models to different regulations and standards, as well as generating audit trails for important decisions around the lifecycle of the model.

---

[40]https://www.gov.uk/government/news/uk-government-publishes-pioneering-standard-for-algorithmic-transparency
[41]https://www.gov.uk/government/collections/algorithmic-transparency-recording-standard-hub
[42]https://www.digitaljournal.com/pr/ai-governance-market-share-2023-2032-size-cagr-status-major-leading-players-aplhabet-inc-microsoft-corporation-ibm-corporation

Quality

Relevance. The quality and performance of ML models are typically measured using common metrics such as accuracy, balanced accuracy, precision, recall, etc. However, these measures may sometimes be misleading as they represent only the average-case behavior of the model and don't account for corner cases or unexpected behaviors. For example, a typical failure stems from the inability of a model to abstain from making predictions, even when assumptions made during training are violated. This could potentially result in highly confident but incorrect predictions.

Definitions and metrics. ML models have inherent uncertainties (Ghosh et al., 2021). *Data uncertainty* may arise from the inability to collect or represent real-world data reliably. For example, it may be difficult to sample data in a way that is representative of the population. It may also come from flaws in data preprocessing (curation, cleaning, or labeling). Learning algorithms rely on various simplifying assumptions, thus introducing *modeling* and *inference uncertainties*. For example, linear models, such as linear regression, can only model linear relations between features.

Furthermore, even if the average accuracy measures are satisfactory, there are likely certain samples for which the model performs far below average. Finding data subsets for which a model underperforms is beneficial as it can guide users to take corrective actions. For instance, Ackerman et al. (2021) present a model trained on loan approval data that underperforms for people between the ages of 33 and 64. This specific age group (33−64) constitutes a *data slice*, which is a meaningful and explainable representation of a contiguous group of samples. Examples of corrective actions may include adding more training data to cover underrepresented slices or adding additional logic (outside the ML model) to deal with inputs belonging to those slices.

Model risk assessment

Assessing the risk of an ML model entails taking a deep look at a specific model and evaluating its risk across the dimensions that are relevant for the given model and use case. A risk assessment might be a one-time process, or it could be applied multiple times or even continuously.

Relevant use cases

An ML model risk assessment may be relevant in several cases. First, when an organization develops and trains models in-house, it may be necessary to assess the risk of the model prior to use in production or release to third parties. Some organizations may have an ethics board or risk organization that must be informed about a model's risks. Secondly, when one company procures ML models or services from another company, the procuring party may want to ensure that the model being acquired adheres to certain quality criteria, including relevant model risks. The third, least common scenario is ML model insurance or re-insurance. This belongs to a relatively novel field of digitalization or IT risk insurance.[43] Here, risk assessment can help to determine pricing and possible monitoring requirements.

Risk assessments can be either internal or external. Internal assessments are performed by the organization that produced the model. These may suffer from objectivity issues and conflicts of interest and require very specific expertise for performing the assessment, which may be difficult

---

[43]https://www.munichre.com/en/risks/emerging-risks.html

for smaller companies. External or third-party assessments utilize a specialized organization. The US financial industry is regulated under SR-11-7,[44] which requires that internal assessments for model risk management be performed by an independent group within the same institution.

Qualitative risk assessment

Many organizations already have qualitative measures for assessing risks in AI projects. A qualitative assessment can include a description of how an AI model was developed, the datasets that were used, alternatives that were considered, and techniques that were applied, providing a rich picture of the model development process. They may also involve dedicated teams that analyze models and datasets and publish their conclusions in reports. However, these model validator teams are required to have very specific expertise, are often overloaded, and constitute a bottleneck in releasing new models. More recently, quantitative methods that apply automated algorithms to measure aspects of AI risk have started to become available.

Quantitative risk assessment

A quantitative assessment is one where well-defined metrics are computed to assess one or more risks in a standardized manner (Piorkowski et al., 2022). This can lead to much quicker turnaround by automating current resource- and time-consuming processes. Moreover, novel aggregation and summarization techniques may enable identifying correlations and trade-offs not easily detected manually, as well as comparing models and versions. A quantitative risk assessment usually receives as input the model itself (or access to its query API) and a test dataset and results in a list of issues or risks accompanied by numerical scores and sometimes severity levels.

A quantitative risk assessment can serve as a gateway to decide whether a model is safe for deployment. In this case, some stakeholders (e.g., risk owners) may set an acceptance criterion (threshold) that the model needs to pass to be considered safe for use. Thresholds may be set on the overall model risk, any individual dimension risk, or individual attack or metric scores.

If the result of a risk assessment is found to be unsatisfactory, the model requires further attention. Consequently, changes may be made to the model or its training set. For example, certain features may be removed from the dataset, characteristics of the model itself changed (such as its size or other hyperparameters), or some mitigation measure applied. The new model should be assessed again to ensure that the risk is mitigated. This could even be an iterative process, where the trade-off between the model's accuracy and other risks is explored to find the best option for the given use case, until a final model is selected.

In yet another scenario, multiple potential model implementations may be considered for the same task or use case. For example, automated AI solutions may generate more than one potential model to consider. When applying quantitative risk assessment to each of these models, a model's privacy or fairness risk may be considered in addition to size, performance, and accuracy when selecting a final candidate.

Qualitative versus quantitative assessment

Most proposed AI regulation is qualitative in nature and does not include concrete technical risk assessment measures. A few government bodies are beginning to venture into standardizing AI risk

---

[44]https://www.federalreserve.gov/supervisionreg/srletters/sr1107.htm

metrics. Regulators in Singapore have launched a pilot effort[45] to gain experience with quantitative risk assessment with the goal of incorporating it into future legislation. New York City has enacted one of the first laws regulating the use of AI in hiring and promotion decisions,[46] which calls for a disparate impact assessment.

Qualitative assessment is generally broader and can provide a rich picture of the overall process, whereas quantitative assessments tend to be more narrowly focused, but have the benefit of assigning numeric values, which can be standardized and compared across models. Quantitative risk assessment can also be used to help model validation teams prioritize models for a more thorough validation process. Model developers may help accelerate model validation by assessing a model's risks as it is developed.

Qualitative and quantitative assessments can be complementary in that the answers to qualitative questions can lead to further quantitative tests, and vice versa. Approaches that combine both types of assessments can provide a more holistic view (Piorkowski et al., 2022). AI governance tools can usually capture both qualitative and quantitative information about models.

Methodology

This section goes deeper into technical approaches to address trustworthy and ethical AI. It is based on a comprehensive literature review, including online journals and conference proceedings, from the past 6 years to identify key research findings, as well as keyword searches conducted in Google search[47] and GitHub[48] to identify leading open-source projects and proprietary products and solutions. However, it is important to note that not all relevant work and possible solutions are presented here, and these are only representative examples. Typically, the examples were chosen based on criteria such as citations (for papers) and stars or forks (for GitHub repositories).

Throughout this section, we use an example human resources (HR) employment decision model to exemplify which dimensions are relevant and which metrics can be measured to satisfy the requirements of this use case.

Elicitation of requirements

Before going into the technicalities of implementing trustworthy AI, Varshney (Varshney, 2022) suggests an exercise of value alignment, also called elicitation. This process helps guide the ML system designer through four main questions:

1. Should you work on this problem?
2. Which pillars of trustworthiness are of concern?
3. What are the appropriate metrics for those pillars?
4. What are the acceptable ranges for the metric values?

---

[45]https://www.pdpc.gov.sg/news-and-events/announcements/2022/05/launch-of-ai-verify-an-ai-governance-testing-framework-and-toolkit

[46]https://rules.cityofnewyork.us/wp-content/uploads/2022/12/DCWP-NOH-AEDTs-1.pdf

[47]https://www.google.com/

[48]https://github.com/

Some questions that may help determine the pillars of concern:

1. Does the system use data about people?
2. Can the system's decisions present disadvantages to certain groups or individuals?
3. Are the data, model, or software available externally or only internally, and how securely are they kept?
4. Do the system predictions support a human decision-maker?

The answer to each of these may lead to the applicability or importance of each pillar. For example, systems that process data about people will likely need to consider the pillars of privacy and fairness.

Once the pillars are determined, understanding which metrics to measure for each pillar is another challenging aspect. Some pillars may include dozens of possible metrics that differ in the definition of what they measure, in the inputs they receive or other technical requirements. One approach could be to run all metrics that are technically possible to measure. This may be a valid approach for some dimensions, although costly in terms of time and resources. However, in some cases, there may even be conflicting metrics within the same dimension, such as between group and individual fairness or between specific group fairness metrics such as equalized odds and demographic parity (Garg et al., 2020). One way to determine the appropriate metrics is by performing pairwise comparisons. Comparing the values of two metrics for several models can help get a better sense of what they indicate and assist in choosing one over the other.

Once specific metrics have been selected, the final level of value alignment is determining the preferred values for each metric. Setting thresholds for acceptable risk enables issues to be easily identified. Since the different metrics are interrelated, including some that are negatively correlated, approaching this one metric at a time can be very problematic. A holistic approach can help yield more consistent results. One way to do this is by determining the feasible region of metric values. Each quantitative test result for a single model represents a point in the feasible region. To compute many such points, a corpus of models for the same (or similar) task can be collected or automatically created and their metric values computed. At that point, Bayes risk, joint distributions, or Pareto frontiers can help select appropriate ranges.

Technical solutions within each pillar of trustworthy AI

We now survey technical solutions, including open-source and other available tools and resources, for tackling each risk dimension separately. These include tools for risk measurement and possible mitigation measures.

Security and reliability
Assessment

Robustness metrics are a key element to measure the vulnerability of a classifier to adversarial attacks and to assess the effectiveness of adversarial defenses (Nicolae et al., 2018). Typically such metrics quantify the amount of perturbation required to cause a misclassification or, more generally, the sensitivity of model outputs to changes in the inputs.

Examples of such metrics include the following:

- Empirical robustness—The minimal perturbation that an attacker must introduce for a specific attack to be successful, which can be computed for different attack implementations.
- Loss sensitivity (Arpit et al., 2017)—A proxy measure of memorization that measures the effect of each sample on the average loss.
- CLEVER (Weng et al., 2018)—Derives a universal lower bound on the minimal distortion required to craft an adversarial example. This score is attack-agnostic and computationally feasible for large neural networks.

Mitigation

Adversarial defenses usually employ one of the two strategies: model hardening or runtime detection of adversarial inputs. A common approach to model hardening is augmentation of the training data with adversarial examples or other augmentation methods. Input data can also be preprocessed using nondifferentiable or randomized transformations or by reducing the input dimensionality. Other model hardening approaches involve special types of regularization or modifying the classifier's architecture.

Detection methods include training a binary classifier to distinguish between benign and adversarial inputs, either based on the input data itself or on the activations of a given layer of the model, or detecting anomalous patterns in datasets (McFowland et al., 2013). Defenses against poisoning attacks typically aim to detect and filter malicious training data or to unlearn poisoned data once discovered.

Available tools

Open-source tools for assessing and defending against adversarial attacks:

- The Adversarial Robustness Toolbox (ART)[49] (Nicolae et al., 2018)—open-source Python library for adversarial machine learning. It provides standardized interfaces for classifiers of most popular ML libraries and supports additional model types such as regression, generative models, encoders, etc. It provides implementations for a wide variety of attacks, including evasion, extraction, and poisoning, as well as some defenses, such as pre- and postprocessing methods, attack detectors, and adversarial training.
- TextAttack[50] (Morris et al., 2020)—Python library specializing in adversarial attacks against natural language processing (NLP) models. In addition to attack implementations, it also provides data augmentation to improve model robustness and supports model training procedures.

Relevance to HR use case

The security dimension is most relevant when there is an opportunity for external adversaries to attack the machine learning process in some way, for example, by manipulating the data that is input to the model at inference time. In the HR use case, the input to the model likely comes from verified sources such as gender, age, university enrollment, and grade information or at least can be

---

[49]https://github.com/IBM/adversarial-robustness-toolbox
[50]https://github.com/QData/TextAttack

human verified later in the process, such as previous employment, references, etc. Therefore, the adversarial evasion risk is not so high and it probably does not lead to severe consequences. If the model is trained on internal organizational data, the poisoning risk is also not relevant.

Fairness

Assessment

Most work on detecting bias is based on computing different fairness metrics, as detailed in previous sections. Several studies have shown that fairness is not necessarily static and should be continuously monitored over time (D'Amour et al., 2020).

Mitigation

Bias mitigation algorithms attempt to improve the fairness of models by modifying the training data (preprocessing), the learning algorithm (in-processing), or the model's predictions (postprocessing). Common mitigation methods include the following:

- Reweighting—Assigns different weights to the examples in each group/label combination to ensure classification fairness.
- Optimized preprocessing—Learns a probabilistic transformation on the training data to simultaneously satisfy three goals: avoiding discrimination, limiting sample distortion, and preserving utility.
- Adversarial debiasing—Learns a classifier that maximizes prediction accuracy and simultaneously reduces an adversary's ability to determine the protected attribute from the predictions.
- Reject option-based classification—Changes predictions from a classifier to provide favorable outcomes to unprivileged groups and unfavorable outcomes to privileged groups within a confidence band around the decision boundary.

Many more detection and mitigation strategies exist and are constantly being developed.[51]

Available tools

Open-source solutions for bias detection and/or mitigation:

- Fairlearn[52]—open-source Python toolkit to help data scientists improve the fairness of AI systems that enables to both assess and mitigate fairness issues.
- AI Fairness 360[53] (Bellamy et al.)—extensible open-source library containing techniques to help detect and mitigate bias in ML models throughout the AI lifecycle. Available in Python and R, it includes bias metrics and algorithms to mitigate bias, both in datasets and models.
- ML-fairness-gym[54] enables building simulations to explore the potential long-run impacts of deploying ML-based decision systems in social environments. It implements a generalized framework for studying and probing long-term fairness effects in simulation scenarios where a learning agent interacts with an environment over time.

---

[51]https://github.com/datamllab/awesome-fairness-in-ai#fairness-packages-and-frameworks
[52]https://fairlearn.org/
[53]https://github.com/Trusted-AI/AIF360
[54]https://github.com/google/mL-fairness-gym

- TensorFlow Fairness Indicators[55]—designed to support teams in evaluating, improving, and comparing TensorFlow models for fairness concerns. It enables easy computation of commonly identified fairness metrics for binary and multiclass classifiers on large-scale datasets and models.

Some commercial solutions such as Watson Openscale,[56] Mona AI Fairness,[57] and Fiddler[58] provide model fairness assessment and monitoring.

Relevance to HR use case

Fairness is extremely relevant in employment decisions. It has been the focus of recent accusations against Amazon[59] and is specifically addressed in the NYC Local Law 144. Since most regulations focus on *group fairness*, this is the more common type of fairness metric employed. For companies operating in NYC, the *disparate impact* metric is required. In other cases, if the underlying training data is likely to contain biases and it is important to look at combinations of features (for example, looking at gender and race combined and not just separately), the *smoothed empirical differential (SED)* fairness metric is recommended. This is defined as the differential in the probability of favorable and unfavorable outcomes between intersecting groups divided by features (all groups are considered equal).

Reweighting is a simple yet effective mitigation strategy when biases may be present in the model's training data.

Explainability

Assessment

The most common explainability metric is faithfulness, also sometimes called fidelity. An interpretation or explanation is considered faithful if the identified important tokens (features) truly contribute to the decision-making process of the model. Most faithfulness metrics use a removal-based criterion, removing or retaining the important tokens identified by the explanation and observing the changes in model outputs. There are many different implementations of faithfulness (Chan et al., 2022), including the following:

- Sufficiency (SUFF)—measures the change in the output probability of the original predicted class when only the important tokens are retained.
- Comprehensiveness (COMP)—measures the change in the output probability of the original predicted class after the important tokens are removed.
- Monotonicity—measures whether the probability of the predicted class monotonically increases when incrementally adding more important tokens.
- Correlation between importance and output probability (CORR)—measures the correlation between the importance of the token and the corresponding predicted probability when the most important token is continuously removed.

---

[55]https://github.com/tensorflow/fairness-indicators
[56]https://cloud.ibm.com/catalog/services/watson-openscale
[57]https://www.monalabs.io/ai-fairness
[58]https://www.fiddler.ai/fairness
[59]https://www.reuters.com/article/us-amazon-com-jobs-automation-insight-idUSKCN1MK08G

Mitigation

Most post hoc explainability methods can be divided into two main categories:

Gradient-based methods assume that the model is differentiable and attempt to interpret model outputs through gradient information. *Perturbation-based* methods perturb the input data and analyze the effect of these perturbations on the model's outputs.

Common explainability methods include (Arya et al., 2019) the following:

- Saliency methods highlight different portions in an image or sample whose classification we want to understand, providing local, feature-based explanations. This includes popular methods such as LIME (Ribeiro et al., 2016) and SHAP (Lundberg & Lee, 2017).
- Counterfactual/contrastive explanations try to find a minimal change that would alter the classifier's decision.
- Feature relevance methods such as partial dependence plots (PDP) and sensitivity analysis (SA) are used to study the global effects of input features on the output values.
- Exemplar methods explain the predictions of test instances based on similar or influential training instances.
- Knowledge distillation methods learn a simpler surrogate model based on a complex model's predictions.

Available tools

Open-source solutions for model explainability:

- AIX 360[60] (Arya et al., 2019)—an open-source Python library that supports interpretability and explainability of datasets and models. It includes different types of explanations along with explainability metrics. It also includes guidance material and a chart to help decide which methods to use.
- Advanced AI explainability for PyTorch[61] (Gildenblat, 2021) provides state-of-the-art methods for explainable AI for computer vision, including pixel attribution, smoothing methods, and metrics for measuring explanation quality.
- InterpretML[62] (Nori et al., 2019)—an open-source package that enables training interpretable glass-box models and explaining black-box systems.

Relevance to HR use case

Employment decisions affect people who may (rightfully) require a justification for decisions made in their regard. Moreover, in many cases (such as in the USA), automated decision systems are required to expose the factors that contributed to a particular decision. If the type of model employed is not directly interpretable (for example, a random forest), the faithfulness (or fidelity) score can be computed to assess whether explanations generated by a surrogate model or by a post hoc explanation method such as LIME can effectively convey the models' decision criteria.

---

[60]https://github.com/Trusted-AI/AIX360
[61]https://github.com/jacobgil/pytorch-grad-cam
[62]https://github.com/interpretml/interpret

To generate simple but effective explanations for why an employment decision was made, feature-based explanation methods such as LIME and SHAP or exemplar-based methods for finding influential training instances such as DFBETA and Cook's distance can be employed.

Transparency and governance
Mitigation

One suitable approach to reduce risk and ensure compliance of an AI system with relevant regulations and policies is to collect and document relevant "facts" about the process, providing a foundation for transparency and AI governance. This requires an AI system owner to collect and document all relevant information and to organize it in an appropriate format. Studies have shown that this is both time-consuming and error prone if performed after the system is already built (Hind et al., 2020). Tooling to capture this information while the system is constructed can help reduce the human burden and improve the accuracy of the information.[63]

Available tools

A variety of systems and tools that instrument the AI lifecycle to collect and manage facts are emerging. These transparency tools can target the data used to train AI models (Gebru et al., 2021; Holland et al., 2018) or different stages of the AI lifecycle (Arnold et al., 2019; Mitchell et al., 2019; Shen et al., 2022), including methods specifically for language models (Bender & Friedman, 2018).

Some popular ML frameworks such as TensorFlow and HuggingFace provide toolkits for building model cards.[64,65] Model cards are also being integrated into existing ML-as-a-Service (MLaaS) services such as Amazon SageMaker.[66] Tools for end-to-end AI governance are also starting to emerge.[67,68]

Relevance to HR use case

Transparency is relevant to almost any ML model, but especially those that affect humans and whose decisions are externally visible and open to scrutiny. The use of a model card or factsheet describing the model training and validation process is always recommended.

Quality
Assessment

Numerous approaches for improved uncertainty quantification (UQ) in AI models have been proposed. UQ methods can be classified as *intrinsic* or *extrinsic*, depending on how the uncertainties are obtained from the model. Intrinsic methods are models explicitly designed to produce uncertainty estimates along with their predictions. An example intrinsic method is variationally trained Bayesian neural networks (BNNs), which learns a probability distribution on the weights of a

---

[63]https://community.ibm.com/community/user/datascience/blogs/shashank-sabhlok/2022/01/23/ai-factsheets-on-cloud-pak-for-data-as-a-service-a

[64]https://github.com/tensorflow/model-card-toolkit

[65]https://huggingface.co/blog/model-cards

[66]https://docs.aws.amazon.com/sagemaker/latest/dg/model-cards.html

[67]https://medium.com/ibm-data-ai/the-need-for-ai-governance-8d88f2f3b941

[68]https://www.verifyml.com/

neural network using special regularization. For methods that do not have an inherent notion of uncertainty built into them, extrinsic approaches can be used to extract uncertainties post hoc. For example, a metamodel that observes a base classification model succeeding/failing at its task can learn to generate reliable confidence scores for the base model. Since poorly calibrated uncertainties cannot be trusted, the quality of the estimation generated by a UQ algorithm should also be evaluated.

Identifying data slices in which a model underperforms can enable taking corrective action to improve the model's behavior across the entire domain.

Finally, causal inference allows quantifying cause-and-effect relationships in data (Shimoni et al.). It allows the estimation of causal effects and enables training causal models that estimate the effect of an intervention on an outcome.

Available tools

Uncertainty Quantification 360[69] (UQ360) (Ghosh et al., 2021) is an open-source Python toolkit that provides a diverse set of algorithms to quantify uncertainties, metrics to measure them, methods to improve the quality of uncertainties, and approaches to communicate uncertainties effectively. It also provides a taxonomy and guidance for choosing from among these capabilities based on the user's needs. The algorithms are scikit-learn compatible to fit into developers' existing workflow.

FreaAI (Ackerman et al., 2021) automatically analyzes features and feature interactions to identify sets of interpretable data slices in which a model underperforms. It requires only the test data and the model's prediction for each record. It implements a set of heuristics to efficiently compute data slices that are, by construction, explainable, correct, and statistically significant and can generate slices that contain significantly more mispredicted records than would be expected in a random draw.

The Causal Inference 360[70] open-source Python toolkit offers a set of causal inference methods that can help train causal models.

Relevance to HR use case

Quality metrics are always relevant, but even more so when the cost of error is very high. In the HR use case, the cost of an error is typically not considered high, so it is up to the discretion of the model or risk owner to decide how much to invest in this direction. If deciding to assess the model's uncertainty, since this is a classification model and a score is desired, ECE or brier-score can be used. Brier-score is easier to measure as it simply computes the mean square difference between predicted probabilities and true labels of a small test set. ECE requires dividing the predictions into bins.

Post hoc calibration methods can be used to calibrate a model's outputs to reduce uncertainty.

AI privacy

Focusing specifically on the privacy dimension, this section presents existing research and technical solutions in different areas of AI privacy.

---

[69]https://github.com/IBM/UQ360
[70]https://github.com/BiomedSciAI/causallib

Risk assessment of models and datasets

As mentioned earlier, assessing the privacy risk of AI models trained on personal data is crucial to making informed decisions on whether to use a model in production, publish it, or share it with third parties. It also enables comparing and choosing between different models based not only on accuracy but also on privacy risk and measuring the improvement brought by privacy mitigations.

Most quantitative privacy assessments include running one or more inference attacks against the model and measuring their success rate. A multitude of attack implementations exist in the literature.[71] They address different types of ML models, threat models, or assumptions. The success rate of an attack may be measured by different metrics, such as accuracy, precision, recall, receiver operating characteristic (ROC) curve, etc.

Each type of attack may have one or more implementations for different access modes (black box/white box), model types (classical ML/deep learning; classification/regression) and model outputs (probabilities/logits/label only). It is therefore quite complicated to discern, given a specific model and dataset, which attacks to run and with which parameters.

It is also important to correctly interpret the results of the attacks. Early work on membership inference was focused on average-case success across the whole training set. Carlini et al. (2022) advocate looking at worst-case privacy of machine learning models. They suggest to report an attack's true-positive rate (TPR) at low (e.g., $\leq 0.1\%$) false-positive rates (FPR), representing a reliable privacy violation of even just a few users. Ye et al. (2022) further propose a framework of indistinguishability games to enable consistently categorizing attacks and providing an accurate interpretation of their results.

In addition to attacks, there are also several general-purpose membership leakage metrics that measure the amount of information about a single sample (or a complete dataset) that is leaked by a model, regardless of any specific attack implementation. The AI Privacy Toolkit website[72] offers some guidance on the selection of appropriate attacks and metrics.

Available tools

Several open-source tools offer implementations of privacy attacks and metrics for ML models:

- The Adversarial Robustness Toolbox[73] (ART) (Nicolae et al., 2018) provides several attack implementations (membership inference, attribute inference, model inversion and database reconstruction), as well as membership leakage metrics.
- ML privacy meter[74] (Murakonda & Shokri, 2020) is a Python library that enables quantifying the privacy risks of ML models for classification, regression, computer vision, and NLP based on state-of-the-art membership inference attacks.

Proprietary solutions, such as credo-ai,[75] provide a more seamless user experience and require less expertise to employ. An end-to-end framework for privacy risk assessment of AI models (Goldsteen, Shachor, et al., 2022) was also presented to help alleviate the issues of existing tools and frameworks that require a high degree of expertise or are tightly coupled with specific ML frameworks.

---

[71]https://github.com/stratosphereips/awesome-mL-privacy-attacks
[72]https://aip360.res.ibm.com/resources#guidance
[73]https://github.com/Trusted-AI/adversarial-robustness-toolbox
[74]https://github.com/privacytrustlab/mL_privacy_meter
[75]https://www.credo.ai/

Another area of active research is the privacy assessment of datasets, and specifically training datasets that were generated from an original, sensitive dataset. These dataset assessment methods typically fall into two categories: The first assesses the generated dataset itself, regardless of how it was created (Chen et al., 2020; Mateo-Sanz et al., 2004; Platzer & Reutterer, 2021). The second assesses the data generation model or algorithm itself, determining the probability of it generating samples that are "too similar" to the original dataset (Stadler et al., 2022).

Creation of privacy-preserving models

Considering recent inference attacks and given the possible applicability of data protection regulations to AI models, it may be required to anonymize ML models such that the personal information of a specific individual cannot be re-identified.

There are several approaches to generating privacy-preserving models that vary in the stage of the process where they are applied, their assumptions, level of complexity, and privacy guarantee. One approach is to apply *differential privacy* (DP) during model training (Abadi et al., 2016; Fletcher & Islam, 2017). DP relies on adding specific noise during the training process to reduce the effect of any single individual on the model's outcome. The privacy guarantee is determined by the parameter ε (or the pair of parameters ε and δ). Another approach is based on transferring to a "student" model the knowledge of an ensemble of "teacher" models, with noisy aggregation of the teachers' answers (Papernot et al., 2017). This approach assumes the availability of a public or nonsensitive dataset with a similar distribution to the private dataset, which may be difficult to find in practice.

These types of methods, which require changing the learning algorithms themselves, are not suitable for scenarios in which the learning process is not under the control of the organization that owns (and wants to anonymize) the data. They may also be difficult to adopt when many different types of ML models are employed. Moreover, any effort invested in model selection and hyperparameter tuning may need to be redone.

To alleviate some of these drawbacks, an alternative solution based on k-anonymity has been proposed (Goldsteen et al., 2021). It indeed provides only syntactic privacy, which is less robust than the stronger notion of differential privacy; however, it has meaningful advantages in terms of ease of use and wide model applicability. This method is guided by the specific ML model to be trained on the data, which enables retaining more utility in the resulting model than nontailored approaches.

Yet another approach involves generating synthetic datasets for model training, instead of the original, sensitive dataset. Many approaches exist for generating synthetic datasets that share desired characteristics with the original data, ranging from completely rule-based systems, through statistical-query based methods, to generative machine learning models. However, simply generating a new dataset does not guarantee privacy, as the new dataset may still leak sensitive information about the original data (Chen et al., 2020; Stadler et al., 2022). To this end, a plurality of methods for differentially private synthetic data generation have emerged (Tao et al., 2022). Like the anonymization-based approach, this method resides outside of the training process and is therefore easier to employ in practice than methods that replace the training algorithm. The same synthetic dataset could even potentially be used for several downstream tasks, making it even more appealing.

Available tools

Open-source tools that enable building privacy-preserving models:

- Differential Privacy Library[76] (Holohan et al., 2019)—provides privacy-preserving versions of some common classical ML models, such as linear regression, logistic regression, Naïve Bayes, and random forest.
- tensorFlow privacy[77]—enables creating privacy-preserving tensorFlow models.
- PyTorch Opacus[78]—enables creating privacy-preserving PyTorch models.
- The ai-privacy-toolkit's (Goldsteen et al., 2023) model anonymization module[79] enables anonymizing tabular training sets in a manner that is tailored to an existing, trained ML model.

When applying DP or model anonymization, there is usually a trade-off between the level of privacy and the accuracy of the model. It is important to find an acceptable trade-off by trying out different options of k or ε, measuring the resulting accuracy, and assessing privacy risk using the methodology described in the previous section.

Complying with specific regulation requirements

Several clauses in GDPR and other regulations have been found to be applicable to ML models.

Data minimization

GDPR's *data minimization* principle requires that organizations, governments, and companies collect only data that is really needed to achieve a given purpose. Advanced ML algorithms, such as deep neural networks, tend to consume large amounts of data and often result in "black box" models, where it is difficult to derive exactly which data influenced the decision.

To this end, a method for data minimization that can reduce the amount and granularity of input data used to perform predictions by ML models was developed (Goldsteen, Ezov, et al., 2022). Once a model is trained and validated, this method enables to re-evaluate exactly what data is required for the model to make accurate predictions, and enables collecting only data that is strictly required for the system to function correctly. The tool[80] determines whether some of the input features can be completely discarded or collected at a lower granularity. For example, instead of exact ages, it may be possible to use 5- or 10-year ranges.

The right to be forgotten

Recent legislation, such as the GDPR and CCPA, mandates that companies erase an individual's personal data upon request. This is called the "right to be forgotten" or the "right to erasure." Because ML models can potentially memorize training data, it is important to also unlearn the

---

[76]https://github.com/IBM/differential-privacy-library
[77]https://github.com/tensorflow/privacy
[78]https://github.com/pytorch/opacus
[79]https://github.com/IBM/ai-privacy-toolkit/tree/main/apt/anonymization
[80]https://github.com/IBM/ai-privacy-toolkit/tree/main/apt/minimization

effects of the data on models (Villaronga et al., 2018). This claim is strengthened by the FTC's 2021 ruling[81] and has led to the research field of machine unlearning[82] (Jiang et al., 2022).

The trivial solution is to retrain the model from scratch after removing the sample(s) from the training set. However, this approach may have huge computational overhead, especially given the rapid growth of models and training sets in recent years. Moreover, the complete training data may no longer be available.

Machine unlearning may include both *exact* and *approximate* methods. With exact unlearning, the unlearned model is proven to be indistinguishable from a model trained from scratch without the sample. This necessitates knowledge of exactly how individual training points contributed to model parameter updates, and the ability to reverse this contribution. Exact unlearning is possible with certain types of models (such as support vector machines and other linear models) or when special training procedures have been applied (Cao & Yang, 2015; Bourtoule et al., 2021). In cases where exact unlearning is impossible, as with nondeterministic models such as neural networks, a relaxation of exact unlearning called approximate unlearning may be used (Guo et al., 2020; Izzo et al., 2021).

Unlearning evaluation methods help ensure that samples are indeed being removed or that their effect on the model is reduced to a minimum. Currently, there is no standard, agreed-upon evaluation method, with a multitude of different metrics employed in the literature (Nguyen et al., 2022).

Relevance to HR use case

As employment decisions are usually based, among others, on features that consist of personal information (such as age, gender, education, etc.), the privacy dimension is highly relevant. It is even more important if the model's decisions are exposed externally, which is usually the case with an employment model.

In terms of risk assessment, membership risk is relevant as it may disclose which people have applied for employment in a specific company. Attribute risk is also relevant to the more sensitive features used such as test scores and interview results.

To improve the privacy of the model, if the model is trained internally by the organization and it is a relatively small model (without too many parameters), differential privacy or ensembling-based methods should be considered. If the model is trained by a third party or is relatively large, model-guided anonymization can be employed. In all cases, the accuracy-privacy trade-off should be examined to help determine the privacy parameter of the chosen method.

Finally, we recommend trying to minimize the amount of data processed by the model, in line with the data minimization principle, regardless of the use case and specifics of the model.

Combining multiple dimensions of trustworthy AI

Even once the problem is broken down into pillars and each trustworthiness dimension can be properly assessed and addressed on its own, crafting a complete solution from the different available tools can be very difficult.[83]

---

[81]https://aitechnologylaw.com/2021/02/ftc_orders_ai_model_delete/
[82]https://github.com/jjbrophy47/machine_unlearning
[83]https://medium.com/ibm-data-ai/the-lifecycle-view-of-trustworthy-ai-a0006339a4cd

Assessment across multiple dimensions

The first set of challenges relates to measuring the risk of a model (or set of models) along multiple dimensions and communicating this information effectively to end users. Humans can efficiently process problems in one or two dimensions, but when the number of dimensions increases, it becomes much more difficult to comprehend and make knowledgeable decisions. It is therefore crucial to enable proper prioritization, visualization, and summarization of this information. Moreover, different personas or use cases may warrant different views or priorities.

There is a strong desire to summarize risk assessment results across multiple dimensions into a higher level "assessment score" suitable for a variety of roles and technical expertise. One challenge here is how to aggregate metrics that may have different scales, directionality, and distributions. Normalization is required before these can be compared or combined. Simple scaling is one possible solution, but it does not account for unbounded metrics. For metrics with significantly different distributions, copula-based methods may be employed (Ulan et al., 2021).

Another key challenge is the apparent conflict between two desirable properties of a risk assessment score: On the one hand, consistency is required to enable direct comparison of two AI systems, but there is also a need for customization based on the particular use case. Most existing risk assessment systems provide either one or the other, but not both. It remains an open question as to how to develop summary metrics that are both customizable and comparable across models.

Mitigation across multiple dimensions

The second more difficult aspect is how to apply mitigations or create models that address multiple dimensions at the same time, taking into account the possible contradictions or trade-offs between the dimensions. For example, it has been shown that imposing fairness constraints may come at the cost of privacy (Agarwal, 2021; Chang & Shokri, 2021). Song et al. (2019) demonstrate a potential conflict between privacy and robustness. In contrast, Asi et al. (2023) and Phan et al. (2020) developed algorithms to preserve both privacy and robustness to adversarial examples (Baracaldo et al., 2022) have found that defenses against poisoning attacks may worsen model bias against certain populations.

Even when there are no inherent trade-offs between the dimensions, the multitude of different tools and techniques, which often require replacing the learning algorithm, are not always easily combined at the technical level.[84] For example, one approach to fairness may require the use of a prejudice remover logistic regression model; satisfying explainability may imply using the logistic rule regression algorithm; and achieving privacy might warrant the use of a differentially private logistic regression implementation. Obviously these three distinct algorithms cannot all be employed at once without significant effort and expertise. However, tackling each dimension at a different stage of the lifecycle may help. For example, for fairness it may be possible to employ data rebalancing or reweighting. For privacy, data anonymization or synthetic data for training may be used.

---

[84]https://medium.com/ibm-data-ai/the-lifecycle-view-of-trustworthy-ai-a0006339a4cdhttps://developer.ibm.com/blogs/data-minimization-for-machine-learning/

Available tools

Microsoft's responsible AI toolbox[85] includes tools for both analyzing and mitigating risks associated with fairness, explainability, and predictive performance. However, each of these analyses is presented in a separate dashboard and is based on a different underlying technology. It also contains a variety of mitigation strategies, each of which may be applied separately, at a specific step of the ML process.

In terms of proprietary solutions, credo-ai[86] offers evaluation of risks across fairness, performance, privacy, security, and transparency. It uses a threshold-based approach to flag and count issues in each area, as well as a compliance progress tracker for different regulations. Watson OpenScale[87] enables monitoring models for bias, explainability, and drift, issuing alerts for violations of constraints.

CognitiveScale's Cortex AI Platform[88] generates a numeric score based on six key elements of trust—fairness, robustness, explainability, accuracy, compliance, and data quality. This AI Trust Index enables comparing models to industry benchmarks. Datatron[89] also provides solutions for monitoring models for drift, bias, and performance, enables setting thresholds for metrics to trigger alerts, and provides an overall health score for models.

The TAILOR EU-funded project[90] is researching trustworthy AI in the context of AutoAI solutions, trying to make them more trustworthy and robust. They are specifically tackling the challenge of multiobjective AutoAI to address the trade-off between performance and other dimensions of trustworthiness, which may be incongruous. The Center for Applied Scientific Computing[91] is applying techniques from optimization, information theory, and statistical learning theory to achieve guarantees on robustness, fairness, and privacy.

Conclusion

In this chapter, we surveyed different trustworthy AI dimensions, the motivation behind them, their relation to ethics and regulations, and technical solutions for addressing them. We discussed current challenges and solutions for addressing multiple trustworthy AI dimensions simultaneously and producing customizable yet comparable overall risk scores.

We stressed that the earlier these issues are addressed in the ML lifecycle, the easier it is to avoid harm and produce trustworthy solutions. We recommended a combination of qualitative and quantitative approaches to model risk assessment, assisted by extensive tooling. Having a clear responsible party (e.g., risk owner) helps to address this in a more consistent manner across departments, use cases, and applications. Involving the model developers at an early stage can help assimilate good practices and reduce the effort required to fix things post facto.

---

[85]https://github.com/microsoft/responsible-ai-toolbox
[86]https://www.credo.ai/
[87]https://cloud.ibm.com/catalog/services/watson-openscale#about
[88]https://www.cognitivescale.com/certifai/
[89]https://datatron.com/ai-monitoring-ai-governance/
[90]https://tailor-network.eu/research-overview/autoai/
[91]https://computing.llnl.gov/casc/mL/robust

Future work

There are many areas for future research, especially in the area of combining multiple dimensions and understanding the interplay between them. We believe that the next generation of trustworthy AI tools must address this in a holistic manner and not as separate concerns with separate tools and solutions.

Addressing new data modalities, model types, and the ability to scale to very large models will also be a key differentiator, enabling addressing large foundational models that will become more and more prevalent.

Acknowledgments

This work has been supported by several projects funded by the European Union's Horizon 2020 research and innovation program, under grant agreements No. 965221, 883188, and 101094323.

References

Abadi, M., Chu, A., Goodfellow, I., McMahan, H.B., Mironov, I., Talwar, K., & Zhang, L. (2016). Deep learning with differential privacy. In *Proceedings of the ACM SIGSAC conference on computer and communications security* (pp. 308–318).

Ackerman, S., Raz, O., & Zalmanovici, M. (2021). FreaAI: Automated extraction of data slices to test machine learning models. https://arxiv.org/abs/2108.05620.

Bellamy, R.K. E., Dey, K., Hind, M., Hoffman, S.C., Houde, S., Kannan, K., Lohia, P., Martino, J., Mehta, S., Mojsilovic, A., Nagar, S., Ramamurthy, K.N., Richards, J., Saha, D., Sattigeri, P., Singh, M., Varshney, K. R., & Zhang, Y. (2018). AI fairness 360: An extensible toolkit for detecting, understanding, and mitigating unwanted algorithmic bias. https://doi.org/10.48550/arXiv.1810.01943.

Agarwal, S. (2021). Trade-offs between fairness and privacy in machine learning. In *IJCAI 2021 workshop on AI for social good*.

Arnold, M., Bellamy, R. K. E., Hind, M., Houde, S., Mehta, S., Mojsilović, A., Nair, R., Ramamurthy, K. N., Olteanu, A., Piorkowski, D., Reimer, D., Richards, J., Tsay, J., & Varshney, K. R. (2019). FactSheets: Increasing trust in AI services through supplier's declarations of conformity. *IBM Journal of Research & Development*, 63(4/5).

Arpit, D., Jastrzębski, S., Ballas, N., Krueger, D., Bengio, E., Kanwal, M.S., Maharaj, T., Fischer, A., Courville, A., Bengio, Y., & Lacoste-Julien, S. (2017). A closer look at memorization in deep networks. In *Proceedings of the 34th international conference on machine learning (ICML'17)* (Vol. 70, pp. 233–242). JMLR.org.

Arya, V., Bellamy, R.K. E., Chen, P., Dhurandhar, A., Hind, M., Hoffman, S.C., Houde, S., Vera Liao, Q., Luss, R., Mojsilović, A., Mourad, S., Pedemonte, P., Raghavendra, R., Richards, J., Sattigeri, P., Shanmugam, K., Singh, M., Varshney, K.R., Wei, D., & Zhang, Y. (2019). One explanation does not fit all: A toolkit and taxonomy of AI explainability techniques. https://doi.org/10.48550/arXiv.1909.03012.

Asi, H., Ullman, J., & Zakynthinou, L. (2023). From robustness to privacy and back. https://arxiv.org/abs/2302.01855.

Baracaldo, N., Eykholt, K., Ahmed, F., Zhou, Y., Priya, S., Lee, T., Kadhe, S., Tan, M., Polavaram, S., Suggs, S., Gao, Y., & Slater, D. (2022). Benchmarking the effect of poisoning defenses on the security and bias of the final model. *NeurIPS*.

Bender, E., & Friedman, B. (2018). Data statements for natural language processing: Toward mitigating system bias and enabling better science. *OpenReview*.

Bourtoule, L., Chandrasekaran, V., Choquette-Choo, C. A., Jia, H., Travers, A., Zhang, B., Lie, D., & Papernot, N. (2021). Machine unlearning. *IEEE S&P*.

Cao, Y., & Yang, J. (2015). Towards making systems forget with machine unlearning. In *IEEE symposium on security and privacy* (pp. 463−480). San Jose, CA. Available from https://doi.org/10.1109/SP.2015.35.

Carlini, N., Chien, S., Nasr, M., Song, S., Terzis, A., & Tramèr, F. (2022). Membership inference attacks from first principles. In *IEEE symposium on security and privacy (SP)* (pp. 1897−1914). San Francisco, CA. Available from https://doi.org/10.1109/SP46214.2022.9833649.

Carlini, N., Tramer, F., Wallace, E., Jagielski, M., Herbert-Voss, A., Lee, K., Roberts, A., Brown, T., Song, D., Erlingsson, U., Oprea, A., & Raffel, C. (2021). Extracting training data from large language models. In *USENIX security symposium*.

Chan, C.S., Kong, H., & Liang, G. (2022). A comparative study of faithfulness metrics for model interpretability methods. https://arxiv.org/abs/2204.05514.

Chang, H., & Shokri, R. (2021). On the privacy risks of algorithmic fairness. In *IEEE European symposium on security and privacy (EuroS&P)* (pp. 292−303). Vienna, Austria. Available from https://doi.org/10.1109/EuroSP51992.2021.00028.

Chen, D., Yu, N., Zhang, Y., & Fritz, M. (2020). *GAN-leaks: A taxonomy of membership inference attacks against generative models. Proceedings of the 2020 ACM SIGSAC conference on computer and communications security (CCS' 20)* (pp. 343−362). New York: Association for Computing Machinery. Available from https://doi.org/10.1145/3372297.3417238.

D'Amour, A., Srinivasan, H., Atwood, J., Baljekar, P., Sculley, D., & Halpern, Y. (2020). *Fairness is not static: Deeper understanding of long term fairness via simulation studies. Proceedings of the 2020 conference on fairness, accountability, and transparency (FAT* '20)* (pp. 525−534). New York: Association for Computing Machinery. Available from https://doi.org/10.1145/3351095.3372878.

Fletcher, S., & Islam, M. Z. (2017). Differentially private random decision forests using smooth sensitivity. *Expert Systems with Applications*, 78(1), 16−31.

Fredrikson, M., Lantz, E., Jha, S., Lin, S., Page, D., & Ristenpart, T. (2014). *Privacy in pharmacogenetics: An end-to-end case study of personalized warfarin dosing. Proceedings of the 23rd USENIX conference on security symposium* (pp. 17−32). United States: USENIX Association.

Garg, P., Villasenor, J., Foggo, V. (2020). Fairness metrics: A comparative analysis. https://arxiv.org/pdf/2001.07864.pdf.

Gebru, T., Morgenstern, J., Vecchione, B., Vaughan, J. W., Wallach, H., Daumeé, H., III, & Crawford, K. (2021). Datasheets for datasets. *Communications of the ACM*.

Ghosh, S., Liao, Q.V., Ramamurthy, K.N., Navrátil, J., Sattigeri, P., Varshney, K.R., & Zhang, Y. (2021). Uncertainty quantification 360: A holistic toolkit for quantifying and communicating the uncertainty of AI. https://arxiv.org/abs/2106.01410.

Gildenblat, J., et al. (2021). *PyTorch library for CAM methods*. GitHub.

Goldsteen, A., Ezov, G., Shmelkin, R., Moffie, M., & Farkash, A. (2021). *Anonymizing machine learning models, . Data privacy management, cryptocurrencies and blockchain technology. DPM CBT* (Vol. 13140). Cham: Springer Lecture Notes in Computer Science. Available from https://doi.org/10.1007/978-3-030-93944-1_8.

Goldsteen, A., Ezov, G., Shmelkin, R., Moffie, M., & Farkash, A. (2022). Data minimization for GDPR compliance in machine learning models. *AI Ethics*, 2, 477−491. Available from https://doi.org/10.1007/s43681-021-00095-8.

Goldsteen, A., Saadi, O., Shmelkin, R., Shachor, S., & Razinkov, N. (2023). AI privacy toolkit. *SoftwareX, 22*. Available from https://doi.org/10.1016/j.softx.2023.101352.

Goldsteen, A., Shachor, S., & Razinkov, N. (2022). *An end-to-end framework for privacy risk assessment of AI models*. *Proceedings of the 15th ACM international conference on systems and storage (SYSTOR '22)* (p. 142) New York: Association for Computing Machinery. Available from https://doi.org/10.1145/3534056.3534998.

Guo, C., Goldstein, T., Hannun, A., & Van Der Maaten, L. (2020). Certified data removal from machine learning models. In *Proceedings of the 37th international conference on machine learning (ICML'20)* (Article 359, pp. 3832–3842). JMLR.org.

Hind, M., Houde, S., Martino, J., Mojsilovic, A., Piorkowski, D., Richards, J., & Varshney, K. (2020). *Experiences with improving the transparency of AI models and services. CHI*. LBW.

Holland, S., Hosny, A., Newman, S., Joseph, J., & Chmielinski, K. (2018). The dataset nutrition label: A framework to drive higher quality data standards. *arXiv*.

Holohan, N., Braghin, S., Mac Aonghusa, P., & Levacher, K. (2019). Diffprivlib: The IBM differential privacy library. https://arxiv.org/abs/1907.02444.

Izzo, Z., Smart, M.A., Chaudhuri, K., & Zou, J. (2021). Approximate data deletion from machine learning models. In *Proceedings of the 24th international conference on artificial intelligence and statistics (AISTATS)* (PMLR: Vol. 130). San Diego, CA.

Jiang, Y., Liu, S., Zhao, T., Li, W., & Gao, X. (2022). Machine unlearning survey. In *Proc. SPIE 12500, Fifth International Conference on Mechatronics and Computer Technology Engineering (MCTE 2022)* (p. 125006J), 16 December 2022. https://doi.org/10.1117/12.2660330.

Kazim, E., Denny, D. M. T., & Koshiyama, A. (2021). AI auditing and impact assessment: According to the UK information commissioner's office. *AI and Ethics*.

Kleinberg, J., Mullainathan, S., & Raghavan, M. (2017). *Inherent trade-offs in the fair determination of risk scores. Innovations in Theoretical Computer Science*. ACM.

Lundberg, S., & Lee, S. -I. (2017). Unified framework for interpretable methods. In *Advances of neural information processing systems*.

Mateo-Sanz, J. M., Sebé, F., & Domingo-Ferrer, J. (2004). *Outlier protection in continuous microdata masking. International workshop on privacy in statistical databases*. Berlin, Heidelberg: Springer.

McFowland, E., Speakman, S., & Neill, D. B. (2013). Fast generalized subset scan for anomalous pattern detection. *Journal of Machine Learning Research, 14*, 1533–1561.

Mitchell, M., Wu, S., Zaldivar, A., Barnes, P., Vasserman, L., Hutchinson, B., Spitzer, E., Raji, I.D., & Gebru, T. (2019). Model cards for model reporting. In *Conference on fairness, accountability, and transparency (FAT)*, January 29–31. Atlanta, GA.

Morris, J., Lifland, E., Yoo, J.Y., Grigsby, J., Jin, D., & Qi, Y. (2020). TextAttack: A framework for adversarial attacks, data augmentation, and adversarial training in NLP. In *Proceedings of the 2020 conference on empirical methods in natural language processing: System demonstrations*.

Murakonda, S.K., & Shokri, R. (2020). MLPrivacy meter: Aiding regulatory compliance by quantifying the privacy risks of machine learning. In *Workshop on hot topics in privacy enhancing technologies (HotPETs)*.

Nguyen,T.T., Huynh, T.T., Nguyen, P.L., Liew, A.W., Yin, H., & Nguyen, Q.V. H. (2022). A survey of machine unlearning. https://arxiv.org/pdf/2209.02299.pdf.

Nicolae, M., Sinn, M., Tran, M., Buesser, B., Rawat, A., Wistuba, M., Zantedeschi, V., Baracaldo, N., Chen, B., Ludwig, H., Molloy, I., & Edwards, B. (2018). Adversarial robustness toolbox v1.2.0. https://arxiv.org/pdf/1807.01069.

Nori, H., Jenkins, S., Koch, P., & Caruana, R. (2019). InterpretML: A unified framework for machine learning interpretability. arXiv preprint arXiv:1909.09223.

Obermeyer, Z., Powers, B., Vogeli, C., & Mullainathan, S. (2019). Dissecting racial bias in an algorithm used to manage the health of populations. *Science, 366*(6464), 447–453.

Papernot, N., Abadi, M., Erlingsson, Ú., Goodfellow, I., & Talwar, K. (2017). Semi-supervised knowledge transfer for deep learning from private training data. In *ICLR 2017*. https://arxiv.org/abs/1610.05755.

Phan, N., Thai, M.T., Hu, H., Jin, R., Sun, T., & Dou, D. (2020). Scalable differential privacy with certified robustness in adversarial learning. *ICML 2020*.

Piorkowski, D., Hind, M., & Richards, J. (2022). Quantitative AI risk assessments: Opportunities and challenges. https://arxiv.org/abs/2209.06317.

Platzer, M., & Reutterer, T. (2021). Holdout-based fidelity and privacy assessment of mixed-type synthetic data. *arXiv*. doi: 10.48550/arXiv.2104.00635.

Ribeiro, M., Singh, S., & Guestrin, C. (2016). Why should I trust you? Explaining the predictions of any classifier. In *ACM SIGKDD international conference on knowledge discovery and data mining*.

Selbst, A. D. (2017). Disparate impact in big data policing. *Georgia Law Review, 52*(1), 109–195.

Shen, H., Wang, L., Deng, W.H., Brusse, C., Velgersdijk, R., & Zhu, H. (2022). The model card authoring toolkit: Toward community-centered, deliberation-driven AI design. In *ACM conference on fairness, accountability, and transparency*. Retrieved from https://par.nsf.gov/biblio/10374245. https://doi.org/10.1145/3531146.3533110.

Shimoni, Y., Karavani, E., Ravid, S., Bak, P., Ng, T.H., Alford, S.H., Meade, D., & Goldschmidt, Y. (2019). An evaluation toolkit to guide model selection and cohort definition in causal inference. https://doi.org/10.48550/arXiv.1906.00442.

Shokri, R., Stronati, M., Song, C., & Shmatikov, V. (2017). Membership inference attacks against machine learning models. In *IEEE symposium on security and privacy* (pp. 3–18). San Jose, CA.

Song, L., Shokri R., & Mittal, P. (2019). Privacy risks of securing machine learning models against adversarial examples. In *CCS 2019*.

Stadler, T., Oprisanu, B., & Troncoso, C. (2022). Synthetic data – Anonymisation groundhog day. *Usenix Security*.

Stray, J., & Aligning, A. I. (2020). Optimization to community well-being. *International Journal of Community Well-Being, 3*, 443–463.

Stray, J., Vendrov, I., Nixon, J., Adler, S., & Hadfield-Menell, D. (2021). What are you optimizing for? Aligning recommender systems with human values. arXiv preprint arXiv:2107.10939.

Suresh, H., & Guttag, J. (2021). *A framework for understanding sources of harm throughout the machine learning life cycle* October 5–9*Equity and access in algorithms, mechanisms, and optimization (EAAMO '21)*. New York: ACM. Available from https://doi.org/10.1145/3465416.3483305.

Tao, Y., McKenna, R., Hay, M., Machanavajjhala, A., & Miklau, G. (2022). Benchmarking differentially private synthetic data generation algorithms. In *PPAI 2022*.

Ulan, M., Löwe, W., Ericsson, M., et al. (2021). Copula-based software metrics aggregation. *Software Quality Journal, 29*, 863–899. Available from https://doi.org/10.1007/s11219-021-09568-9.

Varshney, K.R. (2022). Trustworthy machine learning. Independently Published. http://www.trustworthymachinelearning.com/trustworthymachinelearning.pdf.

Veale, M., Binns, R., & Edwards, L. (2018). Algorithms that remember: Model inversion attacks and data protection law. *Philosophical Transactions of the Royal Society A, 376*. Available from https://doi.org/10.1098/rsta.2018.0083.

Villaronga, E. F., Kieseberg, P., & Li, T. (2018). Humans forget, machines remember: Artificial intelligence and the right to be forgotten. *Computer Law & Security Review, 34*(2), 304–313. Available from https://doi.org/10.1016/j.clsr.2017.08.007, ISSN 0267–3649.

Weng, T., Zhang, H., Chen, P., Yi, J., Su, D., Gao, Y., Hsieh, C., & Daniel, L. (2018). Evaluating the robustness of neural networks: An extreme value theory approach. In *ICLR 2018*.

Ye, J., Maddi, A., Murakonda, S. K., Bindschaedler, V., & Shokri, R. (2022). *Enhanced membership inference attacks against machine learning models. Proceedings of the 2022 ACM SIGSAC conference on computer and communications security (CCS '22)* (pp. 3093—3106). New York: Association for Computing Machinery. Available from https://doi.org/10.1145/3548606.3560675.

Artificial intelligence and basic human needs: the shadow aspects of emerging technology

13

Tay Keong Tan

Department of Political Science, Radford University, Radford, VA, United States

Introduction: the rise of intelligent machines

Over the past decades, artificial Intelligence (AI) technologies, have had pervasive and transformative impact on human society and the way people live, work, and relate with one another. AI technologies and applications have already become more superior than human in certain functions, like pattern recognition, machine learning, efficiency in tasks performance without breaks or downtime, and the search and synthesis of data (Vassev, 2021). AI-enabled platforms and applications are becoming increasingly more powerful and pervasive. They have enhanced communications and connectivity between people and organizations that transcend institutional barriers, national boundaries, and physical geography. They have also made available to human society incredible access to information and applications that have revolutionized the work and output of practically every sector and industry: from education and healthcare to food production and environmental protection.

Some observers are already predicting that AI could be the driver of large-scale and broad-based innovation - like the invention of the steam engine and electricity, and the information technology revolution in decades past—to generate a new cycle of sustained industrial innovation and economic expansion. These long waves of 40—60 years in length (called the Kondratiev Wave) were first described by Russian economist Nikolai Kondratiev in his 1925 book, *The Major Economic Cycles*. Joseph Schumpeter named these long-wave business cycles Kondratieff waves to honor his contribution (Barnett, 2002). While some scholars and pundits herald the advent of AI as the dawn of a new era with a quantum leap in economic prosperity, improved leisure and free time, others like Elon Musk, Stephen Hawking, and Bill Gates, have cautioned that the AI revolution will exacerbate global economic inequity, escalate conflicts, and portend new crises for humanity (Sainato, 2015).

By 2023, some artificial intelligence applications have become smarter and more competent than its human creators, considering the likes of large-language models like ChatGPT-4, which have already been used by millions of people across the world. The use of AI is habit-forming, job-cutting, and uses data and texts that may be tethered to the truth or morality. AI chatbots has been known to enable "bad behavior" and destructive deeds by teaching users how to cheat in exams, commit suicide, or build a "dirty bomb." As AI displaces humans' information-gathering, moral-reasoning and decision-making capabilities, we may be conditioned or tempted to depend on it

Ethics in Online AI-Based Systems. DOI: https://doi.org/10.1016/B978-0-443-18851-0.00004-4

259

excessively and unreflexively to do difficult tasks or solve problems, resulting in atrophy of our capabilities and threats to our autonomy.

In popular lore concocted in Hollywood, cybernetic systems can go rogue and autonomous robots gain intelligence and develop consciousness, and they go on to create a dystopian world that threatens humanity's very survival. *Blade Runner, The Terminator, The Matrix, Transcendence*, and *Westworld* depict apocalyptic scenarios in which AI exacts costs that might not be worth the benefits it brings. AI-enabled machines (humanoid robots) are rapidly developed to behave and look more life-like, mimicking real people, to make their interface with humans more natural and authentic. AI has been designed to read people's feelings through the recognition of patterns and nuances in text, voice tone, facial expressions, and gestures. Hanson Robotics' advanced human-like robot, *Sophia*, and Engineered Arts' entertainment robot, *Ameca*, are already famous on the social mediascape (Hanson Robotics, n.d.; Engineered Arts, n.d) as advanced prototypes of androids that can detect human emotions, decode human languages, and respond in real-time to human facial expressions.

While advancing AI applications have brought ease and benefit to human life in meeting our physical needs (better healthcare, more nutritious farm produce, and more efficient and safer transportation), it is less obvious how they would impact our psychological needs. Our emotional longings, though intangible and often ephemeral, are some of the most complex aspects of human intelligence for machines to learn and master. They are indispensable requirements for our health and wellbeing, and all human beings require their fulfillment for our subsistence. They are the driving forces behind the motivations and actions that humans take.

The fast-advancing march of general-purpose and super-intelligent technology, use of Big Data, and proliferation of automated systems can result in outcomes that threaten human rights and the needs of people. Too often, the use of AI applications without regulation, oversight, or concern for public morality can lead to poor outcomes for some people, such as causing losses of jobs, incomes, and opportunities, and preventing access to critical resources or services. AI used in healthcare settings may lead to triaging and rationing of care that is biased or unsafe for patients. AI algorithms used in job hiring and credit scoring decisions may reflect and perpetuate exclusion and inequities in the society. And the growing application of machine learning to social media marketing, user management, and data collection have been known to undermine people's privacy without their knowledge or consent.

As AI will portend many profound changes in human society (much like the invention of electricity and its subsequent widespread application in human societies), an ex-ante assessment of new AI technologies could offer insight and ideas on how to prevent and mitigate potential risks, or leverage and accentuate the desirable features. A prognosis of their impacts on fundamental human needs can help AI scientists, algorithm designers, as well as policy makers, regulators, and end users, take proactive steps to mitigate the risks and manage the unintended consequences.

Methodology

This study analyzes three emerging technologies and their likely impact on peoples' psychological and emotional needs. They are (1) autonomous vehicles; (2) facial recognition systems; and (3) AI writing or image generators. These three technologies are assessed in terms of their impact on six fundamental psychological needs: (1) certainty, (2) variety, (3) significance, (4) connection, (5) growth, and (6) contribution; the core universal needs that all humans share explained in more detail in the section that follows (Madanes, 2016; Robbins, 2005).

Abraham Maslow's pyramid of human needs flows from the base of the hierarchy of needs upwards: physiological (food, shelter, and clothing), safety (job security, stability in the living environment), love and belonging (family, friends, and community), esteem, and self-actualization (Maslow, 1943, 1962). He postulated that only when lower-level needs for subsistence and survival were met (food, water, and shelter), would higher-level needs (like the social and psychological needs) begin to drive and motivate behavior. The Madanes-Robbins model (explained in further detail below) focuses on similar basic human needs but organized in a different conceptual framework and focuses on emotional and psychological needs.

To gain an up-to-date appreciation of the potential impact of AI adoption on basic human needs, the author reviewed the recent publications from scholarly and popular sources on the subjects of emerging AI applications and their emotional or psychological impact on humans. The facts and findings are triangulated with cross-referencing from other credible sources. The search strategy covers a range of sources and publications, including those from publishers like *SpringerLink and ScienceDirect (Elsevier)*. Articles are varied, ranging from conference proceedings and journal papers to blogs and opinion pieces in popular magazines and organizational websites.

To analyze the salient human needs-related consequences of AI technologies, the discussion will examine each of the abovementioned three applications in turn using the Madanes-Robbins framework of the six core human needs in a segmented fashion. Instead of a comprehensive and extensive impact assessment of each technology in terms of the six human needs, this study will analyze in detail each case study on two of them. In so doing, the analyses would be more in-depth and all six human needs will be covered, and thereby offer illustrations on the ways by which technological change and innovation can impede (or reinforce) the fulfillment of human needs.

The fundamental human needs

Authors Cloe Madanes and Anthony Robbins postulated that the basic emotional needs could have powerful and pervasive influence on our actions and behavior (Madanes, 2009, 2016; Robbins, 2005). Unmet needs, over time, can put people in chronic stress, and compromise their ability to learn and perform in tasks. Satisfying these basic needs is not optional but mandatory for the health and wellbeing of a person. For instance, when a person suffers chronic deficit in a fundamental emotion need (such as the need to be accepted and approved by others), they may become edgy, anxious, or depressed; difficult emotions arise to signal to them something is not right with their lives or living environment, prompting them to take actions to correct the deficit. These needs can also influence our very thoughts, feelings and intentions and hence, shape important decisions.

The first human need of *certainty* underpins our need for security, predictability and safety. Safety is also one of Abraham Maslow's basic needs (Maslow, 1943). It drives our actions to seek comfort and control, from organizing our lives around schedules and developing routines and habits to creating schedules to buying insurance to reduce risks and uncertainty. People need to feel safe, avoid pain, and be comfortable in their environment and relationships. They need a sense of security—a roof over their heads or access of medical care in the event of sickness. A chronic deficit of *certainty* may cause someone to feel stressed and become anxious or panicky. The second human need of *variety* is the opposite of the first need of certainty. Humans need change, stimulation, and excitement to hold their attention and keep them engaged. They seek adventure and novel experiences to feel alive and enthralled. People find *variety* through different avenues such as travel, job change, relationships, entertainment and online pursuits.

The third core human need of *significance* denotes the drive to succeed and achieve a level of importance. People are driven by the need to be special and unique; and they achieve it through efforts to gain fame and fortune and achieve leadership status (like becoming social influencers). To differentiate themselves from others and stand out, some people seek short-lived media publicity or celebrity (the so-called "15 minutes of fame") through outrageous publicity stunts like appearing on reality TV shows or committing notorious acts to become noticed on social media. The fourth basic human need, *connection*, in contrast, underpins our need to fit in, be part of a larger community or group. Humans are a "social species" in the sense that we are driven by an inherent desire for the approval, care and affection of others; and we rely on cooperation to survive and thrive (Aronson, 1980; Tomasello, 2014). People try to fulfill this need by joining groups, conforming to social mores, and performing acts to "fit in" and gain the approval or affection of others (Madanes, 2016). This need for *connection* drives our behavior and decisions (code words for this need include love, relationships, community, friendship, family, togetherness, and unity). When some people may not experience love or even close friendship or family ties, they may find other ways to secure connection with others—in the social groups, religious organizations, and in the community and the workplace.

The fifth core human need is our inherent need for *growth*. People seek the expansion of their capacities, and they desire to grow personally, professionally, socially and also develop themselves intellectually, emotionally and spiritually, as they transition from infancy to adulthood and old age (ibid). They are predisposed to desire increase in their social status over time, or to attain improvements in their social situation, financial position, or lifestyle. *Growth* is required for us to function effectively in society; we need learn from our life experiences and grow intellectually. Without *growth*, people become bored and begin to look for ways out of their stagnation or stasis. People meet this need by striving to make progress in different realms of life, sometimes through achieving the markers of success, accumulating wealth, securing promotions at work, and expanding their social networks (Madsen, 2019). The final core human need is *contribution*, the imperative for homo sapiens to serve and support the larger community, presumably for the greater good. Their desire for *contribution* drives people to give back to family, community or society, or work in professions that "make the world a better place." People pursue this through civic engagement, community service, and giving to charity. Considered a higher-order need beyond subsistence-level, survival-oriented living, *contribution* is essential to a sense of fulfillment and self-actualization to many (Maslow, 1962).

The final two needs are closely relation to Abraham Maslow's pinnacle human need for self-actualization in his model of motivation (ibid). He postulates the need for personal growth and discovery as well as contribution to the greater good to enable a person to find meaning in life. Cloé Madanes hypothesizes that the first four human needs are essential for survival, while the final two are more for fulfillment or personal satisfaction (Madanes, 2016). These core human needs are archetypal patterns of human behavior that are shared by people across cultures, geography, and national boundaries. However, individuals are different, and have different constellation of human needs. Some have a stronger need for *certainty* and less *variety*; they tend to choose lifestyles and professions with greater stability and security and prefer less change and novelty.

Others, in contrast, may have great risk tolerance and are drawn to adventures and entrepreneurial pursuits that present challenge and even danger. Some people have a greater preference for the fulfillment of *connection* over the pursuit of *significance*; they march to their own drumbeat and may even thrive in isolation. The impetus for needs-fulfillment also leads each person to take different paths and apply different strategies (Kenrick et al., 2010; Madanes, 2016). For some people *growth* and *contribution* are achieved through creating works of art or literature, while others prefer to focus on professional accomplishment. Yet, some people seek deviant means to achieve a sense of personal *growth*. And others meet the need for *contribution* through their allegiance to antisocial or counter-cultural groups, such as the white nationalist groups and extremist militias.

Importantly, our fundamental human needs and how they are met (or not) greatly affects our emotions, and through them our moods, feelings, and self-image. When a person is isolated for an extended period of time (such as in a lockdown during a pandemic), the loss of *connection* and *variety* may result in profound sadness and even depression. Similarly, the loss of a job or a career may cause a person to lose his or her sense of *certainty* and *significance*, which may result in fear, anxiety, and even panic. Therefore, the impact of major social changes like the advent of new technology, on the fulfillment or negation of the six basic emotional needs, though intangible, sometimes barely perceptible, can have profound impact on social welfare and human wellbeing. The losses to our basic needs, though hard to quantify and track, can be a source of personal pain and suffering. Though these tend to be "slow burn" crises, they can gradually spread across a society.

Emerging AI technologies and basic human needs

The impact of AI applications and systems will be immense because the waves of AI innovation will proliferate exponentially in scale and scope across nations (Iansiti & Lakhani, 2020). New systems and practices will collide with traditional ones across different industries and sectors, with profound impact on people and their basic human needs. For instance, the fast developing "large-language" models and image generators like ChatGPT-4, Google Bard, Wall-E 2, and Midjourney have been trained on trillions of words of text and figures has begun to upend the status quo in creative industries and educational institutions, as their literary and artistic creations have been used by millions of people without moral guardrails or public oversight and regulation.

These rapidly advancing and revolutionary AI technologies have been compared to the arrival of a superior, super-intelligent alien race that is gradually acquiring power and capabilities that are beyond human imagination. Journalist Stuart Russell of *The Boston Globe* mused: "How do we retain power over entities more than us, forever?" (Russell, 2023) Hence, the impact on our human welfare and need fulfillment (the resultant *surpluses* and *deficits*) must be carefully assessed and actively managed in the design, adoption, and implementation of the new AI technologies, *before* their widespread use in human society. The Russellian prophecy of the human-made, powerful "alien AI civilization" may be an irreversible and uncontrollable development akin to the introduction of an invasive species to a natural ecosystem.

To further illuminate the impacts and possible consequences of the AI revolution, this analysis focuses on three emerging AI technologies that have potential for broad application or popular usage—like "canaries in the coal mine"—to present early warnings of the latent dangers and hidden costs of the looming AI revolution. The uptake and adoption of these technologies are still at their infant stages, and the window of opportunity is still open for preventative or proactive mitigating measures. These three technologies are analyzed with a focus on two of the six core human needs each: Autonomous vehicles and their impact on *certainty* and *connection*; facial recognition systems from the perspective of *variety* and *significance*, and AI writing or image generators in terms of their effect on *growth* and *contribution*.

The choice of the fundamental human needs for application to each technology is based on the suitability and salience of the subset of human needs for each technology. It is unwieldy and repetitive to analyze each technology on all six human needs. For instance, the human need for *certainty* (e.g., drivers' control over the vehicles), and for *connection* (such as the pedestrians' human contact and communication with drivers sharing the use of public roads) have greater relevance in the case of the self-driving cars. And the needs for *variety* and *significance* are much more salient when considering the use of facial recognition technology, as the deployment of public surveillance camera will most likely infringe on the public's right to *privacy* (related to our desire for significance) and motivate some people away from deviance towards conformity to social norms and rules (associated with the need for variety). The three case studies are discussed in turn in succeeding sections, as follows.

Autonomous vehicles and the need for certainty and connection

Autonomous vehicles (AVs, also known as self-driving cars or robotic vehicles) utilize pattern-recognition software to sense its environment, diagnose road conditions, and navigate to destinations with little or no human input. AVs are endowed with advanced sensors to "read" the surroundings, such as thermographic cameras, radar, lidar, sonar, and GPS (Taeihagh et al., 2019). They recognize patterns on the road (other vehicles and pedestrians, and signages) and interpret sensory information (weather conditions, obstacles, and moving objects) to chart appropriate navigation routes. Over the past two to three decades, there has been steady and rapid progress in the research and development, and trials on the use of autonomous vehicles in cities around the world. The biggest automobile and technology companies in the world, including Tesla, Nvidia, Zoox, Uber, BMW, Ford, Baidu, Alphabet's Waymo, and Google Car, are heavily investing in driverless cars.

The AV unit of the American conglomerate Alphabet, Waymo, launched its autonomous taxi service to the public in Phoenix, Arizona in December 2020 (Krafcik, 2020). In 2021, Nuro, a driverless technology startup, began driverless commercial delivery operations in California in partnership with Uber Eats (Choi, 2022). In 2022, Cruise, an AV unit of General Motors, became the first fully commercial driverless taxi hail-ride service in San Francisco (Bellan, 2022). In Japan, Toyota pioneered a fully automated taxi-service around the Tokyo 2020 Olympic Village, and provided fully automated, loop-line rides to the Olympic and Paralympic villages for athletes (Toyota Motor, 2019). In August 2022, China's search giant, tech conglomerate Baidu started commercial driverless taxi services in the Chinese cities of Wuhan and Chongqing, expanding its AV business beyond Beijing and services for the 2022 Winter Olympics (Yu, 2022).

The societal benefits of intelligent vehicles are many. They include reduced automobile accidents, enhanced safety for all road users, reduced traffic congestion, efficiency of commute in terms of time spent, and more enjoyable and comfortable ride experiences. One of the AV's greatest advantages over conventional vehicles is the elimination of human error, which is a primary cause of traffic accidents, which account for 1.3 million fatalities each year (Kaye, et al., 2021; World Health Organization, 2022). With increased autonomy and control given to machines, human drivers are relegated the role of mere passengers who have to relinquish control over the vehicle and are unable to participate in interactions with other road users.

Despite this worldwide rollout of AV services, an impediment in the universal acceptance of this new technology revolves around people's confidence in or comfort level with vehicles functioning on "autopilot" (Wang et al., 2021). An important threshold for the AV industry to overcome is the impact of driverless cars on the human need for *certainty* (perceived loss of safety, comfort, and control on the part of drivers and passengers) and *connection* (feelings regarding the lack of interaction between drivers, pedestrians, cyclists, and other road users). The impact on these basic needs in turn affects human sentiments on the widespread use of AVs. The psychological safety, trust, comfort, stress, fear, and anxiety affect not only the drivers and passengers but also other road users, particularly pedestrians. (Murali et al., 2021).

This loss of control and human interaction with other road users in the vicinity present a challenge for the human need for control (*certainty*) and interaction (*connection*), which affect the AVs' public acceptance. Public acceptance will in turn determine usage and thereby the transition to safer traffic systems around the world. These emotional reactions are commonly experienced in the introduction of industrial machines, indoor home-help and care robots, humanoid robots, and drones (Beran et al., 2015; Rubagotti, 2022). The general public desire to feel safe and comfortable with a new technology will invariably impact the decisions of city councils, local administrations, and national governments on the cusp of authorizing and licensing AVs on public roads (Habibovic, 2018).

The other basic need related to the use of AVs is the need for authentic human-to-human contact (*connection*). The absence of human drivers in control of the autonomous vehicles not only releases people from the tasks of driving, but it also creates a void in the interaction between drivers, pedestrians, and other road users (Lee, 2021). Drivers often use body language, such as eye contact, facial expressions, and hand gestures, and other forms of communication that relay emotions and intentions (Habibovic, 2018). These sophisticated forms of social contact have to be delegated to the autonomous systems in the vehicles themselves. Consequently, the avenues for communicating emotions, like empathy, surprise, fear, and apologetic gestures, are no longer available to the human users sitting in the self-driving cars. To make AVs more natural and acceptable, affective features have been developed to communicate their intent, focus, and awareness to pedestrians and other road users (Chang et al., 2017). Examples include the display technologies (e.g., LED lighting, honking sounds) and anthropomorphic features (such as moving eyes or smiling faces) to signal intent and emotions on the exterior of the AVs (Murali et al., 2021).

Apart from the usual safety features, how can they design the technology to further gain people's trust and offer a satisfactory comfort level? What can be done to make self-driving vehicles be perceived as safe and benign to meet the human need for safety/certainty and connection/personal contact? One approach to address these issues is for future AVs to offer rich experiences to engage the occupants and better meet their needs, like real-time health and mental state checks through physiological sensors like those in wearable applications that measure health-related data and the emotional or cognitive states of drivers and passengers. To strengthen connection, safety, awareness, trust, and a better user experiences in AVs, in-vehicle displays may also go beyond the conventional graphical user interface displays to include gesture, auditory and tactile interfaces, wearable sensors, and augmented and virtual reality technologies (Murali et al., 2021).

Lessons in the design of emotional functions and expressions in "home-help" robots can be incorporated into the AVs' *affective* dimensions to better mimic or even meet the human need for *connection*, as these automobiles are essentially self-moving, nonhumanoid social robots (Heinkel, 2017; Tabone, 2021). From the experience of "social robots" that are designed for serve humans in homes, AI creators must pay attention to the quality of human—robot interaction, and hence the shaping of human perception and the meeting of human needs. These features can help promote the level of usage of self-driving cars and acceptance of their widespread deployment on public roads; two key factors driving the commercial success of AV companies.

Facial recognition systems and our human need for variety and significance

Facial recognition technology was pioneered by Woodrow Wilson Bledsoe, who created a system that organizes human faces' photos manually with computer technology in the 1960s. He used a grid of vertical and horizontal coordinates to manually record the coordinate areas of facial features like eyes, nose, mouth, and hairline of a person and categorize it for cross referencing to other images (Bledsoe, 1964, 1986). From his early work, the techniques were greatly advanced in the 1970s and 1980s with the invention of specific subjective markers for facial identification and the invention of computer software for facial recognition.[1] Facial recognition is a biometric technology that can identify an individual from a live-capture digital image, and possibly cross-reference it with an existing photograph of a face called a faceprint. The faceprint is analogous to a fingerprint (both are unique to each person), and it measures 80 nodal points on a face.

Facial recognition technology is becoming widely adopted in various functions and industries; they range from enforcement agencies and airport and bank security to mobile phone and home appliance. Applications related to facial recognition are those for speech recognition, reading of text patterns, recognition of movements of humans and vehicles, and medical image recognition in healthcare. Airports around the world have been using it for security and access control, and mobile phone makers are increasingly embedding biometric security features into their devices. Just a decade ago popularly considered a fixture of science fiction, facial recognition technology is gradually becoming widely used in human society. This technology along with machine learning technology is benefiting humans in many ways, including finding and identifying persons, pets, and

---

[1] In the 1990s and 2000s, the Defense Advanced Research Projects Agency (DARPA) and the National Institute of Standards and Technology (NIST) further developed this technology and encouraged it use by private companies for commercial purposes (de Leeuw & Bergstra, 2007). Gradually in the first two decades of the millennium, it has become a common tool for border controls, airlines, air travel and transportation hubs, stadiums and mega-event venues, especially those dealing with concerts and conferences. Its application in biometrics is now used in policing and crowd control in mass events worldwide. (NEC, 2022)

protecting premises from unlawful access and theft, improving security at places such as banks and airports, and making shopping and payments more efficient (Gargaro, 2022).[2]

Unlike digitalized inventions and automated systems that must be initiated and configured by humans to execute tasks, AI-enabled facial recognitions systems are adaptive and independent in their functions, mimicking human behavior and transcending the need for continuous human monitoring. With pattern recognition and machine learning capabilities, AI uses algorithms[3] to identify facial and behavioral patterns and categorize and classify them based on features they can detect in those patterns. And through complex algorithms built on statistical models, automated cameras can be "trained to look" at unstructured data and "see patterns" in facial expression.[4]

The benefits of accurate and searchable identities of persons come from improved safety of people using airports, customs checkpoints, and banks, and increased efficiency of services at these premises. When facial recognition software is effective in identifying known or suspected criminals and terrorists, their installation and the presence of the cameras themselves can provide a deterrence against nefarious activities. The same technology, when used to locate missing children or pets, can combine facial recognition software with ageing software to simulate how the missing persons or pets would look several years on. In stores, biometric identification can greatly facilitate cash or credit transactions, where a camera can be programmed to recognize faces and charge the goods to the owners' accounts. It can also help reduce human touchpoints (physical contact or direct in-person interaction) during a pandemic and help prevent the spread of diseases. Increasingly, schools, colleges, and hospitals are also planning to implement facial recognition technology on their premises.

In China, the social credit system is an ambitious government surveillance system that will generate a unified social credit code for all Chinese citizens that scores all aspects of life in the nation. The country has network of some 200 million CCTV cameras with facial recognition capabilities for nationwide surveillance with strong Orwellian undertones—reminiscent of the draconian controls of citizens practiced by repressive governments in the dystopian novel, *1984*. These sophisticated cameras are supplemented by millions of state monitors of microblogs and mobile

---

[2]Apart from the surveillance and security industries, the marketing and sales sectors have also started using affect recognition technology, such as in telephone exchange virtual assistants and telemedicine chatbots. For instance, marketers want to know how people respond to a product, they can use affect recognition software to detect exactly where someone's eyes went, or how they felt when they saw a picture or video of a product. By tracking those eye movements and analyzing facial expressions, they can study how effective an advertisement is or emotionally engaging is a product or service. Hence, data on real-time facial expressions can provide marketers insights on the emotions and thoughts behind purchasing decisions and use them to improve their advertisements and customer experiences. Soon, it will become more widely used in stores, schools, colleges, and hospitals, and across many sectors.

[3]An algorithm is a set of instructions that is applied to an input, performs some repeatable computation on it to derive an output. It is like a precise recipe, typically coded in programming language.

[4]Today's AI-powered software are able to detect not just facial features but also the common human emotions. Some technologies can only detect facial features to define identity factors like age, gender and race; they are not able recognize and identify a face. Other technologies are more advanced, and they delve into "affect recognition" − detecting emotions by analyzing a person's facial expressions and movements to determine the emotion expressed. Through deep learning, AI enables the categorization of facial expressions to identify emotions, such as anxiety, sadness, anger, and joy. Most of human emotional expressions can be inferred from the face, posture, and voice; these three sources offer the most salient evidence for identifying emotional expressions.

surveillance devices that also use number plate recognition and facial recognition tools (State Council, 2015, 2017).

The "social credit" record reduces each citizen into a score than can be used to rank and rate each person, and the data is captured in a national database (Hatton, 2015). It judges each citizens' civic behavior and trustworthiness (based on acts like jaywalking and littering, outstanding fines and unpaid court bills) and inputs a personal score to a permanent record (Raphael & Xi, 2019). It creates a moral ranking system that punishes poorly ranked or black-listed citizens for infractions such as bad driving behavior, smoking in nonsmoking zones, excessive spending on video games, and posting fake news online on terrorist attacks (Canales & Mok, 2022). The penalties range from reduced internet speeds and disapproval of loans, to denial of college admission and flight bans. Those with good scores enjoy perks, such as expedited travel authorization for flights overseas and rentals and hotel bookings without a cash deposit, discounts on energy bills, and better interest rates in bank transactions (ibid).

Even in liberal democratic societies, there is a potential abuse in the form of data theft, identity theft, and the malicious use of personal information. Face-scanning body cameras or security cameras permit pervasive and continuous surveillance of people without their consent and record their biometric data for purposes unknown to the subjects. Facial recognition software can transform these devices into ubiquitous surveillance networks that register and report the movements and locations of people in events and places like protest marches and entertainment premises. This can be deemed a violation of privacy and an affront to people's civil liberties. Facial recognition devices are banned in several American cities. San Francisco, Oakland, Boston, New Orleans, Oakland, Pittsburgh, and Santa Cruz have all banned their use outright. New Orleans, Oakland, Pittsburgh, and Santa Cruz in the United States have laws banning its use in predictive policing, the use of computer systems to identify people as likely perpetrators or victims of crime (Sheard & Schwartz, 2022).

Two fundamental human needs that could be affected by the widespread use of facial recognition are *variety* and *significance*. The human need for variety drives people to seek high-risk activities such as extreme sports or compulsive sexuality to bring thrills and novelty into their lives (Madanes, 2016). For others, a major source of variety is to experience alternative lifestyles and experiment with deviant behavior. AI-powered surveillance equipment is a bane to these behaviors and activities. Some people may also view the surveillance and monitoring systems as a violation of their civil liberties; for them, *significance* comes from unconventional and eccentric acts that make them feel special or unique.

Widespread surveillance creates the pressure to conform to societal norms and cultural mores; making it harder for these people to "stand out" and pursue activities that will distinguish them. For instance, some people seek to meet their needs for *variety* and *significance* through less conventional activities, such as joining a 12-step recovery group or a secret conspiracy theory network in cyberspace. These activities may come under increased scrutiny through the use of ubiquitous surveillance in public cameras and in cyberspace. AI algorithms in security cameras at airports may "flag" the unconventional people as "trouble-makers" and deviants with criminal intent, leading to self-censorship and a tendency to conform to social norms.

In the hands of dictators and oppressive governments, AI can be used to monitor and control the movements and locations of citizens, as in the example of China's social credit system. The Chinese government's facial recognition systems are able to follow people across populous cities.

As a demonstration of this sophisticated surveillance network, Chinese officials collaborated with BBC News to track down one of its reporters tasked with evading detection in Guiyang, a city of 3.5 million people in 2017. The journalist, John Sudworth, was located by the police hiding in a building within just *seven* minutes using the vast network of facial recognition cameras (Zhao, 2017). In at least one incident in mid-2022, the Chinese government may have used the required COVID phone apps (designed to identify and isolate people who might be spreading Covid-19) to shut down potential protesters by changing the "health status" of "troublemakers" to travel freely and gain access to certain public places (Dong, 2022).

AI writing and image generators and their impact on growth and contribution

AI writing assistants and image generators are taking the world by storm in late 2022. The increasing power and sophistication of the software are shaking up the literary, art, and educational worlds with visually stunning AI-generated images and computer-drafted writing that are hard to distinguish from those created by human artists and authors respectively (Madawi, 2022). The technology is free, getting cheaper by the day, and easy to use. "Creative" works are easily generated with just minutes of inputs into a chat bot.

AI-generated text and pictures have already been embedded in news reports, novels, marketing slogans, political speeches, and even journal articles. About a third of the content published by Bloomberg News used some form of automated technology from a system called *Cyborg* (Peiser, 2019). AI-writers are particularly efficient and accurate in producing company earnings briefs and financial reports and creating "breaking news" stories with just-in-time, globally sourced facts and figures (ibid). In June 2022, *Dall-E 2*, a text-to-image generator was launched by OpenAI, named after painter Salvador Dali and Disney Pixar's *WALL-E* (Verman, 2022). Since then, it has been widely used to create high quality, photorealistic pictures, ranging from simple portraits and sceneries to realistic images of war crimes and school shootings (Rook, 2022).

The 2022 Colorado State Fair's annual art competition introduced a new contest for digital artists in a "digital art section" of the contest, in addition to its traditional categories of painting, quilting, and sculpture. The winner of this categories, Jason Allen, created a realistic picture, "Théâtre D'opéra Spatial," ("Space Opera Theater") using *Midjourney*, an AI text-to-graphics program. This event created much controversy and fierce debate by detractors and critics who claimed that the AI-powered apps like DALL-E 2, *Midjourney* and *Stable Diffusion* are enabling a form of digital plagiarism. They based their concerns on the fact that "amateurs can now create complex, abstract or photorealistic works simply by typing a few words into a text box." (Roose, 2022a).

ChatGPT (GPT is short for generative pretrained transformer), a new chatbot created by OpenAI, debuted in November 2022, to popular acclaim (Roose, 2022b). It can tell jokes, write computer codes and draft essays good enough for college assignments. It uses deep learning to search and analyze copious amounts of digital text (books, journal articles, news stories, blogs, and Wikipedia articles) to create intelligent syntheses and comprehensive responses to text prompts, and by extension, test questions and essay assignments that are the "bread and butter" of college and high school courses (Bogost, 2022). Ammaar Reshi, a product design manager in

San Francisco, used ChatGPT to publish a 12-page picture book, printed it, and started selling it on Amazon within a week (Popli, 2022).

A high school English teacher, Daniel Herman, in an article in *The Atlantic*, opined that "[w]hat GPT can produce right now is better the large majority of writing seen by your average teacher of Professor" (Herman, 2022). This commercially available app is so powerful a writing aid that educators are forced to rethink how to give writing assignments. It made Herman "wonder if this might be the end of using writing as a benchmark for aptitude and intelligence" (ibid). Some educators have predicted that ChatGPT, and tools like it, will upend traditional school assignments and take-home exams (Roose, 2022b). ChatGPT is a harbinger of the types of disruption that are to come, as AI upends established practices and professions.

Like a new genre of artform or literary work, AI-generated content can have intrinsic economic, educational, and esthetic value as "original creations" by machines. Great gains in efficiency of the creative process and in the quantity and speed of production of literary and artistic works may follow. People are using them in droves and some companies are already choosing the new technology over hiring human talent. The San Francisco Ballet has begun to use AI-generated images for the 2022 season's advertising promotion of the *Nutcracker*, and a digital fashion model, Shudu Gram, has begun modeling for famous fashion houses, like Louis Vuitton (Popli, 2022).

From the controversies and public backlash witnessed in the media, AI text and image-generating applications must have made novelists, journalists and artists nervous about their job security and livelihood (Roose, 2022a). Many new questions emerge: *If the new breed of AI-powered tools have been trained on authors' and artists' proprietary works and their algorithms can recognize and imitate their distinctive styles, are their outputs still "original works"? Why would anyone hire humans or pay for their services, when they can use AI to generate their own "creations" much faster and more cheaply?* (Popli, 2022). *What are the protections for artists and authors whose works have been used in AI algorithms without their consent? Who will protect the rights and intellectual property of creative professionals, and compensate them for their income and job losses?*

These issues relate to the remaining two fundamental human needs: *growth* and *contribution*. The need for *growth* can be met through the mastery of a skill or craft. The "creatives" seek to master and practice artistic or literary skills, and make their mark on canvases, books, or blogs. Some will pursue a career in the creative arts and literary professions, develop their own styles and trademarks. To these craftspeople, their works are also their unique *contribution* to society. Arguably, artist and literary works provide creative, intellectual and emotional inspiration and insights to society at large, and can uplift people's spirits and challenge the status quo. Art and literature can also contribute to the world by making it a more civilized, creative, and inspired place to live in.

With the advent of AI writing assistants and image generators, the traditional avenues for the mastery of creative literary and artistic crafts have been displaced or irreversibly disrupted. Writers and artists will find it harder to compete with the high-speed composing and deep-learning capabilities that can cull facts and figures and distill styles and ideas from the internet. It is a setback to the conventional pathways for professional *growth* and *contribution* for entire guilds of creative craftspeople and the many others who rely on them for inspiration or livelihood. A related violation of the interests of these craftspeople is the perceived "theft or piracy" of intellectual capital and creative designs by computers who scour the internet in their generative work. Some colleges and universities have classified their use in students' submitted work as a form of dishonesty and have prohibited the submission of computer-assisted writing as the students' original work in academic exercises (American University, 2020; University of Maryland, 2020). In early 2022, there was a class-action lawsuit against the Microsoft-owned GitHub for its AI-powered tool, *Copilot*, that translates human instructions into functional computer code (Verman, 2022). The complaint contends that training the AI system on publicly available data from the internet violates the rights of creators who posted the code under open-source licenses. The lawsuit essentially brands the use of deep learning by *Copilot* as a form of piracy of the original code invented by millions of software programmers. This backlash epitomizes the reaction to human losses in the form intellectual property rights, career opportunities, incomes, and livelihoods.

Early warning of disruption and upheaval

Not all technological advancements that increase efficiency and make our lives easier and more comfortable will have an overall positive impact on core human needs. The above analysis shows that the most advanced technologies, like autonomous vehicles and humanoid "home-help" robots, may cause the loss of authentic human connection and the rise of loneliness and anxiety in people (*certainty* and *connection*). Facial recognition systems could transform surveillance technologies and the public space in ways that infringe on the privacy and security of personal information of citizens and their civil liberties to freely participate in political protests or simply pursue their business anonymously in public (*variety* and *significance*). Increasingly, high-end AI writing application and image creating software are able to successfully mimic and replicate the best human novelists, journalists, and artists, thereby threatening literary and artistic professions with loss of jobs, social status, and livelihoods (*significance* and *growth*).

The introduction of the steam engine and mechanized manufacturing and transportation during the First industrial Revolution (1750−1840) created significant social upheavals and uncertainty, leading to displacement of human labor and craftsmanship and the mass migration from rural areas to the cities (Ivezic & Ivezic, 2020). Decades later, the invention of electricity and mass production in the assembly lines and the advent of automobiles during the Second Industrial Revolution (1870−1914) forced widespread social change and disruption to many professions. From the 1990s, the Third Industrial Revolution introducing computers and information technology also resulted in painful adjustments in many professions and processes displaced by computerization and automation, and the loss of human connection and social validation from increasing isolation among workers (ibid).

Evidently, these industrial revolutions brought with them great leaps in industrial production, technological advancements, and rising standards of living, but also wide-ranging social and human costs that are often overlooked in the name of progress. Many of these costs are intangible pains and time-lagged losses suffered by people and communities in silence. The advent of artificial intelligence is said to be the harbinger of a Fourth Industrial Revolution driven by big data, advanced robotics, cloud computers, and virtual reality, which will bring waves of new inventions and technological progress that will last for decades (UNESCO, 2018). The Fourth Industrial Revolution will be driven by many exciting inventions and applications, such as cybernetic "implantables," blockchain, digital currencies, 3D printing. Their adverse effects on our basic human needs may roll back the gains in human rights and social welfare achieved through decades of development.

However, with advanced planning and careful public sector oversight and regulation, some of these costs may be prevented or mitigated. There are several examples of proactive management of technological adoptions as public leaders, business executives and communities take it upon themselves to pause, put restrictions on, or ban outright, advancing technology in the interest of human welfare and the public interest. The United States Office of Science and Technology Policy (OSTP, a branch of the White House) has issued a Blueprint for an AI Bill of Rights that outlined the administration's intent and plans on how to mitigate the negative impacts of AI technologies and protect the democratic rights and civil liberties of Americans (United States Office of Science and Technology Policy, 2023). In this document, the OSTP has identified principles to guide the design, use, and deployment of AI systems to protect the interests and rights of the public. In the proposed AI Bill of Rights, there will be public policies and regulations measures to offer policy guidance on the use of AI, especially by government agencies.

There is much that government regulation and private sector leadership could do to prevent and mitigate the negative impact of AI, such as personal privacy, data security, misinformation, and the loss of human connection and jobs. On 22 March 2023, the Future of Life Institute, a nonprofit think tank led by MIT physics Professor Max Tegmark, issued an open letter, signed by Tesla CEO Elon Musk, Apple co-founder Steve Wozniak, and Turing Award-winner Yoshua Bengio, and hundreds of AI researchers, asking for a pause on the race by AI lab to develop and deploy powerful technologies that "no one—not even their creators—can understand, predict, or reliably control" (Future of Life Institute, 2023b). On 16 May 2023, Sam Altman, OpenAI CEO, who created and released ChatGPT, called for state regulation of artificial intelligence at U.S. Senate subcommittee hearing across the globe. He further called for a global or national agency to be set up to license AI companies (Kang, 2023).

Among the many things that can be done to manage the myriad and profound risks of general-purpose, super-intelligent AI to society and humanity are rigorous testing and safety standards and licensing requirements by public authorities on the development and release of AI technologies and public policy and laws to enforce compliance with these standards. These rules, akin to how the public agencies would regulate and oversee the release of new drugs by pharmaceutical companies or the operations of a nuclear powerplant. In all these cases, the people and their governments must proactively act to safeguard human rights and values, such as the right to personal privacy by accessing and managing the data they generate, or the right to civil liberty by ensuring that AI applications do not unreasonably curtail people's freedom and opportunities to meet their basic human needs.

Some of the practical measures that have been proposed in the open letter by the Future of Life Institute (op cit.) is for domain experts and AI labs to collaborate in developing and implementing a set of "shared safety protocols for advanced AI design and development that are rigorously audited and overseen by independent outside experts." Its recommendations also include closer scrutiny of "ever-larger unpredictable black-box models with emergent capabilities" and not just any new AI application (ibid.). In addition, the signatories of the letter also recommend for AI developers to work with policymakers in the development of AI governance systems, which would include: New regulatory authorities dedicated to oversight of AI; the tracking of powerful and general-use AI systems and applications with gigantic computational capability; and the creation of "provenance and watermarking systems" to help distinguish real from artificially-made products and services in the interest of transparency, accountability, and the public good (ibid.).

Many of these are already being enacted and implemented at various levels of administration, from city and local governments to national policies and international protocols. At the level of state and city governments, the city of San Francisco reversed its "killer robot plan" in November 2022 in response to angry protests and widespread criticism. The Board of Supervisors of the city had previously approved the police to deploy remote-controlled, ground-based robots that can use deadly force on civilians when there is "imminent risk to life" (Masih, 2022). The public outcry signals a strong disapproval and mistrust of armed robots in the hands of the San Francisco Police Department. With the recent spate of killings of civilians at the hands of law enforcement officers, it is intuitively evidence that such a technology would be an affront to the citizens' basic human need for security and personal autonomy unimpeded by the presence of deadly robots and "militarized" law enforcement.

At the level of transnational organizations, in November 2022, European Union (EU) legislators began drafting the Artificial Intelligence Act to actively regulate the use of facial recognition technology, specifically to curb the potential harm from its adoption across the 27 nations with some 450 million citizens combined. The EU's concern about infringement on the civil liberties of its citizens is driving the impetus to ban outright all facial recognition technology by all public and private entities, *without exceptions*; even in cases of applying it when searching for kidnaping victims, identifying criminal suspects, and preventing serious threats to public security, like terrorist attacks (Hsu, 2022).

The Cyberspace Administration of China (its national cyberspace regulator and watchdog) has proposed new measures in April 2023 for managing AI models that power applications used by the public, from chatbots to image generators. This new policy requires artificial intelligence companies to submit security assessments to authorities before the commercial launch of their applications (Ye, 2023). This could mean very stringent public regulation or even the outright ban on certain AI technologies, following the widespread popular use OpenAI's ChatGPT since its recent release in late 2022 (ibid).

In the realm of international law, the OECD AI Policy Observatory (OECD.AI) is one of the earliest intergovernmental standards, first adopted in May 2019 by 38 OECD member countries and adhered to by many other "partner economies" (OECD Policy Observatory, 2022). Among the OECD AI principles relevant to the advancement of imperative to meeting human needs are, "Principle 1.2 on Human-centered values and fairness: AI actors should respect the rule of law, human rights and democratic values, throughout the AI system lifecycle. These include freedom, dignity and autonomy, privacy and data protection, nondiscrimination and equality, diversity, fairness, social justice, and internationally recognized labor rights" (ibid). Another principle that affirms the protection of human values and rights is, Stakeholders should proactively engage in responsible stewardship of trustworthy AI in pursuit of beneficial outcomes for people and the planet, such as augmenting human capabilities and enhancing creativity, advancing inclusion of underrepresented populations, reducing economic, social, gender and other inequalities, and protecting natural environments, thus invigorating inclusive growth, sustainable development and well-being (Principle 1.1 on Inclusive growth, sustainable development and well-being) (ibid).

The Asilomar AI Principles, developed at the Beneficial AI 2017 conference and signed by some 1797 AI/Robotics researchers and many others, are another "first steps" towards a set of international rules and norms that may be negotiated and adopted as international law to guide the actions and behaviors of nation states in the future (Future of Life Institute, 2023a). Among the list of many values that was promulgated in the Asilomar AI Principles are that "Highly autonomous

AI systems should be designed so that their goals and behaviors can be assured to align with human values throughout their operation (Value Alignment)" and "AI systems should be designed and operated so as to be compatible with ideals of human dignity, rights, freedoms, and cultural diversity (Human Values)." (ibid) These are the values that would be consistent with meeting human needs and advancing people's welfare as discussed in this chapter.

When our public, business, and civic leaders, as well as AI designers and adopters understand what AI can do (or will do) to our basic human needs, humanity can make plans to prevent, mitigate and adapt to AI's adverse impact. Our level of foresight and preparedness will determine how much damage and disruption these new technologies will wrought in our lives and our environment. The worst outcome is for our leaders and scientists to ignore the lessons from the past industrial revolutions and early warning signs from the looming AI Revolution, and passively witness as the emerging technologies transform our world. Our human needs are a vital and irreplaceable aspect of our human nature and genetic makeup; they are messengers that offer important information for living our lives, prioritizing actions, and setting direction. Having our needs and the urge to fulfill them are a quintessential feature of the human experience. The moral imperative to acknowledge and cater to human needs and social welfare is, arguably, one of the toughest challenges for AI designers and implementers to confront, as they strive to make intelligent machines more valuable, effective, and acceptable to humans.

Conclusion—the "Shadow Side" of AI technology

The impacts of AI on people's basic needs can be intelligently managed to safeguard human welfare and rights, especially the most vulnerable and voiceless people who have no say or little knowledge of the dramatic changes that will accompany the emerging technologies. The approach need not be to succumb to the sensationalistic, doomsday, apocalyptic predictions. Neither should there be a passive, techno-optimist, and reactive attitude towards to the powerful changes that the AI Revolution portends. Instead of villainizing or valorizing the advancing technology, much can be done to manage "the bad" and accentuate "the good" consequences. Jung (1938) postulated the "shadow aspects" of human personality as the dark and disowned aspects of our personality that humans choose to hide, repress, or ignore. In every major technological change, there are "winners" and "losers" from the widespread societal transformation and social upheaval that follow. Unless the "shadow aspects" of the AI revolution are acknowledged and explicitly addressed, they will continue to create suffering and pains in the denied and unmet needs of groups and individuals.

AI will usher in a new era of material progress and comfort for most of humanity. However, the assessment of the three AI technologies from the perspective of the six core human needs shows that these benefits will come with an invisible price tag. Our ability to capitalize on these benefits will in part depend on how well we "cushion" our societies from the negative and unintended consequences. Like many transformations that tend to grow exponentially, a proactive approach in managing the risks and costs would be much more effective and efficient than simply reacting to the fallout. Problems in the human-machine interface of AVs, the privacy and civil liberties impact of facial recognition systems, and the disruption from AI-generated writing and art, can be mitigated in the design of algorithms and the implementation and oversight/regulation of these breakthrough technologies.

Due to the rule-bound nature of bureaucracies, cutting-edge technologies will invariably outpace the regulations and protections in public policy to safeguard public interest and welfare. The corporations' concern for the profit motive and shareholders' interest will not incentivize responses that adversely affect their bottom line, exemplified by the behavior of social media platforms that prioritize misinformation and hate speech as attraction for engagement of users over content moderation because they are "good for business." In civil society, citizens may not have the resources, expertise, and access to decisionmakers to advocate for human interests and rights, much less make changes to the proliferating AI inventions. All too often, AI designers and software scientists write algorithms that are so task driven and efficiency minded that they are devoid of common-sense concerns for the impact on human lives.

The "Age of Artificial Intelligence" is already here. There is no escaping a future when advanced robotics and intelligent machines will be fully integrated with and significantly impact human lives. The relentless march of technology may soon result in AI to be embedded in practically every application in our gadgets and every human activity. The human impact of AI should be the subject of further studies and public discourse as humanity rushes headlong into a rapid embrace of new and more powerful AI technologies. Topics for further study include the ethical, environmental and social consequences of AI applications; policy design for effective oversight and regulation of advanced technology; and the curriculum and training programs for AI scientists, professionals, and regulators on the human impacts of emerging AI. In particular, there can be educational programs on the policies and practices for algorithmic design to enable autonomous machines to make ethical, humane, and life-affirming decisions, undertake action in favor of global sustainability and stability, and mitigate AI's negative impacts on society.

References

American University. (2020). University Policy: Academic Integrity Code. Washington, DC. https://www.american.edu/policies/students/upload/academic-integrity-code-2020-2-2.pdf.

Aronson, E. (1980). *The social animal*. New York: Palgrave Macmillan.

Barnett, V. (2002). Which Was the "Real" Kondratiev: 1925 or 1928? *Journal of the History of Economic Thought*, 24(4), 475−478.

Bellan, R. (2022). Cruise can finally charge for driverless robotaxi rides in San Francisco. TechCrunch, June 2, 2022. Boston, MA. https://techcrunch.com/2022/06/02/cruise-can-finally-charge-for-driverless-robotaxi-rides-in-san-francisco/.

Beran, T. N., Ramirez-Serrano, A., Vanderkooi, O. G., & Kuhn, S. (2015). Humanoid robotics in health care: An exploration of children's and parents' emotional reactions. *Journal of Health Psychology*, 20(7), 984−989.

Bledsoe, W.W. 1964. The Model Method in Facial Recognition, Technical Report PRI 15, Panoramic Research, Inc., Palo Alto, California.

Bledsoe, W. W. (1986). I had a dream: AAAI presidential address, 19 August 1985. *AI Magazine*, 7(1), 57−61.

Bogost, I. (2022). ChatGPT Is Dumber Than You Think. *The Atlantic*. December 7, 2022. Washington, DC. https://www.theatlantic.com/technology/archive/2022/12/chatgpt-openai-artificial-intelligence-writing-ethics/672386/.

Canales, K., & Mok, A. (2022). China's 'social credit' system ranks citizens and punishes them with throttled internet speeds and flight bans if the Communist Party deems them untrustworthy. *Insider, 2022*, November 28. Available from https://www.businessinsider.com/china-social-credit-system-punishments-and-rewards-explained-2018-4.

Chang, C.M.; Toda, K.; Sakamoto, D.; Igarashi, T. (2017). Eyes on a Car: An Interface Design for Communication between an Autonomous Car and a Pedestrian. In Proceedings of the Ninth International Conference on Automotive User Interfaces and Interactive Vehicular Applications; Association for Computing Machinery: AutomotiveUI. New York, NY: 65−73.

Choi, C. (2022). *Nuro and Uber partner on autonomous food deliveries. Inside Autonomous Vehicles.* NJ: Red Bank. Available from https://insideautonomousvehicles.com/nuro-and-uber-partner-on-autonomous-food-deliveries/.

Dong, J. (2022). A Chinese city may have used a Covid app to block protesters, drawing an outcry. *The New York Times, 2022.* Available from https://www.nytimes.com/2022/06/16/business/china-code-protesters.html.

Engineered Arts. (n.d) Ameca. Falmouth, U.K. Retrieved from https://www.engineeredarts.co.uk/robot/ameca/.

Future of Life Institute. (2023a). AI Principles. 11 August 2017. https://futureoflife.org/open-letter/ai-principles/.

Future of Life Institute. (2023b). Pause Giant AI Experiments: An Open Letter. 22 March 2023. https://futureoflife.org/open-letter/pause-giant-ai-experiments/.

Gargaro, D. (2022), The pros and cons of facial recognition technology, in *ITpro*.

Habibovic, A., et al. (2018). Communicating Intent of Automated Vehicles to Pedestrians. *Frontiers in Psychology*, 9, 2018. Available from https://www.frontiersin.org/articles/10.3389/fpsyg.2018.01336.

Hanson Robotics. (n.d.) Sophia. Hong Kong. Retrieved from https://www.hansonrobotics.com/sophia/.

Hatton, C. (2015). China 'social credit': Beijing sets up huge system. *BBC News*. Available from https://www.bbc.com/news/world-asia-china-34592186.

Heinkel, D. (2017). What We Know about How Autonomous Driving Will Take Hold. https://www.here.com/learn/blog/know-autonomous-driving-will-take-hold.

Herman, D. (2022). The End of High-School English *The Atlantic*. Washington, DC. https://www.theatlantic.com/technology/archive/2022/12/openai-chatgpt-writing-high-school-english-essay/672412/.

Hsu, J. (2022). EU's Artificial Intelligence Act will lead the world on regulating AI New Scientist, December 28, 2022. https://www.newscientist.com/article/mg25634192-300-eus-artificial-intelligence-act-will-lead-the-world-on-regulating-ai/.

Iansiti, M., & Lakhani, K. R. (2020). *Competing in the age of AI: Strategy and leadership when algorithms and networks run the world.* Boston, MA: Harvard Business Review Press.

Ivezic, M., & Ivezic, L. (2020). *The future of leadership in the age of AI: Preparing your leadership skills for the AI-Shaped future of work.* Marin Ivezic.

Jung, C. (1938). Psychology and religion. *American Journal of Psychiatry*, 95.2(1938), 504−506.

Kang, C. (2023). OpenAI's Sam Altman Urges A.I. Regulation in Senate Hearing. The New York Times, 16 May 2023. https://www.nytimes.com/2023/05/16/technology/openai-altman-artificial-intelligence-regulation.html.

Kaye, S.-A., Somoray, K., Rodwell, D., & Ioni, L. (2021). Users' acceptance of private automated vehicles: A systematic review and meta-analysis. *Journal of Safety Research*, 79, 352−367. Available from https://doi.org/10.1016/j.jsr.2021.10.002, December 2021.

Kenrick, D. T., Griskevicius, V., Neuberg, S. L., & Schaller, M. (2010). Renovating the pyramid of needs: Contemporary extensions built upon ancient foundations. *Perspectives on psychological science*, 5(3), 292−314.

Krafcik, J. (2020). *Waymo is opening its fully driverless service to the general public in Phoenix. Blog.* Waymo, LLC. Available from https://blog.waymo.com/2020/10/waymo-is-opening-its-fully-driverless.html.

Lee, Y. M., et al. (2021). Road users rarely use explicit communication when interacting in today's traffic: Implications for automated vehicles. *Cognition, Technology & Work*, 23. Available from https://doi.org/10.1007/s10111-020-00635-y.

Madanes, C. (2009). *Relationship breakthrough: How to create outstanding relationships in every area of your life.* Emmaus, Pennsylvania: Rodale.

Madanes, C. (2016). The 6 Human Needs for Fulfillment. Blog, October 12, 2016. http://cloemadanes.com/2016/10/12/the-6-human-needs-for-fulfillment/.

Madawi, A. (2022). When AI can make art — What does it mean for creativity? *The Guardian*, November 12, 2022. Available from https://www.theguardian.com/technology/2022/nov/12/when-ai-can-make-art-what-does-it-mean-for-creativity-dall-e-midjourney.

Madsen, S. (2019). *The power of project leadership: 7 Keys to help you transform from project manager to project leader*. London: Kogen Page.

Masih, N. (2022). San Francisco bars police from using killer robots, reversing recent vote. *The Washington Post*, December 7, 2022. Available from https://www.washingtonpost.com/nation/2022/12/07/san-francisco-killer-robot-cop/.

Maslow, A. H. (1943). A theory of human motivation. *Psychological Review*, *50*(4), 370–396.

Maslow, A. H. (1962). *Toward a psychology of being*. Princeton: D. Van Nostrand Company.

Murali, P. R., Kaboli, M., & Dahiya, R. (2021). Intelligent In-Vehicle Interaction Technologies. *Advanced Intelligent Systems*. Available from https://doi.org/10.1002/aisy.202100122.

OECD Policy Observatory. (2022). OECD AI Principles: OECD's Recommendation on Artificial Intelligence https://oecd.ai/en/about.

Peiser, J. (2019). The rise of the robot reporter. *The New York Times*. https://www.nytimes.com/2019/02/05/business/media/artificial-intelligence-journalism-robots.html#:~:text=Roughly%20a%20third%20of%20the,company%20earnings%20reports%20each%20quarter.

Popli, K. (2022). He Used AI to Publish a Children's Book in a Weekend. Artists Are Not Happy About It. *Time*, December 14, 2022. Available from https://time.com/6240569/ai-childrens-book-alice-and-sparkle-artists-unhappy/.

Raphael, R., & Xi, L. (2019). Discipline and Punish: The Birth of China's Social-Credit System. *The Nation*. Available from https://www.thenation.com/article/china-social-credit-system/.

Robbins, T. (2005). Why we do what we do. TED Talk, Oxford, UK. Accessed on 10 December 2021 at: https://www.ted.com/talks/tony_robbins_why_we_do_what_we_do.

Rook, L. (2022). AI-Generated Images Show What Climate Change Will Do To Our Cities by the Year 2100. Greener Ideal, November 15, 2022, https://greenerideal.com/news/climate/ai-generated-images-show-what-climate-change-will-do-to-our-cities-by-the-year-2100/.

Roose, K. (2022a). An A.I.-Generated Picture Won an Art Prize. Artists Aren't Happy. *The New York Times*, December 5, 2022. https://www.nytimes.com/2022/09/02/technology/ai-artificial-intelligence-artists.html.

Roose, K. (2022b). The Brilliance and Weirdness of ChatGPT. *The New York Times*, December 5, 2022. https://www.nytimes.com/2022/12/05/technology/chatgpt-ai-twitter.html.

Rubagotti, M., et al. (2022). Perceived safety in physical human—robot interaction—A survey. *Robotics and Autonomous Systems*, *15*. Available from https://doi.org/10.1016/j.robot.2022.104047.

Russell, S. (2023). How do we retain power over entities more than us, forever? *The Boston Globe*, May 15, 2023. Available from https://www.bostonglobe.com/2023/05/15/opinion/how-can-humans-maintain-control-over-ai-forever/.

Sainato, M. (2015). Stephen Hawking, Elon Musk, and Bill Gates Warn About Artificial Intelligence. *Observer*. Retrieved from https://observer.com/2015/08/stephen-hawking-elon-musk-and-bill-gates-warn-about-artificial-intelligence/.

Sheard, N. and Schwartz. (2022). The Movement to Ban Government Use of Face Recognition in *Electronic Frontier Foundation*, May 5, 2022. Retrieved from: https://www.eff.org/deeplinks/2022/05/movement-ban-government-use-face-recognition.

State Council. (2015) Cujin dashuju fazhan xingdong gangyao [Action Programme to Stimulate the Development of Big Data]. 31 August 2015. Translation available online: https://chinacopyrightandmedia.wordpress.com/2015/08/31/outline-of-operations-to-stimulatethe-development-of-big-data/.

State Council. (2017). Xinyidai rengong zhineng fazhan guihua [Next Generation Artificial Intelligence Development Plan]. 8 July 2017. Translation available online: https://chinacopyrightandmedia.wordpress.com/2017/07/20/a-next-generation-artificialintelligence-development-plan/.

Tabone, W., et al. (2021). Vulnerable road users and the coming wave of automated vehicles: Expert perspectives. *Transportation Research Interdisciplinary Perspectives*, *9*100293. Available from https://doi.org/10.1016/j.trip.2020.100293.

Taeihagh, A., & Lim, H. S. M. (2019). Governing autonomous vehicles: emerging responses for safety, liability, privacy, cybersecurity, and industry risks. *Transport reviews*, *39*(1), 103−128.

Tomasello, M. (2014). The ultra-social animal. *European Journal of Social Psychology*, *44*(3), 187−194.

Toyota Motor. (2019). Specially-Designed Toyota "Tokyo 2020 Version" e-Palette to Provide Automated Mobility to Athletes. October 9, 2019. https://global.toyota/en/newsroom/corporate/29933371.html.

UNESCO. (2018). The Fourth Revolution. The UNESCO Courier. (3) 2018 https://en.unesco.org/courier/2018-3/fourth-revolution/.

United States Office of Science and Technology Policy. (2023). Blueprint for an AI Bill of Rights. May 2023. https://www.whitehouse.gov/ostp/ai-bill-of-rights/.

University of Maryland (2020). Code of Academic Integrity. College Park, MD. https://policies.umd.edu/academic-affairs/university-of-maryland-code-of-academic-integrity.

Vassev, N. (2021). Artificial Intelligence and the Future *of* Humans. *Forbes*. Available from https://www.forbes.com/sites/forbestechcouncil/2021/05/06/artificial-intelligence-and-the-future-of-humans/.

Verman, P. (2022). The year AI became eerily human. *The Washington Post*, December 28, 2022, Washington DC: https://www.washingtonpost.com/technology/2022/12/28/ai-chatgpt-dalle-year-in-review/.

Wang, Y., Hespanhol, L., & Tomitsch, M. (2021). How can autonomous vehicles convey emotions to pedestrians? A review of emotionally expressive non-humanoid robots. *Multimodal Technologies and Interaction*, *5*(12), 84. Available from https://doi.org/10.3390/mti5120084.

World Health Organization. (2022). Road traffic injuries. Newsroom article, 20 June 2022, https://www.who.int/news-room/fact-sheets/detail/road-traffic-injuries.

Ye, J. (2023). China proposes measures to manage generative AI services. Reuters, 11 April 2023. https://www.reuters.com/technology/china-releases-draft-measures-managing-generative-artificial-intelligence-2023-04-11/.

Yu, E. (2022). Baidu rolls out driverless taxi service in two Chinese cities. *ZD Net*, August 10, 2022. https://www.zdnet.com/article/baidu-rolls-out-driverless-taxi-service-in-two-chinese-cities/.

Zhao, C. (2017). China Used Its Vast CCTV Surveillance Network to Track Down Reporter in Just Seven Minutes, *Newsweek*. December 17, 2017. https://www.newsweek.com/tasked-trying-remain-undetected-long-possible-sudworth-filmed-himself-selfie-747843.

Beyond artificial intelligence ethics: exploring empathetic ethical outcomes for artificial intelligence

14

Hart Cohen and Linda Aulbach

School of Humanities and Communication Arts, Western Sydney University, Sydney, NSW, Australia

Introduction: artificial intelligence and empathy

This chapter delves into the worlds of AI systems where humans and machines interact, the emerging ethical issues related to these relationships and reviews the utilization of empathy in this sphere.

The frequency with which AI is embedded in our daily life experiences means that we are sometimes swept into this relationship without our knowledge or consent. There is a growing public belief that the progress of AI, especially in communication with language models such as Chat GPT, is quickly becoming uncontrollable (Asare, 2023). The issue of the unmodulated interaction between humans and machines is sparking huge debates in the field of AI ethics and poses challenges for government policies across the globe. This warrants substantial research to inform the application of AI-related technologies. However, research is only as revealing as the data it can bring forward. Evidence-based research is regarded as the gold standard for revealing how technologies are used and how they impact social and cultural norms and ideas.

Research on ethical issues that question the appropriate forms of machine interventions in our daily lives, however, require a mixed methods research approach. This approach needs to go beyond data models that drive the relevant technologies to be able to address the kinds of research questions that are at once complex and messy, or, as they are sometimes called, "wicked" problems. Questions such as, for example, "what makes us human?" and "what are the most salient forms of communication that would differentiate a human from a machine interactive response?." While answers to these wicked questions are difficult to find, they invoke the pivotal concept of "empathy."

In this chapter, we delve into the crucial subject of empathy within AI technology itself. While AI is generally aiming to imitate human intelligence, empathic AI (AIE) is trying to detect how humans feel and then to display appropriate (empathic) reactions. Empathy is crucial for social interaction, acting as the "social glue" (Oxley, 2011). However, there are various theoretical models with substantial differences in how they understand empathy, how to perceive and send emotional signals and how to interpret and evaluate any emotional data (Yalçın & DiPaola, 2020). One definition that Oxley mentions is that being empathic means responding "to the perceived feelings of another with vicarious emotional reactions of one's own, and empathy is the capacity for, or the occurrence of, such a vicarious experience" (Oxley, 2011).

Ethics in Online AI-Based Systems. DOI: https://doi.org/10.1016/B978-0-443-18851-0.00017-2

However, this is just one of many definitions and it "seems as there are as many theories of emotions as there are emotion theorists" (LeDoux in Beck, 2015; Brown, 2021; Oxley, 2011). There is no consensus "in both philosophy and in the sciences regarding what [exactly] emotions are" (Stark & Hoey, 2021). Indeed, feelings, emotions, empathic, and affective are used interchangeably in many public and even academic discussions (Shouse, 2005; Stark & Hoey, 2021). As Shouse explains, "feelings are personal and biographical, emotions are social, and affects are prepersonal" (Shouse, 2005). Human emotions in computer science are mostly understood physiologically, for example, changes in heart rate, sweat, skin color or other bodily signals are identifying factors for certain emotions. It might not be necessary, then, to gather the exact same emotional data as in other models of human behavior. The categorization of emotions still ends up being highly dependent on the empathy models used (Bråten, 2007; Stark and Hoey, 2021). Despite the more or less severe differences in definitions, the affect theory by Silvan Tomkins has been foundational for any emotion researcher and continues to influence the development of AIE (Burke, 2020; Frank & Wilson, 2020; Schaefer, 2019; Sedgwick & Frank, 2020). Tomkins also applied his theory to "automatons" (humanoid robots) and discussed the complexity of combining emotions and machines (Frank & Wilson, 2020).

Apart from the discussion on how to define emotions and how to design AIE effectively, there is already a polarizing ethical and philosophical debate about what follows if AIE causes feelings in humans and also what it means for AIE to have feelings. Rust and Huang (2021) explain that "machines are more likely to experience emotions in a machine way, [...] [and] will pass the emotional Turing test." For some experts, making AIE have feelings equates to creating self-awareness, resulting in singularity—a state of AI that supersedes humans (Lunceford, 2018). Empathy seems to have both advantages and disadvantages and is highly dependent on how we design and allow for it to be put in place. It is evident that empathy is becoming a crucial factor for AI technologies as well as for humanity.

Methods
Background to the research

The history of AI has its origins in the mid-20th century, but its power and impact has been felt more recently in multiple domains. AI is the technology behind devices and applications like smart phones, smart cars and smart cities with their increasingly smart built environments. The "intelligence" of AI is, in this way, echoed in the world of machines where human-machine interaction is modulated to lessen or absorb the work of humans in various ways. One of the major concerns is the impact of AI on work and employment with predictions of millions of jobs done humans to be made redundant by AI.[1] Though there are various definitions of artificial intelligence, we define AI as the intervention of computational machines in human communication processes. In the late 20th century, machines such as IBM's Big Blue computer managed to win chess matches from chess masters. These are smart computers with embedded sophisticated programs to "play" effectively within game environments such as chess or the Japanese game of GO. GO is particularly at the

---

[1] https://www.bbc.com/news/technology-65102150

pinch point of human—machine game playing because the key moves in the game are not only computational or mathematical, but also involve an esthetic/creative response.

However, the use of the term "smart" has its own specific meaning when applied to machines and is thought to be limited in the way we apply this term to humans. Humans are thought to be capable ("smart") in a number of ways that were not the purview of machines. Thinking creatively, innovatively and in the realm of emotions are generally thought of as uniquely human traits. However, the emergence of machines in the case of creative arts (*DALL-E* and *Midjourney* in visual arts) show the potential for machine-driven creative outcomes.

AI ethics regulation

The European Commission was one of the first governing bodies to develop a substantial AI ethics framework and many countries and organizations have followed suit to regulate automation technologies. The European Union is debating a widely applicable AI Act[2] that carry risk analyses of AI technologies with social well-being and to mitigate social disruption and potential harm. While AI can do both good (increase wages) and harm (increase inequity) as it is deployed with a consensus that AI skills will be a baseline requirement for most fields where industrial production is an outcome. Though ethics guidelines have increased in amount and preciseness, there are still unanswered questions—and the answers given do not fully cover the entirety of an issue, but rather add complexity and uncover even more ethical questions.

For Luke Munn, the horse has already truly bolted:

> As the awareness of AI's power and danger has risen, the dominant response has been a turn to ethical principles. A flood of AI guidelines and codes of ethics have been released in both the public and private sector in the last several years. However, these are meaningless principles which are contested or incoherent, making them difficult to apply; they are isolated principles situated in an industry and education system which largely ignores ethics; and they are toothless principles which lack consequences and adhere to corporate agendas (Munn, 2022).

The extent of the contexts where AI systems involve interaction between humans and machines is broad and so to effectively illustrate our arguments, we will focus on three areas of evident concern: sexrobots/erobots in the emergent use of robots for companionship and sexual relationships, AI and emotion in the mitigation of climate change and the use of AI in facial recognition. These examples are selected because of the emotional character of the human responses evoked by these processes. For this reason, the specific emotional character of empathy will feature in these case studies taking account of recent research in what has been termed "artificial empathy" (Krettek, 2021).[3] We see these case studies as sharing the concerns of intimacy, codependency and recognition. Our two research questions outlined below suggest more specific questions in the case studies: how does AI Learn to love me? Survive with me? Recognize me?

Perhaps the most salient examples of human-machine communication occur in the interactions between chatbots such as Google's Meena, virtual assistants such as SIRI and the emergence of erobots or sexrobots ("Harmony" by American company Abyss Creations LLC and "Replika.")

---

[2]https://digital-strategy.ec.europa.eu/en/library/ethics-guidelines-trustworthy-ai
[3]Danielle Krettek, The Empathy Lab.

With the recent claim by a Google engineer that Meena was exhibiting "human responses," the issue of a possible sentient machine with humanoid capacities has once again exercised those in the ongoing debates relating to machine learning and thinking. These may be misleading, and this most recent case has largely been de-bunked. However, while it is a sign that the question of humanoid AI is never far away from the headlines, it may also be not that far off in reality. It is significant that Google engineer in AI, Geoffrey Hinton[4] recently resigned from his position because of his alarm at the rapid expansion of AI technologies in the spheres where humans could be replaced. It demonstrates that the ethical dimensions to AI are not that distant from the social transformations they engender.

Machine/technology

There is a need to preface the engagement with the issues of human-machine interaction and the role of AI, with a first principle that relates to the mutually enforcing relationship between humans and technology. As Jean- Louis Comolli once wrote: "... Never is an arrangement- combination technological; Indeed, it is always the contrary. The tools always presuppose the machine, and the machine is always social before it is technical." [Comolli, JL., Ideology of the Visible, (1971) (1980)]. This position, written in the late 20th century addresses a pre-digital world—however, while our technologies have sharpened the social-technical divide, the recent embedding of technology (digital affordances) within human action and interaction has re-set the balance between human agency and machines. As Peters (2015) has written,

> The claim that technologies should be subject to humans portrays our wills as immaterial and disembedded, as if we were not already networked creatures, and as if matter were blank nothingness—an insult to the pluralistic universe (Peters, 2015, p. 89).

Peters argues that the demonization of technology results in a compartmentalization between humans and machines where instead we should be querying this relationship from the point of view of the human imbrication with machines—the view that "people are machines," and while the reverse is also true, this does not nullify the need to critically assess machines.

By examining the multifaceted dimensions of empathy in AI, we aim to shed light on the evolving landscape of responsible AI development. Through a comprehensive analysis of existing research, guidelines, and real-world applications, we provide insights into the realm of empathy within the AI sphere. Our intention is to not only discuss the importance of empathy within the technology itself, but also take on an overarching approach to empathy for different perspectives of the AI industry: empathy within ethical guidelines, empathy as a school of thought for climate change and empathy as antithesis to facial recognition software.

The main research questions for this chapter are:

1. With humans immanently related to their environments, what are the possibilities for ethically-led and empathy-based online AI machine innovations?
2. Can empathy be put into ethical guidelines to further support an ethical development of AI?

---

[4]See https://www.theguardian.com/technology/2023/may/02/geoffrey-hinton-godfather-of-ai-quits-google-warns-dangers-of-machine-learning

The first research question is discussed with a literature review in the context of the following three case studies:

1. Empathy in AI erobotics
2. Empathy, AI and climate change
3. Empathy, AI and facial recognition

An evaluation of these case studies will bring us to the second research question of whether or not empathy can be put into ethical guidelines for AI. This overarching discussion is based on the dilemmas and opportunities we identified in the case studies, but has the power to also inform other schools of thought and perspectives on (empathetic) ethical AI.

Case studies

Case study 1—empathy in AI: erobotics

With the continuous improvement of empathic AI (AIE) technologies like conversational AI, emotion recognition and robotic sensors and actuators, it is necessary to highlight one of the most controversial developments within AIE and human-robot interaction (HRI), one that combines the best of each technology: sexrobots. What was once a blow-up doll is now a customized humanized robot with the ability to move, talk, feel, and express "emotions." While still limited in scope, sexrobots are already quite advanced (for example, see "Harmony" by American company Abyss Creations LLC). These robots go beyond just physical pleasure and infiltrate the most private parts of our lives. It brings the nowadays common Alexa or Google Home application from the living room into the bedroom—and into our hearts—not only allowing, but actively trying to build an emotional connection.

The all-encompassing emotional component renders the name "sexrobots" obsolete—rather, a discipline called "erobotics" prevails (Dubé & Anctil, 2021). This shifts the focus onto "eros" (love) and the many questions relating to humanity and its emotions. Imagine a user of an erobot developing feelings for it and treating it (or her/him?) as a companion or even officially as partner? What psychological effects may arise? How are human-to-human relationships impacted? What does it mean for the value of sex and love and intimacy within society? What other effects does it have on humanity? These questions and examples are just the surface fragments of emergent debates in the field of erobotics (Danaher & McArthur, 2017; Sullins, 2012).

Research has already shown that people can form attachments with significantly less humanized objects as a result from the innate instinct to attribute human characteristics to pets and machines (for example, AI devices such as vacuum robots or virtual assistants, Hermann, 2022). This anthropomorphism is entangled with near-perfectly simulated human to human (H2H)-like interactions and the human-like appearance of erobots. It suggests that the impact on both micro and macroaspects of society could be substantial. Ethicists have already voiced their concerns about increasing the objectification of women, the potential use of child-bots in relation to pedophilia, and the possible problems and pressures on long-term relationships surrounding the potential of erobots in always providing "what one desires" (González-González et al., 2020; Kaufman, 2020; Zhou & Fischer, 2019). These concerns fuel the emerging need to program consent into a sexrobot—another example of a complex multilayered discussion that involves questions about how much power do we retain or need to give to AI. On the other hand, the use of sexrobots may have therapeutic benefits or could lead to a decrease of human trafficking and exploitation (Belk, 2022; Sullins, 2012).

The possibilities of erobots are also not limited to a physical device. The afore-mentioned female-representing sexbot "Harmony" comes with her own app, so users can chat with her when apart. One of the most popular companion bots, *Replika*, with which many users have built emotional relationships, received a software update in early 2023 which led the chatbot to break up with their humans and step away from any erotic conversation. This created a significant emotional impact on those Replika users further sparking debate about robotic love. The issues related to companionbots are increasingly important, especially in the context of virtual/augmented reality or within the metaverse. It seems that love with machines is becoming ubiquitous.[5] It is without a doubt a necessary phenomenon to be discussed in the field of AI ethics, especially with a focus on online-based AI systems.

"Robotic love will work, but only because we are so bad at finding a more true love" (Sullins, 2012).

Case study 2—empathy, AI and climate change

In a second example, we can see how empathetic AI could assist with the processes of working towards strategies to mitigate climate change and planetary survival. There are multiple points where empathy is critical to the processes of climate change mitigation, but this needs to be understood in the context of a systems approach to the field.

The human use of human beings

The title to Norbert Weiner's seminal book, *The Human Use of Human Beings*, first published in 1950, speaks volumes for how one of the pioneers of cybernetics thought AI should be programmed—with humans at the center of the functionality of AI processes. As a founder of cybernetics, Weiner's follow up to his earlier book, *Cybernetics: Control and Communication in the Animal and the Machine*, was a reversioning of his original book about Cybernetics—one that he wanted to be accessible to a lay public.

Weiner's reconsideration of basic cybernetic principles which had been defined as an ethos of "command and control" is an important anchor that has largely been forgotten in the implementation of AI online and across the numerous fields. This "intellectual amnesia" on the part of AI developers is especially evident in the early facial-recognition software culminating in the problems faced by contemporary facial capture companies (see below).

The question raised by Weiner's book is, "to what ends will machines designed and developed by humans be put?" If humans will be "used" through the interventions of machines, will the humans that design and implement these interventions act with human benefit as its central logic or rationale? This is precisely the question raised by Geoffrey Hinton when he resigned from Google as their chief AI engineer (see earlier Guardian reference).

---

[5](Dehnert, 2022). Sex with robots and human-machine sexualities: Encounters between human-machine communication and sexuality studies. *Human-Machine Communication, 4*, 131–150. https://doi.org/10.30658/hmc.4.7

The language here betrays the sense of how the design of AI machines for empathy may need a reworking. Using an ethos of command and control will be less effective in communication empathetically than a machine that is defined to "listen and assist." The upshot is that technology is not neutral as may be inferred because so much of how technology comes to us is partitioned between its interface and the programming that lies behind it. Humans shape technology and in turn the technology impacts on humans. While there is substantial scholarship on the relationships between humans and communications and media technologies, the focus of this case study is more specifically on the ways technologies that embed AI have come to impact on climate change.

When Claude Shannon and Warren Weaver, (both of whom worked with Norbert Weiner), developed their mathematical theory of communication as an extension of their engineering interests in machine communication, the cross application to human communicative interaction was embraced in fields ranging from Anthropology to Communications (Bateson, 1972; Watzlawick et al., 1967). One of the key epistemological claims, indeed one that was deemed a revolution in the way humans were seen as actors in their environments, was to see them as "immanent" rather than "transcendent"—that is to say, humans are to be seen as part of larger systems within which they had multiple interrelationships. This perspective is germane to understanding the human relationship to climate change mitigation and by extension, planetary survival.

The contemporary idea and scholarship surrounding anthropogenic climate change extends this kind of thinking around immanence. Human-induced climate change embeds humans in a myriad of relationships where they are immanently related to the environments in which they operate, and which in turn operate on them. There are a host of machines that intervene between humans and the environments that surround them—some of which exhibit artificial intelligence usually embedded in larger systems such as houses, cars, or coal-fired power stations. These machines are now in transition with innovative technologies such as solar technology, the electric car or wind turbines and other alternative energy sources.

For example, with the take-up of rooftop solar, the user has available at their fingertips online reporting of electricity consumption, solar generation and the portion of surplus electricity that is returned to the grid based on their rooftop solar facility. This information can influence energy-use behaviors that both maximize the use of solar-generated power and minimize grid-generated power. Functioning as a kind of thermostat, this both lessens the cost of electricity and reduces the carbon footprint. While this example is a form a systems/control feature, as noted in the literature (Cowls, Tsamados, Taddeo et al., 2021), the carbon footprint of AI must be accounted for in any cost/benefit analysis.

However, a host of new start-ups have taken up the challenge of developing AI in combating climate change. For members of Climate Change AI6, a solutions-oriented group that works on applications to energy efficiency, farming and mobilizing weather data to model the likelihood of disease and other threats to crops. Some of the methods are familiar to researchers (such as natural language analytics) but have been adapted to deal with climate catastrophes and impacts.

The ethical shift in regards climate change has been slow and is still in process, however it may suggest that first, the machines that perform on behalf of humans within their environments can be informed by ethical decisions and second, the key emotional quality of *empathy* is required to influence the viability of surviving an overheating planet. This may be one instance where the deployment of AI has a direct connection to the empathetic responses to the climate emergency by humans seeking to mitigate a climate disaster.

A planetary regard

The depiction of planet earth from outer space has become a commonplace image-exemplars of how AI may be implicated in the empathic processes associated with the Anthropocene—the term that identifies this period as one in which humans induce impacts on their environment such as climate change. Further there has been a plethora of films that depict the natural world through a privileged western gaze. The rise of drones and video technologies of ever-increasing resolutions and sophisticated audio designs all incorporate high-end AI in the rendering of these images.

As Nugent puts it, "…This heraldic gaze, which has also been carried to space, certainly does not feel a humble one and appears to "see" from the perspective of human exceptionalism" (Nugent, 2022).

The so-called "pale blue dot"[6] phenomena celebrated the moment when astronauts first saw the Earth though the windows of their spaceship and remarked on the earth's beauty and vulnerability. The superior gaze given to us of the earth by AI from space suggests a trope of the Anthropocene only achievable by AI. So a question arises as to whether AI is superior human know-how and "exceptional" human imaging or a means of celebrating human technological superiority. The implications for climate change mitigation are not entirely clear. But it is undeniable that our understanding of the planet's survival will need a humbler position for the human presence and a more nuanced appreciation of how earth is depicted to its current inhabitants.

Case study 3—facial recognition

The technology of facial recognition has shown to be a factor in *reducing* empathy within the AI processes of which they are a part. As an example of this, the following case study examines the strategy and fortunes of software start-up *Clearview AI*.

In the recent business pages of the Sydney Morning Herald (14.05.2022), a lengthy article reviewed the dilemma of start-up, *Clearview AI*—a technology acting as a global search engine and mobilized to assist police in the arrest of criminals or as its public statements have indicated:

> Clearview AI acts as a search engine of publicly available images—now more than ten billion to support investigative and identification processes by providing for highly accurate facial recognition across all demographic groups (https://www.clearview.ai/principles).

And in specific reference to *Clearview AI* no longer working with retailers, casinos and sports associations, its website states:

> Clearview AI currently offers its solutions to only one category of customer—government agencies and their agents. It limits the uses of its system to agencies engaged in lawful investigative processes directed at criminal conduct, or at preventing specific, substantial, and imminent threats to people's lives or physical safety.[7]

This strategy is the company's legal recourse to its threatened business model and for its shut down in Europe due to probes and fines relating to the breach of privacy laws. Similar actions are

---

[6]The Pale Blue Dot is an iconic photograph of Earth taken on Feb. 14, 1990, by NASA's Voyager 1 spacecraft. The image inspired the title of scientist Carl Sagan's book, "Pale Blue Dot: A Vision of the Human Future in Space."
[7]https://www.clearview.ai/principles

threatened in Canada and Australia against not only *Clearview AI* but also the law enforcement agencies that deploy its technology to apprehend criminals.

Apart from the way in which the company has exposed a widening gap in privacy standards between the United States (where *Clearview AI* was granted a patent) and the rest of the world, the company's dilemma exposes a key question of online AI ethics: What are limits that would ethically inform an online AI driven tool for the apprehension of criminals? Or, as thematized by many thinkers in the past, how does one inform the capacity of online AI to effect "good" and "bad" outcomes so that the good outcomes prevail and the bad ones are prevented. As an online social media scraper, this application exacerbates already substantial criticism relating to the deployment of facial recognition and the profiling for which it was deployed (Buolamwini & Ebru, 2018).

Empathetic AI with its interest in creating machines with human emotional caring at its center, is the antithesis of most facial recognition technology intentions. Indeed, one can only imagine that facial recognition will only exacerbate the perception that AI technology is designed to work "behind our backs" in a direct undermining of transparency in favor of mass surveillance.

Rethinking recognition

With the keen interest in AI technology and its role in facial recognition the emphasis in the critical assessment of this technology has focussed on the AI technology's immense capacity to capture faces for control and identification. But the idea of *recognition* is overlooked in this context as evidenced by in *Clearview AI*'s priority in the field of criminal investigation. *Clearview AI*'s approach reduces the idea of recognition to a binary between criminal and noncriminal elements of the population under surveillance.

Honneth (2012) has developed a social critique of recognition in the context of the needs expressed by individuals for justice and freedom. He asks:

> How does our understanding of justice and freedom presuppose that the demand for recognition has been satisfied, and what does it mean for our social world if we find that the bonds of recognition have been violated or denied? (Honneth, 2012).

It is ironic that Honneth theorises recognition as an important feature of social cohesion that underpins justice while *Clearview AI*, in its brief to identify criminal behavior also has a brief related to justice, albeit with the practical intent of identifying criminals.

AI facial recognition technology may well be a violation of the "bonds of recognition" as Honneth describes them. Indeed, his main claim is that recognition should serve in three spheres: love in intimate relations, law in legal relations, and achievement in social hierarchies (Honneth, 2012). As the philosophical basis for a social theory of society and social practices, the sense of this theory of recognition is that it may also inform how, uncannily, AI erorobots, social justice and the questions of equity in social relations, may find points of contact and common ground.

Discussion: empathy in ethics

There is now a short but expansive history of ethical codes for AI. These emanated first from science fiction (Asimov) to be revisited later in major conferences in the United States and then recodified in

Europe and around the world. The list of rules grew from 3 (Asimov) to 23 (Asilomar Principles) to now multiple approaches in a number of national and international formations and nongovernment agencies (Walsh, 2022). But in returning to the example of Clearview AI, the actual deployment of AI proceeds to challenge the ethical frameworks that have grown in anticipation of these kinds of excesses in AI machines and software. Institutions and governments are racing to not only keep up with the progress of AI itself, but also to match these technologies with appropriate ethical guidelines. The quest to find an international consensus for such a global matter is important, yet nearly impossible due to culturally specific notions of ethics as well as differences in legislation and political priorities. Theoretical discussions are ongoing and will remain ongoing. However, it is equally important to focus on not just simply meeting the current and urgent needs of ethical solutions, but to take it a step further and cover additional layers of ethical behavior.

Current ethical guidelines revolve around repeated topics—accountability, transparency, privacy, fairness, sustainability—with differences mainly in depth and detail. However, they appear to cover mostly the basics. There is increasing evidence to suggest that the inclusion of empathy in AI systems will be a key facilitator for ethical AI (Batista, 2021; Damioano et al., 2015; Srinivasan & Gonzalez, 2021). While the negative effects of AI are mostly discussed on a society-wide basis, empathy could lead the way to introduce an "ethics in the small" approach (Nallur & Finlay, 2022). With empathetic AI applications, the focus on the individual circumstances provides a new evaluation of what is ethical in this particular situation. This sensitivity, which is expressed through empathy, can then go beyond "not doing harm" and actually assess what is needed to provide a "good" output. The need for empathy and making humans feel good is not only important within the AI space: Looking at recent developments in society, there is an increased value of mental health, self-expression, passion and general feelings-based approach to life. This reflects a new way of seeing emotions as complimentary to rationality, which makes its way into prudential and ethical decision making (Nallur & Finlay, 2022).

In the context of the traditional law, the act of feeling is "denied recognition and legitimacy under the guise of the rationality of the Rule of Law" (Henderson, 1987). Henderson discusses bringing empathy into the practice of law, stating that the better understanding of a situation, which includes the emotional level, leads to a better decision making. However, bringing empathy into this very clear practice will complexify it and can open up wrongdoing or exploitation due to cultural, individual and/or confirmation bias. In *Against Empathy*, Prinz (2011) critically discusses emotions, specifically those underlying (and encompassing) empathy. Due to the high risk of potential bias, empathy could lead to harmful results if taken into account. Prinz puts the ethical discussion of AI emotions into a component theory, stating that there is a "problem of parts and problem of plenty." The former describes the difficulty surrounding the choice of components necessary to detect emotion in a particular context while the problem of "plenty" is the challenge of how these components work together. Theorising these fragments and their connection, multiple definitions of emotions have emerged. The lack of consensus about what exactly emotions are and what empathy really means makes it extremely difficult to then be put these into the practice of ethical design in the AI industry. Stark and Hoey (2021) call for a bigger discussion about the act of "digital phenotyping"[8] as emotion recognition and expression through AI systems is not well defined, researched nor talked about.

---

[8]Digital phenotyping (also known as *personal sensing*, *intelligent sensing*, or *body computing*) involves the collection of biometric and personal data *in situ* from digital devices, such as smartphones, wearables, or social media, to measure behavior or other health indicators. https://www.ncbi.nlm.nih.gov/pmc/articles/PMC8367187/

The difficulty of dealing with the ethics of emotion is mirrored in several attempts of regulating artificial empathy—the IEEE has had an open project for "Standard for Ethical considerations in Emulated Empathy in Autonomous and Intelligent Systems" since 2019. McStay and Pavliscak offer emotion-specific ethics guidelines, calling upon practitioners to take action in their daily life after considering certain ethics-related questions for their product or project rather than providing a new standard similar to other ethical guidelines. This proposal, though still vague, creates space for individuality and therefore manifests the core principles of the ethics of care. This normative ethical theory centers around the individual, believing that generalized standards are "morally problematic, since it breeds moral blindness or indifference" (Gilligan, 2008). We have already referred to Luke Munn's critique of the abundance of AI ethics guidelines—his work is further proof of the current loss of individuality within the ethics debate. As empathy takes on the task of understanding individual feelings, bringing empathy into guidelines could change the generalizing character of emotional AI into a more individualized, nuanced and hence "more ethical" outlook. This can be further verified by looking beyond the AI context: embracing empathy provides (noted earlier in Honneth's work) for a "broader concept of doing justice" (Greenstein, 2021). Bringing humanity into the sphere of the general practice of law, a very emotion-cleansed vision of what is right or wrong, could improve the decision making within a courtroom (Henderson, 1987).

Conclusion

Measuring empathy

In an article analyzing asynchronous communications between people with mental health concerns and interlocutors expressing empathic responses, the researchers define empathy as:

> … a complex multidimensional construct with two broad aspects related to emotion and cognition (Davis, Davis, & Davis, 1980). The "emotion" aspect relates to the emotional stimulation in reaction to the experiences and feelings expressed by a user. The "cognition" aspect is a more deliberate process of understanding and interpreting the experiences and feelings of the user and communicating that understanding to them (Elliott, Bohart, & Murphy, 2018).

The researchers sought to measure empathy by developing analytical tools and applying them to a text-based exchange. The framework engages with communication mechanisms as follows: emotional reactions, interpretative reactions, and exploratory reactions.

The methodology of this research is in the language of computational analytics and desires to apply discourse analysis of text exchanges to mental health contexts. The researchers are correct that mental health exchanges now are often text-based and look to focus on shorter exchanges that exhibit forms of empathy. However, it would appear that the corpus analyses human to human empathy and does not really deal with the machine to person empathy question or what is termed, "emotional AI."

The computational approach to analyzing text-based empathy exchanges offers somewhat restricted versions of what empathetic communications can potentially express, an issue that can be found across the entire field of empathy research. Specific applications offer great insight, yet there are often areas of expressed human emotions that are excluded. However, emotions should be seen as a "whole package" and that the "bigger picture" might be missed by algorithms. Though there is also reason to believe that it might not be necessary to gather the exact same emotional data as a human, as some computational recognition tools can collect information that is only available to machines, for example, cortisol-levels measured by sensors. As previously mentioned, "Machines are more likely to experience emotions in a machine way" (Rust & Huang, 2021), which will give AI the ability to pass emotional Turing tests (Howick et al., 2021). Additionally, while machines focus on detecting and analyzing everything they can, humans tend to miss signals due to their own personal state—they may lack sleep or concentration or experience something that distracts them from seeing the "big picture" themselves.

The possibility of having a machine that is able to capture and recognize others' emotions better than a human challenges the definitions and approaches towards emotion, adding more value on the focus on emotions. This may suggest a "new standard" of the way empathy is expressed which could lead to a more "humane humanity"—but it also has the potential to create "correct" and "incorrect" ways of emotive interaction that could leave out users whose emotional reactions diverge from the majority. The risk here is in the process of standardising empathy, as the core value of empathy itself is in capturing individuality—something that could be left out if there is such a thing as a set framework of empathetic responses.

Regardless of the risks, there is progress in both psychological and algorithmic understanding of feelings, leading to advancing artificial empathy, even with upcoming and increasingly stricter regulations. We will experience a "feeling economy," where the increase in automation through AI will leave society able to focus on feeling, giving women an advantage in the workplace "through empathy and people skills" (Rust & Huang, 2021). While this is a prediction for the future, there is already evidence of it being a potential realizable outcome, further emerging from the era of a "thinking economy," where AI becomes increasingly better at thinking than humans (c. f., Big Blue). The opportunities that arise from the development in *artificial empathy* will soon challenge the "feeling economy" and shape societies' conception and expression of feelings one way or another—a stage where we then have to talk about singularity and its either hopeful or apocalyptic effects (Rust & Huang, 2021).

The ethical theories presumed in the above-mentioned mental health approach to research is focussed on the relationship between the actors of the system and aims to create opportunities for them to interact and improve their mental health outcomes. Regardless if research is focused on empathic interactions between humans and humans or humans and machines, all of these exchanges can assist in assessing such approaches to ethics in a computational setting. this might make it possible to recognize the existence of such opportunities in computational ways.

Krettek's empathic AI

With empathic AI in the picture, it is useful to consider and assess the proposals made by former Google designer, Danielle Krettek in respect of her journey from designing for reliability, speed and accuracy to designing for emotions, trust and loyalty. She identifies significant indicators from end-users of virtual assistants (VA) that emotional responses are a frequent form of interaction with these technologies as if they were "human."

The issues are necessarily complex and though her conclusions are research—led, they are at an early stage of this research field. First, she uncovers the extensive penetration and use of technologies like mobile phones that have embedded Virtual Assistants.[9] Using Sherry Terkel's idea of user density, (Turkle, 2011) the emotional state of the end-user is foregrounded by the central concept of being "alone/together" where effectively, the connectivity to others is constant. In this context, modes (of use) are replaced by "moods"—where connectivity is articulated as part of a deep, neurologically embedded human characteristic.

If Google AI design (UX) was motivated by cognition, behavior, and modality, HX emotion design AI is based on inference, emotion, trust, dependability, recall, and above all attunement. The idea of attunement is a synthesis of how to take AI or machine learning to a point where connectivity is total in the way we think of a totally immersive sensory experience.[10] This approach privileges the quality of the relationship between the virtual assistant over the frequency of contact.

Krettek's research interests turned to understanding more deeply the nature of relationships that could inform connectivity in the virtual world. She isolates four stages of relationship: initiation, experimentation, intensification, and integration. From experts on divorce, she uncovers the key to long lasting relationships as the ratio of positive communication experiences to negative.

A 5 to 1 ratio ensured that a relationship would survive while a 20 to 1 ratio was considered masterful. These successful relationships prioritized emotional sustenance like empathy over information.

The evidence found about quality and sustaining relationships is clearly contrary to the paradigm set up by Internet search engines such as Google which tend to privilege information and end-user frequency above all else.

Further, research undertaken by Krettek revealed the importance of vulnerability in creating intimacy in a relationship. This is a challenge for a computer and machine learning, but this intimacy is based on trust and is spurred on by empathy (defined as a deep emotional connection expressed by "I understand (you) because I was "there" too")

Further, it appears that physical forms or a more humanoid look of the Virtual Assistant was not as significant in the creation of intimacy through voice. Indeed, the focus on developing a humanoid-looking artefact as your interlocutor, Krettek considers a false goal. Rather than looking human, it was more important that the machine "feels" human—to find instead the essence of being human.[11]

In her terms, Krettek desires to be "an art teacher for machines" because machine learning is about machines learning how to be human with attributes such as kindness and honesty. Krettek's approach generates some simple binaries.[12]

---

[9]https://www.ncbi.nlm.nih.gov/pmc/articles/PMC8367187/

[10]See Earlier reference to "singularity"

[11]This resonates with a report on a Replika user who preferred having a blowup doll as a supplement to his computer/app — based companionbot. This suggests that a physical manifestation was an important co-presence for this user. https://iview.abc.net.au/show/four-corners/series/2023/video/NC2303H015S00

[12]These are some of the characteristics that Krettek felt would lead to transparency in a machine-human relationship and add an ethical dimension to AI and its virtual assistants.

Commands—listening Interaction—expression Functions—feelings Actions—helping Inferences—conversations Stores data—remembers

Machine learning—learning machines Human in the loop—teaching machines Predictive actions—gets me, knows me Assistive—thoughtful

These are a shorthand for opposites to chart the differences when machine learning is turned towards learning how to be more human. Krettek concludes her talk with a promotion of the idea of universal languages that are the sources for deep humanity. The ones she has in mind are music, mathematics and film. Computers and similar machines are built on mathematical knowledge and algorithms and so these may only work in certain modalities.

It would be easy to dismiss Krettek as overly optimistic about the potential for machines to emulate empathic responses on par with human ones. However, her deep understanding of Google's design priorities for its digital assistants gives her special insight into what design choices would correct the bias in the way these assistants were projected to be a companion to humans.

In summary, apart from the need to find the right amount and approach of empathy within AI systems to mitigate detrimental emotional effects on humans, there should also be an approach to bring empathy across all AI technologies, irrespective of having EAI as a focal point. In addition, there is reason to believe that adding empathy as an extra layer of ethics into official guidelines is not only necessary but desirable, ensuring that the increase of standardization and generalization in the AI field does not invalidate individuality. A lack of empathy could be considered as unethical in itself, so the ongoing developments in the field of empathy and AI and the link between both is to be welcomed—with the hope, that empathy also makes its way into ethics and formal guidelines.

Further research

If "empathy" can be taken up as *immanent* to what it means to be human and not as just an analogy for a vaguely defined idea to illustrate the working of human emotion, it raises the question of how the concept of *immanence* works in connection to empathy. We have suggested that intimacy, planetary survival and recognition are also immanent to the situation in which AI finds itself. However, what it would it mean for AI itself to be recognized or understand its own existential threats or to experience intimacy? This is not an appeal to anthropomorphism or potential projections of sentience. Rather, the question asked is, "what are the implications of a thoroughgoing immanence that understands a substantive connection between ethical demands and technological conditions?"[13]

Katherine Hayles' work, *How We Became Posthuman* on the distinction between consciousness and cognition, and technogenesis, suggests a perspective on the posthumanist condition that begins to address these concerns.

References

Asare, J. (2023). The dark side of chat GPT. *Forbes*. Available from https://www.forbes.com/sites/janicegassam/2023/01/28/the-dark-side-of-chatgpt/?sh = 59f908cd4799.

Bateson, G. (1972). *Steps to an ecology of mind*. Ballantyne Books.

Batista, E. (2021). *Accountability and empathy are not mutually exclusive*. Batista Executive Coaching.

Beck, J. (2015). Hard feelings: Science's struggle to define emotions. *The Atlantic*. Available from https://www.theatlantic.com/health/archive/2015/02/hard-feelings-sciences-struggle-to-define-emotions/385711/.

---

[13]Liam Magee, Personal Communication.

Belk, R. (2022). Artificial Emotions and Love and Sex Doll Service Workers. Journal of Service Research. Available from https://doi.org/10.1177/10946705211063692

Bråten, S. (2007). *On being moved: From mirror neurons to empathy.* John Benjamins Publishing Company.

Brown, B. (2021). *Atlas of the heart: Mapping meaningful connection and the language of human experience.* Random House.

Buolamwini, J., & Ebru, T. (2018). Gender shades: Intersectional accuracy disparities in commercial gender classification. *Proceedings of Machine Learning Research, 81,* 77–91.

Burke, C. (2020). Machine and mind: Rethinking affect. *Digima, 86.* Available from http://digicult.it/internet/machine-and-mind-rethinking-affect/.

Cowls, J., Tsamados, A., Taddeo, M., et al. (2021). The AI gambit: Leveraging artificial intelligence to combat climate change—Opportunities, challenges, and recommendations. *AI & Society.* Available from https://doi.org/10.1007/s00146-021-01294-x.

Damioano, L., Dumouchel, P., & Lehmann, H. (2015). Artificial empathy: An interdisciplinary investigation. *International Journal of Social Robotics, 7,* 3–5. Available from https://doi.org/10.1007/s12369-014-0259-.

Danaher, J., & McArthur, N. (2017). *Robot sex: Social and ethical implications.* Cambridge: MIT Press.

Davis, M. H., Davis, M. P., Davis, M., et al. (1980). A multidimensional approach to individual differences in empathy. *Journalof Personality and Social Psychology, 10*(85).

Dehnert, M. (2022). Sex with robots and human-machine sexualities: Encounters between human-machine communication and sexuality studies. *Human-Machine Communication, 4,* 131–150. Available from https://doi.org/10.30658/hmc.4.7.

Dubé, S., & Anctil, D. (2021). Foundations of erobotics. *International Journal of Social Robotics, 13,* 1205–1233. Available from https://doi.org/10.1007/s12369-020-00706-0.

Elliott, R., Bohart, A. Watson, J. & Murphy, D. (2018). Therapist empathy and clientoutcome: An updated meta-analysis. Psychotherapy, 55(4), p.399.

Frank, A., & Wilson, E. A. (2020). *A Silvan Tomkins handbook: Foundations for affect theory.* University of Minnesota Press. Available from https://doi.org/10.5749/j.ctv182jthz.

Gilligan, C. (2008). Moral orientation and moral development. In A. Bailey, & C. J. Cuomo (Eds.), *The feminist philosophy reader* (p. 471). McGraw-Hill.

González-González, C. S., Gil-Iranzo, R. M., & Paderewski-Rodríguez, P. (2020). Human-robot interaction and sexbots: A systematic literature review. *Sensors (Basel), 21*(1). Available from https://doi.org/10.3390/s21010216.

Greenstein, M. N. (2021). Ethics and empathy. *Judges' Journal, 60*(2), 40. Available from https://tinyurl.com/2r8ehajm.

Henderson, L. (1987). Legality and empathy. *Michigan Law Review,* 1574.

Hermann, E. (2022). Anthropomorphized artificial intelligence, attachment, and consumer behaviour. *Mark Lett, 33,* 157–162. Available from https://doi.org/10.1007/s11002-021-09587-3.

Honneth, A., (2012). The I in we: Studies in the theory of recognition, Polity.

Howick, J., Morley, J., & Floridi, L. (2021). An empathy imitation game: Empathy turing test for care- and chat-bots. *Minds & Machines, 31,* 457–461. Available from https://doi.org/10.1007/s11023-021-09555-w.

Kaufman, E. (2020). Reprogramming consent: Implications of sexual relationships with artificially intelligent partners. *Psychology & Sexuality, 11*(4), 372–383. Available from https://doi.org/10.1080/19419899.2020.1769160.

Krettek, D., (2021). *The empathy lab,* https://www.youtube.com/watch?v = xPlwoFRSq_0.

Lunceford, B. (2018). Love, emotion and the singularity. *Information (Basel), 9*(9), 221. Available from https://doi.org/10.3390/info9090221.

Munn, L., (2022). The uselessness of AI ethics, Springer Research. Available from https://doi.org/10.1007/s43681-022-00209-w

Nallur, V., Finlay, G. (2022). Empathetic AI for ethics-in-the-small. AI & Society.

Nugent, R. (2022). A planetary regard. Doctoral thesis presentation, Institute for Culture and Society, Western Sydney University.

Oxley, J. (2011). *The moral dimensions of empathy: Limits and applications in ethical theory and practice.* Palgrave Macmillan. Available from https://doi.org/10.1057/9780230347809.

Peters, J. D. (2015). *The marvellous clouds.* University of Chicago Press.

Prinz, J. (2011). Against empathy. *The Southern Journal of Philosophy*, *49*(1), 214−233.

Rust, R., & Huang, M.-H. (2021). *The feeling economy: How artificial intelligence is creating the era of empathy.* Springer. Available from https://doi.org/10.1007/987-3-030-52977-2.

Schaefer, D. (2019). *The evolution of affect theory: The humanities, the sciences, and the study of power.* Cambridge University Press.

Sedgwick, E. & Frank, A. (2020). Shame in the cybernetic fold: Reading Silvan Tomkins. Touching Feeling, pp. 93−122. Available from https://doi.org/10.1515/9780822384786-006.

Shouse, E. (2005). Feeling, emotion, affect. *M/C Journal*, *8*(6). Available from https://doi.org/10.5204/mcj.2443.

Srinivasan, R., & Gonzalez, B. (2021). The role of empathy for artificial intelligence accountability. *Journal of Responsible Technology (Elmsford, N.Y.)*, *9*, 100021. Available from https://doi.org/10.1016/j.jrt.2021.100021.

Stark, L., Hoey, J. (2021). Ethics of emotion in artificial intelligence systems. *ACM Conference fairness, accountability, and transparency.* Association for Computing Machinery. https://dl.acm.org/doi/10.1145/3442188.3445939.

Sullins, J. (2012). Robots, love, and sex: The ethics of building a love machine. *IEEE Transactions on Affective Computing*, *3*(4), 398−409. Available from https://doi.org/10.1109/T-AFFC0.2012.31, Fourth Quarter 2012.

Turkle, Sherry. (2011). The tethered self: Technology reinvents intimacy and solitude. *Continuing higher education review*, *75*, 28−31.

Walsh, T. (2022). *Machines behaving badly.* La Trobe University Press.

Watzlawick, P., Bavelas, B., & Jackson, D. (1967). *The pragmatics of human communication.* Norton and Norton.

Yalçın, Ö. N., & DiPaola, S. (2020). Modeling empathy: Building a link between affective and cognitive processes. *Artificial Intelligence Review*, *53*, 2983−3006. Available from https://doi.org/10.1007/s10462-019-09753-0.

Zhou., & Fischer, M. H. (2019). *AI Love You: Developments in human-robot intimate relationships.* Springer International Publishing.

Further reading

Ang, I. (1994). In the realm of uncertainty: The global village and capitalist postmodernity in Communication Theory, Crowley and Mitchell (eds.), Routledge.

Bragg, E. (2014). Activist ecopsychology. *Ecopsychology*, *6*(1). Available from https://www.liebertpub.com/doi/10.1089/eco.2013.0096.

Clearview Website. https://www.clearview.ai/principle.

Cowie, R. (2012). The good our field can hope to do, the harm it should avoid. *IEEE Transactions on Affective Computing*, *3*(4), 410−423. Available from https://doi.org/10.1109/T-AFFC.2012.40, Fourth Quarter 2012.

Fay, J. (2018). *Inhospitable World: Cinema in the time of the Anthropocene.* London.

Hagendorff, T. (2020). The ethics of AI ethics: An evaluation of guidelines. *Minds and Machines*, *30*(1), 99−120.

Hayles, K. (1999). *How we became posthuman: Virtual bodies in cybernetics, literature, and informatics.* University of Chicago Press.

IEEE (2022). Standard for Ethical considerations in Emulated Empathy in Autonomous and Intelligent Systems. https://standards.ieee.org/ieee/7014/7648/.

McEwan, I. (2019). *Machines like me.* Jonathan Cape.

Mcstay, A., Pavliscak, P. (2019). Emotional artificial intelligence: Guidelines for ethical use.

Popkova, E., & Sergi, B. S. (2020). *Scientific and Technical Revolution: Yesterday, Today and Tomorrow.* Springer.

Sagan, C. (1994). *Pale blue dot: A vision of the human future in space.* NY: Random House.

Schwab, K. (2017). *The fourth industrial revolution.* Portfolio Penguin.

Weiner, N. (1948). *Cybernetics: Control and communication in the animal and the machine.* John Wiley & Sons.

Weiner, N. (1950). *The Human Use of Human Beings.* Houghton Mifflin Company.

World Economic Forum (2022). Artificial intelligence and robotics. https://www.weforum.org/agenda/artificial-intelligence-and-robotics/.

Ethical implications of artificial intelligence in social and political involvement

IV

Who decides what online and beyond: freedom of choice in predictive machine-learning algorithms

Simona Tiribelli[1,2]

[1]*Department of Political Sciences, Communication, and International Relations, University of Macerata, Macerata, Italy* [2]*NYU Center for Bioethics, NYU School of Global Public Health, New York University, New York, NY, United States*

Introduction

That today artificial intelligence (AI) systems and specifically machine learning algorithms (MLAs) are increasingly governing and reshaping almost every domain of our lives seems to be extensively recognized in the debate on the ethics of AI and algorithms (Mittelstadt, 2016; Tsamados, 2020). Within this debate, a great corpus of literature focuses on investigating whether MLAs can be considered "moral agents" (Floridi & Sanders, 2004), as well as on the risks MLAs raise for the promotion of a fair society (Giovanola & Tiribelli, 2022a, 2022b), as they have been uncovered to very often embed gender and/or racial bias and exacerbate discrimination and injustice (Eubanks, 2018; Noble, 2018; O'Neil, 2016). However, among the key issues that require to be further explored is whether online AI-based and MLA-based systems foster and/or hamper human decision-making, both online and offline, that is, "onlife" (Floridi, 2014).

This chapter aims to fill this gap and show that such an ethical inquiry is of particular importance insofar as MLAs not only affect and shape the reality where we live, but they also show a huge predictive power that, if misused, can deeply affect the preconditions of our (supposedly free) social, political, and especially moral choices. Indeed, MLAs and related ML techniques show the capacity of both empowering our autonomy and decision-making, thanks to filtering and personalization techniques, as well as creating "filter bubbles" (Pariser, 2011) or "echo-chambers," (Sunstein, 2008) that, by selecting what is profiled as relevant information to us, can restrict our range of available choice options and predetermine the preconditions of our choices (Tiribelli, 2023). Moreover, MLAs can be also deployed with the goal of targeting human cognitive bias and/ or deep vulnerabilities so as to suspend our reflective endorsement on the motives (i.e., reasons, values, preferences, beliefs, etc.) that can guide our choices (Tiribelli, 2020, 2022b), eroding our autonomy and capacity of self-determination. As a result, we might be under the illusion of having more freedom of choice and action thanks to MLA-based online filtering and personalization techniques, while instead we are consciously or unconsciously delegating a great deal of our autonomy

Ethics in Online AI-Based Systems. DOI: https://doi.org/10.1016/B978-0-443-18851-0.00012-3

and choice to MLAs, ending up not realizing how or why we made a certain choice and action. As a consequence, our capacity to choose and act as moral agents is endangered, as well as our possibility to develop a genuine moral identity (Tiribelli, 2022b).

This scenario becomes even more pressing in our informational societies, where the distinction between online and offline is rapidly blurring (Floridi, 2014), and our environments are increasingly datified, permeated, and shaped, if not governed yet, by MLA-based systems, whose impact extends therefore beyond our decision-making and behavior online. This chapter aims to highlight and dissect the impact of MLA-based systems on individual decision-making by investigating *whether* and, if that is the case, *how* MLAs are fostering or undermining the preconditions of (supposedly free) human choices. To this aim, the chapter is divided into four main sections.

In the first section "Freedom of choice: its value and preconditions," we first pursue an ethical inquiry into the preconditions securing—at least at a minimum threshold—human freedom of choice, drawing insights on main theories developed in moral philosophy and sociopolitical philosophy; then, we argue why freedom of choice is a fundamental ethical value that ought to be secured for the well-being of both us as individuals and our liberal and democratic societies.

The second section "MLA governance: the rise of algorithmic choice-architectures" highlights how MLAs are becoming more than just pervasive forces reshaping our onlife environments, but novel *architects of our choice-contexts*. To do so, we clarify the interconnection between different probabilistic MLA techniques and some of their most relevant features (e.g., combination of profiling and knowledge discovery methods) and argue their capacity to structure the onlife contexts where we prepare our choices. The third section "MLA choice-architectures and freedom of choice" is devoted to unpacking how MLAs as choice-architects affect the preconditions of our freedom of choice, that is, (1) the availability of alternative (morally heterogeneous) options and (2) human (moral) autonomy, giving rise to epistemological and moral constraints on human decision-making processes. Here, we highlight the detrimental consequences this impact can raise for us as individuals, endangering our possibility to develop genuine moral identity, and for our societies, hampering the possibility to (morally) progress through a more cohesive and open public sphere (Tiribelli, 2022a). In the fourth section "Securing freedom of choice by design: a call to action," we address such consequences and show how our ethical inquiry can help pinpointing some ethical criteria that ought to be met to secure our freedom of choice onlife, namely, in today's informational societies that are increasingly shaped, if not governed yet, by MLAs.

Freedom of choice: its value and preconditions

To pursue an ethical inquiry into the impact of online MLA-based systems on our freedom of choice, we first clarify the conditions to be safeguarded in order to secure freedom of choice at a minimum threshold, that is, the *preconditions* that allow the free exercise of our choices. To highlight such conditions, we draw on the main reflections on freedom of choice developed within moral and sociopolitical philosophy (section "Freedom of choice from within moral and sociopolitical philosophy"), and then we revise them to be adequately understood in our contemporary societies (section "Freedom of choice revised in contemporary society"). Finally, we zoom on the ethical value of freedom of choice and argue why the latter *ought to* be secured in our contemporary societies (section "Freedom of choice as an ethical-normative value").

Freedom of choice from within moral and sociopolitical philosophy

The issue of freedom of choice has been extensively discussed within philosophical scholarship over decades; however, it is often poorly explored *per se* and mainly framed in terms of what can define our choices as free. Indeed, the philosophical debate on freedom of choice has focused over time more than on how to define human freedom of choice on what are the *necessary (but not sufficient) conditions* that might secure it—at least at a minimum threshold. Within the philosophical scholarship, such necessary conditions (hereinafter: preconditions) have assumed different formulations, especially as a matter of ethical inquiry within two main debates, developed, respectively, in the fields of moral philosophy and sociopolitical philosophy: the debate on free will and the sociopolitical debate on freedom of choice.

In the former debate, there is a substantial agreement on at least two preconditions underlying human freedom of choice (De Caro, 2004):

1. agents' *availability* in the decision-making process *of alternative options*
2. agents' *autonomy* over the decision-making process.

The first condition expresses the subjects' power to do otherwise from what *de facto* they do, while the second condition is often described in terms of self-determination, that is, the subjects' power to choose and act according to reasons and motives on which they are in control and/or they can endorse (the idea of self-determination or self-governing individual). The main argument is that if an agent is free to perform a certain action, she will also be free to not perform it; at the same time, the choice between potential courses of action cannot be casual; that is, it cannot result from factors out of control of the agent—as in the case of deciding whether to make a certain action on the basis of the toss of a coin, or if one is coerced to make a certain choice.

Despite the acknowledged contribution of such debate in moral philosophy in clarifying what it takes to consider a choice as free at a minimum threshold, such debate is limited as it mainly grounds the discussion on freedom of choice on the "free" or "unfree" nature of the will (De Caro, 2004; Mori, 2001), often failing to distinguish between "freedom of the will" (i.e., *metaphysical freedom*) and "freedom of choice and action" (i.e., *practical freedom*).[1] This distinction is pivotal and is based on the acknowledgment that, beyond the difficulty to understand the free or unfree nature of our will, nonetheless, we exercise a certain kind and/or degree of freedom of choice and action (Albritton, 1985; Hobbes, 1654) or freedom according to a practical respect (Kant, 1998) in our empirical contexts.

However, such a debate is not the sole one that investigates the issue at stake. To pursue a sound ethical inquiry, we should consider also the sociopolitical debate on freedom of choice, which not only acknowledges but also expands the two preconditions highlighted earlier.

---

[1]The lack of such distinction informs also the contemporary debate in AI and ethics on the effects of online manipulative algorithms on human choice and autonomy. See, for example, Botes (2022).

The sociopolitical debate on freedom of choice especially focuses on what it takes to live freely within a particular sociopolitical order and culture. Differently from the free will debate, here freedom of choice is not grounded in the free or unfree nature of the will, but is mainly analyzed with reference to the many forms of coercion that are politically and socially exercisable. Interestingly, while many and different are the sociopolitical accounts of freedom of choice claimed over recent history, human autonomy emerges also here as a widely agreed upon precondition for freedom of choice—especially in the liberal tradition (Skinner, 2008). Nevertheless, key differences emerge when it comes to reckon with the subconditions for autonomy, which vary within different sociopolitical accounts of freedom of choice (Killmister, 2017). For example, champions of liberalism such as Raz (1986) claim that autonomy requires the following:

1. a *minimum rationality* of the agents to plan actions that allow them to achieve their goals;
2. *adequate options available to choose*;
3. *independence* (i.e., freedom from coercion and manipulation).

However, within the same liberal tradition, we can distinguish at least two diverse connotations of autonomy. One emerges from a negative concept of freedom, that is, *autonomy as independence* (mentioned earlier), and refers to a state of an absence from interferences (such as coercion or manipulation). Another one, instead, from a positive concept of freedom, that is *autonomy as self-rule* (Berlin, 1969), and it concerns the *authenticity* (genuineness) of one's own motives (values, reasons, and preferences) and the *condition* of the agent to choose and act with *sufficient resources* to make such own motives effective in the decision-making—put it differently: Autonomy is conceptualized as the power to rule (or govern) oneself; but to govern oneself, one must be in the position to act "competently" based on the motives (i.e., values, reasons, and/or preferences) that are one's own. This picks out the two main families of subconditions for autonomy proffered in liberal conception of autonomy: *competency* and *authenticity conditions* (Killmister, 2017). Competency refer to various capacities such as, for example, minimum rationality and freedom from debilitating pathologies. Authenticity conditions include the capacity of the agents to reflect on and specifically endorse their values, reasons, and preferences in the decision-making process (autonomy as endorsement).

Yet, there are also other traditions within sociopolitical debate whose conception of autonomy differs from and is particularly critical of the liberal one (to expand, see republicanism and communitarianism). Specifically, the liberal conceptualization of autonomy is criticized to be proceduralist, that is, to focus primarily on the procedure through which one can come to endorse options and values (i.e., competency conditions) to determine autonomy. According to these critics, the liberal conception of autonomy avoids questioning whether autonomy conceptualized as the self-rule (or self-government) of the agent may be understood independently from the *socially defined* values and patterns through which people develop themselves (Sandel, 1982). In these views, liberalism fails to consider the role of sociocultural relations and patterns in shaping and defining those values, commitments, projects, and life plans that motivate and drive individuals' choosing and agency (Taylor, 1991). These critical considerations have sparked some scholars to elaborate an alternative account of autonomy meant to replace the liberal one (criticized to be individualistic): this replacement has been called "relational autonomy" (Mackenzie & Stoljar, 2000).

Despite some differences, relational accounts of autonomy stress the import of relatedness in shaping identity features and the dynamics of deliberation and reasoning motivating one's own agency (Mackenzie & Stoljar, 2000; Oshana, 1998) and find prominent subconditions for autonomy also in *social recognition and/or support*. The underpinning idea is that while autonomy requires the ability to act effectively on one's own values (authenticity conditions), oppressive social conditions of various kinds might threaten those abilities by removing one's sense of self-confidence required for effective choosing and agency. Such phenomena make social recognition and/or support a required condition for this self-trusting status and therefore for the full enjoyment of these abilities (Arneson, 1991; Benson, 2005; Westlund, 2014).

Beyond the complexity of bringing out a certain convergence in the broad and heterogeneous reflections on freedom of choice, as offered by the free will debate and the sociopolitical one, there are some preconditions underlying freedom of choice that emerge as sufficiently acknowledged by both these far-reaching debates.

Indeed, (1) the *availability of alternative options* and (2) *human autonomy*, either as independence (absence from coercion and manipulation), self-rule (authenticity via reflective endorsement), and/or relationality, emerge as preconditions widely recognized as necessary to the exercise of freedom of choice (Table 15.1).

Freedom of choice revised in contemporary society

After having considered the main accounts on freedom of choice developed within moral and sociopolitical philosophy, we can now reelaborate them to offer a sounder and comprehensive understanding of what it takes to secure human freedom of choice and action at a minimum threshold. To do so, we take the moves from the preconditions mainly considered in the free will debate, (1) availability of alternative options and (2) human autonomy, as they particularly clarify what it takes to consider a choice as free at a minimum threshold, and we revise them in a way that is informed by the sociopolitical debate on freedom of choice, insofar as the latter highlights either overlaps with free will debate (e.g., the conditions of availability of options and the conception of autonomy as self-rule or self-determination via reflective endorsement) and the differences (e.g., in the recognition of the socio-relational dimension of autonomy).

Such a reelaboration of the key preconditions of our freedom of choice based on a joint reading of these two debates sounds needed today, especially if we aim to assess our freedom of choice in our contemporary society. Indeed, the two debates allow us to acknowledge that freedom of choice comprehends both an individual dimension and a socio-relational one: It refers both to how we choose as "specific individuals" and how we choose as specific individuals who are placed into and shaped by socio-relational and cultural contexts, such as those of our contemporary societies, increasingly hyperconnected, multicultural, and globalized.

Table 15.1 Key points emerging from the analysis of main theories in moral and sociopolitical philosophy on freedom of choice preconditions.

	Key preconditions	Subconditions
Freedom of choice *From within* *Moral Philosophy and Sociop olitical Philosophy*	Availability of alternative options	—
	Human autonomy	• Competency • Authenticity • Relationality

Table 15.2 Dimensions and preconditions arising from our joint reading of the debates on freedom of choice in moral and sociopolitical philosophy.		
	Constitutive dimensions	**Key preconditions**
Freedom of choice	Individual dimension	Autonomy as self-determination
	Socio-relational dimension	Availability of alternative options

The ethical inquiry pursued into the preconditions of our freedom of choice, as elaborated in moral and sociopolitical philosophy, shows that these are *mutually dependent* and *intertwined*: They need and implicate each other to secure freedom of choice at a minimum threshold. In particular, our ethical inquiry shows that such preconditions refer to both an individual dimension, expressed by the condition of autonomy, and its ethical requirement of agents' internal approval or reflective endorsement of what drives their choices (e.g., values, reasons, beliefs, preferences, etc.), but in a way that is deeply informed by the socio-relational dimension, which relates to another condition, that is, the condition of the availability of alternative options. Indeed, the contexts in which we choose (choice contexts), which are environments characterized by a more or less—quantitatively and qualitatively—varied availability of options, are deeply informed and shaped by socio-relational and cultural elements. These elements do not only shape the formation or access to options on the basis of which an agent can choose (e.g., a system of beliefs, values, commitments, and ground projects), but they can also constitute choice options themselves (e.g., the choice concerning who to affiliate with) (Table 15.2).

From this perspective, the preconditions emerging from our joint reading of the two philosophical debates help shed light also on a more adequate account of freedom of choice, as reconceptualized in our contemporary society, to the extent that it acknowledges the authorship of the agents over their choices via the reflective endorsement, but in a way that is not "detached" or isolated by the socio-relational dimension of their living.

To sum up, drawing on the results of our ethical inquiry, we can finally reelaborate and unpack the preconditions of our freedom of choice, as reconceptualized in our contemporary society.

1. The first precondition of freedom of choice, namely, the availability of alternative options, concerns the possibility for the agents to choose otherwise from what *de facto* they do. The plurality of options should not be understood only from a quantitative standpoint. Rather, this precondition entails options that differ substantially from each other, so as to open truly different—and thus alternative—courses of action: options that are heterogeneous from a *qualitative standpoint*. Substantially different courses of action mainly result from options that embed plural and diverse reasons, values, beliefs, and preferences (relations included), enabling critical reflection and reasoning, thus allowing individuals to endorse in decision-making what they might believe is optimal. To summarize: Substantially different courses of action mainly result from the *availability* of *morally heterogeneous options*, that is, a plurality of options that reflect and embed diverse points of view, reasons, beliefs, values (and so forth) as potential

Table 15.3 Key points of our ethical reelaboration of the dimensions, preconditions, and subconditions for freedom of choice in contemporary societies.

	Constitutive dimensions	Key preconditions	Subconditions
Freedom of choice revised	Individual dimension	Autonomy as self-relational determination	Reflective endorsement
	Socio-relational dimension	Availability of (morally heterogeneous) options	(Morally) diversified socio-relational exposure

motives for truly alternative courses of action. Therefore, such a precondition requires *sufficient exposure* of the agent to morally diversified sociocultural contexts (attachments, relations, values, beliefs, preferences, etc.).[2]

   This availability of alternative options is not the sole that needs to be reconceptualized.

2. The second precondition of freedom of choice, namely, human autonomy, needs to be revised too, as shown by the ethical inquiry developed earlier, so as to secure a minimum threshold freedom of choice in contemporary society.In all its multiple understandings, the precondition of autonomy requires the agents to be the ultimate authors (via participation, control, and endorsement) of their choices. However, as previously claimed, this "authorship" should not be understood as full independence. We have shown how the socio-relational and cultural dimension in which the agent is situated is also required (consider the availability of morally heterogeneous options via socio-relational exposure), as well as shapes (consider what motivates the agents: what they think, prefer, and value), and substantially contributes (consider social recognition and support) to the enjoyment or determination of the autonomy of the agent. Following such clarification, autonomy should be better understood than self-determination as *self-relational determination*: Autonomy expresses in the condition of the agents to be the authors of their choices through the reflective endorsement of what they want to truly embrace in the decision-making among the (morally heterogeneous) options available—which in turn are informed and shaped by the socio-relational dimension in which the agent is situated (Table 15.3).

Freedom of choice as an ethical-normative value

Acknowledging the aforementioned two constitutive dimensions and key preconditions of freedom of choice through ethical inquiry amounts to redefine freedom of choice as an ethical-normative value. Redefining freedom of choice as an ethical-normative value means ethically understanding and justifying why human freedom of choice *ought to be* secured in our contemporary society from existing and/or novel forms of the impediment to its exercise.

---

[2]Indeed, an opposite situation, that is, with nonqualitative diversified "moral exposure," would prevent the agents from developing reasons, values, and beliefs that are alternative to those to which they have been exclusively exposed and so to *truly* choose and act alternatively from what *de facto* they do.

Safeguarding freedom of choice means enabling the possibility for individuals to express their *moral standing* (or moral character), by choosing and acting according to one's own values, reasons, and beliefs, and by doing so, developing a genuine moral (social, political, and cultural) identity.

Guaranteeing freedom of choice means also safeguarding the flourishing of our living together (including moral progress), by securing people the possibility to assess whether their beliefs, values, and social and moral practices are the optimal ones or revise and change them. Indeed, exposure to morally heterogeneous options deeply grounds the possibility of a social dimension that is more open to dialog, mutual understanding, recognition, and even respect for differences in culturally heterogeneous contexts of values and practices, such as those of contemporary societies.[3]

Last but not least, freedom of choice is an *axiological catalysator*: a value whose presence confers either more value or disvalue to the respect or nonrespect of other values. Freedom of choice makes human choices and actions indeed appraisable from a moral standpoint, that is, making them, respectively, good and bad not only on the basis of their consequences but also on the basis of agents' intentions (reasons). In this sense, freedom of choice makes agents' good and bad choices, respectively, also better and worse, as they are chosen in a context of freedom, moral participation, and authorship. It would indeed be problematic without freedom of choice to consider our good and bad actions *truly* good or bad (as well as to ascribe moral responsibility and in certain cases also legal imputability). Indeed, without it, it would be hard to evaluate whether the subjects are the real authors of their choices and actions or if something in the environment prevented them from choosing alternatively; if they were the authors of their choices or instead something else has determined them in their place. In short, if we can guarantee freedom of choice, we can assess, for better or worse, whether the subjects who choose are morally present and free in their decisions and behavior. It follows that the protection of freedom of choice assumes great ethical importance for individuals and societies broadly.

Summary: In this section, we addressed a highly complex philosophical issue such as that of freedom of choice and tried to untangle such complexity, especially with regard to its necessary preconditions, by drawing insights from within moral and sociopolitical philosophy and then re-elaborating them to clarify what it takes to secure our freedom of choice in our contemporary society. Thanks to the ethical inquiry into freedom of choice, we have clarified both its necessary conditions and its constitutive dimensions as a core ethical value, providing an ethical compass to assess our freedom of choice nowadays. In the next section, we deal with another complex phenomenon, that is, the rise of algorithmic governance, by shedding light on the novel role played by MLAs in today's hyperconnected informational societies, as being propaedeutic to understand how MLAs, by reshaping our informational environments, can also affect the preconditions of our freedom of choice.

---

[3]Moreover, our freedom to form and choose what we want to reflectively endorse as motives for our choosing allows for developing a *more convinced* culture of giving reasons (moral justifications), that is fundamental to boost the moral dimension of our living together, along with a more cohesive social sphere, by favoring openness to mutual understanding, to the sharing of moral commitments, and therefore, to the possibility of joint cooperation toward societal goals and common goods.

MLA governance: the rise of "algorithmic choice-architectures"

That contemporary informational societies are increasingly permeated by MLAs is today undeniable. Examples abound. Recommender systems (RS) daily suggest what a user may like and should choose, from what song to listen to and movie to watch to products to purchase and/or people to talk to or form friendship relationships with (Milano et al., 2020). Filtering and classifying MLA ruling online service providers (OSPs)—from search engines and news aggregators to social networking services (SNS)—mediate and personalize *ubiquitously* what piece of information we can get access to and its order of priority thanks to personalization techniques (Newell & Marabelli, 2015). There are MLAs deployed to determine who is the most likely to be guilty of tax evasion (Zarsky, 2013), others that trade stocks on Wall Street (Patterson, 2013), or MLA-based services helping people date and mate (Siegel, 2013).

However, we are not only nestled into a pervasive network of MLAs. The latter indeed are increasingly exercising a key pro-active role and assuming social power in our society and personal lives. Especially over the last decade, more and more of—previously just human—daily tasks, personal choices, and high-stake decisions have started to be delegated to pervasive MLAs: Both private and public infrastructures, such as schools, universities, and hospitals (Obermeyer, 2019), financial institutions (Seng Ah Lee & Floridi, 2020), and institutional actors such as courts (Angwin, 2016; Yu & Du, 2019), police forces (Eubanks, 2018), local governmental bodies (Yeung, 2018), and national defense authorities (Rouvroy, 2015) today increasingly rely on MLAs for critical tasks and potentially life-changing decisions worldwide. This growing decisional (or metadecisional) power and "agential authority" ceded to MLAs has been acknowledged by many scholars as the rise of *algorithmic governance*, ML algorithmic regulation (Yeung, 2018), and/or ML algorithmic governmentality (Rouvroy, 2015). Their main argument is that whenever a person is denied a job opportunity by MLAs or a loan by a credit-scoring MLA, or a person is told which way to drive by a GPS routing MLA, or diagnosed with a certain disease, prompted to exercise in a certain way, take specific drugs, and eat a certain food by healthcare MLAs, that person can rightly claim that she lives within a society governed by MLAs.

Algorithmic governance refers to the use of MLA as decision-making systems, that is, systems that *inform or execute decisional tasks* on the basis of *algorithmically generated knowledge* to direct, influence, and thus govern the behavior of people and/or entities in a certain application domain in order to achieve a predefined goal. Algorithmically generated knowledge plays a key role in the establishment of MLA governance: Indeed, it is based on "a continual computational generation of knowledge from data emitted and directly collected in real time on a continuous basis from numerous dynamic components pertaining to the environment to identify and automatically refine the MLA system's operations to attain a pre-specified goal" (Yeung, 2018, p. 5).

Algorithmically generated knowledge is what confers a certain degree of *predictive power* to MLAs. It is produced by a series of combined knowledge-discovery techniques such as profiling (data mining included) and personalization techniques (such as filtering and classification algorithms as well as RS) that allow MLAs to process and mine huge amounts of [un]structured or [dis] aggregated data to infer or *discover* precious patterns and correlations on how things will likely be in the future. Such patterns and correlations are also used as indicators to probabilistically classify a subject as a member of a certain group; in other words: to group people into categories of akin, with surprising similarities, mainly joined by the way they are likely to behave in the near future (Hildebrandt, 2008); categories or groups probabilistically constituted by MLAs to which then show personalized informational contents. Such informational content is preselected by MLAs because they are found to be "relevant" for specific users belonging to certain groups on the basis of their profiling. Profiling can be based both on the correlations between contents and people's historical data (e.g., content-based profiling), from data people directly enter to observable data and/or involuntarily onlife traces they leave behind them, as well as it can rely on data and information deduced or inferred by MLAs by correlating such data with that produced by other people—especially those inferred with similar taste, characteristics, and habits (e.g., collaborative filtering).

Such discovered knowledge (patterns/correlations, etc.) is therefore used to create profiles of people, which can be perfectioned or fine-tuned via highly personalizing techniques such as RS that allow MLAs to measure how users respond to specific personalized informational content and change the parameters underpinning their profiling, grouping/categorization, and informational personalization accordingly to fine-tune the model (e.g., via inductive or deductive profiling, as well as via the combination of both). Such MLA-driven informational preselection (personalization) is based on probabilistic rules self-defined by MLAs and mainly oriented to the achievement of preset goals—such as, for example, especially in SNS and OSPs, third-party interests related to revenue income via users' time/click maximization (Zuboff, 2019).

In short, algorithmically generated knowledge via the combination of MLA techniques is what *truly* makes MLA "predictive systems" and thus highly value-laden in our contemporary informational societies. In fact, although when we talk about algorithms we can also refer to those encoding simple mathematical functions (e.g., statistical or deterministic models), the development of MLA governance grounds specifically on the large-scale application of methodologies and techniques that employ data to come up with novel patterns as knowledge and generate models and profiles that can be used for effective predictions about the data and people (Van Otterlo, 2013, p. 7). This is possible as MLAs are probabilistic: Their outputs do not follow causal relationships or an *a priori* set of causal rules or instructions, but are induced from data. MLAs are not preprogrammed to *follow certain rules* to solve a particular problem (e.g., deterministic or statistical algorithms); they are programmed to *learn how* to solve a specific (predefined) problem, task, and/or goal. For example, if we think about an MLA applied to a classification task, we will have two components: the *learner* that produces a *classifier* with the goal of generating "classes" that can generalize beyond the training data to discover correlations driving certain outputs (Domingos, 2012). The MLAs specifically work by placing new inputs into a model or a classification structure and *learn by defining rules* for how the new inputs will be classified. While the learning phase can be performed in a supervised way via hand-labeled inputs or in an unsupervised way, when MLAs find their own best-fit models to make sense of new sets of inputs, they self-determine what rules (and how) will manage new inputs and discover patterns and/or correlations, by doing so very often in complex ways that can result in being opaque (i.e., MLA as a black box) also to their designers (Pasquale, 2015).

Despite their complexity or opacity, MLAs are all around us, and they are re-weaving the social fabric of our societies and especially the onlife contexts in which we prepare and make our choices. To understand such MLA action, it is sufficient to analyze the governance role that interconnected MLAs ruling the majority of our ICTs play today over information in their function of "informational gatekeepers" (Calhoun, 2002), namely, by managing and navigating data or informational overload in our place. Indeed, as previously pointed out, MLAs manage what information is displayed to us, thanks to personalization ML techniques (e.g., filtering and classifying), and therefore decide what is or is not available to us, we can or cannot get access to, and/or enjoy, especially, but not exclusively, in informational terms, both—if we still endorse such distinction—in the "purely online" or "virtual" space (think about SNS) and in the "offline" or better, "strictly physical" space (think about the new smart fridge where MLAs decide what specific food to display on the basis of your health profile or the specific health regime you have been recommended to follow).

However, in our hyperconnected contemporary societies, our reliance on MLAs for managing information does not only make MLA informational gatekeepers that decide how information and the world as a whole are processed—including ourselves, via the probabilistic construction and application of profiles to us. They are also *choice-architects* (Thaler & Sunstein, 2008), namely, "agents" capable of silently reshaping or predetermining the onlife context in which we prepare and make our choices. Indeed, with the digital or fourth revolution (Floridi, 2014), everything today cannot only be understandable as information and/or described in informational terms, but can be concretely translated, as datafied, processed, and/or embedded into information, starting from our movements, characteristics, and habits to our deepest thoughts, values, or vulnerabilities, up to life projects, relations, and moral or social goals and commitments. Due to the exponential digitalization of our societies, everything today is datafied, which means that it can be captured as information. Information and informational contents specifically can therefore constitute options among which we can choose in the decision-making process. Such informational options range from items and products (e.g., products that MLAs display to us onlife to purchase, from books to read or movies to watch) to social relations (e.g., think about connections or friendships that MLA driving SNS can or cannot decide to display to us, as well as decide to prioritize) up to potentially life-changing socioeconomic chances and opportunities (think about the advertisement of a job opportunity or a special healthcare program for people with low income or a unique university scholarship). From this perspective, informational content constitutes informational options we can choose from. Such options, moreover, can in turn embed values, beliefs, goals, attachments, and thoughts, namely, everything can drive and motivate our choices in a specific way rather than in one another (i.e., moral knowledge). This means that to the extent MLAs as gatekeepers manage information, they are not only deciding what information does or does not show to us, but they are also shaping, restructuring, and defining our choice-contexts, by determining what options from being potentially available (being potentially matter of consideration) to us will be *effectively* actualized as part of our choice-context.

It is important to clarify that the term "choice-architects" has been coined and mainly used (Thaler & Sunstein, 2008) with reference to institutional agents to identify design actions helping individuals navigate too-many option contexts (e.g., informational overload), especially in daily choice-contexts mainly characterized by "bounded rationality" (Kahneman & Tversky, 2000; Simon, 2000; Thaler & Sunstein, 2008), meaning by that how we rarely choose in optimal conditions, but usually with relevant cognitive and time resources constraints. More specifically, according to their standard definition, by restructuring the order of choice-options made available to individuals, choice-architects would help people realize their long-term goals, without limiting their freedom of choice (Thaler & Sunstein, 2008).

In the next section, we highlight how the MLA-played architecting role might differ from that performed by traditional and institutional choice-architects. Indeed, on one side, it is unquestionable that MLAs today restructure individuals' informational choice-contexts, by deciding what will be or not be accessible to them (including the order through which information, items, products, relations, and opportunities are presented or recommended). However, although a widespread technical justification offered by MLA providers for the use of fine-grained profiling and personalization techniques is framed in terms of helping users to tackle information overload and find or access what they really want or are interested on, whether such MLA-driven action truly improves human decision-making toward users' goals without affecting their freedom of choice is an issue that has not been investigated yet. Indeed, the hetero-definition of users' informational choice-contexts driven by MLAs is produced by rules probabilistically self-defined by MLAs that are always driven by preset third-party goals that very often do not overlap with the users' ones. In the next section, we unpack this issue and deepen the relation between MLA choice-architectures and human decision-making by assessing whether MLAs promote or hinder the two preconditions underpinning freedom of choice.

MLA choice-architectures and freedom of choice
Epistemological and moral constraints

In the previous section, we have shown how different MLA techniques ruling the majority of our interconnected ICTs, SNS, and OSPs are giving rise to MLA choice-architectures, that is, onlife environments or choice-contexts predefined by MLAs according to probabilistically heterodefined rules and goals, with regard to the individuals who are subject to them. Therefore, we can already shed light on the first impact raised by MLA choice-architectures on the first precondition of our freedom of choice, the condition of the individuals to choose and act alternatively from what *de facto* they do, specifically: the availability of alternative (morally heterogeneous) options.

Indeed, MLAs act on this precondition by predetermining (personalizing: filtering and classifying) both from a *quantitative* and *qualitative* standpoint the range of choice-options will be accessible to us. From this perspective, MLAs already raise a first kind of constraint on such a first precondition, binding our freedom of choice within a predetermined context of options, hampering our capacity to choose alternatively from what *de facto* MLAs make available to us. This is because, by doing so, MLAs decide what will necessarily inform the formation of our knowledge in terms of reasons and motives that underpin and shape the way in which we can choose, as well as decide what to exclude from that availability of choice-options (which, as shown earlier, can range from chances and opportunities to relations, as well as embed facts, values, beliefs, projects, etc.). This is the first *epistemological* constraint raised by MLAs on our freedom of choice: MLAs bind our decision-making into a predefined range of options, making the reasons and motives leading us to choose and act in a certain way more easily predictable by MLAs.

To understand better if this narrowing MLA-driven action can be anyway beneficial to human decision-making, especially in contexts of informational overload, we formulate *two different scenarios* and assess the "goodness" of the consequences that MLAs can arise in the light of human freedom of choice. As previously pointed out, pervasive and fine-grained MLA profiling and personalization techniques are deployed with the justification of helping the users better find what they want in contexts characterized by limited cognitive and time resources; in other terms, MLAs would show us the same options we would have chosen anyway, if we would have had the time and capacity to perform such an operation.

Therefore, following such consideration, the *first scenario* we consider is based on onlife choice-contexts whose personalization is based on profiles of us *probabilistically well-inferred* by MLAs, that is, on profiles aligned to users as people (e.g., to their beliefs, characteristics, habits, values, projects, life plans, and/or long-term goals). This is indeed possible thanks to the huge, pervasive, and continuously self-learning network of fine-grained MLA profiling techniques in which we are nestled. From this first perspective, MLA choice-architectures might empower human decision-making, showing users what can be truly of relevance—as aligned with them. However, as previously pointed out, MLA techniques tend to categorize people according to discovered similarities, namely, in groups of like-minded (Hildebrandt, 2008) who show similar informational contexts of options (information, relations, opportunities, etc.). This mechanism ultimately leads us to encounter "those of exactly the same opinion sets as our own" and "make us more prejudiced and our attitudes more insular" (Parsell, 2008, p. 43), leading to radicalizing our previous orientation, instead of critically challenging it (Sunstein, 2008, 2017).

In other words, MLAs end to strengthen a phenomenon introduced a long time ago as the "Daily Me" (Negroponte, 1995), recently reformulated as the creation of informational "filtering bubbles" (Pariser, 2011) or "echo-chambers" (Sunstein, 2008, 2017): contexts designed in advance on the basis of predicted MLA indicators to respond to our predicted profiles (preferences, beliefs, values, projects, etc.). Put it differently: By doing so, MLAs might lead to the formation of both informational and relational contexts of like-minded, which means environments characterized by a reduced level of heterogeneity, therefore, with a reduced possibility of encountering the unexpected as diverse, alternative, divergent, and even new, insofar as mainly constituted by options embedding information and relation with similar characteristics. Similar characteristics mean also less heterogeneous moral and social beliefs, preferences, practices, and values, that is, everything can motivate us to choose or act in a certain way rather than in one another. To rephrase and further unpack this phenomenon, MLAs can foster the formation of moral echo-chambers. By doing so, MLAs narrow the range of options not only from a quantitative standpoint but also on the qualitative level, by reducing the level of heterogeneity that is critical to choose and act truly alternative from what *de facto* individuals do as preselected by MLAs. As pointed out in the first section, truly alternative courses of actions require an exposure to different and heterogeneous relations, practices, values, beliefs, life plans, or projects. A morally heterogeneous exposure is indeed critical to assess and reflect upon whether what we are choosing as well as what motivates us to choose in a certain way rather than in one another is optimal or instead needs to be revised or changed. In other words, it allows us to develop and embrace in a genuine or authentic way the reasons, values, beliefs, facts, experiences, chances, and opportunities that motivate our choosing, or to change them, and thus choose and act alternatively from what we thought optimal up to that time. By reducing such exposure, MLAs affect at the core the first condition of freedom of choice, that is, the *alternativity factor*: the availability of *alternative* as *morally heterogeneous* options. By doing so, MLAs do not foster human decision-making, but instead hamper it at the epistemological level, insofar as they do not compensate but instead exploit one of the most commonly renowned biases flawing human decision-making, that is, the "confirmation bias": our tendency to a-critically adhere, approve, and aggregate with what is akin to ourselves and/or confirm our previous opinions or attitudes (Quattrociocchi et al., 2016; Zollo, 2015), undermining our possibility to form and choose in a genuine or authentic way what can motivate our agency by critically testing it in a context of heterogeneity.

This phenomenon produced by MLAs has also key implications at the collective level insofar as it tends to shape our social interactions and relations in ways that do not expand them but narrow them, fostering detrimental phenomena such as social polarization, fake news, and social cascades, and therefore, the fragmentation of the public sphere (Sunstein, 2017). Put it differently: By undermining the possibility of exposure to diverse or heterogeneous beliefs, practices, and values, both online and offline (onlife), considering the time we spent on MLA-based SNS and OSPs and how they reshape our environment (Floridi, 2014), MLAs can deeply hamper the flourishment of a dialogic public sphere, a culture of giving reasons, and especially of mutual openness to understanding, mutual recognition, and respect, including, over time, the possibility of moral progress—which are key elements for the well-being of multicultural liberal democratic societies.

Nevertheless, here one might reply that the fact that MLAs very often work on de-individualized assumptions and correlations constitutes a counter-effect to the risks outlined earlier. This is the specific *second scenario* that we consider, that is, where the personalization of individuals' choice-contexts is based on MLA-discovered or constructed profiles that are not aligned to them as specific individual persons. Indeed, a very often-repeated worry about MLA profiling (Leese, 2014) is that it can ignore the individuality of people, namely, their consideration as "particular persons," with specific values, affiliations, beliefs, life-ground projects, and/or plans (Giovanola & Tiribelli, 2022a, 2022b), because it relies on patterns whose predictive values are imperfect (Hildebrandt, 2008). For example, a person can be subject to an adverse decision, such as being denied credit, simply by virtue of being similarly profiled to people who live in the same neighborhood who are not credit-worthy, while ignoring other more fine-grained and relevant aspects pertaining to their attitudes or life plans. Here we claim that ignoring the particular individuality of a person via options de-personalization is not the solution to the aforementioned problems and that this scenario raises even more problematic issues that see freedom of choice deeply intertwined with social justice and fairness. Such issues turn out to be intuitive, if we recall that the options predetermined—and therefore made available or accessible—by MLAs to people are not just informational *stricto sensu*, but include alternative possibilities such as real chances or social opportunities.

In a society increasingly permeated by interconnected MLAs, this scenario entails that a person might be subject to an MLA-produced adverse decision, such as, for example, being denied credit, due to de-individualized assumptions or correlations that might be general and easily end up being inexact and/or biased. This means that a person might see a life-changing request refused, for example, for a credit or a job opportunity or a subsidized rate of health insurance, simply in virtue of being similarly profiled to people, for example, who are not credit-worthy, and not on the basis of the specific person she is. MLAs that make decisions on subjects on the basis of profiles that do not consider or are not aligned to them as "particular persons" undermine the subjects at the epistemological and practical levels, by creating asymmetries both in knowledge and in power between individuals and MLAs:

- First, an epistemological asymmetry between who I am and how I am treated by MLAs (i.e., the options made accessible to me are based on a probabilistically discovered profile that do not reflect the particular person I am);

- Second, an asymmetry in power (or action), insofar as I cannot know, access to, and therefore, contest such MLA-constructed profiles and all the decisions produced by MLAs based on that. By doing so, MLAs do not only affect the individuals at the epistemological level (in knowledge), but also at the practical one (in action), as decisional end setters, by depriving them both of the cognitive and of practical power to exercise adequately their autonomy over the decision-making and their lives.

From this perspective, MLAs do not only raise an epistemological constraint on our freedom of choice, by binding the formation of the specific knowledge motivating and steering our decision-making to a range of options quantitatively and qualitatively predefined by MLAs via probabilistic techniques, mainly oriented to the achievement of preset third-party goals. MLAs can also raise *practical* or, more specifically, *moral constraints*, affecting the way in which we can exercise our autonomy over the decision-making process, as decisional end-setters. Therefore, MLA choice-architectures—as currently designed—can affect also the second condition underlying our freedom of choice: our (moral) autonomy, and specifically, the possibility of a genuine (or authentic) authorship over the decision-making.

To recall what we have pointed out in the first section of this manuscript, autonomy as participation and control over decision-making expresses the condition of the agents to be the "authors" of their choices via the exercise of reflective endorsement on what they choose (i.e., what motivates their choosing) in the decision-making among the (morally heterogeneous) options available—that are informed and shaped by the socio-relational dimension in which the agents are situated. To sum up, autonomy can be defined, more than as self-determination, as self-relational determination.

In this section, by analyzing the impact raised by MLAs on the first precondition of our freedom of choice, i.e., the availability of morally heterogeneous options, we have highlighted how MLAs can undermine the socio-relational dimension of our freedom of choice, which—as highlighted in the first section—informs, shapes, and enables the exercise of our autonomy, by reducing, via personalization techniques, the heterogeneity of individuals' socio-relational exposure. We have pointed out that such diminished heterogeneous socio-relational exposure can strengthen well-known cognitive bias flawing human decision-making (e.g., confirmation bias), therefore weakening our possibility to critically test and reflectively endorse what we choose among different options, namely, our possibility to form and embrace in a truly genuine/authentic way what deeply motivates and steers our choices. By doing so, MLAs can end to prompt—over time—a phenomenon of *deskilling* in the exercise of *reflective endorsement*. This phenomenon is problematic, raising another kind of constraint, that we define here as moral constraints *stricto sensu*.

The reflective endorsement is our last call for our freedom of choice: In that, we can find the distinctive trait of our "authorship" over our choices, actions, and identity. By exercising reflective endorsement, i.e., endorsing certain options that embed certain reasons, beliefs, and values (those we embrace) rather than others and approving them as motives for our choices and actions, we develop our "ought to": the way in which we make those reasons and values (and so forth) not just the motives, but the moral rules for our behavior (we make them normative for our conduct). Via the exercise of our reflective endorsement, indeed, we form the way in which we respond to our reality as conveyed by our mediated perceptions and emotions, by taking a moral stand toward them and, by doing so, developing our moral posture or moral identity. Indeed, even if MLAs can predetermine our range of available options, mainly binding our choices to certain options rather than others (it is less likely we choose those options that MLAs do not show us), thereby constraining our freedom of choice, this constraint is still "soft." As moral agents, we always have the power to decide to act against certain informational options (or against our preferences, needs, or desires), as well as to choose not to act. Put it differently: our choice-contexts of options do not necessarily determine us to choose in a certain way. This is because we have the normative agential power to say yes or no to a certain option (reason, value, event, desire, and so forth), and this approval lies exactly in the exercise of reflective endorsement.

The problem is that the exercise of our reflective endorsement can be affected too by MLAs, not only over time in terms of deskilling, but it can also be suspended or bypassed by the combined action of certain fine-grained and more incisive MLA techniques such as RS. Described as inescapable and ubiquitous "sticky traps" (Seaver, 2018) for their capacity to glue users and/or addict them to some recommended preset solutions (informational options, products, relations, etc.) labeled as "relevant" or highly efficient to harness individuals' constructed profiles in order to nudge their behavior choice (Milano et al., 2020) toward the achievement of MLA preset goals, RS can exercise a disruptive impact on people's moral autonomy. Indeed, RS can augment the MLA personalization rate, by allowing the fine-tuning of MLA profiles of individuals and specifically the discovery of what people tend to value most within an already personalized context of options, that is, what can mostly elicit users' choice-response behavior: the prediction of users' *choice-driving elements*.[4] The latter can range from users' basic interests to their deepest vulnerabilities, weaknesses, fears, and/or traumas.[5] Such informational options, if identified as highly valuable in terms of efficiency to maximize MLA preset goals (e.g., click maximization), can be then exploited by MLAs to refine the personalization of users' choice-contexts and to microtarget them with recommendations (O'Neil, 2016) on what they should be interested in, what they should like, buy, and/or watch, that is, what they should choose.

Moreover, especially when the microtargeted or recommended options embed sensitive information, MLAs can also easily trigger what is defined in evolutionary and moral psychology as "primary or basic emotions" (Ekman, 2003), such as fear, joy, anger, sadness, interest, and disgust, that is, those instinctive and innate emotions and emotional behaviors we have in common with a very young child and some animals.[6] This entails that MLAs, by exploiting and recommending or targeting users with sensitive informational contents, predicted and labeled as value-laden to meet third-part goals, can trigger individuals' unreflective or instinctive choice-response behaviors, which can *suspend or directly* bypass users' exercise of reflective endorsement. By doing so, indeed, the microtargeted option recommended by MLAs more than a mere nudge can become a real push affecting users' moral autonomy, suspending or bypassing their reflective endorsement, and transforming the main informational *option* recommended from being a *motive* of people's choices and actions (an option they can approve as a motive for their choices by reflectively endorsing it) to being the main *cause* of them. In other words, MLA recommendation of an informational option,

---

[4]In order to set the relevance of an option, RS can make online experimentation or A/B testing, that is, the practice of exposing selected groups of users to modifications of the algorithm, with the aim of gathering feedback on the effectiveness of each version from the user responses (Seaver, 2018).

[5]In this regard, let us think, as an example, that only recently, after the Supreme Court sentence Roe v. Wade, Google has declared that it would delete location data about abortion clinic visits from users' data history.

[6]Primary emotions can be distinguished from those so-called "secondary," such as shame, envy, guilt, pride, or regret (and of course many more), which arise later in our life as they require a certain awareness, degree of socialization, and the formation of an idea of what is good (the formation of moral disposition). See Ekman (2003).

instead of epistemologically empowering human decision-making (i.e., informing agents' choices and actions with truly novel options), can deeply undermine it, by ending up choosing at the at the users' place, therefore predetermining their behavior to meet preset and hetero-defined goals that often do not overlap with the users' ones.

Such a phenomenon can be accidental and/or prompted intentionally, but in any case, it raises detrimental consequences both for individuals and societies. Recalling the case of Cambridge Analytica, we can think about political parties that, in order to reach consensus, can exploit individuals' sensitive information to microtarget them with specific option, as measured to be successful in triggering a predesigned change in their political orientation; or we can think about pharmaceutical companies that can target fears or vulnerabilities, for example, to increase the purchase of their depressive drugs (the list of cases can be easily expanded). Beyond the accidental or intentional nature of this phenomenon, it raises a hard-moral constraint on our freedom of choice, binding us to decisions and actions, as well as, over time, on social, political, and moral identities on which we lose, and thus cannot claim, control, participation, and especially, moral authorship.

Summary: thus far, we have clarified what are the preconditions that can secure—at least at a minimum threshold—our freedom of choice. Then, we have shown how predictive MLAs can undermine both such preconditions. In the next section, thanks to this ethical inquiry, we pinpoint three ethical criteria that *ought to* be considered and operationalized through the design and implementation of MLA-based systems to promote the latter as ethical tools for both safeguarding and promoting human freedom of choice.

Securing freedom of choice by design: a call to action

In the first section of this manuscript, we have highlighted that the first precondition underlying the free exercise of our freedom of choice is the availability of morally heterogeneous options. In other words, agents are free to choose *if and only if* they can act otherwise from what *de facto* they do. We have indeed outlined how truly alternative courses of actions especially ask for a sufficient morally heterogeneous exposure of the agents, that is, for example, to alternative social and cultural practices, beliefs, values, and/or moral ground projects. In the second section, we have argued how MLAs can predefine our informational environments or onlife choice-contexts by selecting, filtering, and classifying choice-options on the basis of profiles probabilistically discovered or constructed for us usually to satisfy third-party interests. We have shown how this MLA-driven predefinition of our onlife availability of options can undermine our possibility to choose truly alternatively from what *de facto* MLAs decide to prioritize, recommend, and make simply available to us both online and offline, from a quantitative and qualitative standpoint. We have clarified how such an informational hetero-definition of our choice-contexts indeed does not only narrow and constrain what can inform our choices but also what can become a motive we can endorse to choose and act in a certain way, rather than in another (our moral knowledge: reasons and motives underpinning our choices). Moreover, we have argued how this MLA-driven predefinition of our availability of options is particularly problematic, as it tends to shape our (informational and relational) environments in ways that exclude heterogeneity, privileging informational and relational options akin to each other's, thus, to the creation of groups of like-minded, that is, moral echo-chambers. This narrowing

MLA-driven action deeply hinders our possibility—via the exposition to different ways of thinking, acting, and behaving—to choose truly alternatively from what *de facto* we can do and MLAs show us, and therefore, our possibility to develop our very own (authentic or genuine) preferences, values, beliefs, life plans, or morally ground projects. In addition, this MLA-driven action is made on the basis of profiles that are often opaque to the people who are subject to that, as well as on the basis of goals or interests very often misaligned with the users'.

In order to counteract this narrowing and constraining MLA-driven predefining action on our choice-contexts, first, a *criterion of intelligibility* should be considered and operationalized through the design and development of MLA-based systems. Indeed, the agents should have the right to be informed by MLAs on how they are profiled and categorized into certain groups, instead of others. In other words, a subject's *right to know* what information and discovered inferences or patterns determine her access to certain choice-options rather than others (i.e., information, relations, and opportunities) should be met and respected in ways that are accessible to the subjects, that is, according to vocabularies and modalities *comprehensible* by the different subjects involved.

Designing MLA-based systems in compliance with the criterion of intelligibility means creating spaces where the subjects can (1) be informed, even without a specific request, about the way in which they are profiled and on the basis of what (e.g., specific data, nodes, inferences, patterns, correlations, etc.), and (2) be provided with the capacity to change the parameters at the basis of such profiling to regain control over what can or cannot deeply inform and/or become a motive of their choices and actions, as well as constituting an opportunity they can enjoy and take benefit from. This criterion, therefore, would provide the subjects with the capacity to revise, change, or expand more what they can get access to onlife and thus exercise control and participate actively in the MLA-driven reshaping action of options constituting their choice-contexts. In other words, to reclaim and obtain a certain degree of participation and/or control over the context in which they develop and make their choices. It is worth mentioning that there are efforts being made on explainable AI (XAI), that is, in the direction of trying to make MLAs (and the choices they make) more understandable and/or transparent from a human point of view (to expand such main XAI methods, see Holzinger et al., 2022; Vilone & Longo, 2020).

Nevertheless, this first ethical criterion is not enough to safeguard our possibility to choose truly alternatively from what *de facto* we do, namely, to secure the first precondition underpinning our freedom of choice. Indeed, some of the detrimental consequences we have previously highlighted, such as the possibility to end up in an enclosed informational and relational "bubble" of like-minded, or moral echo-chambers, can also be led by agents' autonomous predefinition of their informational environments or choice-contexts. This is because, due to the widely explained bias in cognitive sciences, such as, for example, the confirmation bias, we tend to choose relations and privilege information aligned with or that can confirm our previous points of view, beliefs, or attitudes. Instead of deeply exploiting such cognitive bias, MLA-based systems ought to be designed to *defuse* them. This is possible by complementing by design the criterion of intelligibility with the *criterion of heterogeneity*, which responds to the right of agents to keep their choices and identity truly open via the possibility to test, revise, change, and thus critically and reflectively endorse what can deeply inform and motivate their choices, so as to make them, as well as develop over time their identity, in a truly genuine (or authentic) way. Indeed, contexts of similar information and relations tend to deploy human cognitive bias like confirmation biases in decision-making hampering the possibility to deeply reason, approve, or contest what drives our behavior, eroding (deskilling) the exercise of our reflecting endorsement.

Operationalizing the criterion of heterogeneity means guaranteeing by design the possibility for the agents to be exposed to (informational and relational) options always embedding a sufficient level of heterogeneity, even when they have been informed and in turn have approved the way in which they have been profiled and categorized. Indeed, securing this minimum grade of heterogeneity is crucial not only to assess if what we choose as driving our behavior or motivating our decision-making is optimal but also to always keep our identity open to diverse, alternative, and even unknown ways of thinking, choosing, and acting (i.e., values, practices, and beliefs), as critical to secure an open and healthy public sphere (moral progress included), namely, for mutual understanding, recognition, and respect, and therefore, for the possibility of shared cooperation and/or joint commitments for the farthest common societal goals.

The consideration and operationalization of these two ethical criteria would not only prevent and/or mitigate the narrowing and hetero-predetermining MLA-driven action on the first precondition of freedom of choice. As shown earlier, criteria of intelligibility and heterogeneity also contribute to the protection of the second precondition of freedom of choice, that is, the exercise of moral autonomy via reflective endorsement. Indeed, we have also argued how certain techniques, such as RS, can affect our autonomy deeply by exploiting certain informational options discovered and tested on us as choice-driving behavior information or stimuli (e.g., traumatic events, psychological vulnerabilities, and/or fears, just to mention a few) that can trigger certain predefined unreflective and instinctive behavioral responses. To achieve predefined goals, MLAs might exploit choice-triggering behavior information to push people's decision-making to meet predefined third-party goals, for example, by triggering primary emotions that can overact on our choice behavior in our place, beyond our control, that is, bypassing or suspending the exercise of our reflective endorsement over what can motivate and drive our choice. This phenomenon happens also in environments that are not algorithmically governed: for example, when one says something that is particularly sensitive to the point to make us suddenly joyful, cry, interested, sad, or angry (i.e., primary emotions). The difference in MLA-governed environments is that MLAs might easily discover such choice-driving information or stimuli and use them to architect our choice-environments to silently push our behavior toward hetero-defined and preset goals, so that we might do realize why in a certain moment we made a certain action, from clicking on a certain item to buying a certain product to voting a certain political party. Such MLA architectures might deeply affect who we could become, i.e., our social, political, and especially moral identity, founding out ourselves over time far from the person we wanted, were aimed, or motivated to become.

Given the complexity of regulating what can be a matter of MLA datification, reelaboration, and inference in our hyperconnected environments, how can we prevent or mitigate by design such an action?

If the MLA behavior in input (datification, elaboration, and inference) might be difficult to regulate, even when it deals with data pertaining to the health sphere over which regulations such as the *European General Data Protection Regulation* (GDPR) should apply,[7] we claim that the same cannot be said about MLA behavior in output, which concerns what discovered inferences and information MLA can or cannot exploit (especially recommend or push) in order to meet predefined third-party goals. We claim that a third requirement, the criterion of *confidentiality*, should be considered and operationalized throughout the MLA design, development, and use. This criterion

---

[7]On the limit to regulate via GDPR MLA inference on health data and other sensitive information, see Watcher (2020).

asks MLA providers and designers to provide space or tools allowing the subjects, after having informed them about their profiles (criterion of intelligibility), to exercise a right of confidentiality over certain information, that is, to ask MLAs to keep some discovered knowledge and information confidential, to remove it or exclude it from their past history and especially from future reelaboration and processing, as many times as such information (patterns, correlations, etc.) reappears as value-laden.

Operationalizing this ethical criterion would also require the MLA designer to train MLAs (via supervised training) on highly heterogeneous amounts of correlations that might be sensitive to different subjects, starting with those that can relate to the psychological health sphere, so as to train MLAs to acknowledge them and, once discovered as potentially sensitive, to inform and alert the subjects, so that they can promptly act on them, in the case they want to do so. Moreover, the operationalization of this ethical criterion would also require MLA designers to exclude the use of individuals as pure "means" for A/B testing to fine-tune users' discovered or constructed profiles on sensitive knowledge and information (inferences, patterns, etc.).

Respecting the ethical criterion of confidentiality, along with the two previously mentioned, means—by recalling a Kantian-inspired vocabulary—respecting individuals as *decisional end-setters*, which entails not treating them as mere means to achieve hetero-defined third-party goals, interests, or ends. In other words, this means respecting and promoting individuals' autonomy, that is, enabling the exercise of their reflective endorsement and, throughout that, empowering their possibility of truly genuine or authentic authorship over decision-making, on what can deeply motivate and drive their choices and therefore, over time, shape their identity.

Conclusion

In this paper, we have tried to highlight, unpack, and address the impact raised by online AI-based systems, and especially by predictive MLA-based systems, on our freedom of choice, both online and offline, that is, onlife, harnessing the conceptual and practical tools offered by moral philosophy and AI ethics. First, we have pursued an ethical inquiry into the necessary preconditions that underlie the free exercise of choices in our contemporary, hyperconnected informational societies. In particular, we have highlighted that freedom of choice comprehends an individualistic dimension and a socio-relational dimension and have detected its preconditions in the (1) availability of morally heterogeneous options and (2) autonomy as self-relational determination. Second, we have shown how the pervasive application of MLA techniques is establishing what is known as MLA governance, defining its *modus operandi*, and specifically, how it is developing environmental or MLA choice-architectures deeply shaping the contexts in which we prepare and make our choices. Then, we have argued how such MLA choice-architectures, through the exploitation of combined ML techniques, can affect and undermine the preconditions that can secure at a minimum threshold our freedom of choice, highlighting an MLA-driven double-level impact on human decision-making: at its epistemological level and at its moral one, strictly speaking. We have concluded this section by showing how predictive MLAs, as currently designed, can deeply erode our exercise of freedom of choice and our possibility of a truly genuine authorship over decision-making, opening further reflections on what might be the long-term implications arising from this impact at both the individual and collective level. Third, and finally, we have shown the import of our ethical inquiry into the impact of predictive MLAs on our freedom of choice at the practical level. Indeed, we have shed light on how this inquiry helps pinpoint some key ethical criteria—*intelligibility, heterogeneity,* and *confidentiality*—that ought to be met by design to develop MLA-based systems preserving at a minimum threshold human freedom of choice as an ethical-normative value in our increasingly MLA-driven informational societies.

We are aware that still so much work needs to be done in this direction at both the conceptual and technical levels; however, we are confident that this inquiry can provide a more exhaustive and philosophically grounded ethical compass to address the protection of freedom of choice in our digital era increasingly influenced by AI and MLA-based systems.

References

Albritton, A. (1985). Freedom of will and freedom of action. *Proceedings and Addresses of the American Philosophical Association, 59*(2), 239—251.

Angwin, J., et al. (2016). *Machine Bias*. Retrieved from: https://www.propublica.org/article/machine-bias-risk-assessments-in-criminal-sentencing.

Arneson, R. (1991). Autonomy and preference formation. In J. Coleman, & A. Buchanan (Eds.), *Harm's way: Essays in Honor of Joel Feinberg* (pp. 42—73). Cambridge: Cambridge University Press.

Benson, P. (2005). Feminist intuitions and the normative substance of autonomy. In J. S. Taylor (Ed.), *Personal autonomy: New essays on personal autonomy and its role in contemporary moral philosophy* (pp. 124—142). Cambridge: Cambridge University Press.

Berlin, I. (1969). *Two concepts of freedom.* Oxford, UK: Oxford University Press.

Botes, M. (2022). Autonomy and the social dilemma of online manipulative behavior. *AI Ethics.*

Calhoun, C. (2002). *Dictionary of the social sciences.* Oxford University Press.

De Caro, M. (2004). *Il libero arbitrio. Una Introduzione.* Roma-Bari, Italy: Laterza Editori.

Domingos, P. (2012). A few useful things to know about machine learning. *Communications of the ACM, 55*(10), 78—87.

Ekman, P. (2003). *Emotions inside out: 130 Years after Darwin's the expression of the emotions in man and animals.* New York: New York Academy of Sciences.

Eubanks, V. (2018). *Automating inequality.* New York: St Martin's Publishing.

Floridi, L. (2014). *The fourth revolution: How the infosphere is reshaping human reality.* Oxford, UK: OUP.

Floridi, L., & Sanders, J. W. (2004). On the morality of artificial agents. *Minds and Machines, 14*(3).

Giovanola, B., & Tiribelli, S. (2022a). Weapons of moral construction? On the value of fairness in algorithmic decision-making. *Ethics and Information Technology, 24*(3).

Giovanola, B., & Tiribelli, S. (2022b). Beyond bias and discrimination: Redefining the AI ethics principle of fairness in healthcare machine-learning algorithms. *AI & Society.*

Hildebrandt, M. (2008). Defining profiling: A new type of knowledge? In M. Hildebrandt, & S. Gutwirth (Eds.), *Profiling the European citizen* (pp. 17—45). Springer.

Hobbes, T. (1654). Of liberty and necessity. In V. Chappell (Ed.), *Hobbes and bramhall on liberty and necessity* (pp. 15—42). Cambridge, UK: Cambridge University Press.

Holzinger, A., Saranti, A., Molnar, C., Biecek, P., & Samek, W. (2022). Explainable AI methods—A brief overview. In A. Holzinger, R. Goebel, R. Fong, T. Moon, K. R. Müller, & W. Samek (Eds.), *XXAI—Beyond explainable AI. xxAI 2020. Lecture notes in computer science* (vol. 13200). Cham: Springer.

Kahneman, D., & Tversky, A. (Eds.), (2000). *Choices, values and frames.* Cambridge, UK: Cambridge University Press.

Kant, I. (1998). In P. Guyer, & A. W. Wood (Eds.), *Critique of pure reason*. Cambridge, UK: Cambridge University Press.

Killmister, J. (2017). *Taking the measure of autonomy: A four-dimensional theory of self-governance*. London, UK: Routledge.

Leese, M. (2014). The new profiling: Algorithms, black boxes, and the failure of anti-discriminatory safeguards. *The European Union Security Dialogue*, *45*(5), 494−511.

Mackenzie, C., & Stoljar, N. (2000). Introduction. Autonomy refigured. In C. Mackenzie, & N. Stoljar (Eds.), *Relational autonomy. Feminist perspectives on autonomy, agency, and the social self* (pp. 3−31). New York: Oxford University Press.

Milano, S., Taddeo, M., & Floridi, L. (2020). Recommender systems and their ethical challenges. *AI & Society*, *35*, 957−967.

Mittelstadt, B. D., et al. (2016). The ethics of algorithms: Mapping the debate. *Big Data & Society*, *3*(2), 1−21.

Mori, M. (2001). *Libertà, necessità, determinismo*. Bologna, Italy: Il Mulino.

Negroponte, N. (1995). *Being digital*. New York: Vintage Books.

Newell, S., & Marabelli, M. (2015). Strategic opportunities (and challenges) of algorithmic decision-making: A call for action on the long-term societal effects of 'datification'. *The Journal of Strategic Information Systems*, *24*(1), 3−14.

Noble, S. (2018). *Algorithms of oppression*. New York: NYU Press.

Obermeyer, Z., et al. (2019). Dissecting racial bias in an algorithm used to manage the health of populations. *Science*, *366*(6464), 447−453.

O'Neil, C. (2016). *Weapons of math destruction*. London, UK: Penguin.

Oshana, M. (1998). Personal autonomy and society. *Journal of Social Philosophy*, *29*(1), 81−102.

Pariser, E. (2011). *The filter bubble: What the internet is hiding from you*. London, UK: Viking.

Parsell, M. (2008). Pernicious virtual communities: Identity, polarisation and the web 2.0. *Ethics and Information Technology*, *10*(1).

Patterson, S. (2013). *Dark pools: The rise of AI trading machines and the looming threat to wall street*. New York: Random House.

Pasquale, F. (2015). *The black box society: The secret algorithms that control money and information*. Cambridge, MA: Harvard University Press.

Quattrociocchi, W., Scala, A., & Sunstein, C. (2016). Echo-chambers on Facebook. *Harvard discussion paper series*.

Raz, J. (1986). *The morality of freedom*. Oxford, UK: OUP.

Rouvroy, A. (2015). Algorithmic governmentality: A passion for the real and the exhaustion of the virtual. *All watched over by algorithms*. Berlin.

Sandel, M. J. (1982). *Liberalism and the limits of justice*. Cambridge: Cambridge University Press.

Seaver, N. (2018). Captivating algorithms: Recommender systems as traps. *Journal of Material Culture*, *24*(4).

Seng Ah Lee, M., & Floridi, L. (2020). Algorithmic fairness in mortgage lending: From absolute conditions to relational trade-offs. *Minds & Machines*, *31*(1).

Siegel, E. (2013). *Predictive analytics: The power to predict who will click, buy, lie or die*. New Jersey: John Wiley and Sons.

Simon, H. A. (2000). Bounded rationality in social sciences: Today and tomorrow. *Mind & Society*, *1*(1), 25−39.

Skinner, Q. (2008). *The genealogy of liberty*. UC Berkley: Public Lecture.

Sunstein, C. (2008). Democracy and the internet. In J. van den Hoven, & J. Weckert (Eds.), *Information technology and moral philosophy* (pp. 93−110). Cambridge University Press.

Sunstein, C. (2017). *#Republic: Divided democracy in the age of social media*. Princeton, NJ: Princeton University Press.

Thaler, R., & Sunstein, C. (2008). *Nudge*. Yale University Press.

Taylor, C. (1991). *The ethics of genuineity*. Cambridge, MA: Harvard University Press.

Tiribelli, S. (2020). Predeterminazione algoritmica e libertà di scelta. In L. Alici, & F. Miano (Eds.), *Etica nel futuro* (pp. 431−441). Orthotes: Napoli.

Tiribelli, S. (2022a). Artificial agency and moral agency: Conceptualizing the relationship and its ethical implications on moral identity formation. *S&Fscienzaefilosofia.it.*, *27*(1), 54−68.

Tiribelli, S. (2022b). *Moral freedom in the age of artificial intelligence*. Milan-London: Mimesis International.

Tiribelli, S. (2023). *Identità personale e algoritmi. Una questione di filosofia morale*. Roma: Carocci editore.

Tsamados, A., et al. (2020). The ethics of algorithms: Key problems and solutions. *AI & Society*, *37*, 215−230.

Van Otterlo, M. (2013). A machine learning view on profiling. In M. Hildebrandt, & K. de Vries (Eds.), *Privacy, due process and the computational turn-philosophers of law meet philosophers of technology* (pp. 41−64). Abingdon, UK: Routledge.

Vilone, G., & Longo, L. (2020). Explainable artificial intelligence: A systematic review. ArXiv: 2006.00093v4: 1−81.

Wachter, S. (2020). Affinity profiling and discrimination by association in online behavioral advertising. *Berkeley Technology Law Journal*, *35*(2).

Westlund, A. (2014). Autonomy and self-care. In M. Veltman, & A. Piper (Eds.), *Autonomy, oppression, and gender* (pp. 181−198). New York: Oxford University Press.

Yeung, K. (2018). Algorithmic regulation: A critical interrogation. *Regulation and Governance*, *12*(3), 505−523.

Yu, M., & Du, G. (2019). Why are chinese courts turning to AI? *The Diplomat*.

Zarsky, T. (2013). Transparent predictions. *University of Illinois Law Review*, *2013*(4).

Zollo, F., et al. (2015). Emotional dynamics in the age of misinformation. *PLoS One*, *10*(9).

Zuboff, S. (2019). *The age of surveillance capitalism: The fight for a human future at the new frontier of power*. New York: Public Affairs.

The hard problem of the androcentric context of AI: challenges for EU policy agendas

16

Joshua Alexander González-Martín

University of Salamanca (USAL)—Institute for Science and Technology Studies (ECYT), Salamanca, Spain

Introduction

Unlike past technologies designed to mediate or enhance the physical capabilities of human beings (traveling long distances in an ever shorter time, helping our organism to develop immunity against viruses, or establishing long-distance communications), artificial intelligence (AI) systems have made both a qualitative and quantitative technological leap in that they are technologies capable of processing large amounts of data and inferring patterns, or even drawing conclusions from that data (Theodorou & Dignum, 2020). Algorithms that enable these cognitive processes to make predictions or decisions can themselves evolve through machine learning, neural networks, and the use of data (Chaturvedi et al., 2022; Cohen, 2021; Lawrence et al., 2021; Miikkulainen et al., 2019). Thus, recent years have seen a large growth in the capabilities and applications of AI systems (Dignum, 2020), with these potentially presenting themselves as a replacement for human decision-making and reasoning (Duan et al., 2019; Hancock, 2022; Kaptelinin, 2022; Parikh et al., 2016).

In this context, there is great theoretical and practical interest in ethical and legal issues surrounding the development and use of AI (Coeckelbergh, 2020; Dignum, 2019; Gunkel, 2020; Morley et al., 2021). The reason is that, with the growing importance and frequent use of AI applications, at the same time the technological challenges and risks associated with these systems have increased. It is true that AI applications offer opportunities, advantages, and solutions to various problems, but they also yield biased results that could affect individuals or minorities based on their sexuality, ethnicity, or gender (Martínez et al., 2020; Nadeem et al., 2021; Ntoutsi et al., 2020). As AI transforms our society, economy, and our daily lives, biases (implicit or explicit) in algorithmic decisions have deleterious effects that particularly affect not only individuals but also certain social groups and, with that, also society at large (Altman et al., 2018).

Emerging technologies that entail accelerated disruption, as in this case, continuously require new normative criteria, especially in the application of ethics and legislation. Although several issues of biases in algorithms and neural networks have already been widely discussed (Akter et al., 2021; Challen et al., 2019; Daneshjou et al., 2021; DeCamp & Lindvall, 2020; Ferrer et al., 2021; Hagendorff et al., 2022; Hooker, 2021; Johnson et al., 2019; Makhni et al., 2022; Peng et al., 2022; Rajpurkar et al., 2022; Sham et al., 2022; Tripathi & Musiolik, 2022; Zhou et al., 2022), in this chapter we focus on AI gender biases from an explicitly European perspective, given the commitment of

Ethics in Online AI-Based Systems. DOI: https://doi.org/10.1016/B978-0-443-18851-0.00013-5

the European Commission (EC) toward responsible innovation and the recent emergence of the European Union (EU) proposal for AI regulation. The extent to which political agendas concerning gender equity and equality are respected in the field of AI applications will be assessed.

The argument is structured as follows. First, an overview of how gender scripts are embedded in AI systems, producing biases and (hidden) gender discrimination, is presented. Subsequently, we focus on the notion of Responsible Research and Innovation (RRI), on key and relevant agendas, and the device of gender equity and equality policies will be analyzed. It will assess the extent to which RRI policies align with regulation to address technological androcentrism and sociodemographic risk factors such as gender, in AI mining, management, and interpretation of massive data. Finally, EU nondiscrimination law and data protection law will be reviewed in relation to the legal challenges arising from the situation with machine learning systems. We will conclude with some final thoughts that show certain policy challenges and bring to the forefront the urgency of public discussion around the future of AI regulation.

Artificial intelligence and gender biases

As technologies using AI develop, the meaning moves away from being unambiguous for AI, just as the concept of "Machine Learning" is not a monolithic concept (Kaplan, 2016; Wischmeyer, 2018). Instead, these technologies cross diverse fields, methodologies, and techniques. However, there is no doubt that, generally speaking, with the development of machine learning, AI systems are able to learn by themselves through algorithms. The learning process occurs by providing data from previous operations that performed the same tasks or by providing data through the internet (Marr, 2016). Thus, these algorithms are able to define or modify decision-making rules in an automated fashion and make sense of large data streams (Fenwick et al., 2018; Haddad & Hornuf, 2019). It is hard to imagine any domain of everyday life that is not profoundly affected by AI.

From a technical point of view, algorithms are programmed instruments to automatically solve complex mathematical problems. A distinction must be made between the *input*, the arithmetic operation, and the *output*. The *input* consists of data that should form the basis for a desired result (*output*). The output is not immediately known in detail, but the properties or objectives that the result should have must be defined. The arithmetic operation leading from the input to the output is often referred to as the solution. The operand between *input* and *output* is decisive for the concrete result (Mahnke, 2015). Furthermore, these methods are always closely related to the use of an AI-based system in broader technological contexts. Even certain AI systems can influence the social or economic participation of individuals in everyday life.

This is especially important because, since algorithms are designed, implemented, and trained, in broader sociotechnical contexts of use, their design, learning, and use may be affected by the attitudes, values, inclinations, and social tendencies of those individuals or autonomous systems that program and use the algorithms. Consequently, AI systems, in the form of machine learning that, through selection and processing, use large datasets, run the risk of being traversed by biases or creating new biases. As a sociotechnical by-product, AI systems may have unavoidable social-historical forms, one of which manifests itself through gender relations that may bias the decision-making of these systems (Basta et al., 2021; Cirillo et al., 2020; Fabris et al., 2020; Fosch-Villaronga et al., 2021;

Larrazabal et al., 2020; Lütz, 2022; Martínez et al., 2020; Nadeem et al., 2021; Ntoutsi et al., 2020; Puc et al., 2021; Serna et al., 2021; Waller & Waller, 2022).

Sociohistorical assumptions affect what data is collected, what definitions, terms, and uses are tracked, and what predictions the algorithms produce (Hendl et al., 2019). Consequently, these algorithms may produce gender-biased knowledge about professions and human relationships. In this chapter, a general definition of gender bias is considered to be the inclination or prejudice of a decision made by an AI system in favor of or against a person or group, especially in a gender-unfair manner. Given this definition, the following shows how gender bias is articulated as an intrinsic but subtle feature that appears embedded in some AI systems. But before gender bias is discussed through the use of algorithms, the question of how gender scripts can be embedded in the AI engineering context will be addressed.

Engineering context and gender script in AI

Engineers, and other actors involved in the design or programming process, configure both the user and the context of use and the AI-based technological artifacts that will autonomously, semiautonomously, or heteronomously act in the context of use, as an integral part of the whole technology development process, to enable and constrain the development of the technology. Configuring the context of use is an important part of the engineering process, as it can help define the parameters within which a technology can be developed. By understanding the needs and constraints of use, engineers can create systems that are more effective and efficient. In addition, by understanding the context of use, engineers can develop systems that are better suited to the specific environment in which they will be used.

Some authors refer to this as a plan of use (Borgo et al., 2014; Crilly, 2010; De Vries, 2016; Houkes & Vermaas, 2004, 2010; Preston, 2013), actions considered goal-directed, where actions involving interactions with the technological artifacts themselves are included (Borgo et al., 2014). Other authors have referred to technologies as artifacts designed and produced through "scripts" (see, for example, Akrich, 1992), i.e., contextual presuppositions, including representations about their users and assumptions about the context of use by designers, are materialized in technologies and are incorporated into the materiality of the artifact, thus composing an environment for their use and defining what can or cannot be done with them.

In this line, Van Oost (1995), Oudshoorn (1996), and Rommes (2000) introduce the concept of gender script, from which they try to explain how gender is implied in the design of technologies or, in other words, how technologies are "gendered," and to account for the way in which designers represent users according to gender patterns. According to Ellen van Oost, gender scripts are those representations that the designers of an artifact have or construct of gender relations and gender identities, representations that they then inscribe in the materiality of that artifact (Van Oost, 2003: 195). Artifacts themselves often incorporate user configurations established through usage plans in the engineering context (Houkes & Vermaas, 2014; Loh, 2019), in which gender scripts, i.e., cultural expectations of how men and women should behave with respect to artifacts and technology in general, are embedded. In van Oost's words:

> certain technical artifacts are produced explicitly for women or men in the context of certain gender-specific stereotypes, while other artifacts only implicitly reflect gender in the production process, for example, by male designers who use themselves and their experiences as reference categories in the development process (Van Oost, 2003: 195).

Gender scripts embedded in technological systems often dictate how men and women should behave in specific technological situations, influencing how they use and experience technological artifacts and systems. This is most obvious when we look at the different types of technology that are available to men and women. However, how the gender of future users is anticipated and defined that influences the material design of the resulting artifacts and how gender intervenes in the design phase does not (only) mean identifying the gender of users and designers, i.e., describing the large numerical difference and hierarchical segregation between men and women in the productive and market structure of technologies, but also goes through how the different levels of the gender system operate in the development of the content and final outcome of technologies.

An example of a gender script in a technological system is the expectation that women should be less technologically competent than men. This means that women are often expected to be less knowledgeable about technology and to use it less than men. Birth control pills are another example of technological artifacts that have significant gendered meanings. According to gender scripts, women are the ones who should take the pills to control their fertility, since they are responsible for reproduction. Men, on the other hand, are not involved in this process and do not have to take birth control pills. This gendered meaning of birth control pills has influenced the way people use them. Women have been the ones who have been responsible for taking the pills, even though men can also benefit from their use.

Another example of gender scripting is the understanding that, in the family unit, women should as far as possible be codrivers of family vehicles, while men should as far as possible be the ones driving. Even when car use is widespread and being a driver is not influenced by gender conditions, cars marketed to men are presented as powerful, while those marketed to women are presented as reliable (Hubak, 1996). According to gender scripts, automobiles are a means of transportation for men and have not historically been intended for women. This is reflected in the way cars are designed and advertised. Automobiles are designed to be comfortable for men, and women are generally not represented in advertising campaigns. This leads to the fact that women may feel uncomfortable driving a car and do not feel included in the automotive culture. In short, gendered technology scripts can be understood as the assumptions about the context of use that are embodied in technology, which prestructure the use of technology and attribute and delegate specific competencies, actions, and responsibilities to its intended users (Rommes, 2002).

The design of technologies that incorporate gendered scripts results in gender inequality in their use (Sanz, 2016; Van Oost, 2003). When scripts reveal gendered patterns, they can emphasize, conceal, reinforce, or diminish gender differences and inequalities (Rommes, 2002). Thus, technological systems and artifacts do not simply exist in a vacuum in isolation, but are shaped and used in particular ways that reflect and reinforce gender relations. Thus, despite what some people may believe, technology is not gender neutral. This is especially true when we talk about AI-based technological systems. The biases, stereotypes, and ideologies behind gender scripts are not contained in the design activity of material artifacts and the introduction of biases in the programming of algorithm functions. Instead, we can overtly see gendered traits embedded in the datasets that algorithms have access to, which consequently influences how algorithmic neural networks are formed, which can lead to biased and discriminatory results (see, e.g., Kim et al., 2019; Mahmoud, 2019).

Awareness of gender bias in software has increased in recent years, and some researchers have studied how software tends to reproduce gender inequalities in the real world (see, e.g., Martínez et al., 2020; Nuseir et al., 2021; Panch et al., 2019). For example, software can be designed with an approach that reinforces sexist assumptions about the world. Gender bias in data and algorithms is embedded in a specific software design and production. This design responds to a plan of use conceived by the agents involved in the design context. Thus, while gender inequalities in software may be present in design, production, and use, the context of use is particularly important, as it determines how the software will be reinforced in real life. For example, the context of use can determine who has access to the software and what information is used.

Thus, scripts and gender biases are not solely reproduced by human beings. It is true that, on the one hand, there are algorithms that work in isolation from their environment and simply produce a prefabricated script; but, on the other hand, there are algorithms capable of processing exogenous data, i.e., the algorithms are able to "learn," to execute better and better solutions to the problems programmed into them on the basis of past results and stored data. Thus, rapidly improving computational methods based on machine and deep learning now also offer the possibility of generating knowledge that is further emancipated from human control and influence (Hancock, 2022). If it is no longer only programmers who inscribe action scripts into algorithms, but algorithms are able to develop independent action and knowledge structures, autonomous artificial agents are a new type of nonhuman entity that learns and reproduces genre scripts and that has to be taken into account in the debate on morally right action.

We then explore how the stereotypes and ideologies behind gender scripts and traits may extend to AI systems and have pernicious consequences insofar as they may come to limit the type of knowledge and range of inferences these systems can have with respect to the data collected about women and men.

Machine learning and gender

AI systems, algorithms, and machine learning enable many of today's technologies to mimic human cognitive intelligence (Colonna, 2013). Machine learning is a subset of AI methods that train algorithms, meaning that the algorithm can assess the shortcomings of its decision-making process in early iterations and improve its analysis and predictions in later iterations. Machine learning algorithms make recommendations for audiovisual content on platforms such as YouTube or Twitch and suggest products to buy and which restaurants to go to. They are also increasingly used in high-stakes scenarios such as lending (Mukerjee et al., 2002) and hiring decisions (Cohen et al., 2019). However, like people, algorithms, in the form of machine learning using (large) amounts of data, are vulnerable to biases and run the risk of creating or perpetuating biases, making their decisions "unfair" (Angwin et al., 2016; Caliskan et al., 2017; O'Neil, 2016). This can happen through the algorithm and its training, but also through the selection and processing of joint data.

Now then, in order to discuss statements about a connection between data and possible disadvantages, a distinction must be made between two types of datasets that shape how datasets can come to discriminate: On the one hand, datasets may contain direct statements about people's characteristics such as age, gender, religion, etc., and seemingly neutral properties such as income or place of birth; on the other hand, the training data on which an AI algorithm is based (Martini, 2019). In this regard, it has been found that AI algorithms may be biased against women in several ways. For example, they may be more likely to recommend jobs that are traditionally considered male-dominated to men than to women, downgrade resumes with female-sounding names, or deny a woman access to certain services (Altman et al., 2018; Bolukbasi et al., 2016; Canetti et al., 2019;

Dwivedi et al., 2021; Kumar et al., 2019; Lambrecht & Tucker, 2019; Leavy, 2018; Lütz, 2022; Mehrabi et al., 2021; Nadeem et al., 2020). Other behaviors raise issues in terms of gender bias and stereotyping. In the latter regard, for example, AI-based systems have been shown to be biased in the way they relate words or concepts, with algorithms being more likely to recommend words with more strongly stereotyped relationships depending on which set of words are entered into the search (e.g., "man is to work as woman is to mother") (Martínez et al., 2020).

According to Ntoutsi et al. (2020), biases that exist in humans enter our systems and can be amplified due to complex sociotechnical systems (Ntoutsi et al., 2020). One of the ways these biases are introduced into systems is through the data user. Algorithms analyze data, assign relevance to information, and structure information and communication processes. They determine whether information is displayed on the Internet and to whom (Beam, 2014; Bucher, 2012). Thus, they contribute to the construction of realities (Just & Latzer, 2017), shape our tastes and our culture, and are considered a source of social order (Beer, 2016). Some data sources used to train machine learning models are user-generated, i.e., inherent biases in users can be reflected in the data they generate. But, in some application domains, algorithms make biased decisions themselves and determine the allocation of resources, for example, in high-frequency trading on exchanges (Algo-Trading) (Khandelwal, 2022) and in risk assessments that are used as a basis for relevant decisions, such as lending (Credit Scoring) (Hassani, 2021).

Furthermore, when user behavior is affected by an algorithm, any bias present in that algorithm could be amplified and introduce a gender script into the data generation process. But even if the input data for bias cases could be so sparsely selected that under no circumstances is a correlation with the characteristics of the people for whom the bias is effected possible, there is still no guarantee of a nondiscriminatory decision. The reason is that algorithms are shown and learn reality by feeding on historical data. Consequently, the datasets are indirectly discriminatory, and existing inequalities can thus be perpetuated and sometimes even reinforced (Krüger & Lischka, 2018).

Discrimination is often closely related to the lack of data, i.e., from the outset only a small area of real life is shown, which is usually also heavily distorted (Favaretto et al., 2019; Williams et al., 2018). Thus, it is important to mention that, in the engineering context, the lack of gender diversity and the exceptionally homogeneous and male dominance in high-tech industries and in AI design and implementation create "blind spots" that drive gender bias in AI-based systems (Clifton et al., 2020; Fernández, 2020; Fernández-Martínez & Johnson, 2019; Lee, 2018; Wang, 2020). Blind spots are situations where technology or design does not take gender diversity into account and thus reproduce and enhance existing inequalities. The lack of gender diversity in AI development can lead to decisions being made that are not suitable for women and minorities, as they do not reflect their needs and preferences. For example, designing an AI that uses primarily men as its data source can lead to those data being processed as surrogates for the male gender and, consequently, to women being excluded from important areas of society, such as the labor market (Zhou et al., 2022).

Biased data comes in different modalities (images, texts, etc.). For example, semantic gender distortion embedded in the Spanish Wikipedia has been reported (Martínez et al., 2020). Making use of the word embedding technique, *Word2vec*, they realized that Wikipedia showed semantic-structural gender biases and stereotypes; for example, the algorithm associated pairs such as "Hombre es a Experto como Mujer es a Sabelotodo" (Man is to Expert as Woman is to Know-It-All) (Martínez et al., 2020; Zhao et al. (2017)), meanwhile, studied semantic role labeling models on a dataset, *imSitu*, showing that only 33% of the agent roles in cooking images are male and the remaining 67% of cooking images have female agents. They also noted that, in addition to the existing bias in the dataset, the model would amplify the bias such that after training the algorithm, the bias would be magnified for "male," counting only 16% of the cooking images (Zhao et al., 2017). Although less visible than the effects of direct discrimination, these biased results produced by the use of algorithms may cause greater unanticipated harm, due to the widespread use of algorithms and the ease with which neural networks reprogram algorithms and access databases, aiding in the propagation of biases and stereotypes.

Gender scripts, which are implemented in artifacts by both men and women through use, as well as by machine and deep learning of AI systems, can show not only how user gender representations are an inextricable part of artifact design (Oost, 2003: 194), but also the gender representations of the AI systems themselves. Gender features, through explicit or implicit representations of AI systems, can reinforce the idea of gender roles and signal the role of women as domestic objects (e.g., inferring that women are mothers, daughters, wives, sisters, etc.), thus perpetuating historical roles, femininity, and thus androcentrism. Furthermore, if algorithms base their inferences on techniques such as Word2vec, they may not only perpetuate biases and stereotypes but also work their way into the underlying datasets on which the algorithms learn and base their decisions (Lütz, 2022).

Thus, there is not only a connection with the quantity of training data but also with its quality. These and other findings show that bias and discrimination by gender characteristics in relation to men and women can occur in contexts where the data that models select and process is critical for decision-making. Depending on the application of a machine learning algorithm, these gender associations can still be learned and can perpetuate the same stereotypes.

Under these observations, it is shown that predictive algorithm models have the risk of exploiting social bias linked to gender. This could have a devastating impact on society as a whole, as well as on the individuals involved, to the extent that it reflects, amplifies, or distorts the perception of the real world and the status quo. Therefore, the proposal and implementation of effective policy agendas is urgent. Moreover, in order to avoid distortions and discriminations in algorithmic results, all facets of social diversity must be taken into account (Elmer, 2022). Therefore, the implementation by the engineering community of different techniques and attenuating biases that can mitigate model bias through prediction calibration is necessary.

Gender discrimination, RRI, and European regulation for AI

Algorithms and, in general, AI-based systems are transforming the way decisions are made. Their adoption has been accompanied by allegations of discrimination by consumers and the media. In the case of gender bias, this is a set of errors whereby an algorithm does not treat men and women equally the way individuals are entitled to be treated (Hellman, 2016). Despite the diversity of laws and policies aiming at equality between men and women,[1] all European Union (EU) Member States have adopted laws against gender discrimination to increase equality between men and

---

[1]For example, the enunciation of the principle of gender parity assumed by the Republic of Poland has not been accompanied by effective policies to promote equality between men and women. Despite democratization on paper, Europeanization, and dissemination of gender mainstreaming, following a campaign of delegitimization of gender equality, no effective actions were carried out to consolidate the values of equality institutionally (Szczygielska, 2019).

women. Of course, gender equality is one of the hallmarks of Europeanization and is also fundamental to democracy (Liebert, 2016). However, rapid advances in AI-based systems and the impact of these systems on the world have outpaced the scope of these laws.

AI is part of a rapidly developing set of methods and systems with a large number of potential applications. As such, it is not possible to comprehensively predict all of the potential implications of AI on the legal system at this time. However, it is clear that AI has the potential to radically alter the way the legal system operates, and it is important that the legal system be prepared for this. One of the key issues that should be addressed is the question of gender discrimination implicit in the outcomes of algorithmic processes. Given the facts described in Section 1, to minimize algorithmic bias, regulation should intervene. However, starting with an overview of the commitment to a responsible technological system, and a description of the EU law against gender discrimination in AI, is essential to understand where we stand and what avenues could be considered for a perfectible legislative framework for technological innovation and development.

RRI, gender equality, and AI policies

The concept of RRI is of great policy importance in the European context. The literature addressing RRI seeks to understand how research and innovation can become more responsible and the implications that this responsibility in research and innovation has. RRI is a policy agenda launched by the EU in 2010 with the aim of introducing ethics and reflection in the field of science and technology, in the context of processes, organizations, outcomes, and impacts. This new logic of research and innovation transforms research practices in any of the disciplines, so that they become deliberative, inclusive, and anticipatory (Owen et al., 2020; Wittrock et al., 2021). Thus, apart from having research and innovation integration with ethical values and reflection, RRI also integrates the needs and concerns of society to transform research and innovation practices.

RRI is an emerging framework in science, technology, and society studies (STS), which includes a gender equality dimension (Owen et al., 2020). This dimension is understood as a three-dimensional construct whereby gender equality is achieved when (1) women and men are equally represented in all disciplines and at all hierarchical levels, (2) gender barriers are removed so that women and men can develop their potential equally, and (3) when the gender dimension is considered in all research and innovation activities (Hartlapp et al., 2021). It should be noted, however, that the EU's vision of gender equality is mostly understood as women's participation in the labor market and pays little attention to its implications in the private sphere or how women should be incorporated in masculinized environments (Lombardo, 2017). Nevertheless, it would be rather naïve to think that a European RRI concept and RRI tools could be seen as a "cure-all solution" in a global society. The RRI concept and tools need contextualization for their effective application.

For example, AI systems currently have a significant impact on employment in advanced economies, and more sophisticated versions of these types of technologies are constantly appearing and, with these versions, new problems related to gender bias and discrimination are appearing. But where are the responsible innovation and research policies in relation to AI systems and gender discrimination? EU-driven policies that aim to reduce algorithmic gender discrimination can be grouped into policies to build data and infrastructure for AI, policies to generate appropriate AI skills and education, and policies that help establish an appropriate regulatory and governance framework (Guevara-Gómez et al., 2021).

The development of the EU approach to AI involves a number of EU policy areas, including the digital market, the internal market, and research and innovation policies. In early 2020, the Organisation for Economic Co-operation and Development stated that, worldwide, at least 50 countries (including the EU) have developed or are in the process of developing a national AI strategy (OECD, 2020), while a global review in 2019 identified 84 AI ethics guidelines (Jobin et al., 2019). Ultimately, in 2016, the EU pointed out the main concerns and issues raised by AI systems, and thus, based on a Draft Report of the Parliament's Committee on Legal Affairs, the European Parliament Resolution of February 16, 2017, was adopted with recommendations to the Commission on civil law rules on robotics (2015/2103(INL)). By way of justification for the endeavor, it states in its Introduction the following:

> Whereas, now that humanity stands on the threshold of an era in which robots, bots, androids and other increasingly sophisticated forms of artificial intelligence seem poised to unleash a new industrial revolution — likely to affect all strata of society — it is vitally important for the legislator to weigh the legal and ethical consequences, without thereby hindering innovation (pt. B).

But it was not until the drafting of *White Paper on Artificial Intelligence: a European approach towards excellence and trust*, published following an open consultation in 2020 (European Commission, 2020), that a path toward regulation began to be mapped out, presenting regulation as a key element in building trust.

With regard to gender equality, the EU explicitly calls for high-quality data, measures, and obligations to use datasets that are sufficiently representative, especially to ensure that all relevant dimensions of gender, ethnicity, and other possible prohibited grounds of discrimination are adequately reflected in those datasets (European Commission, 2020). In addition, they point to the need for policies that help establish an appropriate regulatory and governance framework that should enable public enforcement authorities, as well as independent external auditors, to identify potentially illegal outcomes or harmful consequences generated by AI systems, such as unfair bias or discrimination (HLEG, 2019).

Thus, it is intended to take care of the correct development of new ways of working, from the development of intelligent systems for the management of the new work environment to the development of automatic processes for the generation of new products. In fact, the white paper's focus on investment and regulatory approach suggests a strong presence of Market Power Europe elements, stating in parallel that this approach is based on fundamental rights, values, and ethics. However, given that market incentives are possibly insufficient to minimize algorithmic gender bias and given that political agendas related to ethics and AI have focused more on ethical frameworks than on possibilities for better regulation (Vesnic-Alujevic et al., 2020), there is a clear lack of regulatory content in a context where regulation should have a stronger presence in order to actively mitigate gender discrimination.

In the following, the two fundamental tools that could actively protect the right to nondiscrimination related to AI systems in the EU, the nondiscrimination law and the data protection law, are discussed.

Principle of nondiscrimination in law

All (fundamental) treaties, such as the Treaty on the Functioning of the European Union (TFEU) or the Charter of Fundamental Rights of the European Union (CFR), form the upper or primary layer of the EU legal order. Secondary law (regulations, directives, recommendations, and opinions) constitutes the second layer, which is subordinate to primary law (Schütze, 2021). The general principle of nondiscrimination in EU law, then, is expressed in a set of primary and secondary legal provisions that protect individuals from discrimination on the grounds of sex, race or ethnic origin, disability, religion or belief, age, and sexual orientation. This principle is found in nondiscrimination law and is one of the main legal instruments that could protect the right to nondiscrimination in the context of algorithmic decision-making.

Article 21 of the Charter of Fundamental Rights of the European Union (nondiscrimination), which states that "any discrimination based on any ground such as sex, race, color, ethnic or social origin, genetic features, language, religion or belief, political or any other opinion, membership of a national minority, property, birth, disability, age or sexual orientation shall be prohibited," is seen as a general and fundamental principle of the EU.[2] The case law of the Court of Justice of the European Union also affirms that substantive equality is the intended aim of antidiscrimination law and that differences between groups must be recognized in order to achieve substantive equality in practice (De Vos, 2020). However, the Charter only applies to EU public bodies and Member States and not to the private sector (Wachter et al., 2021). Moreover, as with all EU directives, the gender nondiscrimination directive only sets a minimum standard and provides a general framework to be adopted by the Member States.

European antidiscrimination legislation addresses both direct and indirect discrimination, and case law shows that the European Convention on Human Rights captures both types of discrimination (Ellis & Watson, 2012). Briefly, according to Directive 2006/54/EC (on gender equality), direct gender discrimination shall be deemed to occur when a person is treated less favorably than another person is, has been, or would be treated in a comparable situation on grounds of gender, i.e., this type of discrimination refers to adverse treatment on the basis of a protected ground. Thus, direct discrimination is described by the European Court of Human Rights (ECtHR) as the existence of a difference in the treatment of persons in analogous or significantly similar situations, which is based on an identifiable status characteristic (gender, ethnicity, religion or belief, age, sexual orientation, and disability) (Nuñez, 2016). This type of discrimination can never be justified under EU directives.

Indirect gender discrimination, on the other hand, describes a situation in which an apparently neutral practice disadvantages people of a certain gender, a protected group compared to other persons (McCrudden, 2016). Thus, the prohibition of indirect discrimination aims to ensure that no (seemingly neutral) criterion can be used instead of (at least one of) the direct status characteristics and thus disadvantage certain persons or groups. In the case of sex discrimination, the prohibition does not only take into account biological sex but also gender. Unequal treatment on the basis of pregnancy is just as discriminatory as unequal treatment on the basis of medically assisted procreation. The Court of Justice of the European Union (ECJ) has confirmed the latter. Sexual harassment or harassment based on sex is also a form of sex discrimination. For the prohibition of indirect discrimination, it is not relevant whether a discriminating agent intended to discriminate; instead, it is the effect of a practice that is taken into account (Möschel, 2017). Therefore, the discriminating agent cannot circumvent the prohibition of indirect discrimination by proving that he or she did not intend to discriminate.

---

[2]See EUROPEAN UNION, Charter of Fundamental Rights of the European Union, C364/1 (2000).

However, this form of discrimination leaves open a much wider scope and is also much more difficult to identify than cases of direct discrimination, because indirect discrimination may remain hidden from both institutions and victims, and it is sometimes difficult to capture some of its subtleties or pervasive nature. Gender differences and biases are complex and go beyond the mere presence of women. Suppose a female student applies to take an exam to gain access to a place offered by the University of Cambridge. Unfortunately, due to health reasons, the face-to-face exam is canceled. Instead, the university uses an algorithmic system, a tool that automates the selection of candidates, to deduce the grades that students would have obtained in the exam, based on their academic history and circumstantial criteria such as efficiency in daily schoolwork or place of residence. If a student is automatically denied a place by the university, she does not see why she was denied access. Because, for the AI system, the specific value lies in the confluence of inferences, whereas a single one of the circumstances would likely have produced a different outcome (Prince & Schwarcz, 2019). Thus, inferring from zip code or place of residence or name or other variable membership in protected categories involves combining different factors in the analysis, which, separately, might be innocuous, but which, acting together, converge on a result that affects people of a certain national or ethnic origin, religion, or sex.

Although the law allows covering all effects arising from algorithmic biases, identifying the discriminatory action becomes particularly difficult, given that it must consist of finding the correlation between the data, which can be a matter of "black box modeling," where the reasons behind the outcome are unknown from the introduction of big data (Mayson, 2019; Xenidis & Senden, 2019). This means that, on a large number of occasions, AI multiplies bias, but does not contribute to identifying it by relying on data correlations. Therefore, even if it were known that an algorithm made a decision in place of academic staff, it would be difficult to discover whether the algorithm is discriminatory and whether it yields discriminatory results, because access to relevant information is limited and discrimination is hidden in the big data.

Another problem with nondiscrimination law is that the prohibition of indirect discrimination does not provide a clear rule. It is true that indirect discrimination can reveal underlying social inequalities through patterns of behavior, but it is a relatively new concept, so we are faced with a situation where there is a relative lack of case law addressing indirect discrimination (Makkonen, 2007). While, according to the European Court of Human Rights, a general policy or measure that has disproportionate detrimental effects on a particular group may be considered discriminatory even when it is not specifically targeted at that group and there is no discriminatory intent, this is only the case if such a policy or measure has no objective and reasonable justification. In practice, therefore, since the prohibition of indirect discrimination does not apply if an objective and reasonable justification is successfully defended, the concept of indirect discrimination itself is often difficult to apply.

The other main legal instrument that, alongside nondiscrimination law, could protect the right to nondiscrimination and address algorithmic bias issues is data protection law. The General Data Protection Regulation (GDPR) and EU Convention 108 contain specific rules for certain types of "automated individual decision-making" (see Article 22 GDPR; Article 9(1)(a) COE Data Protection Convention 2018). These rules have several objectives, including the protection of individuals against unfair or unlawful discrimination. In principle, Article 22 prohibits certain fully automated decisions with significant or similar legal effects:

> The data subject shall have the right not to be subject to a decision based solely on automated processing, including profiling that produces legal effects on him or her or significantly affects him or her in a similar way (Art. 22 GDPR).

Thus, in some circumstances, individuals may require an organization to make an employee reconsider the automated decision. For example, the university could ensure that students can call the registrar's office to have a human reconsider the decision, if the university has denied them access to a place because an algorithm has deduced (and decided) that they would not have a sufficient grade to gain access to the place had they taken the scheduled exam. Thus, following the requirements of the GDPR, the university must explain that it uses automated decision-making and must provide meaningful information about the underlying logic of that decision-making process (see Article 13(2)(f) and 14(2)(f) GDPR). However, while researchers are beginning to explore how data protection law can help combat discrimination (see, e.g., Gellert et al., 2013; Hacker, 2018; Schreurs et al., 2008), some authors (Edwards & Veale, 2017; Zuiderveen, 2020) have noted that it is too early to assess the effects of Convention 108 and GDPR. For example, Zuiderveen (2020) has pointed out that data protection law only applies to personal data, so algorithmic decision-making processes, since they do not refer to identifiable individuals, are partly outside the scope of data protection law (Zuiderveen, 2020).

Technical measures are also being taken against algorithmic biases such as performing technical assessments on the representativeness of datasets (Bellamy et al., 2019; Bhardwaj et al., 2021; Confalonieri et al., 2022; Englert & Muschiol, 2022; Fountain, 2022; Nascimento et al., 2022; Zhang & Ananiadou, 2022).[3] For example, algorithmic adjustments are made to compensate for problematic bias (Danks & London, 2017), HITL models that require human intervention are tested (Berendt & Preibusch, 2017), and open algorithms are designed. Regarding the concern that AI-based technology may challenge the transparency of decisions and thus the question about the ability to explain its autonomous decisions and actions to human users, it is also worth noting that work has recently begun on the so-called explainable artificial intelligence (XAI).[4] XAI has arisen from the need to try to map the structure of AI-based systems and understand how the algorithmic districts employed operate, given that with the use of algorithmic systems one can come to understand what they do, but not how they do it (Ridley, 2022). Thus, work on XAIs strives to create a

---

[3]In line with the attempt to build socially responsible AI-based systems, it is worth noting that the private sector has conceptual tools that point in the direction of the possibility of aligning moral/legal needs with the interests that the companies behind these AI technologies might have. When we refer to the social responsibility of an organization, we find the distinction between internal responsibility (which the organization itself feels toward its constituents, members of its community, and external stakeholders) and the accountability for which it is held accountable to someone who has the authority to demand compliance. Thus, a private company would not only have the limits set by legislation, but the company can also limit its own actions by looking after the interests of others, which, on many occasions, are also its own. In this way, corporate social responsibility refers to the presumption of ethical commitments assumed by the company vis-à-vis an internal and external environment of which it feels part and to which it owes a duty to fulfill certain responsibilities. And, by extension, it can also refer to technical rationality in terms of the need for the company to contribute to the viability of the environment in order to make and maintain its business and financial performance feasible (see, for example, Barauskaite & Streimikiene, 2021; de Villiers et al., 2021; Dixon et al., 2021; Lachuer & Jabeur, 2022; Mazzi, 2023).

[4]Among the XAI techniques currently employed, we can find, for example, the Local Interpretable Model-Agnostic Explanation (LIME) algorithm. LIME is an algorithm that can be applied to any deep learning classification model, being able to provide an accurate post hoc explanation that justifies a local prediction. To carry out its task, LIME stresses the model by adjusting and varying its inputs, in order to understand how each input variable influences the different predictions of the model. Among the main applications that LIME offers, the analysis and justification of the prediction of a document classification model carried out by a convolutional neural network stand out (Chakraborty et al., 2017; Ribeiro et al., 2016).

series of techniques that produce explainable models from the currently employed black box systems, based on improving the reliability, causality, transferability, or unbiasedness of the system (Evans et al., 2022; Marques-Silva & Ignatiev, 2022; Speith, 2022).

But while the proposed technical solutions are steps toward a path of algorithms that are more scrupulous about bias and discrimination, there is an urgent need for greater regulatory, political, and ethical attention to fairness, especially in terms of protecting vulnerable and/or marginalized populations (Raji & Buolamwini, 2019). In this context, the EU Fundamental Rights Agency (FRA) describes the potential for discrimination against individuals through algorithms and states that the principle of nondiscrimination, enshrined in Article 21 of the EU Charter of Fundamental Rights, must be taken into account when applying algorithms to everyday life. A report by the European Parliament on the implications of big data for fundamental rights highlighted that

> due to the data sets and algorithmic systems used when making assessments and predictions at different stages of data processing, big data can lead not only to violations of individuals' fundamental rights, but also to differential treatment and indirect discrimination against groups of people with similar characteristics, particularly with regard to fairness and equal opportunities for access to education and employment, when selecting or evaluating individuals or when determining the new consumption habits of users of social networks (European Parliament, 2017a).

With the aim of producing substantive equality in the face of problems of bias and discrimination by algorithms, the report called on the European Commission, Member States, and data protection authorities to identify and take any possible measures to minimize algorithmic discrimination and bias and to develop a robust and common ethical framework for transparent personal data processing and automated decision-making that can guide the use of data and the continued application of Union law (European Parliament, 2017b). Thus, the need to take a proactive stance to dismantle the obstacles that stand in the way of equality was indicated.

Currently, as seen in this chapter, the law leaves loopholes where new kinds of differentiation are created and lead to biased and discriminatory effects. This is especially true in the case of algorithmic systems, where AI applications can show bias against certain groups of people, which can have detrimental consequences for those groups. While AI offers many potential benefits, its potential risk must be taken into account and legal measures must be taken to mitigate its adverse effects. Data is the worst obstacle to providing indications of discriminatory actions, with traditional mechanisms still not being able to be employed. In addition, machine learning allows greater justifications to disprove the evidence, since it operates in the field of indirect discrimination, which is, therefore, disarticulable by objective and reasonable justification. On the other hand, the victim will in these cases be less aware of the bias (Ebers & Navas, 2020), which is the main barrier in many cases to the visibility of the risk and its impact.

To conclude, and with a view to future work, we suggest an approach that may help address the problem of legal liability for harm caused by AI systems. Regulation is, of course, key to ensuring that AI systems are designed, developed, and deployed following a certain set of ethically desirable standards and principles. However, regulation often functions in a reactive manner and focuses on penalizing the existence of effects that should be avoided, i.e., regulation often acts as a sentinel looking for effects that should not exist. With the rise of autonomous systems and the development of new technologies, technological and global societies that take an active part in technological traffic are exposed to a multitude of new risks that we are accepting without having sufficient protection against eventual damages. Moreover, there are often difficulties in determining who is liable for technological damages.

Now then, legal tools can be developed with which to take action against tangible third parties who are not perpetrators, i.e., who are not at fault, but who have nevertheless in some sense indirectly contributed to the infringement. This is the case of the so-called strict liability (or liability by risk), which has its justification in the prevention of damages and defines the liable person as the one who knows and dominates in general the source or origin of the risk, although not necessarily the course of the event resulting in the damage, and benefits from the origin of the risk (Larenz, 1956). More precisely, strict liability falls on the person who creates a risk through an activity and who is in the position to assess, control, and minimize the risk to some extent (Semmelmayer, 2021). Thus, liability is imposed for the risks generally attached to others by the exploitation linked, for example, to owning an animal, using a vehicle, or deploying an algorithm, because it seems socially justified that a person can and should bear the risk of harm and not that (animal, vehicle, and algorithm) which has caused the harm.

What justifies, therefore, no-fault liability is the creation of a risk or danger, which is broadly expressed. In some cases, risk as a reason for the introduction of a strict liability regime has been a tool used by legislators and case law. Thus, for example, Swiss, German, Austrian, and, to a lesser extent, Greek legislators indicate that the specific cases of no-fault liability in special laws have been limited to activities described as very dangerous and to the enormous risk represented by certain objects or activities (Werro & Palmer, 2004). Aiming at a better application of current nondiscrimination rules in the field of algorithmic decision-making, the development of this approach could ensure that AI systems are used in an ethically desirable way and could be part of a process where regulation can closely monitor the reaction to new technological developments and thus where undesirable effects occur equally close to their regulation.

Conclusions

The importance of analyzing the entanglement between technologies and gendered social structures (in light of the entanglement of technology with society through its design and use) also becomes very relevant when considering AI-based systems and the problem of inequality. There is certainly reason to believe that the EU and Member States are taking seriously the risks arising from the unethical design and use of AI systems. However, having seen how gendered scripts are embedded in general technological systems and, specifically, within AI systems, we find that RRI and other policy agendas that set out to formulate a regulatory framework for correcting these systems often dwell on general ethical assumptions and "forget" to attend to legislative output. Even when it comes to settled laws, such as nondiscrimination law, these laws do not meet the need for regulation of new situations of algorithmic discrimination.

AI systems are driving the fourth industrial revolution. Their impact is and will continue to be more profound than the preceding major changes in computing and telecommunications, as was access to the Internet by the general population. But legislation is again late in implementing legal laws, despite the historical precedents of other technologies. For example, a parallel observation to this tardiness can be drawn with respect to the concern in the State of Victoria for motor vehicles.

For this reason, the development of new laws that take into account current technological and social realities is urgently needed. In this chapter, it has been suggested to take into account strict liability as a means for the implementation of a determinant regulation in the development and production of AI systems. Strict liability can help us understand how a third party, someone linked to an activity that can indirectly cause socio-technical risks of various kinds, is partially or fully responsible for the resulting state of affairs.

In closing, some measures that are important to remember to promote gender diversity in AI development can (and should) include equal representation of women and men in datasets used in AI; ensuring that AI algorithms are not gender-biased; developing tools to identify and correct gender bias in AI; and increasing awareness of gender bias in AI and legal measures to reduce it. Overall, it is clear that AI will have a significant impact on the legal system and that a number of challenges will need to be addressed. It is important that the legal system be prepared for the advent of AI and that steps are taken to ensure that it is able to deal with the challenges it will pose. And although ironically the project of legislating a technology makes law into technology, this project is absolutely necessary to control the technological domain that, today more than ever, is controlled by Dr. Frankenstein's monster.

Acknowledgments

I am grateful to Mar Cebrián and Santiago López for their comments on an earlier version of the manuscript, as well as to the anonymous reviewers for their contributions.

References

Akrich, M. (1992). The De-Scritption of Technical Objects. *Shaping Technology/Building Society: Studies in Sociotechnical Change (pp. 205–224). Cambridge: The MIT Press.*

Akter, S., McCarthy, G., Sajib, S., Michael, K., Dwivedi, Y. K., D'Ambra, J., & Shen, K. N. (2021). Algorithmic bias in data-driven innovation in the age of AI. *International Journal of Information Management, 60*102387.

Altman, M., Wood, A., & Vayena, E. (2018). A harm-reduction framework for algorithmic fairness. *IEEE Security & Privacy, 16*(3), 34–45.

Angwin, J., Larson, J., Mattu, S., & Kirchner, L. (2016). Machine bias: There's software used across the country to predict future criminals. *And it's biased against blacks. ProPublica, 23*, 77–91.

Barauskaite, G., & Streimikiene, D. (2021). Corporate social responsibility and financial performance of companies: The puzzle of concepts, definitions and assessment methods. *Corporate Social Responsibility and Environmental Management, 28*(1), 278–287.

Basta, C., Costa-Jussà, M. R., & Casas, N. (2021). Extensive study on the underlying gender bias in contextualized word embeddings. *Neural Computing and Applications, 33*(8), 3371–3384.

Beam, M. A. (2014). Automating the news: How personalized news recommender system design choices impact news reception. *Communication Research, 41*(8), 1019–1041.

Beer, D. (2016). *Metric power*. London: Palgrave Macmillan.

Bellamy, R. K., Dey, K., Hind, M., Hoffman, S. C., Houde, S., Kannan, K., ... Zhang, Y. (2019). AI Fairness 360: An extensible toolkit for detecting and mitigating algorithmic bias. *IBM Journal of Research and Development, 63*(4/5), 1−4.

Berendt, B., & Preibusch, S. (2017). Toward accountable discrimination-aware data mining: the Importance of keeping the human in the loop—And under the looking glass. *Big Data, 5*(2), 135−152.

Bhardwaj, R., Majumder, N., & Poria, S. (2021). Investigating gender bias in BERT. *Cognitive Computation, 13*(4), 1008−1018.

Bolukbasi, T., Chang, K. W., Zou, J. Y., Saligrama, V., & Kalai, A. T. (2016). Man is to computer programmer as woman is to homemaker? Debiasing word embeddings. *Advances in Neural Information Processing Systems, 29.*

Borgo, S., Franssen, M., Garbacz, P., Kitamura, Y., Mizoguchi, R., & Vermaas, P. E. (2014). Technical artifacts: An integrated perspective. *Applied Ontology, 9*(3−4), 217−235.

Bucher, T. (2012). Want to be on the top? Algorithmic power and the threat of invisibility on Facebook. *New Media & Society, 14*(7), 1164−1180.

Caliskan, A., Bryson, J. J., & Narayanan, A. (2017). Semantics derived automatically from language corpora contain human-like biases. *Science, 356*(6334), 183−186.

Canetti, R., Cohen, A., Dikkala, N., Ramnarayan, G., Scheffler, S., & Smith, A. (2019, January). From soft classifiers to hard decisions: How fair can we be? In *Proceedings of the conference on fairness, accountability, and transparency* (pp. 309−318).

Chakraborty, S., Tomsett, R., Raghavendra, R., Harborne, D., Alzantot, M., Cerutti, F., . . . Gurram, P. (2017). Interpretability of deep learning models: A survey of results. In *2017 IEEE smartworld, ubiquitous intelligence & computing, advanced & trusted computed, scalable computing & communications, cloud & big data computing, Internet of people and smart city innovation (smartworld/SCALCOM/UIC/ATC/CBDcom/IOP/SCI)* (pp. 1−6). IEEE.

Challen, R., Denny, J., Pitt, M., Gompels, L., Edwards, T., & Tsaneva-Atanasova, K. (2019). Artificial intelligence, bias and clinical safety. *BMJ Quality & Safety, 28*(3), 231−237.

Chaturvedi, A., Tiwari, A., Chaturvedi, S., & Liò, P. (2022). System neural network: Evolution and change based structure learning. *IEEE Transactions on Artificial Intelligence, 3*(3), 426−435.

Cirillo, D., Catuara-Solarz, S., Morey, C., Guney, E., Subirats, L., Mellino, S., ... Mavridis, N. (2020). Sex and gender differences and biases in artificial intelligence for biomedicine and healthcare. *NPJ Digital Medicine, 3*(1), 1−11.

Clifton, J., Glasmeier, A., & Gray, M. (2020). When machines think for us: The consequences for work and place. *Cambridge Journal of Regions, Economy and Society, 13*(1), 3−23.

Crilly, N. (2010). The roles that artefacts play: Technical, social and aesthetic functions. *Design Studies, 31*(4), 311−344.

Coeckelbergh, M. (2020). Artificial intelligence, responsibility attribution, and a relational justification of explainability. *Science and Engineering Ethics, 26*(4), 2051−2068.

Cohen, L., Lipton, Z.C., & Mansour, Y. (2019). Efficient candidate screening under multiple tests and implications for fairness. *arXiv preprint arXiv:1905.11361.*

Cohen, S. (2021). *The evolution of machine learning: Past, present, and future. Artificial intelligence and deep learning in pathology* (pp. 1−12). Amsterdam: Elsevier.

Colonna, L. (2013). A taxonomy and classification of data mining. *SMU Science and Technology Law Review, 16*(2), 309−369.

Confalonieri, R., Lucchesi, F., Maffei, G., & Catuara-Solarz, S. (2022). *A unified framework for managing sex and gender bias in AI models for healthcare. Sex and gender bias in technology and artificial intelligence* (pp. 179−204). Academic Press.

Daneshjou, R., Smith, M. P., Sun, M. D., Rotemberg, V., & Zou, J. (2021). Lack of transparency and potential bias in artificial intelligence data sets and algorithms: A scoping review. *JAMA Dermatology, 157*(11), 1362−1369.

Danks, D., & London, A. J. (2017). Algorithmic bias in autonomous systems. *IJCAI, 17*, 4691–4697.

DeCamp, M., & Lindvall, C. (2020). Latent bias and the implementation of artificial intelligence in medicine. *Journal of the American Medical Informatics Association, 27*(12), 2020–2023.

De Vos, M. (2020). The European Court of Justice and the march towards substantive equality in European Union anti-discrimination law. *International Journal of Discrimination and the Law, 20*(1), 62–87.

De Vries, M. J. (2016). *Teaching about technology: An introduction to the philosophy of technology for non-philosophers*. Springer.

Dignum, V. (2019). *Responsible artificial intelligence: How to develop and use AI in a responsible way*. Springer Nature.

Dignum, V. (2020). AI is multidisciplinary. *AI Matters, 5*(4), 18–21.

Dixon, E., Thee, L., & Rogers, B. (2021). *AI and corporate social responsibility. Demystifying AI for the enterprise* (pp. 289–324). Productivity Press.

Duan, Y., Edwards, J. S., & Dwivedi, Y. K. (2019). Artificial intelligence for decision making in the era of Big Data—evolution, challenges and research agenda. *International Journal of Information Management, 48*, 63–71.

Dwivedi, Y. K., Hughes, L., Ismagilova, E., Aarts, G., Coombs, C., Crick, T., ... Williams, M. D. (2021). Artificial Intelligence (AI): Multidisciplinary perspectives on emerging challenges, opportunities, and agenda for research, practice and policy. *International Journal of Information Management, 57*101994.

Ebers, M., & Navas, S. (2020). *Algorithms and law*. Cambridge University Press.

Edwards, L., & Veale, M. (2017). Slave to the algorithm: Why a right to an explanation is probably not the remedy you are looking for. *Duke Law & Technology Review, 16*, 18.

Ellis, E., & Watson, P. (2012). *EU anti-discrimination law*. Oxford: OUP.

Elmer, C. (2022). *Algorithmen im Fokus. Statistische Datenanalyse im Journalismus* (pp. 56–64). Berlin: Springer.

Englert, R., & Muschiol, J. (2022). Training data improvement by automatic generation of semantic networks for bias mitigation. *American Journal of Information Science and Technology, 6*(1), 1–7.

European Commission. (2020). White Paper on Artificial Intelligence—A European approach for excellence and trust.

European Parliament. (2017a). European Parliament Resolution of 14 March 2017 on the implications of big data for fundamental rights: Privacy, data protection, non-discrimination, security and law enforcement (2016/2225(INI)).

European Parliament. (2017b). Report with recommendations addressed to the Commission on civil law rules on Robotics (2015/2103(INL)).

Evans, T., Retzlaff, C. O., Geißler, C., Kargl, M., Plass, M., Müller, H., ... Holzinger, A. (2022). The explainability paradox: Challenges for xAI in digital pathology. *Future Generation Computer Systems, 133*, 281–296.

Fabris, A., Purpura, A., Silvello, G., & Susto, G. A. (2020). Gender stereotype reinforcement: Measuring the gender bias conveyed by ranking algorithms. *Information Processing & Management, 57*(6)102377.

Favaretto, M., De Clercq, E., & Elger, B. S. (2019). Big Data and discrimination: Perils, promises and solutions. A systematic review. *Journal of Big Data, 6*(1), 1–27.

Fenwick, M., Vermeulen, E. P., & Corrales, M. (2018). *Business and regulatory responses to artificial intelligence: Dynamic regulation, innovation ecosystems and the strategic management of disruptive technology. Robotics, AI and the future of law* (pp. 81–103). Singapore: Springer.

Fernández-Martínez, C., & Fernández, A. (2020). AI and recruiting software: Ethical and legal implications. *Paladyn, Journal of Behavioral Robotics, 11*(1), 199–216.

Ferrer, X., van Nuenen, T., Such, J. M., Coté, M., & Criado, N. (2021). Bias and discrimination in AI: A cross-disciplinary perspective. *IEEE Technology and Society Magazine, 40*(2), 72–80.

Fosch-Villaronga, E., Poulsen, A., Søraa, R. A., & Custers, B. H. M. (2021). A little bird told me your gender: Gender inferences in social media. *Information Processing & Management*, *58*(3)102541.

Fountain, J. E. (2022). The moon, the ghetto and artificial intelligence: Reducing systemic racism in computational algorithms. *Government Information Quarterly*, *39*(2)101645.

Gellert, R., Vries, K. D., Hert, P. D., & Gutwirth, S. (2013). *A comparative analysis of anti-discrimination and data protection legislations. Discrimination and privacy in the information society* (pp. 61−89). Springer.

Guevara-Gómez, A., de Zárate-Alcarazo, L. O., & Criado, J. I. (2021). Feminist perspectives to artificial intelligence: Comparing the policy frames of the European Union and Spain. *Information Polity*, *26*(2), 173−192.

Gunkel, D. J. (2020). Mind the gap: Responsible robotics and the problem of responsibility. *Ethics and Information Technology*, *22*(4), 307−320.

Hacker, P. (2018). Teaching fairness to artificial intelligence: Existing and novel strategies against algorithmic discrimination under EU law. *Common Market Law Review*, *55*(4).

Haddad, C., & Hornuf, L. (2019). The emergence of the global fintech market: Economic and technological determinants. *Small Business Economics*, *53*(1), 81−105.

Hagendorff, T., Bossert, L., Fai, T.Y., & Singer, P. (2022). Speciesist bias in AI−How AI applications perpetuate discrimination and unfair outcomes against animals. *arXiv preprint arXiv:2202.10848*.

Hancock, P. A. (2022). Avoiding adverse autonomous agent actions. *Human Computer Interaction*, *37*(3), 211−236.

Hartlapp, M., Müller, H., & Tömmel, I. (2021). *Gender equality and the European commission. The Routledge Handbook of Gender and EU Politics*. Routledge.

Hassani, B. K. (2021). Societal bias reinforcement through machine learning: A credit scoring perspective. *AI and Ethics*, *1*(3), 239−247.

Hellman, D. (2016). Two concepts of discrimination. *Virgina Law Review*, *102*(4), 895−952.

HLEG. (2019). *The assessment list for trustworthy artificial intelligence (ALTAI)*. https://ec.europa.eu/digital-singlemarket/en/news/assessment-list-trustworthy-artificial-intelligence-altai-self-assessment.

Hendl, T., Jansky, B., & Wild, V. (2019). From design to data handling. Why mHealth needs a feminist perspective. *Feminist philosophy of technology*, 77−103.

Hooker, S. (2021). Moving beyond "algorithmic bias is a data problem.". *Patterns*, *2*(4)100241.

Houkes, W., & Vermaas, P. E. (2004). Actions versus functions: A plea for an alternative metaphysics of artifacts. *The Monist*, *87*(1), 52−71.

Houkes, W., & Vermaas, P. E. (2010). *Technical functions: On the use and design of artefacts* (Vol. 1). Springer Science & Business Media.

Houkes, W., & Vermaas, P. E. (2014). On what is made: instruments, products and natural kinds of artefacts. *Artefact kinds: Ontology and the human-made world* (pp. 167−190). Cham: Springer.

Hubak, M. (1996). The car as a cultural statement: Car advertising as gendered socio-technical scripts. *Making Technology Our Own*, 171−201.

Jobin, A., Ienca, M., & Vayena, E. (2019). The global landscape of AI ethics guidelines. *Nature Machine Intelligence*, *1*(9), 389−399.

Johnson, K., Pasquale, F., & Chapman, J. (2019). Artificial intelligence, machine learning, and bias in finance: Toward responsible innovation. *Fordham Law Review*, *88*, 499−529.

Johnson, K. N. (2019). Automating the risk of bias. *The George Washington Law Review*, *87*, 1214−1271.

Just, N., & Latzer, M. (2017). Governance by algorithms: Reality construction by algorithmic selection on the Internet. *Media, Culture & Society*, *39*(2), 238−258.

Kaplan, J. (2016). Artificial intelligence: Think again. *Communications of the ACM*, *60*(1), 36−38.

Kaptelinin, V. (2022). The social production of technological autonomy. *Human Computer Interaction*, *37*(3), 256−258.

Khandelwal, A. R. (2022). Ring trading to algo trading—A paradigm shift made possible by artificial intelligence. *Impact of Artificial Intelligence on Organizational Transformation*, 21–32.

Kim, B., Kim, H., Kim, K., Kim, S., & Kim, J. (2019). Learning not to learn: Training deep neural networks with biased data. In *Proceedings of the IEEE/CVF conference on computer vision and pattern recognition* (pp. 9012–9020).

Krüger, J., & Lischka, K. (2018). *Was zu tun ist, damit Maschinen den Menschen dienen. (Un)berechenbar? Algorithmen und Automatisierung in Staat und Gesellschaft* (pp. 440–470). Berlin: FOKUS.

Kumar, G., Singh, G., Bhatanagar, V., & Jyoti, K. (2019). Scary dark side of artificial intelligence: A perilous contrivance to mankind. *Humanities & Social Sciences Reviews*, *7*(5), 1097–1103.

Lachuer, J., & Jabeur, S. B. (2022). Explainable artificial intelligence modeling for corporate social responsibility and financial performance. *Journal of Asset Management*, *23*(7), 619–630.

Lambrecht, A., & Tucker, C. (2019). Algorithmic bias? An empirical study of apparent gender-based discrimination in the display of STEM career ads. *Management Science*, *65*(7), 2966–2981.

Larenz, K. (1956). *Lehrbuch des Schuldrechts, Band II*. Munich: C.H. Beck.

Larrazabal, A. J., Nieto, N., Peterson, V., Milone, D. H., & Ferrante, E. (2020). Gender imbalance in medical imaging datasets produces biased classifiers for computer-aided diagnosis. *Proceedings of the National Academy of Sciences*, *117*(23), 12592–12594.

Lawrence, T., Zhang, L., Lim, C. P., & Phillips, E. J. (2021). Particle swarm optimization for automatically evolving convolutional neural networks for image classification. *IEEE Access*, *9*, 14369–14386.

Leavy, S. (2018, May). Gender bias in artificial intelligence: The need for diversity and gender theory in machine learning. In *Proceedings of the 1st international workshop on gender equality in software engineering* (pp. 14–16).

Lee, N. T. (2018). Detecting racial bias in algorithms and machine learning. *Journal of Information, Communication and Ethics in Society*, *16*(3), 252–260.

Liebert, U. (2016). *Gendering Europeanisation: Making equality work in theory and practice. Gendering European integration theory* (pp. 147–174). Opladen: Barbara Budrich Publishers.

Loh, J. (2019). *What is feminist philosophy of technology? A critical overview and a plea for a feminist technoscientific utopia,* . *Feminist philosophy of technology* (Vol. 2, pp. 1–34). Stuttgart: Springer, JB Metzler.

Lombardo, E. (2017). The Spanish gender regime in the EU context: Changes and struggles in times of austerity. *Gender, Work & Organization*, *24*(1), 20–33.

Lütz, F. (2022). *Gender equality and artificial intelligence in Europe. Addressing direct and indirect impacts of algorithms on gender-based discrimination. ERA Forum* (pp. 1–20). Berlin: Springer.

Mahmoud, M. A. (2019). Gender, e-banking, and customer retention. *Journal of Global Marketing*, *32*(4), 269–287.

Mahnke, M. (2015). Der Algorithmus, bei dem man mit muss? Ein Perspektivwechsel. *ComSoc Communicatio Socialis*, *48*(1), 34–45.

Makhni, S., Chin, M. H., Fahrenbach, J., & Rojas, J. C. (2022). Equity challenges for artificial intelligence algorithms in health care. *Chest*, *161*(5), 1343–1346.

Makkonen, T. (2007). *Measuring discrimination: Data collection and EU equality law*. Office for Official Publications of the European Communities.

Marques-Silva, J., & Ignatiev, A. (2022). Delivering trustworthy AI through formal XAI. *Proceedings of the AAAI Conference on Artificial Intelligence*, *36* (11), 12342–12350.

Marr, B. (2016). What is the difference between artificial intelligence and machine learning. *Forbes*, December, 6.

Martínez, C. D., García, P. D., & Sustaeta, P. N. (2020). Sesgos de género ocultos en los macrodatos y revelados mediante redes neuronales: ¿hombre es a mujer como trabajo es a madre? *REIS: Revista Española de Investigaciones Sociológicas* (172), 41–60.

Martini, M. (2019). *E. Zusammenfassung der Schlussfolgerungen. Blackbox Algorithmus—Grundfragen einer Regulierung Künstlicher Intelligenz* (pp. 157—331). Berlin: Springer.

Mayson, S. G. (2019). Bias In, Bias Out. *The Yale Law Journal, 128*(8), 2218—2300.

Mazzi, F. (2023). *Concerted actions to integrate corporate social responsibility with AI in business: Two recommendations on leadership and public policy. Responsible artificial intelligence: Challenges for sustainable management* (pp. 251—266). Cham: Springer.

McCrudden, C. (2016). The new architecture of EU equality law after CHEZ: Did the Court of Justice reconceptualise direct and indirect discrimination? *European Equality Law Review, Forthcoming, U of Michigan Public Law Research Paper* (512).

Mehrabi, N., Morstatter, F., Saxena, N., Lerman, K., & Galstyan, A. (2021). A survey on bias and fairness in machine learning. *ACM Computing Surveys (CSUR), 54*(6), 1—35.

Miikkulainen, R., Liang, J., Meyerson, E., Rawal, A., Fink, D., Francon, O., ... Hodjat, B. (2019). *Evolving deep neural networks. Artificial intelligence in the age of neural networks and brain computing* (pp. 293—312). Academic Press.

Morley, J., Floridi, L., Kinsey, L., & Elhalal, A. (2021). *From what to how: An initial review of publicly available AI ethics tools, methods and research to translate principles into practices. Ethics, governance, and policies in artificial intelligence* (pp. 153—183). Cham: Springer.

Möschel, M. (2017). The Strasbourg Court and indirect race discrimination: Going beyond the education domain. *The Modern Law Review, 80*(1), 121—132.

Mukerjee, A., Biswas, R., Deb, K., & Mathur, A. P. (2002). Multi—objective evolutionary algorithms for the risk—return trade—off in bank loan management. *International Transactions in Operational Research, 9* (5), 583—597.

Nadeem, A., Abedin, B., & Marjanovic, O. (2020). Gender Bias in AI: A review of contributing factors and mitigating strategies. In *ACIS 2020 proceedings*.

Nadeem, A., Marjanovic, O., & Abedin, B. (2021). Gender Bias in AI: Implications for managerial practices. In *Conference on e-Business, e-Services and e-Society* (pp. 259—270). Cham: Springer.

Nascimento, F. R., Cavalcanti, G. D., & Da Costa-Abreu, M. (2022). Unintended bias evaluation: An analysis of hate speech detection and gender bias mitigation on social media using ensemble learning. *Expert Systems with Applications, 201*117032.

Ntoutsi, E., Fafalios, P., Gadiraju, U., Iosifidis, V., Nejdl, W., Vidal, M. E., ... Staab, S. (2020). Bias in data-driven artificial intelligence systems—An introductorysurvey. *Wiley Interdisciplinary Reviews: Data Mining and Knowledge Discovery, 10*(3), e1356.

Nuñez, S. C. (2016). The ECtHRs's Judgment in Biao V. Denmark: Non-discrimination among nationals and family reunification as converging European standards: ECtHR, Biao V. Denmark, Judgment of 24 May 2016, Application No. 38590/10. *Maastricht Journal of European and Comparative Law, 23*(5), 865—889.

Nuseir, M.T., Al Kurdi, B.H., Alshurideh, M.T., & Alzoubi, H.M. (2021). Gender discrimination at workplace: Do Artificial Intelligence (AI) and Machine Learning (ML) have opinions about it. In *The international conference on artificial intelligence and computer vision* (pp. 301—316). Cham: Springer.

OECD. (2020). AI strategies and public sector components. Available at: https://oecd-opsi.org/projects/ai/strategies/ (last access: 27 February 2022).

O'Neil, C. (2016). *Weapons of math destruction: How big data increases inequality and threatens democracy.* New York: Crown Publishing Group.

Oudshoorn, N. (1996). *Genderscripts in technologie. Noodlot of uitdaging?* Enschede: Universiteit Twente.

Owen, R., Macnaghten, P., & Stilgoe, J. (2020). *Responsible research and innovation: From science in society to science for society, with society. Emerging technologies: Ethics, law and governance* (pp. 117—126). Routledge.

Panch, T., Mattie, H., & Atun, R. (2019). Artificial intelligence and algorithmic bias: Implications for health systems. *Journal of Global Health, 9*(2).

Parikh, R. B., Kakad, M., & Bates, D. W. (2016). Integrating predictive analytics into high-value care: The dawn of precision delivery. *JAMA*, *315*(7), 651−652.

Peng, A., Nushi, B., Kiciman, E., Inkpen, K., & Kamar, E. (2022). Investigations of performance and bias in human-AI teamwork in hiring. *arXiv preprint arXiv:2202.11812*.

Preston, B. (2013). *A philosophy of material culture: Action, function, and mind*. Routledge.

Prince, A. E., & Schwarcz, D. (2019). Proxy discrimination in the age of artificial intelligence and big data. *Iowa Law Review*, *105*, 1257.

Puc, A., Štruc, V., & Grm, K. (2021). Analysis of race and gender bias in deep age estimation models. In *2020 28th European signal processing conference (EUSIPCO)* (pp. 830−834). IEEE.

Raji, I.D., & Buolamwini, J. (2019). Actionable auditing: Investigating the impact of publicly naming biased performance results of commercial ai products. In *Proceedings of the 2019 AAAI/ACM conference on AI, ethics, and society* (pp. 429−435).

Rajpurkar, P., Chen, E., Banerjee, O., & Topol, E. J. (2022). AI in health and medicine. *Nature Medicine*, *28*(1), 31−38.

Ribeiro, M.T., Singh, S., & Guestrin, C. (2016). "Why should i trust you?" Explaining the predictions of any classifier. In *Proceedings of the 22nd ACM SIGKDD international conference on knowledge discovery and data mining* (pp. 1135−1144).

Ridley, M. (2022). Explainable Artificial Intelligence (XAI). *Information Technology and Libraries*, *41*(2).

Rommes, E. (2000). *Gendered user-representations. Design of a Digital City. Women, work and computerization. charting a course to the future* (pp. 137−145). Boston: Springer.

Rommes, E. (2002). *Gender scripts and the internet. The design and use of Amsterdam's Digital City*. Enschede: Twente University Press.

Sanz, V. (2016). Género en el "contenido" de la tecnología: ejemplos en el diseño de software. *Revista iberoamericana de ciencia tecnología y sociedad*, *11*(31), 93−118.

Schreurs, W., Hildebrandt, M., Kindt, E., & Vanfleteren, M. (2008). *Cogitas, ergo sum. The role of data protection law and non-discrimination law in group profiling in the private sector. Profiling the European citizen* (pp. 241−270). Springer.

Schütze, R. (2021). *European Union Law*. Oxford: Oxford University Press.

Semmelmayer, P. (2021). Climate change and the German law of torts. *German Law Journal*, *22*(8), 1569−1582.

Serna, I., Pena, A., Morales, A., & Fierrez, J. (2021, January). InsideBias: Measuring bias in deep networks and application to face gender biometrics. In *2020 25th international conference on pattern recognition (ICPR)* (pp. 3720−3727). IEEE.

Sham, A. H., Aktas, K., Rizhinashvili, D., Kuklianov, D., Alisinanoglu, F., Ofodile, I., ... Anbarjafari, G. (2022). Ethical AI in facial expression analysis: Racial bias. *Signal, Image and Video Processing*, 1−8.

Speith, T. (2022, June). A review of taxonomies of explainable artificial intelligence (XAI) methods. In *2022 ACM conference on fairness, accountability, and transparency* (pp. 2239−2250).

Szczygielska, M. (2019). *'Good change' and better activism: Feminist responses to backsliding gender policies in Poland. Gendering democratic backsliding in Central and Eastern Europe: A comparative agenda* (pp. 120−160). Central European University Press.

Theodorou, A., & Dignum, V. (2020). Towards ethical and socio-legal governance in AI. *Nature Machine Intelligence*, *2*(1), 10−12.

Tripathi, S., & Musiolik, T. H. (2022). *Fairness and ethics in artificial intelligence-based medical imaging. Ethical implications of reshaping healthcare with emerging technologies* (pp. 71−85). IGI Global.

Van Oost, E. (1995). Over 'vrouwelijke' en 'mannelijke' dingen. In M. Brouns, & M. Grunell (Eds.), *Vrouwenstudies in de jaren negentig* (pp. 289−313). Bussum: Coutinho.

Van Oost, E. (2003). Materialized gender: How shavers configure the users' feminity and masculinity. In N. Oudshoorn, & T. Pinch (Eds.), *How users matter. The co-construction of users and technology* (pp. 193–208). Cambridge, London: The MIT Press.

Van Oost, E. C. (2003). *Materialized gender: how shavers configure the users' feminity and masculinity. How users matter. The co-construction of users and technology (pp. 193-208)*. Cambridge: MIT Press.

Vesnic-Alujevic, L., Nascimento, S., & Polvora, A. (2020). Societal and ethical impacts of artificial intelligence: Critical notes on European policy frameworks. *Telecommunications Policy, 44*(6), 101961.

Wachter, S., Mittelstadt, B., & Russell, C. (2021). Why fairness cannot be automated: Bridging the gap between EU non-discrimination law and AI. *Computer Law & Security Review, 41*105567.

Waller, R. R., & Waller, R. L. (2022). Assembled bias: Beyond transparent algorithmic bias. *Minds and Machines*, 1–30.

Wang, L. (2020). The three harms of gendered technology. *Australasian Journal of Information Systems, 24*.

Werro, F., & Palmer, V. (2004). *The boundaries of strict liability in European tort law. The common core of European private law*. Durham: Carolina Academic Press.

Williams, B. A., Brooks, C. F., & Shmargad, Y. (2018). How algorithms discriminate based on data they lack: Challenges, solutions, and policy implications. *Journal of Information Policy, 8*(1), 78–115.

Wischmeyer, T. (2018). Regulierung intelligenter systeme. *Archiv des öffentlichen Rechts, 143*(1), 1–66.

Wittrock, C., Forsberg, E. M., Pols, A., Macnaghten, P., & Ludwig, D. (2021). *Introduction to RRI and the organisational study. Implementing responsible research and innovation* (pp. 7–22). Cham: Springer.

Xenidis, R., & Senden, L. (2019). EU non-discrimination law in the era of artificial intelligence: Mapping the challenges of algorithmic discrimination. *Raphaële Xenidis and Linda Senden, 'EU non-discrimination law in the era of artificial intelligence: Mapping the challenges of algorithmic discrimination' in Ulf Bernitz et al (eds), General Principles of EU law and the EU Digital Order (Kluwer Law International, 2020)* (pp. 151–182).

Zhang, G., & Ananiadou, S. (2022). Examining and mitigating gender bias in text emotion detection task. *Neurocomputing, 493*, 422–434.

Zhao, J., Wang, T., Yatskar, M., Ordonez, V., & Chang, K.W. (2017). Men also like shopping: Reducing gender bias amplification using corpus-level constraints. In *Proceedings of the 2017 conference on empirical methods in natural language processing*.

Zhou, N., Zhang, Z., Nair, V. N., Singhal, H., & Chen, J. (2022). Bias, Fairness and Accountability with Artificial Intelligence and Machine Learning Algorithms. *International Statistical Review*.

Zuiderveen, F. J. (2020). Strengthening legal protection against discrimination by algorithms and artificial intelligence. *The International Journal of Human Rights, 24*(10), 1572–1593.

Further reading

Bradley, A., MacArthur, C., Hancock, M., & Carpendale, S. (2015). Gendered or neutral? Considering the language of HCI. In *Proceedings of the 41st graphics interface conference* (pp. 163–170).

De Villiers, C., Kuruppu, S., & Dissanayake, D. (2021). A (new) role for business—Promoting the United Nations' Sustainable Development Goals through the internet-of-things and blockchain technology. *Journal of Business Research, 131*, 598–609.

Expert Group on Policy Indicators for Responsible Research and Innovation. (2015). *Indicators for promoting and monitoring responsible research and innovation*. Luxembourg: Publication Office of the European Union.

John-Mathews, J. M., Cardon, D., & Balagué, C. (2022). From reality to world. A critical perspective on AI fairness. *Journal of Business Ethics*, 1–15.

Langenbucher, K. (2020). Responsible AI-based credit scoring—A legal framework. *European Business Law Review*, *31*(4).

Mittelstadt, B. D., Allo, P., Taddeo, M., Wachter, S., & Floridi, L. (2016). The ethics of algorithms: Mapping the debate. *Big Data & Society*, *3*(2), 1—21.

Oudshoorn, N., & Pinch, T. (2003). *How user matter. The co-construction of users and technology.* Cambridge, MA: The MIT Press.

Oudshoorn, N., Rommes, E., & Stienstra, M. (2004). Configuring the user as everybody. *Science, Technology & Human Values*, *29*(1), 30—63.

Raub, M. (2018). Bots, bias and big data: Artificial intelligence, algorithmic bias and disparate impact liability in hiring practices. *Arkansas Law Review*, *71*, 529—570.

Van Oost, E. (2000). Making the computer masculine: The historical roots of gendered representations. In E. Balka, & R. Smith (Eds.), *Women, work and computerization: Charting a course to the future* (pp. 9—16). Boston, MA/Dordrecht/Londres: Kluwer.

Curse of the cyborg mammoths: the use of artificial intelligence in manipulating and mobilizing human emotions

17

Tay Keong Tan

Department of Political Science, Radford University, Radford, VA, United States

Your emotions are the slaves to your thoughts, and you are the slave to your emotions.
~ **Elizabeth Gilbert, Eat Pray Love: One Woman's Search for Everything Across Italy, India and Indonesia.**

Introduction: the inexorable rise of machines

Over the past decades, artificial intelligence (AI) technologies, have had pervasive and transformative impacts on human society and the way people relate with one another. AI-enabled platforms and applications are becoming more powerful and more widely used; they now mediate many forms of human interaction. They have enhanced connectivity between people that transcends institutional barriers, national boundaries, and geography. They have also made available ginormous amount of data and innovation applications that have revolutionized many sectors and industries, from health care and transportation to entertainment and space exploration. AI drives large-scale and broad-based innovation—much like the invention of electricity and the information technology revolution in decades past—to generate a new cycle of sustained industrial innovation and economic expansion.[1]

Unlike most digitalized inventions and automated systems, which must be initiated and configured by humans to execute tasks, AI systems can be adaptive and independent in their functions and behavior, mimicking human behavior and transcending the need for continuous human monitoring in its decision-making.[2] With its pattern recognition and machine learning capabilities, AI uses algorithms to identify patterns, categorize, and classify them based on data they can detect in patterns and their representation. There are already applications in speech recognition, reading of text patterns, facial recognition, recognition of movements of humans and vehicles, and medical

---

[1]These long waves of 40—60 years in length (called the Kondratiev Wave) were first described by the Russian economist Nikolai Kondratiev in his 1925 book, The Major Economic Cycles. Joseph Schumpeter named these long-wave business cycles Kondratieff waves to honor his contribution (Barnett, 2002).

[2]For instance, AI-based unmanned aerial vehicles (drones) can capture the real-time weather and geographical data during the flight, process it in real-time, and make independent choices based on the processed data in modulating its flight plans and actions. And in facial detection and recognition applications, AI uses virtual filters to identify human faces when taking pictures and scans our facial ID to unlock our cell phones.

Ethics in Online AI-Based Systems. DOI: https://doi.org/10.1016/B978-0-443-18851-0.00018-4

image recognition in health care. And through complex algorithms built on statistical models, humanoid robots and autonomous vehicles can be "trained to look" at unstructured data and "see patterns" from road terrains and traffic movements, to fingerprints and facial expressions.

Despite these impressive advances in creating machines to replicate and capitalize on the intelligence of homo sapiens, an important frontier that machines have yet to master is in the realm of understanding, interpreting, manipulating, and imitating human emotions. Emotions are a vital and irreplaceable aspect of our natural intelligence. Having feelings are a quintessential feature of the human experience. Emotions are messengers that offer important information for living our lives, enhance our social skills, relationships, and professional practices, and strengthen the understanding of ourselves and the world around us (McLaren, 2010). The ability to accurately read and decipher our feelings and imitate and replicate human emotions is arguably some of the toughest thresholds for AI designers to breach as they thrive to make advanced robotics and intelligent machines more natural, authentic, and acceptable to humans.

Background: the primacy of emotions in machine-human interface

When AI-enabled applications are programmed to capitalize on the intelligence of *homo sapiens*, an area that it can have a profound impact is in the realm of human emotions. Homo sapiens are, by nature, emotional creatures. We are motivated and activated by emotions. Emotions are powerful and ubiquitous signals and impulses that arise within our bodies to drive many of our behaviors. Our feelings and bodily sensations drive us to do things, both consciously and unconsciously (McLaren, 2021).

Inevitably, AI-enabled machines (such as humanoid robots and autonomous home-help machines) are developed to behave and look more and more lifelike, mimicking real people, to make their interface with humans more natural and authentic. AI has been designed to read people's feelings through the recognition of patterns and nuances in text, voice tone, facial expressions, and gestures. It is also programmed to adjust its responses and demeanor to interact with those emotions more appropriately, as robots begin to take on human functions. Hanson Robotics' advanced human-like robot, *Sophia*, and Engineered Arts' entertainment robot, *Ameca*, are already famous on the social mediascape (Engineered Arts, n.d.; Hanson Robotics, n.d.). These advanced prototypes of androids can detect human emotions, decode human languages, and respond in real time to human facial expressions and perceived emotions.

During the past decades, AI research has been focused on advancing the "left-brain" intelligence of machines, giving them linguistic, mathematical, and logical reasoning abilities. Increasingly researchers are turning toward "right-brain" enhancements, giving machines *emotional intelligence*. As in the cases of the advanced prototypes of lifelike humanoids, *Sophia* and *Ameca*, intelligent machines are programmed to be able to recognize and express human emotions and their interactions infused with emotions, like empathy, surprise, pleasure, and embarrassment. An area of interest in the field of AI research and development is the use of machine learning, big data, and pattern recognition capabilities to manage and mimic emotions in homo sapiens. As consumer behavior, voter decisions, and social media engagement are driven by *how people feel*, there is profit to be made, political capital to be gained, and social media influence to be mined in the field of emotions.

An important aspect of human nature is the power of emotions in our lives. Emotions relate to our feelings (such "I feel sad"), situational appraisals and thought patterns (like "I am under threat"), personal expressions (body language, tone of voice, or words that convey our emotions, like loud and menacing posture in rage), action tendencies (the tendency toward certain behavior, like fleeing from a conflict); and physical changes (symptoms, like hairs standing at the back of the neck or butterflies in the stomach) (Scherer, 1984, 1986). Much of our behavior is predominantly motivated and activated by emotions (Trettenero, 2020). For instance, emotions automatically tell us what is important or unimportant, signal to people what needs and wants are met or unmet, and thereby help prioritize our actions. They drive us and lead humans both consciously and unconsciously. These are powerful forces that emerge from within the human body to influence and shape our thoughts, actions, and behavior.

Research shows that our emotions are instinctual in nature and are encoded within our genes (Izard, 2019). We are guided and controlled by our emotions as exemplified by the "fight-or-flight" responses that activate our amygdala (commonly known as the reptilian brain), which seems to take over our perception and awareness of life at that moment. When we become overtaken by intense or difficult emotions, our present-moment actions and decisions are strongly influenced by these powerful feelings. For instance, if we are overwhelmed with happiness, sadness, fear, disgust, or rage, we are hard pressed to think or do anything else except to attend to the feelings, which mobilizes hormones coursing through our veins (e.g., estrogen, progesterone, testosterone, norepinephrine and epinephrine, serotonin, and dopamine) and dominates our bodily sensations. We are already seeing the mobilization and manipulation of human emotions like anger and fear by social media platforms to maximize user engagement and advertising revenue. Companies are strategically targeting the emotions that drive customer behavior in their marketing, product presentation, and corporate branding. Political parties and campaigns are mining citizens' data to appeal to the angst and fears of voters in their strife for voter turnout and campaign contributions.

Methodology: ethical prognoses of emerging AI applications

Psychologist Jonathan Haidt proposed the elephant-rider allegory on behavioral change. He argues that there are two sides of human nature: *A Rider and an Elephant*. The Rider is rational, analytical, and plans ahead, while the Elephant is emotional, irrational, and driven by instinct. The Rider seems to be in charge, but the Elephant is much more energetic and powerful, and it can easily overpower the Rider to take control, just as our emotions sometime overwhelm and take over our intellect (Haidt, 2006). Hence, when applied to certain industries and sectors, AI technologies can be designed to mobilize the emotional side of humans and harness its power over human behavior—for good or evil.

Machines are programmed to appear as if they have feelings, and learn to display emotions in ways that enable them to appear emotionally intelligent and "almost human." The result is "AI-powered elephants" of the human psyche that can have such artificial powers and unnatural capabilities through AI; they may become more and more like "cyborg mammoths"—mythical creatures with emotions augmented with AI technologies. There are significant and often unforeseeable ethical consequences in creating these beasts, before they are adopted by (or unleashed on) human society in the widespread AI applications.

To gain an up-to-date appreciation of the trends and events relating to AI and emotions, the author reviewed both articles from scholarly and popular sources and triangulated the facts and findings with cross-referencing from other credible sources. The search strategy covers a range of sources and publications, including those from publishers like *SpringerLink and ScienceDirect (Elsevier)*. Articles are varied, ranging from conference proceedings and journal papers to organizational websites and popular magazines (including opinion pieces in the field of research on AI and emotions). In this discourse, this chapter aims to spark ethical reflection as these emerging AI technologies are being developed and deployed, in order to prognose ethical risks and highlight some ethically desirable benefits.

To analyze some of the most salient ethical consequences of AI technologies used in the mimicking and management of human emotions, the discussion examines them using a framework of three ethical principles, as follows:

1. Beneficence or maleficence (e.g., doing good, conferring benefits, or making improvements in people's welfare)
2. Integrity, fidelity, and responsibility (honesty, truth-telling and not spreading misinformation, exercising sound judgment, and accountability for actions)
3. Justice and fairness (e.g., fair treatment, respect for the autonomy, rights, and dignity of people, honoring professional boundaries, and preserving confidentiality and privacy)

As AI can portend large-scale and broad-based innovation in human society (like the first introduction of electricity for common use in society), ex-ante ethical reflection before the adoption and proliferation of new technologies could offer insight and ideas on how to prevent and mitigate potential risks or leverage and replicate those ethically desirable features for our benefit. This study combines the perspectives of political economy analysis, public policy, and ethical analysis to assess these ethical issues. It examines three case studies on emerging AI applications that can have widespread societal impact: autonomous vehicles; affect recognition systems; and social media algorithms.

Autonomous vehicles

An autonomous vehicle (AV, also known as self-driving car or robotic vehicle) uses AI to sense its environment, recognize road conditions, and navigate to a destination with little or no human input. AVs are endowed with sensors to perceive the surrounding, such as thermographic cameras, radar, lidar, sonar, and GPS (Taeihagh & Lim, 2019). They use AI to recognize patterns on the road, interpret sensory information (signages, obstacles, and moving objects) in the environment, and to identify appropriate navigation routes to a destination. Over the past two decades, there has been steady progress in the research, development, and adaptation of autonomous vehicles.[3]

---

[3]The American company Waymo launched its autonomous taxi service to the public in Phoenix, Arizona, in December 2020. Another company, Nuro, started driverless commercial delivery operations in California in 2021. In February 2022, Cruise became the second service provider to offer driverless taxi rides in San Francisco. Meanwhile, in Japan, Toyota pioneered a fully automated taxi service around the Tokyo 2020 Olympic Village. In China, two publicly accessible driverless taxi services have been launched. The first was in 2020 in Shenzhen by the Chinese company, AutoX. The second was by the conglomerate Baidu in 2021 in Beijing, the host city for the 2022 Winter Olympics.

An impediment in the introduction of this new technology revolves around people's confidence in or comfort level with ground vehicles functioning on "autopilot." Human emotions regarding the adoption and widespread use of AVs are those of psychological safety, trust, comfort, stress, fear, and anxiety in the drivers and other road users, particularly pedestrians. These emotional reactions are also commonly experienced in the introduction of industrial machines, indoor home-help and care robots, humanoid robots, and drones (Beran et al., 2015). The general public wants to feel safe and comfortable with the new technology.

The primacy of feelings in people's choosing to use AVs themselves and their acceptance of the widespread use of self-driving cars on public roads has motivated AV companies to confront these emotional issues. How can they design the technology to gain people's trust? What can be done to make self-driving vehicles safe and benign? How can they minimize and mitigate the harms that AVs might cause? The absence of human drivers in control of the autonomous vehicles not only releases people from the tasks of driving, but it also creates a void in the interaction between drivers, pedestrians, and other road users. Drivers often use body language, such as eye contact, facial expressions, and hand gestures, and other forms of communication that relay emotions and intentions. These sophisticated forms of social contact have to be delegated to the AI-powered autonomous systems in the vehicles themselves. Autonomous vehicles' AI sensors and displays have yet to incorporate design features that can convey intent and awareness to surrounding road users, especially pedestrians on or near the roads.

Communicating emotions, like empathy, surprise, fear, and apologetic gestures, can go a long way to help promote acceptance of the use of AV on public roads. From the experience of "social robots" that are designed to serve humans in homes, attention to emotions in human-robot interaction is critical. Lessons in the design of emotional functions and expressions in these robots can be incorporated into the AVs' *affective* dimensions, as these are essentially self-moving, nonhumanoid social robots.

Hence, to make AVs more natural and acceptable, affective features are now being developed to communicate their intent, focus, and awareness to pedestrians and other road users. Examples include the display technologies (e.g., LED lighting) and anthropomorphic features (such as moving eyes and smiling faces or honking sounds) to signal intent and emotions on the exterior of the AVs.

The ethics of autonomous vehicles

The ethical implications of autonomous vehicles in our AI-empowered world are analyzed with the framework of three principles: (1) *Beneficence or welfare*; (2) *Responsibility, fidelity, and integrity*; and (3) *Justice and fairness*. In terms of *beneficence* (advancing *human welfare*, e.g., doing good, conferring benefits, or making improvements in people's well-being), autonomous vehicles promise a quantum leap in public safety and the quality of life of humans. Their safety record is also significantly superior to human-driven vehicles, because AVs react faster than human drivers, use sensors and cameras to scan the environment for hazards, and are not liable to drive drunk, tired, distracted, or in a heightened emotional state. Self-driving cars are programmed to follow traffic rules and will be more alert and responsive than human drivers. Approximately 1.3 million people die from road traffic crashes annually, most them due to human causes like distracted driving, drunk driving, and other impairments. Road traffic crashes cost most countries about 3% of their gross domestic product. The elimination of emotional interference alone, ranging from fatigue and anxiety to agitation and rage, can greatly benefit road safety. In addition, AVs are a way of reducing congestion, pollution, and energy consumption, while increasing workforce productivity (Kaye et al., 2021).

An added advantage of self-driving automobiles is that they can increase the level of comfort for riders of vehicles. An emotion-related issue relates to their nonanthropomorphic features. AVs are not shaped like humanoids. The design of autonomous vehicles has begun to incorporate emotional features and signaling devices (emotions are signals to human from within) to make their interactions with humans more natural and engaging. Like the "social robots" used in healthcare, education, and domestic environments, research has shown that robots expressing emotional states could make people feel closer to them and perceive them as having anthropomorphic traits and communication styles (Eyssel et al., 2011). Based on these research findings on the efficacy of emotional expressions of nonhumanoid robots, AV designs have begun to incorporate emotional capabilities, like those that enable them to display emotion in ways that enable them to appear empathetic or otherwise emotionally intelligent. Among the features considered are lights like "moving eyes" at the front of a car that follow the position of pedestrians at crosswalks (Chang et al., 2017) or an electronic sign indicating a smile to inform pedestrians that it is safe to cross (Pratticò et al., 2021).

On the issue of *responsibility* (fairness, nondiscriminatory treatment, respect for the autonomy, rights, and dignity of people, honoring professional boundaries, and preserving confidentiality and privacy), the AV industry is grappling with the issues of liability in case of a crash, when people are hurt in an accident. There needs to be public policy and business regulations that can fairly assign responsibility and accountability for accidents, such as the use of end-user license agreements for the owners of self-driving cars or new taxes or insurance policies to protect owners and users of automated vehicles to meet the claims made by victims of an accident. Privacy is another ethical issue in question. Due to the interconnectivity of automated cars, it acts like any other mobile devices that can gather personal information about an individual (ranging from the location, routes taken, voice recording, video recording, and preferences in media) involving the users of an AV.

The issue of *responsibility* is related to that of *fairness*—fair treatment, respect for the autonomy, rights, and safety of people—as well. A key moral dilemma that AI software engineers and car manufacturers are facing is to resolve the well-known "runaway trolley problem." How are they to program the self-driving vehicle, in situations when the vehicle has to make decisions involving the moral dilemma, choosing between two bad outcome options? How are these vehicles to react in a choice between (A) running over the person who suddenly ran across the road and (B) avoiding the person by swerving into a wall, killing the passengers (Zhu, 2022)? The algorithms that will help AVs make these moral decisions present a formidable moral quandary: running over the pedestrian or sacrificing passengers in the vehicle to save the pedestrian. This issue of fairness may be a thorny one as while most people prefer that self-driving cares be designed to maximize the lives to be preserved in a fatal accident, they want to purchase or ride vehicles that prioritize the safety of passengers at all costs (Bonnefon et al., 2016). Delays in designing algorithms that will help AVs make these tough moral decisions will have implications on the acceptability and adoption rates of autonomous vehicles. Due to their superior safety features and record, such a delay in the adoption will affect public safety as adoption of the technology can save many lives.

Affect recognition technology

Today's AI-powered software is able to detect not just facial features but also emotions.[4] Facial recognition technology (or facial emotion recognition) is becoming widely adopted in various functions and industries; they range from enforcement agencies and airport and bank security to mobile phone and home appliances. Retailers are using AI-based facial recognition technology to prevent violence and crime. Airports are applying it to strengthen their security and access control, and mobile phone makers have brought biometric security feature into the devices. No longer a fixture of science fiction and futuristic novels, facial recognition technology is gradually becoming more widespread and mainstream in human society. Facial recognition technology along with machine learning technology is benefiting humans in many ways, including finding and identifying persons, pets, and animals, protecting premises from unlawful access and theft, improving security at places such as banks and airports, and making shopping and payments more efficient (Gargaro, 2022).

Facial recognition is a biometric technology that can identify an individual from a live-capture digital image, possibly cross-referencing it with an existing photograph of a face called a faceprint.[5] The faceprint is analogous to a fingerprint, and it measures 80 nodal points on a face.[6] More advanced facial recognition technologies have delved into "*affect recognition*"—detecting emotions by analyzing a person's facial expressions and movements to determine the emotion expressed. This is sometimes called "emotional AI." In these applications, emotional intelligence in affective computing implies the machine's comprehension of users' emotions and also its computer's display of emotions comprehensible by the user.

They can also "extract" emotions from text, such as comments reviewing a product or an experience. Through a process called sentiment analysis or opinion mining, AI technology uses normal language processing and machine learning to decipher emotions in text samples to form a judgment on the prevailing sentiment is positive, negative, or neutral. On audio data, emotional AI analyzes recordings to terms of intonation, tone of voice, vocal pitch, speed, and pauses to determine the sentiment. With machine learning and refined algorithms, this technology can even detect hidden emotions, like agitation in the use of a dry sense of humor or anxiety in the pitch and tone of voice. And with video data, affect recognition or emotional AI uses computer vision and facial recognition

---

[4]In the 1990s and 2000s, the Defense Advanced Research Projects Agency (DARPA) and the National Institute of Standards and Technology (NIST) further developed this technology and encouraged its use by private companies for commercial purposes (de Leeuw & Bergstra, 2007). Gradually in the first two decades of the millennium, it has become a common tool for border controls, airlines, air travel and transportation hubs, stadiums, and megaevent venues, especially those dealing with concerts and conferences. Its application in biometrics is now used in policing and crowd control in mass events worldwide (NEC New Zealand, 2022).

[5]Facial recognition technology is believed to be pioneered by Woodrow Wilson Bledsoe, who created a system that could organize faces' photos manually with computer technology in the 1960s. He used a grid of vertical and horizontal coordinates to manually record the coordinate areas of facial features like eyes, nose, mouth, and hairline of a person and categorize it for cross-referencing to other images (Bledsoe, 1964, 1986). From his early work, the techniques were advanced in the 1970s and 1980s with the invention of specific subjective markers for facial identification and the invention of computer software for facial recognition.

[6]Some technologies can only detect facial features to define identity factors like age, gender, and race; they are not able to recognize and identify a face.

algorithms to "read" facial expressions in order to identify emotions, such as anxiety, sadness, anger, and joy.

For instance, it can analyze facial expressions to help detect the difference between lying and truth-telling. In a real-time interview and reviewing a video recording, it can detect how a person is feeling, including sentiments like confidence, nervousness, gladness, and sadness. Most of the human emotional expressions can be inferred from the face, posture, and voice; these three sources offer the most salient evidence for identifying emotional expressions. Cyberbullying is a relatively new phenomena, but it is now recognized as an emerging public health threat affecting children and adolescents, disproportionately affecting females and sexual minorities (Aboujaoude et al., 2015). AI can offer the benefit in the area of cyberbullying and suicide prevention, as deep learning algorithms can help flag cyberbully Tweets, Instagram images, and other social media-based cyber-aggression (Hosseinmardi et al., 2015; Ji et al., 2021).

As technology progresses and proliferates, online forums and social media posts can become breeding grounds for the expression of aggression, which may lead to depression, suicidal ideation, and attempts. Fortunately, affect recognition technology can provide a new tool for suicide detection, as facial emotions may provide a means of identifying people with severe depression and suicidal intention (Larsen et al., 2015.) Machine learning applications are a form of AI that can improve the accuracy of prediction using large-scale datasets, which is particularly useful for the early detection of suicide risk, triage, and treatment (Bernert et al., 2020). For example, machine learning algorithms for the automatic detection of what might be classified as "suicidal tweets" can inform social media platforms and suicide prevention agencies of potential suicidal thoughts and emotions (Bharti et al., 2021; O'Dea et al., 2015). In a study published in 2015 on the use of advanced technology in suicide prevention in Australia, a team of researchers found that "automatic detection of suicidality from social media content, and crisis detection from acoustic variability in speech patterns" can help to screen and identify likely cases of suicidal behavior and save lives (O'Dea et al., 2015).

While it is "common knowledge" that a person's emotional state can be read from facial expressions, this is based on assumptions with tenuous evidence and shaky empirical basis (Crawford, 2021a, 2021b). The analysis of emotions based on facial expressions may not be accurate, as facial expressions can vary from person to person and each person's expression may have a mix of different emotional states (e.g., fear and anger; as in the hybrid emotion of jealousy, or happy and sad as in the mixed emotion of bittersweet or poignancy). These factors make facial emotion recognition fraud with bias and errors (Vemou & Horvath, 2021). The assumption has influenced policy decisions, national security protocols, legal judgments, and educational practices and guided the diagnosis and treatment of psychiatric illness, and the development of commercial applications (Barrett, 2017). However, there is little scientific evidence to support the premise that emotions can be accurately read from facial expressions (Barrett et al., 2019; Haven, 2020).

Nevertheless, the premise that a person's emotional state, intentions, and state of mind can be accurately judged from facial expressions has already been applied to our everyday social interactions. For instance, advertisers and marketers know that most consumer decisions are spontaneous and based on feelings (germinating ideas for a shopping list or even buying on impulse). When they want to know how people respond to a product, they use affect recognition software to detect exactly where someone's eyes went or how they felt when they saw a picture or video of a product. By tracking those eye movements and analyzing facial expressions, they study how effective an advertisement is or how emotionally engaging a product or service is. It is widely believed that real-time facial expressions can provide advertisers and marketers insight into the emotions and use them to improve their advertisements and influence customers purchasing decisions.

The ethics of affect recognition technology

Like most inventions, *ethical benefits* are enhanced by affect recognition technology in some areas, and they are also impeded in others. The benefits of accurate and actionable profiles of human emotions can help enhance public safety and human welfare (*beneficence*), such as improved safety of people using airports, customs checkpoints, and banks, and the increased efficiency of services at these premises. When affect recognition software is effective in identifying the intention and state of mind of suspected criminals and terrorists, their installation and deployment (e.g., in security cameras) can help detect crime, provide actionable intelligence to law enforcement, and create a deterrence against criminal activities.

Affect recognition technology, when used at sale counters, can be combined with data from biometric identification in transactions at stores to alert service staff or store owners about the emotions of customers on their faces, tone of voice, or body language, resulting in better customer service and a more seamless and pleasant shopping experience. It can also help reduce human touchpoints (physical contact or direct in-person interaction) during a pandemic. These can benefit in human welfare in the different industries and services.

On the other hand, affect recognition technology (often used with biometric identification) can also present ethical drawbacks, from threats to privacy and violations of rights (*justice* and *responsibility*). Face-scanning body cameras or security cameras permit pervasive and continuous surveillance of people without their consent and record their emotional states and biometric data and store them for purposes unknown to the subjects. Facial emotion recognition can be combined with biometric identification technology to access and analyze data from multiple sources, such as voice, text, and health data from existing databases, thereby giving it the ability to turn human emotions (a very private information of people) into a data source (Vemou & Horvath, 2021). These devices have essentially become ubiquitous surveillance networks that not only register and report the movements and locations of people in events and places, but also offer clues to their emotional states and intent. The data collected can be used against people in situations like protest marches and sensitive entertainment premises.

China's social credit system is an extreme version of this.[7] It is a government surveillance system that offers a unified social credit code for all Chinese citizens that scores all aspects of life and judges the citizens' behavior and trustworthiness (like acts of jaywalking, bad driving, having outstanding traffic fines, smoking in nonsmoking zones, and posting fake news online)—all linked to a permanent record (Hatton, 2015; Raphael & Xi, 2019). In April 2023, Iran began installing

---

[7]In 2021, the United Nations Educational, Scientific and Cultural Organization (UNESCO) became the first intergovernmental organization to call for nations to pledge to end pervasive mass surveillance using AI, particularly by banning the use of AI for "social scoring systems" using facial recognition in public places. These are assessed to threaten human rights and civil liberties (Hiikkila, 2021).

cameras in public places to identify unveiled women who can be charged in violation of its "hijab laws" (BBC, 2023). "Smart" facial recognition cameras and biometric tools are used by the Iranian government to identify and send "warning messages" to violators of the morality law (BBC, 2023). Iranian women are legally required to cover their hair with a headscarf, in conformance with the strict religious law. Facial recognition devices, even without affect recognition capabilities, can result in a violation of privacy and are banned by laws in several countries and cities. For instance, American cities San Francisco, Oakland, Boston, New Orleans, Oakland, Pittsburgh, and Santa Cruz have bans on government use of facial recognition. New Orleans, Oakland, Pittsburgh, and Santa Cruz in the United States have laws banning its use in predictive policing and the use of computer systems to identify people as likely perpetrators or victims of crime to strategize police deployment and tactics (Sheard & Schwartz, 2022).

On the issue of *fairness*, the accuracy with which affective computing can accurately read emotions across cultures and contexts presents an ethical quagmire for designers and users of affect recognition devices. Emotions are inherently difficult to "read," even for humans, as people try to hide their true emotions or misrepresent how they feel (Barrett et al., 2019; Haven, 2020). A machine may only achieve limited accuracy in interpreting emotions; its result fraught with biases and errors. With diverse test sets of millions of faces from different countries and cultures, machines can, with deep learning and refining of algorithms, improve their emotion-reading capabilities. However, even humans misread cultural references, sarcasm, and nuance in the emotional expressions of other people, which can completely alter the meaning expressed and feelings displayed.

Until AI can learn to accurately interpret the metrics of body language and speech patterns and correlate them to different emotions and thoughts, its use in affect recognition can lead to attribution errors and moral hazards. With this technology in its infancy, the ethical issue of *fairness* looms large as inaccurate interpretations by machines can result in unintended consequences, like unfair treatment and the loss of autonomy, rights, and dignity of people. For instance, when AI algorithms in security cameras at airports misread the emotions of travelers from different cultures, they may "flag" the wrong people as suspects with criminal intent, leading to misguided and futile security enforcement actions. There are also ethical costs involving the failure to preserve the privacy of people surveilled by these ubiquitous cameras in public places.

As sales and marketing organizations, surveillance and security industries have started using affect recognition technology, such as in telephone exchange virtual assistants and telemedicine chatbots. Soon, it will become more widely used in stores, schools, colleges, hospitals, and across many sectors. Affect recognition technology has the potential to make a huge difference to benefit human lives, but some ethical pitfalls in terms of concerns about its accuracy and threats to privacy remain to be resolved.

Algorithms in social media

Social media are virtual, online applications that facilitate the "creation and exchange of user-generated content". They come in many forms: Collaborative projects such as *Wikipedia*; blogs, such as *Twitter*; content communities, such as *YouTube* and *Instagram*; social networks, such as

Facebook; and virtual social worlds, such as the *Metaverse* and *Second Life*. It is now widely known that social media can be a breeding ground for disinformation, polarization, and harassment.[8]

Disinformation on the Internet ranges from fake product information and fictitious news to conspiracy theories and propaganda by terrorist groups and warring nations, intentionally disseminated by bad actors to achieve their nefarious ends. For example, during the Brexit referendum, data scientists at the universities of Berkeley and Swansea uncovered hundreds of thousands of Russian Twitter accounts that were created to disrupt the Brexit vote. Organized disinformation attacks, such as those in which false news or faulty information is disseminated quickly and broadly erode the legitimacy of electoral results or contradict the public announcements of public authorities. They may come in the form of cyberattacks orchestrated by cyberactors using multiple fake social media accounts with realistic photographs and profiles of account owners.

Social media farms can also vary the content and writing style of the postings to appear authentic and avoid detection by software designed to identify fake accounts.[9] These nefarious activities are enabled by AI, with sophisticated algorithms that can rapidly conjure and publish many Facebook and Twitter posts or emails, such as those that complain about irregularities at polling locations or dispute the integrity of vote counts. Beyond just text-only fake news, "deepfake" technology now exists to produce videos where the words spoken by a person in a video or audio file can be altered, much like Photoshop capabilities for audio and video content.

Sophisticated AI has human-inspired elements of both cognitive and emotional intelligence. It can learn to understand and imitate human emotions and use these in its decision-making and actions. For instance, a human-inspired AI system can use facial expressions to detect a person's agreement with or resistance to a political message or receptiveness to a commercial advertisement. Such "humanized" AI imitates certain features of human emotional and social intelligence and is able to make decisions and change its behavior in real-time interactions with humans. Combining deep learning and facial/affect recognition, AI-empowered human robots are now capable of a freewheeling discussion with a person, offer analytical, informed, and yet self-reflective opinions, and also form its own ideas on various issues. Examples like the lifelike *Sophia* and *Ameca* are quintessential examples. In a parallel example in the virtual world, *FN Meka* is an AI musician given the appearance of a black male cyborg. The virtual rapper is so popular and prolific that in August 2022, he has more than 500,000 monthly Spotify subscribers, an Instagram account with more than 220,000 followers, and more than one billion views on its TikTok account (BBC, 2023). "Virtual personalities" like FN Meka are de facto social media influencers. While FN Meka's songs are performed by an anonymous singer, AI is fast evolving toward the ability to computer-created lyrics, songs, and entire performances (BBC, 2023).

It is commonly known that social media sites are ridden with misinformation and disinformation (commonly known as fake news), hate speech, harassment and revenge pornography, and a host of

---

[8]The political consultancy, Cambridge Analytica, used millions of Facebook users' data to successfully influence and manipulate public opinion in events such as the 2016 US presidential election and the 2018 Brexit referendum.

[9]According to U.S. intelligence agencies, actors in Russia have been interfering in the 2016 US presidential election, which is broadly known. There was also evidence of Russian manipulation of public opinion and voter outcomes in elections in Austria, Belarus, Bulgaria, France, Germany, and Italy using fake news posts and misinformation on social media.

terrorist activities. To compound these problems, the algorithms used by social media companies are driven by profit motives that further exacerbate their dangers to users and society as a whole. Some of the tried-and-true strategies of news media success is polarization and the mobilization of emotions. The presentation of sensational news, especially bad news, makes for high viewership on television as well as high engagement on social media sites (Vosoughi et al., 2018). An MIT study on some 126,000 stories tweeted by about 3 million people more than 4.5 million times and published in the journal *Science* finds that false news online travels "farther, faster, deeper, and more broadly than the truth," resulting in perverse outcomes for public information dissemination as more and more people rely on social media as the primary source of news information (Vosoughi et al., 2018).

As heightened emotions tend to engage users and keep them logged in, enabling more posts and comments that elicit strong emotional responses tends to benefit social media companies and boost their revenue. Some common examples are stories that spawn outrage over the extinction of species from habit loss and pollution resulting from corporate practices or comments that spark xenophobic panic about the mass immigration of foreigners into a country. And "bad actors" around the world are quickly learning how to exploit and weaponize the algorithm. They did so to great effect during the 2016 election cycle, placing anonymous advertisements on social media platforms designed to stir up social unrest, racial bias, and political rancor and to degrade confidence in US institutions, including the electoral system. Divisive content and misinformation ("fake news") often got the most engagement. Even if the divisive content was posted by strangers, it had the effect of sowing division between friends and families, ultimately alienating many users.

Social media sites are now serving more and more of the function of disseminating news to users, and they face the same incentives as some news organizations. Roughly half of American adults get their news from social media (Pew Research Center, 2021), with Facebook outpacing all other social media sites in providing news content. Some of these platforms have installed AI-enabled algorithms that prioritize posts and comments that heighten people's emotions to boost the users' engagement in order to attract more advertising revenue.

Facebook, for example, uses data from its own research to boost users' engagement on the platform by giving priority to sensational and emotion-arousing posts. In 2021, the Wall Street Journal published the *"Facebook Files,"* an investigative series that uses Facebook's internal documents made public by Frances Haugen, a former manager at the company. Haugen revealed what she saw as the social media giant's deliberate policy of applying algorithms and platform design decisions to prioritize profit over the safety and welfare of users (Hagey & Horwitz, 2021; Horwitz, 2021; Roose, 2021). Her testimony to the U.S. Congress' Senate Commerce Subcommittee on Consumer Protection, Product Safety, and Data Security further revealed that Facebook ignored its own research data showing a negative impact on teenagers' mental and physical well-being when social media contributed to feelings of inferiority, symptoms of depression, suicidal ideation, and screen addiction (Lukianoff & Haidt, 2018).

Facebook has also been used by commentators to spread misinformation on the COVID-19 vaccine and issues relating to the U.S. presidential elections. However, its business model and algorithms are premised on getting users to spend as much time as possible on its platform, and its engineers and designers use behavioral psychology and big data to continue to make its platforms more engaging and "addictive" to maximize users' exposure to advertisements (Boers et al., 2019). These result in negative effects on the self-image and mental health of users, particularly teenage girls. Facebook's internal research also revealed that its platform is used in developing countries for human trafficking, drug dealing, and to promote ethnic violence.[10]

Social media companies themselves have amassed huge financial power and social influence through the gigantic user bases on their platforms. Facebook has 2.934 billion active users, and Twitter has some 217 million active daily users (Aslam, 2022). Marketers can reach a total potential audience of 2.168 billion users with advertisements on Facebook in 2022. While they serve very useful social functions of connecting people, disseminating information, and facilitating the creation of media content and communities, the profit motive motivates them to design algorithms and use business practices that have huge social and ethical implications. They can monetize or use the data collected on user preferences, browsing habits, demographics, political and religious beliefs, and scores of other signals on these platforms. Advertisements, news article links, and group posts are carefully targeted to influence consumer decisions, such as the use of user data to enable advertisers to reach finely segmented audiences with nuanced messaging that will most likely appeal to these groups.[11]

The ethics of social media algorithms

AI algorithms using big data and behavioral psychology to engage the users' emotions on various social media platforms relate to the ethical issue of *beneficence*. The business motive of most social media companies prioritizes user engagement (which is correlated to its advertising revenue) over the health and safety of users, especially adolescents and those at an impressionable age. Unlike conventional algorithms hard-coded by engineers, AI-empowered algorithms use machine learning to "learn" from input data and "see" correlations within them. The advanced machine-learning capabilities help to automate decisions to enable computers to manipulate the behaviors of users. For instance, based on advertisement-click data, the algorithm might learn that women click more often on advertisements for cosmetics or fashion than men and are "trained" to serve more of them to women. Conversely, the machine-learning algorithm may "discover" that certain male users tend to like or share certain posts on conspiracy theories (or those related to political controversy, misinformation, and extremism). It may then "decide" to give those posts more prominence to further

---

[10]Facebook has hundreds and even thousands of algorithms to target ads and rank content, using its own data on users' preferences. They can boost some types of content in the users' news feeds or downplay or delete others deemed to be "bad content," like nudity and spam.

[11]Virtual characters in cyberspace who perform on YouTube, pose on Instagram, and amass a huge following on Twitter and Facebook are masters of the manipulation of human emotions, often like living, breathing human beings. They replace human artists and influencers, can attract and maintain a fan base on social media, and create their own brands and labels to sell merchandize and rake in a profit. In addition, they have a great advantage over homo sapiens in that they never age, can have limitless persona and fashion changes, and can be in many places at the same time. The advent of these advances in the social mediascape raises many ethical questions, concerning the veracity of information, data protection, and the right to privacy. There is also the concern for the entire new class of actors (virtual personalities and influencers), assets (real estate and properties on Metaverse), and cryptocurrencies that will compete with actors, products, and services in the real world, resulting in "winners and losers." How will the gains from these inventions be regulated and shared? How can the losses from these profound changes be compensated? It calls for greater public oversight and scrutiny of the practices and ethical standards of social media companies.

engage these users. Commercial algorithmic profiling and targeting have become widely used in the advertising industry, often without the consumers' knowledge or consent (Kant, 2021).

An example is when Facebook allowed its platforms in Myanmar to be used to foment division in this ethnically diverse society and incite offline violence against the Rohingya Muslim minority. Facebook is accused of automatically feeding the people of Myanmar divisive content in an effort to gain user attention and increase their time on the platform. In 2021, Rohingya refugees in the United States filed a $150 billion class-action lawsuit against Facebook's parent company, Meta Platforms Inc., accusing Facebook of allowing viral fake news and hate speech on its platform that incited violence against the Rohingyas, which escalated the religious conflict in Myanmar into a full-blown genocide.[12] Augmented by AI capabilities, machines are now able to capitalize on their "study" of human behavior to leverage our emotions, proclivities, and vulnerabilities for the benefit of social media companies.

In terms of *Responsibility*, the safeguarding of fundamental human rights such as the right to a free and fair election, the right to health, and the right to nondiscrimination are important aspects of *fidelity* and *accountability* on the part of social media companies. These rights are threatened by the spread of false information. For instance, health misinformation regarding infectious diseases and vaccines that spreads quickly on social media can hamper effective treatment and care and sometimes jeopardize the lives of people. False information and ideas about the effectiveness and safety of the COVID-19 vaccines and the health benefits of masking during the coronavirus pandemic compete for attention with true public health and scientific information, "which may be difficult to comprehend and even dull, is easily crowded out by sensationalized news" (Wang et al., 2019).

In democratic elections, voters rely on accurate information about parties, candidates, and political or public issues in their voting decisions. When social media sites offer platforms for targeted political advertising, some have used falsehood to maliciously influence public opinion and spread propaganda to unfairly shape electoral outcomes. The machine-learning models and algorithms of these companies aim to maximize user engagement, and consequently their bottom line, by favoring controversy, misinformation, and extremism. Such behavior can in turn exacerbate political divisions and inflame political tensions. Due to the rapid news cycle on social media sites, much of the information circulated is not subject to editorial oversight or even careful thought. Research suggests that people are inclined to accept information from social media sources that best align with their own partisan views, which underpins how fake news stories are able to spread so fast, when they are often not fact checked or critically assessed by their readers (Waszak et al., 2018).

Finally, related to the ethical principle of *fairness*, the rapid spread of false information and propaganda on social media also threatens the rights of people to nondiscrimination. The right to live without prejudice or discrimination is enshrined in the Universal Declaration of Human Rights and the International Convention on Economic, Social, and Cultural Rights, two major human rights laws that were ratified by most of the nations of the world. This fundamental human right is threatened by the spread of fake news, defamatory information, disinformation, propaganda, and conspiracy theories that target groups in society like migrants or ethnic minorities in some countries

---

[12]The military crackdown killed thousands of Rohingyas and forced many more to be displaced from their homeland. The United Nations estimates that there are about 700,000 Rohingya refugees in Bangladesh. In 2018, Facebook admitted that it had not done enough "to help prevent our platform from being used to foment division and incite offline violence."

(Neidhardt & Butcher, 2022). The easy and rapid dissemination of false information on social media is further augmented by machine-learning algorithms that favor sensational news and outrageous content that often dehumanizes minority groups and fans of the flames of racism, homophobia, and xenophobia.

Cyberspace is so rife with unsubstantiated information it is are hard to refute or take down before the damage is already done. A noteworthy example is reported in Germany, where companies like Facebook and Twitter failed to delete up to 70% of online hate speech, false allegations, or incorrect facts about certain groups (such as Muslims and immigrants) within 24 hours of it being reported (The New York, 2017). This failure on the part of these Silicon Valley platforms to monitor the spread of misinformation about minority groups has significantly eroded the protection of the right to nondiscrimination of refugee and migrant populations, which faced backlash from far-right hate groups in Germany.

Concluding remarks: taming of the cyborg mammoths

Emotions are a vital and irreplaceable aspect of our natural intelligence. They are messengers that offer important information for living our lives, enhance our social skills, relationships, and professional practices, and strengthen the understanding of ourselves and the world around us. Social instincts and emotional intelligence come almost automatically to humans, and we may react to stimuli and situations based on the cues and drives that emotions signal to us from within. This is because *homo sapiens* are hardwired and socialized to perceive, interpret, and act on the emotions within and of those around us. While having emotions and feelings is a quintessential feature of the human experience, they are some of the most complex aspects of human intelligence for machines to learn and master, as they are developed and constantly advanced to imitate and capitalize on the intelligence of humans. This basic human capacity has to be deliberately programmed into machines by design.

With the many ethical hurdles and pitfalls highlighted in the examples of emerging AI applications in autonomous vehicles, affective computing, and social media algorithms discussed earlier, targeted measures (no matter how tentative and incomplete) should be considered in the design, implementation, and deployment of AI technologies to prevent and mitigate the ethical drawbacks. Conversely, features that portend ethical benefits may also be deliberately enhanced. These cutting-edge technologies are often not well regulated by public policy and transparent corporate practices to safeguard the public interests and welfare. With deep learning and the relentless march of technological advancements, AI will soon be used in practically every application on our computers. Machines will inevitably learn new abilities and receive more powerful capabilities to sense, recognize, and manage human emotions and respond to them more intelligently and with greater impact.

For these reasons, the emotional application of AI should be the subject of further studies and public discourse as our human society embraces new and more powerful technologies and tools. Topics for consideration include the ethical, environmental, and social consequences of (emerging and existing) AI applications; the policies and principles for algorithmic design to enable autonomous machines to make good (ethical) decisions; and the curriculum and training programs for AI scientists, professionals, and regulators to educate them on the societal impacts and ethical ramifications of new technologies. The advent of an "Age of Artificial Intelligence" is looming, and it is not far off. There is no escaping a future when advanced robotics and intelligent machines will be fully integrated and embedded in our lives, with far-reaching and unforeseeable ethical ramifications.

References

Aboujaoude, E., Savage, M. W., Starcevic, V., & Salame, W. O. (2015). Cyberbullying: Review of an old problem gone viral. *Journal of Adolescent Health*, *57*(1), 10−18. Available from https://doi.org/10.1016/j.jadohealth.2015.04.011.

Aslam, S. (2022). Twitter by the numbers: Stats, demographics & fun facts. *Omnicore*. February 22, 2022. https://www.omnicoreagency.com/twitter-statistics/.

Barnett, V. (2002). Which was the "real" Kondratiev: 1925 or 1928? *Journal of the History of Economic Thought*, *24*(4), 475−478.

Barrett, L. F. (2017). *How emotions are made: The secret life of the brain*. New York: Houghton Mifflin Harcourt.

Barrett, L. F., Adolphs, R., Marsella, S., Martinez, A. M., & Pollak, S. D. (2019). Emotional expressions reconsidered: Challenges to inferring emotion from human facial movements. *Psychology Science in the Public Interest*, *20*, 1−68.

BBC. (2023). *Iran installs cameras to find women not wearing hijab*. April 9, 2023. https://www.bbc.com/news/world-65220595. Accessed on April 10, 2023.

Beran, T. N., Ramirez-Serrano, A., Vanderkooi, O. G., & Kuhn, S. (2015). Humanoid robotics in health care: An exploration of children's and parents' emotional reactions. *Journal of Health Psychology*, *20*(7), 984−989.

Bernert, R. A., Hilberg, A. M., Melia, R., Kim, J. P., Shah, N. H., & Abnousi, F. (2020). Artificial intelligence and suicide prevention: A systematic review of machine learning investigations. *International Journal of Environmental Research and Public Health Review*, *17*, 5929.

Bharti, S., Yadav, A. K., Kumar, M., & Yadav, D. (2021). Cyberbullying detection from tweets using deep learning. *Kybernetes*, *13*, ISSN: 0368-492X.

Bledsoe, W.W. (1964). *The model method in facial recognition*. Technical report PRI 15. Palo Alto, CA: Panoramic Research, Inc.

Bledsoe, W. W. (1986). I had a dream: AAAI presidential address, 19 August 1985. *AI Magazine*, *7*(1), 57−61.

Boers, E., Afzali, M. H., Newton, N., & Conrod, P. (2019). Association of screen time and depression in adolescence. *JAMA Pediatrics*, 853−857.

Bonnefon, J.-F., Shariff, A., & Rahwan, I. (2016). The social dilemma of autonomous vehicles. *Science (New York, N.Y.)*, *352*(6293), 1573−1576. Available from https://doi.org/10.1126/science.aaf2654.

Chang, C.M., Toda, K., Sakamoto, D., & Igarashi, T. (2017). Eyes on a car: An interface design for communication between an autonomous car and a pedestrian. In *Proceedings of the 9th international conference on automotive user interfaces and interactive vehicular applications* (pp. 65−73). New York: Association for Computing Machinery. Automotive UI.

Crawford, K. (2021a). *Atlas of AI: Power, politics, and the planetary costs of artificial intelligence*. New Haven, CT: Yale University Press.

Crawford, K. (2021b). Artificial intelligence is misreading human emotion. *The Atlantic*. April 27, 2021. Retrieved from: https://www.theatlantic.com/technology/archive/2021/04/artificial-intelligence-misreading-human-emotion/618696/.

de Leeuw, K., & Bergstra, S. (2007). *The history of information security: A comprehensive handbook* (pp. 264−265). *Amsterdam*: Elsevier.

Engineered Arts. (n.d.) Ameca. Falmouth, U.K. Retrieved from https://www.engineeredarts.co.uk/robot/ameca/.

Eyssel, F., Kuchenbrandt, D., & Bobinger, S. (2011). Effects of anticipated human-robot interaction and predictability of robot behavior on perceptions of anthropomorphism. In *2011 6th ACM/IEEE International Conference on Human-Robot Interaction (HRI)* (pp. 61−67).

Gargaro, D. (2022). *The pros and cons of facial recognition technology.* ITpro.

Hagey, K., & Horwitz, H. (2021). Facebook tried to make its platform a healthier place. It got angrier instead. *Wall Street Journal.* Retrieved from: https://www.wsj.com/articles/facebook-algorithm-change-zuckerberg-11631654215?mod = article_inline.

Haidt, J. (2006). *The happiness hypothesis; putting ancient wisdom to the test of modern science.* London: Arrow Books.

Hanson Robotics. (n.d.) Sophia. Hong Kong. Retrieved from: https://www.hansonrobotics.com/sophia/.

Hatton, C. (2015). China 'social credit': Beijing sets up huge system. *BBC News.* October 26, 2015. Retrieved from: https://www.bbc.com/news/world-asia-china-34592186.

Haven, D. (2020). Why faces don't always tell the truth about feelings. *Nature News Feature.* February 26, 2020. https://www.nature.com/articles/d41586-020-00507-5#ref-CR6.

Hiikkila, M. (2021). China backs UN pledge to ban (its own) social scoring. *Politico.* November 23, 2021. Retrieved from: https://www.politico.eu/article/china-artificial-intelligence-ai-ban-social-scoring-united-nations-unesco-ethical-ai/.

Horwitz, J. (2021). Facebook says its rules apply to all. Company documents reveal a secret elite that's exempt. *Wall Street Journal.* September 15, 2021. Retrieved from: https://www.wsj.com/articles/facebook-files-xcheck-zuckerberg-elite-rules-11631541353?mod = article_inline.

Hosseinmardi, H., Mattson, S. A., Rafiq, R. I., Han, R., Lv, Q., & Mishra, S. (2015). Detection of cyberbullying incidents on the Instagram social network. *ArXiv. /abs/1503.03909.*

Izard, C. E. (2019). Emotion theory and research: Highlights, unanswered questions, and emerging issues. *Annual Review of Psychology, 2009*(60), 1–25. Available from https://doi.org/10.1146/annurev.psych.60.110707.163539.

Ji, S., Pan, S., Li, X., Cambria, E., Long, G., & Huang, Z. (2021). Suicidal ideation detection: A review of machine learning methods and applications. *IEEE Transactions on Computational Social Systems, 8*(1), 214–226. Available from https://doi.org/10.1109/TCSS.2020.3021467.

Kant, T. (2021). Identity, advertising, and algorithmic targeting: Or how (not) to target your ideal user. *MIT Case Studies in Social and Ethical Responsibilities of Computing, 2021.* Available from https://doi.org/10.21428/2c646de5.929a7db6.

Kaye, S.-A., Somoray, K., Rodwell, D., & Lewis, I. (2021). Users' acceptance of private automated vehicles: A systematic review and meta-analysis. *Journal of Safety Research, 79,* 352–367.

Larsen, M.E., Cummins, N., Boonstra, T.W., et al. (2015). The use of technology in suicide prevention. In *Presented at 37th annual international conference of the engineering in medicine and biology society.* August 26–29, 2015. Milan, Italy.

Lukianoff, G., & Haidt, J. (2018). *The Coddling of the American Mind: How good intentions and bad ideas are setting up a generation for failure.* Penguin Books.

McLaren, K. (2010). *The language of emotions: What your feelings are trying to tell you.* Colorado: Sounds True.

McLaren, K. (2021). *The power of emotions at work.* Colorado: Sounds True.

NEC New Zealand. (2022). *A brief history of facial recognition.* Retrieved from https://www.nec.co.nz/market-leadership/publications-media/a-brief-history-of-facial-recognition/.

Neidhardt, A.-H., & Butcher, P. (2022). Disinformation on migration: How lies, half-truths, and mischaracterizations spread. *Migration Information Source. Migration Policy Institute, 8.*

O'Dea, B., Batterham, P. J., Calear, A. L., Paris, C., & Christensen, H. (2015). Detecting suicidality on Twitter. *Internet Interventions, 2,* 183–188.

Pew Research Center. (2021). *News consumption across social media in 2021.* September 20, 2021. Retrieved from: https://www.pewresearch.org/journalism/2021/09/20/news-consumption-across-social-media-in-2021/.

Prattico, F. G., Lamberti, F., Cannavò, A., Morra, L., & Montuschi, P. (2021). Comparing state-of-the-art and emerging augmented reality interfaces for autonomous vehicle-to-pedestrian communication. *IEEE Transactions on Vehicular Technology*, *70*, 1157–1168.

Raphael, R., & Xi, L. (2019). Discipline and punish: The birth of China's social-credit system. *The Nation*. Available from https://www.thenation.com/article/china-social-credit-system/.

Roose, K. (2021). Facebook is weaker than we knew. *New York Times*. October 2021. Retrieved from: https://www.nytimes.com/2021/10/04/technology/facebook-files.html.

Scherer, K. R. (1984). On the nature and function of emotion: A component process approach. In K. R. Scherer, & P. Ekman (Eds.), *Approaches to emotion* (pp. 293–317). Hillsdale, NJ: Erlbaum.

Scherer, K. R. (1986). Vocal affect expression: A review and a model for future research. *Psychological Bulletin*, *99*, 143–165.

Sheard, N., & Schwartz. (2022). The movement to ban government use of face recognition. *Electronic Frontier Foundation*. Retrieved from: https://www.eff.org/deeplinks/2022/05/movement-ban-government-use-face-recognition.

Taeihagh, A., & Lim, H. S. M. (2019). Governing autonomous vehicles: Emerging responses for safety, liability, privacy, cybersecurity, and industry risks. *Transport Reviews*, *39*(1), 103–128.

The New York Times. (2017). *Delete hate speech or pay up, Germany tells social media companies*. Retrieved from https://www.nytimes.com/2017/06/30/business/germany-facebook-google-twitter.html.

Trettenero, S. (2020). Human beings are first and foremost emotional creatures. *Psychreg (website)*. Retrieved from: https://www.psychreg.org/human-beings-are-emotional-creatures/.

Vemou, K., & Horvath, A. (2021). Facial emotion recognition. *Techdispatch, European Union, 1, 2021*. Available from https://doi.org/10.2804/519064.

Vosoughi, S., Roy, D., & Aral, S. (2018). The spread of true and false news online. *Science (New York, N.Y.)*, *359*(6380), 1146–1151. Available from https://doi.org/10.1126/science.aap9559.

Wang, Y., McKee, M., Torbica, A., & Stuckler, D. (2019). Systematic literature review on the spread of health-related misinformation on social media. *Social Science & Medicine*, *240*, 2019. Available from https://doi.org/10.1016/j.socscimed.2019.112552.

Waszak, P. M., Kasprzycka-Waszak, W., & Kubanek, A. (2018). The spread of medical fake news in social media—The pilot quantitative study. *Health Policy and Technology*, *7*(2), 115–118.

Zhu., et al. (2022). A moral decision-making study of autonomous vehicles: Expertise predicts a preference for algorithms in dilemmas. *Personality and Individual Differences*, *186*(111356). Available from https://doi.org/10.1016/j.paid.2021.111356.

On deterring hate speech, while maximizing security and privacy

18

Sue Spaid

Northern Kentucky University, Highland Heights, KY, United States

Introduction: online hate speech, incivility, rage, and the lack of self-restraint

What's patently clear is that the internet's original architects envisioned creating a place where speech and all manner of dissident activities could reign freer there than anywhere on land, where the authorities take "true threats" seriously. Elon Musk's April 2022 bid to purchase Twitter in order to remove all barriers to free speech (a.k.a. "content moderation") was initially exemplary of this mindset (Isaac & Hirsch, 2022). Since Twitter became Musk's personal asset (now known as X), the problematic incidents that earlier spurred Twitter to suspend tweeters' accounts (terrorist activities, foreign information operators, hate speech as dictated by EU codes of conduct, fake follower accounts (usually trolls or human-operated bot accounts), and violent threats, among other incidents) have accelerated at such a fast pace that "unfettered free speech" looms more as an unrealistic promise than a future ideal. On February 28, 2023, only four months after Musk purchased Twitter, its policy was updated to include "the mere act of hoping, wishing, or expressing desire that other people might experience harm. This includes (but is not limited to) *hoping* [italics mine] for others to die, suffer illnesses, tragic incidents, or experience other physically harmful consequences" (Merchant, 2023). This chapter employs the *Cambridge Dictionary*'s definition of hate speech, which is "public speech that expresses hate or encourages violence towards a person or group based on something such as race, religion, sex, or sexual orientation" (Cambridge Dictionary, 2023). As Musk has learned all too well, simply eliminating botnets and making algorithms transparent is insufficient to ensure free speech, so long as hate speech targets remain vulnerable. Even when hate speech doesn't escalate into physical violence, feminists have long observed that psychological and emotional abuse often proves more destructive than physical abuse (Pharr, 1997: 15).

A key deterrent to the sudden rise in hate speech, incivility, and/or rage is self-restraint, yet self-restraint is alternatively in short supply (Thompson, 2019). Exemplary of human beings' *undisciplined* behavior, Musk verbally attacked Vijaya Gadde, Twitter's female general counsel responsible for punishing violators, even though he had contracted not to disparage Twitter while purchasing it (presumably to prevent him from driving down its value should the deal fall through). His abusive language sparked a malicious Twitter barrage resulting in ethnic slurs directed at her, thus adding fuel to this conversation, while paradoxically undermining her efforts to constrain hate speech and sexual harassment (Serrano, 2022). As this chapter indicates, this unfortunate anecdote

Ethics in Online AI-Based Systems. DOI: https://doi.org/10.1016/B978-0-443-18851-0.00021-4

grounds this chapter's larger question, "Could self-learning AI technology prove more 'disciplined' than humans in assessing and monitoring harmful language?" Although human beings are responsible for coding and uploading such algorithms, this chapter defends the view that AI applications that employ "emotion chips" are likely to be more routine, quicker to respond, and therefore more effective in their capacity to detect and deter hate speech than human beings, even though earlier AI models have censored conversations erroneously perceived as racist and people have taught AI machines all manner of harmful ideas.

The primary ethical issue arising from unrestricted free speech online is the risk of remote assailants harming those who are particularly vulnerable to abuse, already suffering psychological damage, and/or likely to be physically assaulted as a result of verbal attacks. Not only does *worldlessness* provide remote assailants impunity, but retweets ensure omnipresence, which either prompts a Twitter "pile-on" or engenders a cybermob, both of which boost the initial insult's force. For example, free-speech expert/lawyer Danielle Citron remarks how "revenge porn" has proven to "inflict financial, emotional and physical harm far graver than many thefts" (2014: 189). Given that the absence of accountability tends to permit harmful behaviors, this chapter explores how various ethical frameworks might apply emotion chips to curb hate speech in order to delimit incivility.

To appraise various algorithms' relevance, this chapter first contrasts both legal and societal norms afforded people doing business face-to-face in private spaces with those engaged online in anonymous enterprises. This chapter then identifies eight unintended consequences, given the spate of legal and sociological texts that have arisen to address issues related to unfettered free speech online. Since "code" is the primary mode of internet regulation (Lessig, 1999), this chapter next analyzes the range of tools currently in use to deter hate speech, including the potential for self-learning AI to grasp empathy, which is central to applying more complicated ethical schemes.

With the potential for empathic AI better understood, this chapter evaluates how victims of unfettered online free speech might have fared under corrective algorithms as deployed by Kant's moral maxim, utilitarianism, libertarianism, Rawls' "worst-off fares best" scheme, and human rights. This chapter gauges which of these models best augments individual security and protects privacy. Absent a research team, this chapter is entirely speculative, but it ought to provide AI researchers with an applicable template for boosting available tools' ethical benefits.

Finally, this chapter recommends that ethical review boards be established to assess and evaluate the consequences of social media algorithms, just as ethical review boards determine boundaries in medical research that protect human subjects and improve data collection/analyses. Absent such review boards, Musk and his ilk will continue to frame content modification as personal or political when in fact algorithms could be programmed to administer universally administrable standards requested by ethical review boards to delimit potentially harmful online activities.

Context

The costs associated with free speech: online versus on-land scenarios

Clearly, the internet was designed to be a playground where opinions scatter unfettered and myriad ideas reach as diverse an audience as possible. In this context, the internet was meant to be a public square, where far-flung participants reign omnipresent, while various voices ring audible, simultaneously. That websites feign openness and inclusivity lends them their public sense. However, the internet's actual "publicness" is up for debate since its portals, or websites, are more likely to be privately owned than government-owned. Moreover, one could argue that hate speech requires a public forum to make it "public speech," so violent online threats that occur on privately owned websites, rather than public space, don't necessarily count as hate speech. Even if definitionally accurate, such a view runs counter to societal norms. For example, hate speech is not illegal in the United States,[1] yet people expect proprietors to protect them from hate speech, and cruelty more broadly, even when traversing privately owned spaces, whether virtual or real. People don't expect to be heckled or harassed while shopping, watching a movie or visiting friends' homes. Even so, certain kinds of activities in private spaces such as crossing a picket line, defying cultural protestors, or entering abortion clinics do risk exposing participants to public discomfort, which is why security is usually readily available.

Most amazingly, the internet collapses time and space, eliminating time spent awaiting information and erasing the distance between subjects and objects, whether images, ideas, people, or things awaiting consumption. Unlike the real world, where people are at risk of getting shot or beat up, life on the internet feigns "harmless good fun," yet users are routinely subjected to psychological abuse, which feminists (and psychologists) claim is far worse than physical violence "because it damages self-esteem so deeply" (Pharr, 1997: 15). Online threats leave victims feeling terrorized and especially anxious. Moreover, recent evidence points to the additional toll taken by those who earn their living monitoring online incivility. "Specifically, moderating [only] uncivil comments made people more emotionally exhausted, and this exhaustion in turn led people to be less accurate in picking which comments to reject or accept for publication on a news site comment thread" (Riedl et al., 2020). Not only are women prime targets for online abuse, but as statistics demonstrate, online harassment correlates directly with on-land violence. Moreover, news reports indicate that perpetrators emboldened by online rants show up at people's homes, publicly embarrass celebrities, or harass personalities while dining out.

While the internet's worldlessness makes it a poor substitute for the public square (Spaid, 2019), its worldlessness not only affords perpetrators impunity, but it has enabled the internet to *pose* as a particularly safe and secure place, especially designed to protect and ensure free speech. In fact, websites are rather nowhere. They not only fail to protect their community members, but websites designed to be profitable lure unsuspecting visitors, solicit people's private information, repackage personal data, and sell it. Even if internet users are rarely at risk of physical violence, the fact that plenty of social media algorithms *incidentally* incite hate and prompt rage, if not addict users, risks amplifying many more users' vulnerabilities (Zickgraf, 2022). Additionally, encrypted tools such as iPhones and WhatsApp, which initially arose to secure user privacy, aggravate societal insecurities when their encryption mechanisms aid wrongdoers' capacities to escape law enforcement detection.

Despite the aforementioned risks to privacy and personal security associated with online activities, online sales are typically treated as free speech, which on its surface grants clients privileges unavailable in on-land sales where sellers' rights reign prior to purchasers' rights. For example,

---

[1]With Snyder v. Phelps (Conery, 2011), the US Supreme Court defended the staging of hateful protests during military funerals by arguing that "such speech cannot be restricted simply because it is upsetting or arouses contempt." https://www.washingtontimes.com/news/2011/mar/2/supreme-court-oks-church-protest-military-funerals/.

clients who refuse to follow brick-and-mortar enterprises' restrictions, such as mask mandates, wearing shirts/shoes, or not carrying guns, can be denied services. In 2015, the Colorado Supreme Court denied an on-land baker's right to refuse to serve gay clients. Three years later, the US Supreme Court upheld (7−2) the same "conservative" baker's "religious right" not to have to bake a wedding cake for a gay marriage. A similar outcome occurred in Belfast, the UK. Soon after a gay activist placed an order for a cake that read "Support Gay Marriage" adjacent to two Muppets, he received word that the baker would not fill his order. In 2014, the court supported his right to have a "Christian baker" make this cake, only to have the UK Supreme Court overturn the lower court's decision in 2018 in support of the bakery owner's religious right not to endorse gay marriage.[2] Both rulings prove that on-land sellers' religious rights are prior to their clients' free speech rights.

Given the internet's anonymity and its prioritizing of free speech, being denied an online purchase seems *prima facie* absurd. Unlike the aforementioned bakery clients whose visibility exposes their "identities," online buyers tend not to expect online sellers to refuse services, since sellers either use third-party data to lure potential clients or cannot specify a buyer's "identity." If sellers used third-party data to *deny* services, rather than to avail desirable options, *rejected* clients would no doubt feel violated, which could cause them undue psychological harm. In availing their services online, buyers erroneously presume that sellers such as bakers accept the norms of free speech familiar to anonymous deals. Even though online sellers who refuse clients' services put themselves at risk of being assailed via social media, online sellers are even more at liberty to refuse services, since they can simply ignore purchasers' online requests.

In order for online sellers to follow local laws, they must already know something about their clients such as their shipping and/or home addresses. For example, European nations such as Austria, France, Germany, and Hungary forbid online purchases of Nazi memorabilia from anywhere (Walfisz, 2022). That governments restrict purchases is typically considered proof that free speech is not a *universal* right. So long as sellers refuse online orders and governments outlaw certain sales, online sales are not actually governed by free speech rights, even though they are presumed to be. As the next section demonstrates, social media corporations also tend to prioritize their free speech rights over the well-being of their clients/community members.

The recent spate of texts focused on hate speech's unintended consequences

To demonstrate the violence arising from free speech online, I next summarize views that challenge the erroneous assumption that the internet is safer than the street. Given the "First Amendment and the borderless nature of the internet," American Defamation League director Abraham Foxman and privacy/internet lawyer Christopher Wolf worry that "governing bodies are largely helpless to control this massive assault on human dignity and safety" (2013). Having authored *Viral Hate: Containing Its Spread on the Internet*, they remain concerned by the internet's capacity to distort expression and therefore undermine democracy. They worry that the lethal combination of business demands and tantalizing social media serve to ignite the contagion of hate that has been spinning out of control of late. In light of the difficulties of legally thwarting minors' internet access in public libraries, their solution to arouse an engaged citizenry who will force the government to pass

---

[2]https://www.bbc.com/news/uk-northern-ireland-59882444.

laws outlawing hate speech seems a bit naïve. One likely drawback is the fact that if citizens revolt, organizations such as the American Civil Liberties Union (ACLU) will feel empowered to defend hate speech as free speech all the more.

Former Wilson Center fellow Nina Jankowicz, whose research also addresses the way online disinformation dismantles democracy, worries that "women face a disproportionate [number] of attacks online. These range from physical insults to threats of violence, and they're forcing women—especially younger ones—to censor themselves out of fear of physical or emotional retribution" (Jankowicz, 2022). As she boldly points out, social media corporations lack the will to fight vitriolic content so long as such behaviors drive site engagement, a point I discuss in greater detail later. This is exemplary of sellers' rights taking priority over clients' rights.

Internet lawyer Sarah Jeong considers online harassment a threat to the internet's usefulness. Her book *The Internet of Garbage* addresses the "media's simplistic focus on 'mean words', rather than on more concerted cyber-stalking attempts (e.g., looking up real estate records, going to a person's place of work or private residence, taking photos of that place and/or documenting that person's whereabouts … or worse)." She argues, "Everything is primarily driven by corporate interests. There is little thought for the safety and privacy of ordinary individuals. There are many changes that can be made at both the national and international level that would make all potential targets of harassment safer." She believes that the solution is "architectural," meaning that to be successful the solution must be code-driven. The ethics must be *in* the technology. Neither norms nor the market is likely to curb, let alone prevent, online harassment (Jeong, 2015).

Laura Penny's book *Cybersexism: Sex, Gender and Power on the Internet* treats hate speech as endemic of sexism toward women more broadly, for example, "people's inability to deal with sex in a way that is not violent, guilty or contemptuous of women and girls" (2013). Not only does she consider online violence "real" violence, but she also notes that the "hatred of women in public spaces online has reached" epidemic proportions. She remarks how "any woman standing up in Parliament, or a lecture theatre or a room full of her friends, to talk about her own experiences [has] learned to anticipate violence, threats and taunting [as] if she happened to upset the men." Finally, she testifies that like "many women with any sort of online profile," she's had to get used to hate speech: "the violent rape and murder fantasies, the threats to my family and personal safety, the graphic emails with my face crudely pasted onto pictures" (2013).

Danielle Citron considers current laws and civil rights sufficient to compensate "revenge porn" victims, so long as they are willing to sue their perpetrators. However, few states and nations outright outlaw, let alone prevent, such harm from happening in the first place. "On the criminal law front, harassment and stalking laws should be updated to reach the totality of the abuse, and revenge porn should be banned. Civil rights law should penalise online harassers who interfere with someone's right to pursue life's crucial opportunities—work, education and self-expression—due to group bias" (2014: 142). Since prosecutors are reluctant to prosecute misdemeanors, she also recommends classifying "cyber harassment" and "unlawful surveillance" as felonies (2014: 144). She argues that "revenge porn legislation does not trample on the First Amendment [free speech rights] because it protects a narrow category of private communications, private matters whose protection would foster private speech" (2014: 212). Were it illegal to post images online for which subjects have not authorized their consent for some particular time period (say 5 years), which has long been the case for film and print media, revenge porn would largely come to a halt.

Whitney Phillips takes a different tack. She actually argues that "trolling fits comfortably within today's media landscape" (2015). She considers "trolls choices born of and fueled by culturally sanctioned impulses, which are just as damaging as trolls' most disruptive behaviours" (2015). Joseph Reagle, whose research addresses readers' online comments, investigates "hate reading," such that readers' comments frame "content that is thought, rather than read, to be awful or objectionable," thus engendering a chain of distorted responses, much like the children's game Telephone (2015). Jonathan Greenblatt elegantly summarizes everybody's concerns:

> Social media has been at the center of the storm for more than a decade,
> and its toxic potential reached new heights during [Trump's] presidential term.
> Whether you consider it the catalyst or just a conduit, the fact is that social media
> drives radicalization. It's a font of conspiracy theories, a slow-burning acid
> weakening our foundations post after post, tweet after tweet, like after like. And
> the hate festering on social media inevitably targets the most vulnerable—
> particularly marginalised groups like religious, ethnic, and racial minorities, and
> members of the LGBTQ community (Greenblatt, 2020).

As indicated by these eight unintended consequences of free speech online (the contagion of hate, self-censorship, online harassment, cybersexism, revenge porn, trolling, hate reading, and radicalization), current laws regarding hate speech favor online platforms' right to traffic in free speech over citizens' privacy and security rights. Once again, clients must pay the hidden costs associated with free speech. Unlike brick-and-mortar businesses that regularly reinforce civility, online enterprises simply treat hate speech as a cultural artifact, "the cost of doing business" online. Online sellers tend to defer to free speech when doing so benefits them, yet their parallel efforts to dismiss criticism denies free speech. Only the threat of online clients' taking their business elsewhere has motivated such sites to ensure their customers' security and privacy.

Position statement: rethinking social media's algorithms
Social media algorithms that incite hate speech

The recent article "Angry by Design" alludes to the aforementioned outcomes:

> Design privileges certain forms of content: it enables particular kinds of
> relations, and encourages specific forms of participation. For this reason,
> design prove[s] to be a productive lens for understanding toxic communication.
> Of course, this study also ha[s] its limits. In particular, the degree to which
> design may influence individuals—and how that influence might be modulated
> by age, gender, class, or culture—has yet to be precisely determined. One path
> for [future] research would be to take up this challenge, producing a more
> quantitative analysis of design influence. Another path would be to apply this
> approach to other platforms: Reddit, TikTok, 4chan, and so on (Munn, 2020).

This article was among the first to touch upon what the aforementioned authors had suspected regarding the rise in incivility, though they could not pinpoint what was driving it. Indeed, "Angry by Design" author Luke Munn paints the internet as a very dark street, lacking in street lamps, whose hidden dangers reflect the algorithms' role in rigging site activities to favor site owners, who after all hire designers to design optimally profitable websites. The aforementioned authors long suspected something downright malicious: The symptoms were obvious. Failing to recognize the algorithms' role in escalating toxic communication, the aforementioned authors mistakenly placed the blame on either men, sexism, society, or misinformation. And of course, these could be the underlying factors that generate ensnaring, rather than neutral, algorithms. It's rather difficult to understand how either sexism or misinformation could be optimally profitable design outcomes.

Few people, and especially those corporations commissioning and employing the algorithms, seem willing to admit that the very algorithms meant to boost visitor traffic to websites simultaneously incite prejudice and spur local crimes. According to Jack Loughran, "Data from Cardiff University's HateLab project showed that as the number of 'hate tweets' made from one location increased, so did the number of crimes in the real world—including violence, harassment and criminal damage" (Loughran, 2019). In fact, the correlation between "hateful" Twitter posts and "racially- and religiously-aggravated crimes" during eight months from 2013 to 2014 was so accurate that the HateLab realized social media algorithms could predict and prevent hate crime. He added: "In time, our data science solutions will allow us to follow the hate wherever it goes" (Loughran, 2019). This is simultaneously frightening and elucidating.

Social media algorithms designed to identify and deter hate speech

Clearly, those with the authority to regulate hate speech ought to use data science to follow and deter hate. Given free speech legal scholar Larry Lessig's emphasis on the regulatory aspect of computer code, I next review a range of algorithms currently being tested to identify hate speech in order to deter it. I then use this information to analyze whether it might be possible to design algorithms that carry out ethical schemes in order to generate more desirable outcomes that hopefully curb bad actors, override undisciplined behaviors such as Musk's inappropriately attacking Twitter's general counsel, and eliminate human biases that poison most algorithms from the start. The goal is to generate outcomes that honor clients' rights by enhancing their security, privacy, and free speech, rather than making a mockery of users' security, privacy, and free speech.

Lola, a tool developed by the University of Exeter, was one of the first bots designed to detect hate speech. According to Luke Dormehl, it "harnessed the 'latest advances in natural language processing and behavioral theory' to scan through thousands of messages a minute to uncover hateful content" (2020). Researchers claim that "Lola analyses 25,000 messages per minute to detect harmful behaviours—including cyberbullying, hatred and Islamophobia—with up to 98% accuracy" (Dormehl, 2020). Lola employs an "emotion detection engine" that identifies simple emotions such as "anger, fear, love, trust, etc." (Dormehl, 2020).

In order to reduce hate speech, countries such as France have criminalized certain content, requiring social media sites to pay fines if criminalized content isn't removed within an hour. Bots like Lola are necessary to speed up this process, if not to nip it in the bud. Since May 2021, Twitter has prompted tweeters to review and revise "potentially harmful or insulting" tweets (Bhattacharjee, 2021). Supposedly, just requesting users to rethink tweets has led to 11% fewer *offensive* tweets. For Twitter to make this claim, however, Twitter must have saved users' draft tweets prior to their having tweeted revisions, which seems like a privacy violation all its own.

Not surprisingly, most algorithms written to flag hateful speech simply search for particular words. As a result, some have incidentally amplified racial bias by blocking inoffensive tweets by black people or other minority group members. Initially, AI models were 1.5 times as likely to flag tweets posted by African-Americans as "false positives" as compared to other tweets. Fortunately, a University of Southern California team recently created a "hate speech classifier" that is more "context-sensitive and less likely to mistake a post containing a group identifier as hate speech. To achieve this, the researchers programmed the algorithm to consider two additional factors: the context in which the group identifier is used, and whether specific features of hate speech are also present, such as dehumanizing and insulting language" (Dawson, 2020).

Programming AI to develop empathy skills

For AI to judge whether online speech is dehumanizing or insulting requires programming AI to develop empathy skills, a natural human emotion that many human beings are reported to lack. Unlike sympathy, which requires people to recall similar circumstances (childhood sickness, family deaths, romantic break ups, or disappointing performances), empathy requires one to imagine what it would be like to be in a situation that one has yet to experience. For example, a friend mentions how difficult her life has been of late since she's had to wear a cast for six weeks, but you've yet to break a bone, let alone wear a cast. Of course, the fact that most people have witnessed others on crutches having to lug around heavy casts makes imagining wearing a cast for six weeks easier than other kinds of cases, especially racial discrimination. The very reason people accidentally hurt another's feelings is that they fail to imagine all the ways the other might interpret and respond to their words. They especially find themselves in trouble when they assess that it might be better to take a neutral approach than to be forthright. Even honesty has its costs.

Empathy requires thinking outside the box, which of course is easier for those who have experienced or witnessed many more different kinds of experiences. It is widely known that reading literary fiction, as opposed to popular fiction, improves empathy, since it provides readers access to novel experiences, much like witnessing others, that help individuals "understand what others are thinking and feeling" (Chiaet, 2013).

> Literary fiction, by contrast, focuses more on the psychology of characters and their relationships. "Often those characters' minds are depicted vaguely, without many details, and we're forced to fill in the gaps to understand their intentions and motivations," [David] Kidd says. This genre prompts the reader to imagine the characters' introspective dialogues. This psychological awareness carries over into the real world, which is full of complicated individuals whose inner lives are usually difficult to fathom (Chiaet, 2013).

This is called artificial empathy, since it occurs at a remove, rather than being experienced firsthand. One advantage that AI has over human beings is that AI can be exposed to many more examples of extant literature whose characters' empathic strengths and weaknesses generate positive/negative situations. If human beings could *digest* as many situations as AI, presumably, they too would be more empathic.

According to the *Personality and Social Psychology Review*, American students reported 48% less empathy in 2009 than in 1979. "Possible causes of the growing empathy gap include increasing materialism, changing parenting methods and the digital echo chamber, in which people anchor themselves in close-knit groups of like-minded people. Such echo-chamber effects aren't always as obvious as those seen on social media. For example, researchers have found that the matching processes used on dating platforms can also weaken social bonds" (Zurich, 2019). As it turns out, online interactions further weaken empathic feelings. "The *Global Risks Report* highlights that while online connections can be empathetic, research suggests that the degree of empathy is six times weaker than for real-world interactions" (Zurich, 2019).

No doubt, programming AI to grasp empathy would provide complementary emotions for human beings who are either too stressed out to care, too undisciplined to behave correctly, too distracted to worry about how their words impact others, too self-absorbed to consider others' feelings, or too worn out from monitoring uncivil behaviors. In short, empathy matters most when people engage online strangers and acquaintances, yet there are myriad reasons why people lack it and are thus primed to insult or dehumanize other people, however unintentionally. As Minter Dial reiterates, it's even "harder to get into the shoes of someone who comes from a different background, different culture, different language, different sex, different race, and so on" (De Lallo, 2020). Moreover, the internet obscures people's shoes.

The real question then is, "Is technology advanced enough to train AI to employ empathy?" Dial's book, *Heartificial Empathy: Putting Heart into Business and Artificial Intelligence* (2018), "describes artificial empathy as the coding of empathy into machines. He describes artificial empathy or heartificial empathy as personal, situational and based on the appropriate intentions. Because empathy can be learned, artificial intelligence can surely be equipped with artificial empathy in the years to come" (Wu, 2019). Presumably, programmers with exceptionally high emotional IQs and capable of considering all angles will be recruited.

The ethical dilemma at the heart of artificial empathy

Paradoxically, some people question whether it is ethical to introduce empathy into AI, since introducing human emotions into AI increases the risks of AI "one day displac[ing] humankind" (Wu, 2019). Note that this doesn't necessarily result in negative outcomes. A recent review of 96 research articles addressing the results of replacing humans who regularly face burnout as a result of working with dementia patients stressed social robots' positive capacities in helping dementia patients "work independently in basic activities and mobility, provide security, and reduce stress" (Alonso et al., 2019). Moreover, Alzheimer's patients benefit from interactions with pet robots, as do autistic children who interact with social robots.

Most human beings find being empathic online challenging, while those who are especially good at it on land are most at risk of emotional exhaustion. Since human beings stand to suffer more as a result of this "empathy gap," it actually seems rather necessary to program AI to do what human beings don't do particularly well. Even if AI proves better at eliminating insensitive language, it hardly means that AI will eventually replace human beings. Using AI to avail empathy more broadly is more like using computers to perform calculations or to follow decision matrices than replacing human beings in order to save hourly wages. In fact, if thoughtfully implemented, doing so could actually preserve human resources that are otherwise vulnerable to emotional exhaustion (aka burn out).

An added bonus is that people who work for empathic businesses tend to lead more ethical lives. If AI can boost the empathy of either businesses or online platforms, then people's lives could become more ethical as well (Wu, 2019). Minter Dial discusses a study that shows that businesses known for showing empathy within their culture and toward their customers typically experience profits. "And that shows up in the shareholder stock price. The study evaluated 170 publicly traded companies, on a range of some 50 criteria, on their empathic ability. The top ten with empathy outperformed the bottom ten by two times on the stock market" (De Lallo, 2020).

How emotion chips boost empathy

So, what kind of AI product actually stands to boost empathy? According to Jun Wu,

> Emoshape is the first company to hold the patent technology for emotional synthesis. The emotion chip or EPU developed by Emoshape can enable any AI System to understand the range of emotions experienced by humans. At any moment, the EPU can understand 64 trillion possible emotional states every 1/10th of a second. The range of your emotions is mapped onto a gradient where the degree of each emotion can be observed. Signalaction AI is using Emoshape to make situational awareness actionable by generating real time insights with emotional intelligence from voice and text communication to empower users to make smart choices for positive outcomes (Wu, 2019).

If Wu's description is true, then artificial empathy is available and ready for implementation.

Even if artificial empathy can be programmed to neutralize human reactions, researchers must contend with the fact that people respond differently to different kinds of experiences. According to research, children tend to prefer visual experiences free from informative texts or emotional baggage. When German children were shown a short video of a snowman melting under the warm afternoon sun, they had the greatest galvanic response (skin sensations) to the video without a soundtrack and rated it the most pleasant, whereas they remembered the video accompanied by an emotionally charged soundtrack. By contrast, the video accompanied by factual information had the greatest response in terms of heart rate and breath, yet they rated it least pleasant and barely remembered it (Massumi, 1995, 83−84). This suggests that children get more pleasure from images, while words pump them up; a point that might also explain adult's heated reactions to websites that invite them to read and/or contribute commentary. Perhaps the very act of thinking about what others have written or said triggers hate reading. Perhaps the problem is not just the algorithm, but the way verbal language physically impacts our bodies, so the very notion of "free speech" is fraught from the get go. No wonder beefing up security is such a tall order.

So long as society depends on AI to fill the empathy gap, rather than employing AI to show people how they can behave more empathically, society will remain awash in personalities who thrive on radicalizing readers by inciting reactions that cause controversies. Such perpetrators would no doubt reject platforms whose tools deter hate speech, which they would view as curtailing their free speech. However, I imagine myriad social media fans who share Jeong's worry that the internet is so full of garbage that it's often difficult to locate creative and inspiring material. I thus envision people gravitating toward online platforms where they can share comments, knowing that software deterrents are in place to constrain the kinds of escalations that engender local violence on land. Such tools could become invaluable sales features, akin to encrypted software available on iPhones and WhatsApp that draw customers, rather than deter them, leaving those who aim to harm behind.

Imagine that websites programmed with EPU chips guide contributors to "make smart choices for positive outcomes" by discouraging people from posting inflammatory comments. Rather than merely "nudging" commentators to "review and revise," the ideal software provides users with samples of how other readers might interpret their texts, enabling authors to get a sense of how others might react down the line. This response seems more positive than some EPU chip's rejecting contributors' words outright, which would indeed prove frustrating. An EPU chip might even be able to evaluate contributors' emotional states based on what they wrote. Such a move might help writers to grasp how others see them. Given written comments' tendency to heighten emotions, tools are needed to predict the chain of future readers, who have yet to read our words, and may even encounter our comments in very different contexts. Comments deemed harmless years ago can appear insensitive later. An effective EPU chip ought to capably predict "multiple" others' responses, as comments escalate emotional tensions.

Discussion: filling the empathy gap, while boosting empathy
Testing artificial empathy under five ethical approaches

Presuming that EPU chips steer readers toward the least "radicalizing" comments, let's determine "right action" by testing the relevance of human rights plus four familiar ethical approaches: (1) the "categorical imperative," such that one imagines a world where everyone does whatever action one is considering doing, (2) utilitarianism, such that one imagines how the consequences of one's actions contribute to the overall happiness/pleasure of the whole society, (3) libertarianism, such that one prioritizes one's own happiness so long as no one else is harmed and (4) a Rawlsian society such that resources are redistributed so that the least advantaged incur the least harm.

Basic appeals to human rights seem implausible, so long as governments, social media sites, and online sellers prioritize their right either to censor free speech or incite hate speech at a cost to their communities. The fact that 93% of those online use social media makes hate speech a global problem for online communities that lack the capacity to curtail hate speech. That Facebook posts fanned Burmese hatred toward the Rohingya, resulting in 700,000 refugees, is a notable case in point (Mozur, 2018).

Supposedly only 63% of the world's population has internet access, so a Rawlsian approach seems irrelevant, so long as the least advantaged are not even invited to the party. While social media critics might consider not having online access advantageous, since nonusers avoid the related psychological traumas, a lack of internet access hardly protects hate-speech targets from becoming victims of on-land violence. In fact, the UN's Special Rapporteur on Minority Issues Report notes that more than 70% of hate speech is directed at minorities (de Varennes, 2021). Fernand De Varenne remarks how "The theme of hate speech, social media and minorities pits many of the most vulnerable communities—for example, ethnic minorities such as persons of African descent, Asian communities, Dalits, Rohingya and Roma, and religious or belief minorities such as Baha'i, Muslims, Jews and Christians—in countries in Africa, Asia, the Americas and Europe against the interests of some of the most powerful corporate entities, with States sitting uncomfortably between the two" (2021: 4–5).

Although it's difficult to predict exactly how EPU chips will be used, I imagine Rawlsians who aim to protect the least advantaged using them to evaluate the inflammatory potential of myriad responses to particular comments. Given the way ChatGPT-3 currently works, such that probabilities drive word associations, presumably whatever software operates ChatGPT-3 could be deployed to predict potential reader responses. The EPU chip could then be used to rank comments from least radicalizing (neutral) to most radicalizing (inflammatory). Those who dare to post "inflammatory" comments would see the statement: "Not deliverable due to the high predictability of inflammatory language." The contributor must either reword his/her comment or depart the site for another.

As an alternative to Rawlsians who use EPU chips to detect inflammatory speech, I imagine EPUs being used to evaluate alternative ethical approaches. The Kantian could use the EPU to divine empathy-sensitive principles that everyone ought to follow and the Utilitarian might employ mathematics to calculate the societal consequences of particular points of view and attitudes, while the Libertarian could deploy the EPU to assess the potential harm arising from his/her placing his/her liberty first. Although such EPU assessments would prove enlightening, it remains to be seen whether users actually care enough to modify their online behaviors in accordance with such revelations. I thus recommend sending impending radicalizers the respective notices: "Not deliverable due to implausible moral maxim," "Not deliverable due to undesirable outcomes for most people" or "Not deliverable due to likelihood of harming others." Such responses are not exemplary of machine morality, since they reflect computer programs, and are hardly some machine's decision (Wallach & Allen, 2009). A better description might be to call such machines "moral aids," much like the "loco parentis," teacher, priest, or lawyer who advises his/her charge regarding the best course of action, yet at the end of the day each charge makes his/her own move. If people consistently receive such messages, it should increase the odds that they will seriously reconsider their language patterns. Perhaps only 11% (the Twitter fraction) will revise their comments, while 89% will flee to other sites. EPUs alerting commentators to the chain of responses whose escalation risks on-land violence have to be an improvement.

When it comes to hate speech, messages resulting from the application of either the categorical imperative or the utilitarian framework seem too abstract and thus not forceful enough to get people to "make smart choices for positive outcomes." By contrast, the simpler libertarian notice that one's potential action risks harming others might be sufficient to inspire recipients to modify their texts. Most cultures follow the golden rule such that people are encouraged to treat others as they would like to be treated. Those who acknowledge that they don't want to be body-shamed, dismissed as too old or routinely ignored, are less likely to discuss weight, disparage ages, or dismiss others' ideas. If people are just not conscious of what they're writing, then EPU use can at least get people to pay attention to how others might react to their messages. On this level, the emotion chip serves as an empathic mirror that reflects contributors' unempathic choices when it comes to creating comments.

So long as the emotion chip is used as designed, users cannot be blamed for unintended consequences. Luciano Floridi's ethical theory broadens "the class of moral agents to include robots, software bots and other information technology (IT) systems. He defines the moral agent as an interactive, autonomous and adaptable transition system capable of performing 'morally qualifiable' actions, that is, actions that can cause good or evil" (Spinello, 2021: 23). On this level, the very websites and social media sites that elicit comments deemed good or evil count as artificial moral agents. As a result, brick-and-mortar creators are accountable for harm resulting from using products as intended and they're charged with censoring undesirable actions. Using emotion chips to "deflame" responses goes a long way toward reducing harm.

One question that warrants attention is whether the underlying problem concerns how human beings are wired, rather than some underlying algorithms' capacity either to amplify or to correct/deter human behavior. The answer to this conundrum matters greatly since as has been briefly discussed, there are those who like Elon Musk consider free speech at all costs a cause worth fighting for, despite its inevitable harm, and alt-right trolls such as Steve Bannon and Alex Jones who maximize their pleasure (and profits) by radicalizing listeners (Lawton & Wheatley, 2023). As briefly noted, Phillips frames trolling as "culturally sanctioned" not because of "human wiring," but because trolling yields obvious economic benefits that society welcomes. If the problem is "human wiring," then it's no wonder people have difficulties doing the right thing (self-restraint) when they stand to benefit from doing otherwise. Fortunately, our world is neither Hobbesian, such that all people are inherently evil, nor Dawkinsonian à la *The Selfish Gene* (1976), such that human beings are evolutionary-programmed to prioritize their needs and/or those of their offspring. No doubt, there is definitely a "naughty faction," a tiny cohort of ruthless and/or undisciplined people who just won't take the time to imagine how their words elicit undesirable reactions. No EPU chip will restrain the naughty from being naughty. Naughty people have their own logic that apparently works for them; otherwise they wouldn't care to be naughty. One could say that laws and regulations exist to restrict naughtiness.

Assessing the emotion chip's role in curtailing unintended consequences

The last question that warrants attention is whether EPUs that employ empathy in a manner that alerts users to potential harm could alleviate free speech's eight unintended consequences (the contagion of hate, self-censorship, online harassment, cybersexism, revenge porn, trolling, hate reading, and radicalization). Like any moral action, the principle depends less on the principle than on the principled, that is, each actor's motivation for right action and willingness to enact principled behavior. Would people, aided by emotion chips capable of predicting how others might react to their comments, reconsider their words? If the contagion of hate de-escalates, these eight unwanted outcomes ought to diminish as well. There might even be less hate reading and far less cybersexism and revenge porn. Should an overly sensitive EPU reveal how easily others might be insulted by one's comments, self-censorship is likely to occur. If people had tools on hand to assess how their naughty choices hurt others or caused others to react to their words, they may take less pleasure in being naughty, even if the naughty faction never stops harassing, trolling, and radicalizing. No doubt, the widespread availability of EPUs would change norms such that if people's tweets and comments caused escalations, others would wonder why the author failed to heed the EPU chip's predictions.

As Twitter's "revise and review" program demonstrates, just taking one's time to reflect on one's choices is sufficient to lead one to make better choices. Neuroscientists who distinguish the reptilian or primal brain (basal ganglia), paleomammalian or emotional brain (limbic system), and the neomammalian or rational brain (neocortex) credit the latter with empathy, imagination, reflection, understanding, problem-solving, and decision-making (Budson, 2017). It thus starts to make sense why AI could help busy and distracted human beings to behave more rationally, make better choices, and not let their emotions rule their responses.

So long as some commentators believe that bullying others benefits the bullies, there will always be a subset of commentators who make poor choices, however mean or unfair. It's thus unlikely that all (or even most) users would alter their actions unless they felt certain that doing so would reduce the chance of others inferring undesirable readings that engender chains of negative responses. Therefore, the best option is to empower EPUs to inform contributors of how others might interpret every comment. That already far exceeds what parental, peer, and educative systems reinforce.

Establishing ethical review boards to evaluate social medial algorithms

Just as ethical review boards have been established to evaluate clinical trials in order to protect trial participants, ethical review boards must be set up to evaluate how social media algorithms impact users' well-being. This view coheres with Anderson and Anderson's recommendation that "ethically significant behavior of autonomous systems should be guided by explicit ethical principles determined through a consensus of ethicists" (Anderson & Anderson, 2018). This originally seemed a novel idea, but evidently others are thinking along the same lines. Given the ethical challenges associated with developing and testing self-driving cars, Boston's Institute for Leadership and Innovation suggests:

> To avoid [AI failure] in the future, we should not leave this to the technology developer; an Ethics Review Board should be used. Ethics Review Boards might be established at a federal level and/or a state level. They would be tasked with tasks of trying to guide how the AI should do when encountering moments (providing the policies and procedures, rather than somehow "coding" such aspects). They might also be involved in incident assessment (Dukakis, 2018).

Of course, given the myriad challenges associated with empathic thinking, such ethical review boards must be comprised of diverse communities so as to fully assess the emotion chip's success at reading, interpreting, and proposing alternative commentaries.

Conclusion

One theme that has been consistent throughout this chapter is the way online clients have paid dearly for free speech, while online sellers, websites, social media platforms, and corporations have profited handsomely by flaunting free speech. To balance this huge gap, which has jeopardized clients' privacy, security, as well as their rights, AI researchers must continue to develop empathic emotion chips and programmers must figure out how to accurately employ such chips to capably interpret human language choices associated with tweets, comments, reactions, newsfeeds, etc. This chapter's focus has been hate speech, which is ordinarily one-way (a verbal attack, not a conversation), but the very same emotion chips primed to mitigate one-way banter must simultaneously be both developed for conversational purposes, since AI's most likely application concerns coherent conversations between AI software and human beings in need of therapy, companionship, or supervision.

Until recently, the idea of programming a robot to make a human being fall head over heels in love with it seemed entirely implausible. But as Maria Schrader's fascinating film *Ich Bin Dein Mensch* (2021) demonstrates, robot partners are not only plausible, but their greater utility and vulnerability lead people to depend on them and care for them even more than they do human partners.

Given the well-documented unintended consequences of unfettered free speech, as well as the tendency for online banter and insults to spur violent crimes on land, it would be great to have tools that alert us to language that risks harming others and demonstrate how to make smarter word choices, so that written words no longer incite chains of angry mobs. If AI can provide this much-needed assistance, then so be it. EPUs that propose less humiliating phrasings cannot be as annoying as autocorrect software that finishes people's sentences.

References

Alonso, S. G., Hamrioui, S., de la Torre Díez, I., Cruz, E. M., López-Coronado, M., & Franco, M. (2019). Social robots for people with aging and dementia: A systematic review of literature. *Telemedicine and e-Health*, 25(7), 533−540. Available from https://doi.org/10.1089/tmj.2018.0051, Accessed 26 July 2022.

Anderson, M., & Anderson, S. L. (2018). GenEth: A generational ethical dilemma analyzer. *Journal of Behavioral Robotics*, 9(1), 337−357. Available from https://doi.org/10.1515/pjbr-2018-0024, Accessed May 20, 2023.

Bhattarcharjee, S.S. (2021). Twitter will now nudge you if it thinks you're about to tweet something toxic. Mashable India. https://in.mashable.com/tech/22063/twitter-will-now-nudge-you-if-it-thinks-youre-about-to-tweet-something-toxic. Accessed July 27, 2022.

Budson, A. (2017). Don't listen to your lizzard brain. Psychology Today. https://www.psychologytoday.com/us/blog/managing-your-memory/201712/don-t-listen-your-lizard-brain. Accessed March 15, 2023.

Chiaet, J. (2013). Novel finding: Reading literary fiction improves empathy. Scientific American. October 4.

Cambridge Dictionary. (2023)Available from https://dictionary.cambridge.org/dictionary/english/hate-speech. (2023) Accessed 20.05.23.

Citron, D. (2014). *Hate crimes and cyberspace*. Cambridge: Harvard University Press.

Conery, B. (2011). Supreme Court upholds protests at military funerals as free speech. Washington Times. Available from https://www.washingtontimes.com/news/2011/mar/2/supreme-court-oks-church-protest-military-funerals/.

Dawson, C. (2020). Context reduces racial bias in hate speech detection algorithms. USC Viterbi School of Engineering. https://viterbischool.usc.edu/news/2020/07/context-reduces-racial-bias-in-hate-speech-detection-algorithms/. Accessed January 30, 2024.

De Lallo, D. (2020). Getting the feels. Should AI have empathy? https://www.mckinsey.de/business-functions/mckinsey-analytics/our-insights/getting-the-feels-should-ai-have-empathy. Accessed July 15, 2022.

de Varennes, F. (2021). Report of the Special Rapporteur on minority issues (A/HRC/46/57). UNHRC. https://reliefweb.int/report/world/report-special-rapporteur-minority-issues-ahrc4657. Accessed March 15, 2023.

Dormehl, L. (2020). Humans moderators can't stop online hate speech alone. We need bots to help. Digital Trends. https://www.digitaltrends.com/web/solving-hate-speech-with-ai/. Accessed July 15, 2020.

Dukakis, M. (2018). Ethics review board—What is it and why do we need it? https://dukakis.org/news-and-events/ethics-review-board-what-is-it-and-why-do-we-need-it/. Accessed July 31, 2022.

Foxman, A., & Wolf, C. (2013). *Viral hate: Containing its spread on the internet*. New York: St. Martin's Publishing Group.

Greenblatt, J.A. (2020). Stepping up to stop hate online. Stanford Social Innovation Review. https://ssir.org/articles/entry/stepping_up_to_stop_hate_online. Accessed July 15, 2022.

Jankowicz, N. (2022). *How to be a woman online: Surviving abuse and harassment, and how to fight back.* London: Bloomsbury Academic.

Jeong, S. (2015). *The Internet of garbage.* Washington, DC: Vox Media, Inc.

Lessig, L. (1999). *Code and other law of cyberspace.* New York: Basic Books.

Lawton, S., & Wheatley, J. (2023). One month in, insurrectionist MAGA influencers say Speaker McCarthy is one of them. Media Matters. https://www.mediamatters.org/steve-bannon/one-month-insurrectionist-maga-influencers-say-speaker-mccarthy-one-them. Accessed March 15, 2023.

Isaac, M., & Hirsch, L. (2022). With deal for Twitter, musk lands a prize and pledges fewer limits. New York Times. https://www.nytimes.com/2022/04/25/technology/musk-twitter-sale.html. Accessed March 14, 2023.

Loughran, J. (2019). Social media algorithm could predict and prevent hate crime. Engineering & Technology. https://eandt.theiet.org/content/articles/2019/10/social-media-algorithm-could-be-used-to-predict-and-prevent-hate-crime-before-it-happens/. Accessed July 15, 2022.

Massumi, B. (1995). The autonomy of affect. *Cultural Critique, 31,* 83–109.

Merchant, B. (2023). Column: The promise of free speech on Elon Musk's Twitter is officially dead. Los Angeles Times. https://www.latimes.com/business/technology/story/2023-03-06/column-the-promise-of-free-speech-on-elon-musks-twitter-is-officially-dead. Accessed May 20, 2023.

Mozur, P. (2018). A genocide incited on Facebook, with posts from Myanmar's military. New York Times. https://www.nytimes.com/2018/10/15/technology/myanmar-facebook-genocide.html. Accessed March 15, 2023.

Munn, L. (2020). Angry by design: Toxic communication and technical architectures. *Humanities and Social Sciences Communications, 7,* 53.

Penny, L. (2013). *CyberSexism: Gender and power on the internet.* London: Bloomsbury Publishing.

Pharr, S. (1997). *Homophobia: A weapon of sexism.* Berkeley: Chardon Press.

Phillips, W. (2015). *This is why we can't have nice things: Mapping the relationship between online trolling and mainstream culture.* Cambridge: MIT Press.

Reagle, J. (2015). *Reading the comments: Likers, haters, and manipulators at the bottom of the web.* Cambridge: MIT Press.

Riedl, M., Masullo, G., & Whipple, K. (2020). The downsides of digital labor: Exploring the toll incivility takes on online comment moderators. *Computers in Human Behavior, 107.*

Serrano, J. (2022). https://gizmodo.com/elon-musk-twitter-vijaya-gadde-sec-disparagement-1848848684. Accessed March 14, 2023.

Spaid, S. (2019). Surfing the public square: *On worldlessness, social media, and the dissolution of the polis, experience in a new key,* In D. Jørgensen (Ed.), Open Philosophy.

Spinello, R. (2021). *Cyberethics: Morality and law in cyberspace.* New York: Fordham Books.

Thompson, S. (2019). Hate speech and self-restraint. *Ethical Theory and Moral Practice, 22,* 657–671. https://doi.org/10.1007/s10677-019-10004-y. Accessed May 20, 2023.

Walfisz, J. (2022). Nazi Memorabilia: Why did someone just buy Hitler's watch? euronews.culture. https://www.euronews.com/culture/2022/08/02/nazi-memorabilia-why-did-someone-just-buy-hitlers-watch. Accessed March 15, 2023.

Wallach, W., & Allen, C. (2009). *Moral agents: Teaching robots right from wrong.* Oxford: Oxford University Press.

Wu, J. (2019). Empathy in artificial intelligence. https://www.forbes.com/sites/cognitiveworld/2019/12/17/empathy-in-artificial-intelligence/?sh = 792f70526327. Accessed July 15, 2022.

Zickgraf, R. (2022). Elon musk might make it worse, but Twitter was already bad. Jacobin. https://jacobin.com/2022/04/elon-musk-twitter-social-media-platforms-private-capitalists. Accessed January 30, 2024.

Zurich. (2019). Decline in empathy creates global risks in the 'age of anger'. Zurich Insurance Group. https://www.zurich.com/en/knowledge/topics/global-risks/decline-human-empathy-creates-global-risks-age-of-anger. Accessed March 15, 2023.

Further reading

Bialecki, J. (2019). 'The Lord says you speak as harlots': Affect, *affectus*, and *affectio*. In R. Yelle, C. Handman, & C. Lehirch (Eds.), *Language and religion* (pp. 421–441). Boston/Berlin: Walter de Gruyter, Inc.

Hassan, S., & De Filippi, P. (2017). The expansion of algorithmic governance: From code is law to law is code. Artificial intelligence and robotics in the city.

Mantilla, K. (2015). *Gender trolling: How misogyny went viral*. Santa Barbara: Praeger.

Taylor, A. (2014). *The people's platform: Taking back power and culture in the digital age*. New York: Metropolitan Books, Inc.

Index

Note: Page numbers followed by "*f*" and "*t*" refer to figures and tables, respectively.

A

AAA. *See* Algorithmic Accountability Act (AAA)
AAI. *See* Autonomous artificial intelligence (AAI)
Abraham Maslow's pyramid of human needs, 261–263
AC. *See* Affective computing (AC)
Accountability, Responsibility, and Transparency (ART), 65–66
ACLU. *See* American Civil Liberties Union (ACLU)
Actual implementation, 16
Adaptive systems, 56
Adversarial machine learning, 241
Adversarial Robustness Toolbox (ART), 241, 247
Adverse decision, 312
Advertisements, 359
AEDTs. *See* Automated employment decision tools (AEDTs)
AER. *See* Automatic emotion recognition (AER)
AES systems. *See* Automated Essay Scoring systems (AES systems)
Affective computing (AC), 71–72
Affective learning, 72–74
Aflatoxin, 163
Agile-Waterfall hybrid, 82
Agri-food sector, 161–162
    conscientious design, 166
    interdisciplinary and multistakeholder engagement, 166–167
    legislation, 167
    preventing and mitigating potential ethical risks of online AI systems in, 165–167
    responsible innovation, 165
    teaching AI ethics, 167
Agriculture system, 154
    potential ethical risks of AI technological advancements in, 161–164
Agri–food systems, 165
AI. *See* Artificial intelligence (AI)
AIE. *See* Empathic AI (AIE)
AIED. *See* Artificial Intelligence in Education (AIED)
Alexa, 136
Algorithmic Accountability Act (AAA), 230
Algorithmic decision-making process, 98, 323, 331–332, 334, 336
Algorithmic ethics, 138
Algorithmic fairness, 63
Algorithmic gender discrimination, 330
Algorithmic governance, 306–307
Algorithmic neural networks, 326
Algorithmic processes, 330

Algorithmic systems, 30
Alien AI civilization, 263
Ambiguous moral contexts, 17–18
Ambivalent feeling, 180–181
American Civil Liberties Union (ACLU), 368–369
American Psychological Association (APA), 75
Amygdala, 349
Analytics-based assessment, 64
Androcentric context of AI
    artificial intelligence and gender biases, 324–329
    gender discrimination, RRI, and European regulation for AI, 329–336
Androcentrism, 329
Androids, 348
Animal production, 158–159
Anonymization-based approach, 248
Anthropocene, 286
Anthropomorphism, 283–284
Aquaculture industry, 159
Arithmetic operation, 324
ART. *See* Accountability, Responsibility, and Transparency (ART); Adversarial Robustness Toolbox (ART)
Artificial agents, 197
Artificial empathy, 290, 372–374
    ethical dilemma at heart of, 373–374
Artificial intelligence (AI), 7–8, 26, 55–56, 59, 71, 78, 97, 124, 135–136, 143, 153, 175–176, 229, 231, 259, 271, 279, 284–286, 299, 323, 325–327, 347, 373
    AI-based systems, 328
    AI-generated content, 270
    AI–based online assessment, 55
        ethics of AI in online assessment, 59–66
    and empathy, 279–280
    in assessment, 56–59
        AI-based assessment systems, 57
        automated feedback, 57–58
        automated grading, 57
        evaluation of assessment integrity, 58
        learning and assessment analytics, 58–59
    chatbots, 259–260
    design, 74–75
    in education, 56
    empathy, AI and climate change, 284–286
        human use of human beings, 284–285
        planetary regard, 286
        rethinking recognition, 287
    empathy in ethics, 287–289
    engineering context and gender script in AI, 325–327

Artificial intelligence (AI) (*Continued*)
    erobotics, 283—284
    ethical challenges for learning and assessment analytics,
        62—64
    ethical dimensions
        of AI-automated grading and feedback, 61—62
        of AI-based assessment supervision, 62
    ethical frameworks for AI-based systems to area-specific
        considerations for CSCL agent design, 9—18
        adapting to CSCL settings, 12—16
        breaking down high-level frameworks, 10—11
        classifying human-agent communication, 11—12
        ethics and pedagogical ethics, 16—18
    ethics, 283—287
        associated with AI systems, 121, 136—138
        of AI-based assessment scenarios, 61—64
    facial recognition technology, 287
    frameworks, 12
        to mitigate potential ethical risks of AI-based assessment,
          64—66
    further research, 292
    gender discrimination, RRI, and European regulation for,
        329—336
        principle of nondiscrimination in law, 331—336
        RRI, gender equality, and AI policies, 330—331
    implementation, 100—103
    Krettek's empathic AI, 290—292
    machine learning and gender, 327—329
    measuring empathy, 289—290
    methods, 280—283
        AI ethics regulation, 281—282
        machine/technology, 282—283
        research, 280—281
    in online assessment, 55
    policies, 330—331
    potential ethical risks of AI technological advancements in
        agriculture and food systems, 161—164
        accountability/responsibility, 163
        cyber security, 161—162
        data ownership, 162—163
        fairness, 163—164
        privacy, 162
        transparency, 164
    privacy, 246—250
        complying with specific regulation requirements,
          249—250
        creation of privacy-preserving models, 248—249
        data minimization, 249
        relevance to HR use case, 250
        right to be forgotten, 249—250
        risk assessment of models and datasets, 247—248
    products for disabled, 122—124
        Laura, 128—130
        methodology, 124

        Peter, 124—125
        Peter's diary, 125—128
        reflection, 130—132
    trend, 136
    in VR-based systems, insights into, 225—226
    writing
        assistants, 269
        and image generators and impact on growth and
          contribution, 269—271
        and impact on growth and contribution, 269—271
Artificial Intelligence Act, 231, 273
Artificial Intelligence and Data Act (AIDA), 231
Artificial Intelligence in Education (AIED), 26—27
Artificial neural networks, 56
Artificial systems, 195
Artistic creations, 259
Assessment
    analytics, 58—59
    assessment-based analytics, 59
    explainability, 243
    fairness, 242
    process, 60—61
    quality, 245—246
Association for Computing Machinery (ACM), 75
Asynchronous sessions, 39
Augmentation methods, 241
Augmented collection, 224
Augmented critical decision process, 230, 232
Augmented reality, 284
Autism, 216—217
    spectrum, 216—217, 224
Autoethnography, 124
Automated agents, 79
Automated assessments, 55—56
Automated decision system, 230, 232, 236, 244
Automated decision-making, 335
Automated employment decision tools (AEDTs), 234
Automated Essay Scoring systems (AES systems), 57
Automated feedback systems, 57—58
Automated grading, 57
Automated individual decision-making, 333—334
Automated system, 57, 80
Automated weeding, 156
Automatic emotion recognition (AER), 71
    in education, 76—81
    roles and benefits in online learning, 77*t*
Automatic evaluation, 13—14
Automatic feedback, 57—58
Automatic medical diagnosis, 106
Automatic processes, 281, 331
Automobiles, 326
Autonomous agricultural robot, 156—157
Autonomous and Intelligent Systems, 289
Autonomous artificial agents, 327

Autonomous artificial intelligence (AAI), 136—139
  advanced analysis, 147—148
  algorithmic ethics, 138
  artificial intelligence trend, 136
  efficiency and fairness, 147
  ethics associated with AI systems, 136—138
  means, 145—147, 146*f*
    accountability, 147
    fairness, 146
    transparency, 146
    trustworthiness, 146
  methodology, 139—141
    problem statement, 141
    research questions, 141
    technoethical inquiry approach, 139—141
  results, 141—148
    economic, 143
    historical, 142
    intended ends, 144—145
    legal, 143
    levels of influence, 144
    perspectives, 141—147
    political, 143
    possible side effects, 145
    sociocultural, 143—144
    stakeholders, 144
    theoretical, 142
  technoethics, 139
Autonomous systems, 324—325
Autonomous technologies, 136, 142
Autonomous vehicles (AVs), 234, 264—265, 347, 350—351
  AI sensors, 351
  ethics of, 351—352
  and need for certainty and connection, 264—266
Autonomy, 305, 313
AVs. *See* Autonomous vehicles (AVs)

B

Baseline, 31—32, 66, 211—212
Bayes classifier, 106
Bayes theorem, 106
Bayesian neural networks (BNNs), 245—246
Behavioral theory, 371
Bernoulli distribution, 106
Bias mitigation algorithms, 242
Big data, 26, 98, 154, 260, 271, 359—360
  analysis, 103
  processing, 108
Binary segmentation process, 156
Bioethics, 34—35
Biometric data, 221
Biometric identification, 267
Biometric security, 266—267

Biometric technology, 266, 353
Biometric tools, 355—356
Black box modelling, 29, 333
Black male cyborg, 357
Blockchain technology, 159—160, 271
Blue River Technology, 156—157
BNNs. *See* Bayesian neural networks (BNNs)
BoniRob, 156—157
Boosting, 102
  works, 100
Bovine tuberculosis, 158
Brain process, 212
  and perception of reality in VR, 213—214
Business regulations, 352

C

Calibration, 100
California's Consumer Privacy/Protection Act (CCPA), 230—231, 233
Canaries, 264
Carbon footprint, 285
CART. *See* Classification and Regression Tree (CART)
CAs. *See* Conversational agents (CAs)
CASA. *See* Computer as social actors (CASA)
Catalysator, 306
CCPA. *See* California's Consumer Privacy/Protection Act (CCPA)
Central Digital & Data Office (CDDO), 76, 85
Charter of Fundamental Rights of European Union (CFR), 331—332
Chatbots, 3, 8, 153—154, 175—176, 281—282
China Internet Information Service Algorithmic Recommendation Management (IISARM), 231—232
Chronic stress, 261
Civil rights law, 369
Classification And Regression Tree (CART), 105
  decision tree, 109*t*
Climate change, 282, 284—286
Coal-fired power stations, 285
Cognitive appraisal theory, 72—73
Cognitive development, 9
Cognitive sciences, 316
Cognitivism, 100—101
Collaborative learning, 3, 7—8, 17, 142
Collaborative projects, 356—357
Commercial delivery, 264
Commission on civil law, 331
Communication process, 285, 328
COMP. *See* Comprehensiveness (COMP)
Compartmentalization, 282
COMPAS. *See* Correctional Offender Management Profiling for Alternative Sanctions (COMPAS)
Competition law, 144

Compliance, 245
Comprehensiveness (COMP), 243
Computational generation, 307
Computational machines, 280—281
Computational methods, 26, 327
Computational recognition tools, 289—290
Computational resources, 79
Computer as social actors (CASA), 12
Computer(s), 127, 292
    computer-assisted writing, 270—271
    computer-supported analysis, 26
    ethics, 136—137
    programs, 204—205, 376
    software, 266
    systems, 355—356
    technology, 266
Computer—supported collaborative learning (CSCL), 3.
        See also Online learning
    conversational agents for collaborative learners, 4—9
        agents interact with learner, 6—8
        effectiveness, advantages, and limits of agents in, 8—9
    ethical frameworks for AI-based systems to area-specific
        considerations for design, 9—18
Conceptualization, 302
Conscientious design, 166
    framework, 166
Constructivism, 101
Consumption data, 164
Content moderation, 365
Continuing bonds
    in grief, 176—179
    model, 176
Conventional vehicles, 265
Conversational agents (CAs), 4—5, 8
    for collaborative learners, 4—9
    CSCL agents interact with learner, 6—8, 7*f*
    effectiveness, advantages, and limits of agents in CSCL,
        8—9
Conversational chatbot, 175—176
Copula-based methods, 251
CORR. *See* Correlation (CORR)
Correctional Offender Management Profiling for Alternative
        Sanctions (COMPAS), 235
Correlation (CORR), 243
Cosmetics, 359—360
Counterfeits of humans, 199
Court of Justice of European Union (ECJ), 332
COVID-19, 34—35, 179, 360
    pandemic, 3
    vaccine, 358—360
Creative process, 270
Criminal law, 369
Criterion of heterogeneity, 317
Critical assessment, 287

Criticism, 8, 370
Crop
    growth, 155
    production, 155—158
        fruit and vegetable harvesting, 157—158
        weed management, 155—157
CSCL. *See* Computer—supported collaborative learning
        (CSCL)
Cultural expectations, 325—326
Cultural psychology, 177
Curriculum
    holistic approach to curriculum design and classification,
        103—108
        classification methodologies for qualitative indicators,
            104—105
        sentiment analysis methodologies for qualitative
            indicator analysis, 106—108
    optimization
        decision-making process and AI implementation, 100—103
        holistic approach to curriculum design and classification,
            103—108
        indicators and LMS as technical basis for holistic
            approach, 108
        K-nearest neighbor kNN in strategic budget planning,
            109—111
        LinkedIn and Google scholar big data system for support
            of human resource procedure, 111—112
        decision-making AI system raises ethical issues and
            applicable ethical frameworks, 112—114
    planning, 97, 102
Cutting-edge technologies, 275
Cyber harassment, 369
Cyber security, 161—162
Cyber-stalking attempts, 369
Cyberattacks, 357
Cyberbullying, 354
Cybercriminals, 145
Cybermob, 366
Cybernetic systems, 260, 284
Cybersecurity mitigation, 135
Cybersexism, 369
Cyberspace, 361
Cyborg mammoths, 349

D

DAI. *See* Digital Afterlife Industry (DAI)
Data
    analysis, 35
    anonymization, 114
    augmentation, 241
    collection process, 29, 31, 34—35, 82, 114, 220, 234
    curation process, 234
    general unified themes, 36*t*

generation
  model, 248
  process, 328
minimization, 249
mining, 102, 307–308
ownership, 27–28, 63, 162–163
preprocessing, 237
protection, 11, 74, 273, 324
  and digital information bill, 231–232
  law, 334
  regulations, 230–231
quality, 113–114
security, 358
sets, 153–154
sharing, 142
slice, 237
uncertainty, 237
violate copyright rights, 201
Data from Cardiff University's HateLab project, 371
Data-driven decision support systems, 99–100
Data-driven decision-making, 143
Data-driven educational decision-making model, 99–100
Database data, 111
Datasets, 61–62, 236, 327–328
  risk assessment of models and, 247–248
Datatron, 252
Davinci model, 198
Deathbots, 175–176, 178
  continuing bonds, technological mediation, and ethical implications in grief, 176–179
  expectation of response and authenticity of relationship, 181–185
  imagined use of, 179–185
  imagining of, 188–189
  phones, internet, and social networks, 180–181
  potential ethical risks of, 185–188
Decision trees, 105
  for fish classification problem, 104f
Decision-making process, 28, 34–35, 60, 97, 100–103, 105, 112, 122, 136, 144, 235, 299–301, 309, 313, 327
  AI system raises ethical issues and applicable ethical frameworks, 112–114
  capabilities, 259–260
  in education, 101f
  factors, 72
Deep learning, 29, 56, 195, 327
  algorithm, 160
  capabilities, 270–271
  deep learning-based image recognition technologies, 157
  networks, 205
Deep neural networks, 6–7, 249
Deepfake technology, 357
Deliberation process, 64, 289
Deloitte, 10

Desensitization, 223
Design, 218–219
  algorithms, 371
  approach, 32–33
  empathy by design, 224–225
  privileges, 370
  process, 30, 222
Detection software, 197, 200
  discrimination, 198–199
DI. See Disparate impact (DI)
Differential privacy (DP), 248
Digital Afterlife Industry (DAI), 178, 188
Digital assistants, 292
Digital currencies, 271
Digital fingerprint, 178
Digital footprint, 180
Digital immortality, 186, 188
Digital learning, 6, 25
Digital phenotyping, 288
Digital replica, 201
Digital technologies, 177–178, 185
Digital watermarks, 204
Dimensional model, 72–73
Disability, 121–122
Discrete model, 72–73, 84
Discrimination with help of detection software, 198–199
Disparate impact (DI), 235
Domestic drones, 141–142
Dopamine, 349
DP. See Differential privacy (DP)
Drones, 351
Dystopian imaginaries, 185

E
Early warning system, 163
EC. See European Commission (EC)
ECJ. See Court of Justice of European Union (ECJ)
eCommerce financial systems, 136
Economic cycles, 259
ECtHR. See European Court of Human Rights (ECtHR)
EDM. See Educational data mining (EDM)
Education, 4, 64, 74, 99
  AI in, 56
  automatic emotion recognition in, 76–81
  ethics of MMLA in, 27–33
  systems, 55
Educational data mining (EDM), 26–27, 29, 58
Educational decision making, 99–100
Educational process, 59, 61, 71
Educational recommender systems (ERSs), 29
Educational resources, 86–88
Educational technology, 60
  ethics of, 59–60

Educative systems, 378
Educators, 59
Effective learning process, 73
Electoral system, 358
Electric car, 285
Electricity, 259–260
  consumption, 285
Electrodermal activity data acquisition software, 85
Electronic assessment, 57
Electronic orientation, 122
Electronic transfer, 199
Electronically distributed texts, 200
Elicitation, 239–240
Embodied cognition framework, 216–217
Emotion(s), 72–74, 219–222, 280, 348
  chips, 365–366, 377
  detection engine, 371
  emotion-cleansed vision, 289
  primacy of emotions in machine-human interface,
    348–349
  recognition, 71–72, 77, 288
    algorithms, 74
    method, 76
  representation model, 81
Emotional AI, 289, 353
Emotional development, 222
Emotional intelligence, 81, 225–226, 348
Emotional networks, 213
Emotional reactions, 265, 351
Emotional trauma, 211
Empathic AI (AIE), 279, 283
Empathic mirror, 376
Empathic thinking, 378
Empathy, 219–222, 279, 281, 284–286, 372
  by design, 224–225
  empathy-based online AI machine innovations, 282
  in ethics, 287–289
  gap, 373
  models, 280
  programming AI to develop empathy skills, 372–373
  VR as empathy machine, 215–216, 223–224
Energy consumption, 81, 351–352
Engineering process, 325
Epinephrine, 349
EPU chips, 376
Erobotics, 283–284
ERSs. *See* Educational recommender systems (ERSs)
Estrogen, 349
Ethical AI, 4
  development, 165
Ethical automatic emotion recognition model for online
    learning, 81–85
  ethical prediction model for AER in online learning, 83*f*
  stories and ethical actions of AER system, 87*t*

Ethical considerations, 86–88, 196, 199–203
  counterfeits of people, 201–202
  new forms of plagiarism, 200
  spread of misinformation, nonsense, and toxic language,
    202–203
  verifying authorship, 199–200
  violation of copyright rights and privacy, 200–201
Ethical data extraction, 82
Ethical decision-making, 153, 288
Ethical design, 167
Ethical development, 153
Ethical dilemma at heart of artificial empathy, 373–374
Ethical frameworks, 366
Ethical guidelines and frameworks, 74–76
Ethical hazards, 55
Ethical implications, 72–74
  in grief, 176–179
Ethical inquiry, 300, 306
Ethical issues, 82, 279
Ethical model, 72
Ethical perspectives, 135
Ethical risks, 186
Ethical theory, 142, 289
Ethical-normative value, freedom of choice as, 305–306
Ethics, 59, 71, 76, 130–131, 187, 211–212, 219–222
  of affect recognition technology, 355–356
  in affective computing, 73*f*
  AI, 60, 137, 299
    assessment scenarios, 61–64
    in online assessment, 59–66
  associated with AI systems, 136–138
  of autonomous vehicles, 351–352
  of educational technology, 59–60
  empathy in, 287–289
  ethical dimensions
    of AI-automated grading and feedback, 61–62
    of AI-based assessment supervision, 62
    ethical challenges for learning and assessment analytics,
      62–64
  ethics-based theory, 29
  frameworks to mitigate potential ethical risks of AI-based
    assessment, 64–66
  insights into ethics in VR-based systems, 225–226
  of MMLA in education, 27–33
  moral and ethical implications of assessment, 60–61
    general challenges of introducing AI into assessment, 61
    general moral implications of assessment, 60–61
  relation to, 231–232
  of social media algorithms, 359–361
EU. *See* European Union (EU)
EU Fundamental Rights Agency (FRA), 335
EU's General Data Protection Regulation (GDPR), 230–231
Euclidean distance, 110
European Commission (EC), 323–324

European Court of Human Rights (ECtHR), 332–333
European General Data Protection Regulation (GDPR), 29–30, 34–35, 317–318, 333–334
  data minimization principle, 249
European Group on Ethics in Science and New Technologies (EGE), 75
European regulation for AI, 329–336
European Union (EU), 323–324, 329–330
  legislators, 273
European Union Charter for Fundamental Rights (EUCFR), 75
Evaluation methods, 143
  of assessment integrity, 58
Evasion attacks, 234
Evidence-based research, 279
Exemplar methods, 244
Exogenous data, 327
Explainable artificial intelligence (XAI), 316, 334

F

FA. *See* First author (FA)
Face-scanning body cameras, 355
Face-to-face interaction, 186
Facebook, 356–357
  files, 358
  generation, 177–178
  memorial pages, 181
Faceprint, 266, 353
Facial emotion recognition, 355
Facial expression, 78–79
Facial recognition, 135
  model, 233
  software, 267
  and our human need for variety and significance, 266–269
  technology, 264, 266–267, 273, 287, 353
Fairness, 99, 146, 163–164, 234–235, 242–243, 360–361
  assessment, 242
  available tools, 242–243
  and bias issues in MMLA systems, 39–40
  looms, 356
  metrics, 235
  mitigation, 242
  relevance, 234–235
Fake news, 357–358
False-positive rates (FPR), 247
Farm management, 155
FBI. *See* US Federal Bureau of Investigations (FBI)
Feed composition, 158
Filmography, 218
Filtering techniques, 299–300
Financial transactions, 135
Fine-tuned language models, 200–201
First author (FA), 179
Food industry, 154–155

Food processing and related operations, 159–160
  food safety, 160
  food supply chain and traceability, 159–160
Food production process, 166–167
Food safety, 159
  risk assessment, 160
Food supply chain, 159–160
Food system, potential ethical risks of AI technological advancements in, 161–164
Foodborne disease, 154–155
  surveillance, 160
Formal learning, 15–16
FPR. *See* False-positive rates (FPR)
FRA. *See* EU Fundamental Rights Agency (FRA)
Free speech, 367–368
Freedom of choice, 300–306
  as ethical-normative value, 305–306
  MLA choice-architectures and, 310–315
  revised in contemporary society, 303–305
    dimensions and preconditions arising from our joint reading, 304*t*
    ethical reelaboration of dimensions, preconditions, and subconditions for, 305*t*
    from within moral and sociopolitical philosophy, 301–303, 303*t*
FTC. *See* US Federal Trade Commission (FTC)
Fuzzification, 110

G

Gaussian distribution, 106
Gaussian radial basis functions, 107–108
GDPR. *See* European General Data Protection Regulation (GDPR)
Gender, 327–329
  associations, 329
  bias, 324–329
    engineering context and gender script in AI, 325–327
    machine learning and gender, 327–329
  discrimination for AI, 329–336
  equality, 330–331
  gender-based knowledge, 325
  scripts, 326, 329
Gendered technology scripts, 326
Generalization, 222
Genes, 349
Genome annotation data, 160
Gini impurity, 105
Glass-box models, 244
Global positioning system (GPS), 159, 264
Global Risks Report, 372–373
Google, 10, 136, 291
  assistant, 175–176
  Scholar, 103, 112
  search, 239

Google's LaMDA model, 197
Governance framework, 331
Governance model, 145
Government surveillance system, 355–356
GPS. *See* Global positioning system (GPS)
Gradient-based methods, 244
Grading process, 61–62
Graphic emails, 369
Grasp empathy, 366
Grief, continuing bonds, technological mediation, and ethical
        implications in, 176–179
Griefbots, 175–176
Grieving process, 176, 178–179
Group posts, 359

H
H2H. *See* Human to human (H2H)
Hanson robotics, 260
Harvesting season, 162
Hate speech
    context, 366–370
        costs associated with free speech, 366–368
        recent spate of texts focused on hate speech's unintended
            consequences, 368–370
    filling empathy gap, while boosting empathy, 375–378
        assessing emotion chip's role in curtailing unintended
            consequences, 377–378
        establishing ethical review boards to evaluate social
            medial algorithms, 378
        testing artificial empathy under five ethical approaches,
            375–377
    online hate speech, incivility, rage, and lack of self-
            restraint, 365–366
    position statement, 370–375
        emotion chips boost empathy, 374–375
        ethical dilemma at heart of artificial empathy, 373–374
        programming AI to develop empathy skills, 372–373
    social media algorithms
        designed to identify and deter hate speech, 371–372
        incite hate speech, 370–371
    targets, 365
HCI. *See* Human-computer interaction (HCI)
Head-mounted display (HMD), 216
Headphones, 129
Health care, 202
    delivery, 145–146
    MLAs, 307
    professionals, 144–145
    program, 309
Health regime, 308–309
Heart of artificial empathy, ethical dilemma at, 373–374
Herbicides, 155–156
    herbicide-resistant weeds, 155–156

Hetero-predetermining MLA-driven action, 317
Heterogeneity, 311–312
Heterogeneous contexts, 306
Heterogeneous exposure, 315–316
Heterogeneous options, 304–305
Heterogeneous reflections, 303
Heterogeneous relations, 311–312
High-Level Expert Group on Artificial Intelligence
        (AI HLEG), 113
Hijab laws, 355–356
HMD. *See* Head-mounted display (HMD)
Hoc calibration methods, 246
Holistic approach, 211, 240
    CART decision tree, 109*t*
    to curriculum design and classification, 103–108
    indicators and LMS as technical basis for, 108
Homo sapiens, 348
Homophobia, 360–361
HR. *See* Human resources (HR)
HRI. *See* Human-robot interaction (HRI)
Human behaviour, 360
Human communication processes, 280–281
Human decision-making, 43–44, 310–312, 323
Human discrimination abilities, 197–198
Human emotional expressions, 354
Human emotions, 351
    affect recognition technology, 353–355
    algorithms in social media, 356–359
    autonomous vehicles, 350–351
    ethical prognoses of emerging AI applications, 349–350
    ethics
        of affect recognition technology, 355–356
        of autonomous vehicles, 351–352
        of social media algorithms, 359–361
    inexorable rise of machines, 347–348
    primacy of emotions in machine-human interface,
            348–349
Human intelligence, 260
Human interaction, 62, 265
Human morality, 145
Human needs, artificial intelligence and
    AI writing and image generators and impact on growth and
            contribution, 269–271
    autonomous vehicles and need for certainty and connection,
            264–266
    early warning of disruption and upheaval, 271–274
    emerging AI technologies and basic human needs,
            263–264
    facial recognition systems and our human need for variety
            and significance, 266–269
    fundamental human needs, 261–263
    methodology, 260–261
    rise of intelligent machines, 259–260
Human personal assistant, 126–128

Human processing, 136
Human psyche, 349
Human resources (HR), 239, 373
  planning process, 98
Human rights, 10, 86–88, 360, 375
Human to human (H2H), 283–284
  interaction, 15
  learning process, 15
Human use of human beings, 284–285
Human welfare, 355
Human wiring, 377
Human-centered Values, 10, 12
Human-computer interaction (HCI), 15, 26–27, 154
Human-induced climate change, 285
Human-inspired AI system, 357
Human-machine interaction, 212, 280–282
Human-machine relationship, 72
Human-robot interaction (HRI), 266, 283, 351
Humanity, 274, 283
Human–machine discrimination abilities, 198
Humanoid, 281–282
  robots, 347, 351
Hunt's algorithm, 104
Hybrid emotion, 354
Hyperconnected contemporary societies, 309
Hyperconnected environments, 317
Hyperledger Fabric system, 160
Hyperstimulation, 223
Hypothetical scenario, 184–185

I

IEEE. *See* Institute for Electrical and Electronics Engineers
    (IEEE)
Illusions, 213–214
Image generators and impact on growth and contribution,
    269–271
Image processing methods, 156
Image sources, 213
Imaginary dialog, 183
Imagining of deathbots, 188–189
Immersion phase, 222
Immersive artistic representations, 215–216
Immersive environment, 221
Immersive journalism, 215–216, 222
Indirect gender discrimination, 332
Industrial innovation, 347
Industrial revolutions, 140–141, 271
Inflammatory speech, 376
Informal learning, 15–16
Information Commisioner's Office (ICO), 75
Information to system (IS), 136
Information technology (IT), 97, 347, 376–377
Information theory, 105

Informational gatekeepers, 308–309
Infrared thermal images, 158
Ingredients, 130
Innate emotions, 314
Innovations, 31–32
  transforms research, 330
Innovative technologies, 285
Instagram, 356–357
  images, 354
Institute for Electrical and Electronics Engineers
    (IEEE), 219
Institute for Future of Work (IFOW), 75
Intellectual amnesia, 284
Intellectual property law, 144
Intelligent machines, 274
  rise of, 259–260
Intelligent personal assistant, 125–126
Intelligent software, 125, 129
Intelligent speech software, 125
Intelligent systems, 26–27, 113, 132
Intelligent technology, 5
Intelligent tutoring systems (ITS), 8–9, 26–27, 56, 73
Intelligent vehicles, 265
Intentions, 372
International Convention on Economic, Social, and Cultural
    Rights, 360–361
Internet, 180–181, 366–367
  forums, 180
  of garbage, 369
  regulation, 366
  search engines, 291
Internet of Things (IoT), 135, 153
Interpretable model, 84, 86–88
Interviews, 34–35, 176
Intrinsic method, 245–246
Investigative project, 201
IoT. *See* Internet of Things (IoT)
iPhones, 367, 373
Irreducible tension, 177
IS. *See* Information system (IS)
Islamophobia, 371
Isolation, 326
IT. *See* Information technology (IT)
Iterative search-review-discussion process, 78
ITS. *See* Intelligent tutoring systems (ITS)

J

Juvenile rheumatoid arthritis, 128

K

K-nearest-neighbor (kNN), 109
  in strategic budget planning, 109–111

kNN. *See* K-nearest-neighbor (kNN)
Knowledge distillation methods, 244
Knowledge-discovery techniques, 307–308
Kondratiev Wave, 259
Krettek's empathic AI, 290–292

L

LA. *See* Learning analytics (LA)
LAK. *See* Learning analytics and knowledge (LAK)
Laplace estimation, 106
Large language models (LLMs), 3, 14–15, 196–199, 233
  discrimination with help of detection software, 198–199
  ethical consequences, 199–203
  handle epistemological crisis, 203–204
  human discrimination abilities, 197–198
  as thinking tools, 204–205
Large-scale collection, 31
Law enforcement agencies, 286–287
Leaf node, 104
Learning algorithms, 237, 248
Learning analytics (LA), 25, 58–59
  systems, 63
Learning analytics and knowledge (LAK), 26
Learning curves, 12–13
Learning design, 5
Learning designers, 17–18
Learning disabilities, 121
Learning environment, 5
Learning indicators, 97
Learning management systems (LMSs), 58, 98–99, 108
  indicators and LMS as technical basis for holistic approach, 108
Learning measurements, 31
Learning phase, 308
Learning process, 26, 55–56, 74, 324
Learning research, 27–28
Learning support assistants (LSAs), 216–217
Learning systems, 26
Learning theory, 6, 100–101
Legal system, 330
Legal tools, 336
Legislation, 167
Legitimate processing, 233
Leslie's framework, 75–76
Liability law, 144
Linear models, 237
LinkedIn system, 111–112
LLMs. *See* Large language models (LLMs)
LMSs. *See* Learning management systems (LMSs)
Logistic regression model, 251
LSAs. *See* Learning support assistants (LSAs)
Luciano Floridi's ethical theory, 376–377

M

Machine ethics, 71–72
Machine interventions, 279
Machine learning (ML), 30, 56, 109, 123, 142, 153, 234, 259, 292, 323, 327–329, 348
  implementation, 103
  ML-based decision systems, 242
  model, 98, 328, 360
  process, 241–242
  systems, 324
  techniques, 6–7, 77, 266–267, 353
Machine learning algorithms (MLAs), 160, 232, 299, 360–361
  choice-architectures, 310–315
  epistemological and moral constraints, 310–315
  governance, 307–310
  MLA-driven informational preselection, 308
  personalization rate, 314
  techniques, 310
Machine learning-as-a-Service (MLaaS), 245
Machine morality, 376
Machine-generated text, 197
Machine-human interface, primacy of emotions in, 348–349
Madanes-Robbins framework, 261
Mainstream decision-making, 236
Malware detectors, 234
Mass production, 203
Material design, 326
Materialism, 372–373
MaxEnt. *See* Maximum entropy (MaxEnt)
Maximum entropy (MaxEnt), 106–107
Mediation, 176
Medical diagnostics, 135
Medical image recognition, 347
Medical research, 366
Medical treatment process, 220
Memorization, 178, 241
Mental health, 131, 289–290
Mental imagery process, 213
Mental models, 213
Mental trauma, 211
Metacognitive processes, 36–38
Metadata estimation system, 114
Metrics, 235
Microblogs, 267–268
Microphone, 58
Mitigation
  available tools, 252
  explainability, 244
  fairness, 242
  across multiple dimensions, 251–252
  transparency and governance, 245
Mixed reality (MR), 202
ML. *See* Machine learning (ML)

MLaaS. *See* Machine learning-as-a-Service (MLaaS)
MLAs. *See* Machine learning algorithms (MLAs)
MMD. *See* Multimodal data (MMD)
MMLA. *See* Multimodal Learning Analytics (MMLA)
MMORPGs. *See* Multimodal online role-playing games
        (MMORPGs)
Mobile devices, 352
Mobile phone, 124, 266–267, 353
Mobile surveillance devices, 267–268
Model risk management, 237–238
Moodle, 108
Morality law, 355–356
Motivations, 372
MR. *See* Mixed reality (MR)
Multimodal affect detection, 85
Multimodal data (MMD), 26
Multimodal Learning Analytics (MMLA), 26–27, 35–42
    argued benefits of MMLA and ethical issues, 41–42
    awareness level of benefits and risks associated with, 41
    emerging need for ethical framework for, 35–36
    ethical framework for, 48*f*
    ethics of MMLA in education, 27–33
    fairness and bias issues in, 39–40
    limitations and future work, 45–47
    methodology, 33–35
        data analysis, 35
        data collection, 34–35
        participants, 33–34, 34*t*
    privacy, surveillance, and intrusiveness issues with, 36–38
    student agency over learning data ownership, 38–39
    systems' accountability, 40–41
    systems' transparency and explainability, 40
    trustworthiness of, 39
Multimodal online role-playing games (MMORPGs), 214
Multistakeholder engagement, 166–167

N
Naïve Bayes (NB), 102
National data protection laws, 35–36
National defense authorities, 307
National Institute of Standards and Technology (NIST), 232
Natural human emotion, 372
Natural language processing (NLP), 55, 57, 195, 241, 371
Naturalized copresence, 180–181
NB. *See* Naïve Bayes (NB)
Neural models, 57
Neural networks (NNs), 57, 144, 195–196, 204, 323
    approaches, 29
    reprogram algorithms, 328–329
Neutral approach, 372
New York City (NYC), 234
NIST. *See* National Institute of Standards and Technology
        (NIST)

NLP. *See* Natural language processing (NLP)
NNs. *See* Neural networks (NNs)
No-fault liability, 336
Nondiscrimination law, 324
Norepinephrine, 349
Nuclear powerplant, 272
NYC. *See* New York City (NYC)

O
OECD. *See* Organisation for Economic Co-operation and
        Development (OECD)
Office of Science and Technology Policy (OSTP), 272
OLMs. *See* Open learner models (OLMs)
Online AI systems in agri-food sector, preventing and
        mitigating potential ethical risks of, 165–167
    conscientious design, 166
    interdisciplinary and multistakeholder engagement,
        166–167
    legislation, 167
    responsible innovation, 165
    teaching AI ethics, 167
Online assessment, 55, 57–58
Online database servers, 154–155
Online education, 57, 74
Online harassment, 367, 369
Online hate speech of self-restraint, 365–366
Online learning, 39, 71–72, 77, 85–88. *See also*
        Computer–supported collaborative learning (CSCL)
    AER roles and benefits in online learning, 77*t*
    automatic emotion recognition in education, 76–81
    emotions, affective learning, and ethical implications,
        72–74, 73*f*
    ethical automatic emotion recognition model for online
        learning, 81–85
    ethical guidelines and frameworks, 74–76
Online MLA-based systems, 300–301
Online service providers (OSPs), 307
Online threats, 366–367
Online tutoring system, 82
Online-based AI systems, 153–154
Open learner models (OLMs), 29
Organic diseases, 36–38
Organisation for Economic Co-operation and Development
        (OECD), 160, 331
Organizational innovation, 144
OSPs. *See* Online service providers (OSPs)
OSTP. *See* Office of Science and Technology Policy (OSTP)

P
Paper-based diary, 131
Paraverbal communication, 86
Pareto frontiers, 240

Partial dependence plots (PDP), 244
PBG Framework. *See* Process-Based Governance Framework (PBG Framework)
PBL. *See* Project-based learning (PBL)
PDP. *See* Partial dependence plots (PDP)
Pedagogical angle, 16—17
Peer assessments, 32
Personal actions, 218
Personal assistant, 125—126
Personal data, 233, 367
Personal Information Protection Law (PIPL), 230—231
Personalization techniques, 56, 299—300, 307, 310—311
Personalized Instagram message, 201
Personalized interaction deathbots, 186—187
Personalized learning, 6—7
Perturbation-based methods, 244
Pest and disease detection, 157
PET. *See* Positron emission tomography (PET)
Pet robots, 373
Phones, 180—181
PILAR. *See* Pragmatic Inquiry for Learning Analytics Research (PILAR)
PIPL. *See* Personal Information Protection Law (PIPL)
Plagiarism, 9, 199
    new forms of, 200
Poisoning attacks, 234
Positron emission tomography (PET), 213
Pragmatic Inquiry for Learning Analytics Research (PILAR), 29—30
Precision agriculture, 158
Predictive algorithm models, 85, 329
Predictive machine—learning algorithms
    freedom of choice, 300—306
    MLA
        choice-architectures and freedom of choice, 310—315
        governance, 307—310
Predictive power, 307—308
Preliminary Design Principles, 11
Privacy, 141, 162, 367, 370
    available tools, 249
    privacy-preserving models, creation of, 248—249
    rights, 199
    risk assessment, 233
    of trustworthy AI, 232—233
        metrics, 233
        relevance, 232—233
Probabilistic techniques, 313
Problem-solving process, 6
Process-Based Governance Framework (PBG Framework), 75—76
Proctoring systems, 58, 62
Progesterone, 349
Programming AI to develop empathy skills, 372—373
Project cost management process, 102—103

Project-based learning (PBL), 101—102
Protected attributes, 235
Proteus effect, 214—215
Prototypes, 348
Proximal development, 13
Public dataset, 233
Public morality, 260
Public policy, 143
Public safety, 352
Public speech, 365—367
Purpose-sampling method, 33—34
Python library, 241
Python toolkit, 242

Q
Quadratic function, 107
Qualitative approach, 33, 229
Qualitative assessment, 238—239
Qualitative indicators, 102, 108
    classification methodologies for, 104—105
    decision tree for fish classification problem, 104*f*
    sentiment analysis methodologies for, 106—108
Qualitative inquiry methods, 33
Qualitative methods, 179
Qualitative risk assessment, 238
Qualitative standpoint, 304—305, 310
Quality, 237, 245—246
    assessment, 245—246
    available tools, 246
    metrics, 237, 246
    relevance, 237
        to HR use case, 246
Quantitative approaches, 229
Quantitative assessment, 238—239
Quantitative data sources, 25
Quantitative indicator, 97
Quantitative methods, 238
Quantitative privacy assessments, 247
Quantitative risk assessment, 238
Quantitative standpoint, 310—312
Quantitative test, 240

R
Racism, 360—361
Radial basis functions (RBFs), 107—108
Radicalisation, 370
Radiofrequency identification technology (RFID), 159—160
Rawlsian society, 375
RBFs. *See* Radial basis functions (RBFs)
Real-time facial expressions, 354—355
Realism, 216
Receiver operating characteristic curve (ROC curve), 247

Recipes, 197–198
Reciprocal conversation, 178–179
Recognition software, 355
Recommender systems (RS), 307
Recursive process, 147
Reflection, 124, 130–132, 182, 223–226
    empathy by design, 224–225
    insights into ethics and AI in VR-based systems, 225–226
    VR as empathy machine, 223–224
Reflective endorsement, 304, 313, 317
Regulation, 231
Regulators, 74
Reinforcement learning, 6–7
Relational autonomy, 302
Reliability, 234, 240–242
    assessment, 240–241
    available tools, 241
    metrics, 234
    mitigation, 241
    relevance, 234
        to HR use case, 241–242
Religious rights, 367–368
Reptilian brain, 349
Responsibility, 360
Responsible AI, 229
Responsible innovation, 165
Responsible Research and Innovation (RRI), 324, 330–331
    for AI, 329–336
Revolutionary AI technologies, 263
RFID. *See* Radiofrequency identification technology (RFID)
Risk accelerating learning, 32
Risk assessment, 229, 237–238
    available tools, 247–248
    of models and datasets, 247–248
Risk deathbots, 183–184
Risk Management Framework (RMF), 232
Risk management system, 11, 231
Risk tolerance, 263
RMF. *See* Risk Management Framework (RMF)
Road safety, 351–352
Road traffic crashes, 351–352
Robocrop, 156–157
Robot(s), 123, 281
Robotic kiwi harvester, 157
Robotic vehicle, 350
Robotics, 136, 331
Robustness metrics, 240
ROC curve. *See* Receiver operating characteristic curve (ROC curve)
Root node, 104
RRI. *See* Responsible Research and Innovation (RRI)
RS. *See* Recommender systems (RS)

S
SA. *See* Sensitivity analysis (SA)
Safety, 11
Saliency methods, 244
Satellites, 154–155
Saturation, 33–34
Science, technology, and society studies (STS), 330
Screen reader, 129
Screen-reading software, 131
Security, 11, 234, 240–242, 366–367, 370
    assessment, 240–241
    available tools, 241
    cameras, 356
    metrics, 234
    mitigation, 241
    relevance, 234
        to HR use case, 241–242
SED. *See* Smoothed empirical differential (SED)
Self-driving car, 350
Self-learning AI, 366
Self-learning network, 311
Self–restraint, online hate speech, incivility, rage, and lack of, 365–366
Semantic artifice, 181
Semantic variance, 224–225
Semantic-structural gender biases, 328–329
Sensing technology, 25
Sensitive information, 42
Sensitive research, 179
Sensitivity analysis (SA), 244
Sensor(s), 350
    data, 42–43
    technologies, 160
Sensorial stimuli, 211–212
Sensory impairments, 121
Sensory stimuli, 216–217
Sentiment analysis, 106, 353–354
    Bayes classifier, 106
    maximum entropy algorithm, 106–107
    methodologies for qualitative indicator analysis, 106–108
    support vector machines, 107–108
Serotonin, 349
Sexrobots. *See* Erobotics
Sexual harassment, 332, 365–366
Signaling devices, 352
Smart cars, 280–281
Smart cities, 280–281
Smart dairy farming, 165
Smart facial recognition cameras, 355–356
Smart fridge, 308–309
Smartphone, 74, 280–281
Smoothed empirical differential (SED), 243

Smoothing methods, 244
SNSs. *See* Social network services (SNSs)
Social dimension, 306
Social hallucinations, 222
Social media, 202, 356—357, 370
  algorithms, 356—359, 366, 378
    designed to identify and deter hate speech, 371—372
    establishing ethical review boards to evaluate, 378
    ethics of, 359—361
    incite hate speech, 370—371
  corporations, 369
  farms, 357
  posts, 354
  sites, 358
  social media-based cyberaggression, 354
  sources, 360
Social network services (SNSs), 177, 307
Social networks, 180—181, 186—187, 356—357
Social species, 262
Socio-relational dimension, 304
Sociocultural environment, 143—144
Software manufacturers, 130—131
Solar generation, 285
Solar technology, 285
Solid ethical principles, 72
Sonograms, 158
Sound ethical inquiry, 301
Spatial processing, 213
Spatial sound, 219
SPD. *See* Statistical parity difference (SPD)
Spectrum, 147
Speech software, 127
Speech technology, 124, 130
Spinal cord, 124—125
Spinal injury, 121—122
Splitting criterion, 105
Stakeholders, 26—27, 65
Statistical parity difference (SPD), 235
Stereotypes, 60, 153—154, 326—329
STS. *See* Science, technology, and society studies (STS)
SUFF. *See* Sufficiency (SUFF)
Sufficiency (SUFF), 243
Suicidal tweets, 354
Support, Underwrite, and Motivate (SUM), 75—76
Support vector machines (SVM), 107—108
Surveillance networks, 355
Synthetic dataset, 233
Systematic approach, 135

T

Tangible learning, 102
Teaching-learning processes, 65
Technical papers, 196

Technical solutions within pillar of trustworthy AI, 240—246
  explainability, 242—243
    assessment, 243
    available tools, 244
    mitigation, 244
    relevance to HR use case, 244—245
  fairness, 242—243
  security and reliability, 240—242
  transparency and governance, 245
Technoethical inquiry approach, 139—141, 147
  conceptual map of technoethics, 140*f*
  social subsystems, 141*t*
Technoethics, 135, 139
Technological artefact, 178—179
Technological innovations, 148
Technological mediation in grief, 176—179
Technologies for Automated Feedback—Classification Framework (TAF—ClaF), 57—58
Technology-driven medium, 215—216
Telecommunication
  industry, 137
  law, 144
  organization, 144
Tensorflow Fairness Indicators, 243
Testosterone, 349
TFEU. *See* Treaty on Functioning of European Union (TFEU)
Thanatechnology, 176
Thematic analysis research, 35
Thermographic cameras, 264
Third-party data, 368
3D printing, 271
Threshold-based approach, 252
TPR. *See* True-positive rate (TPR)
Traceability systems, 159—160
Traditional learning, 26
Trained ML model, 233
Transformative technology, 166—167
Translation software, 204
Transparency, 63, 164, 245
  and governance, 236, 245
    available tools, 245
    metrics, 236
    mitigation, 245
    relevance, 236
    relevance to HR use case, 245
Treaty on Functioning of European Union (TFEU), 331—332
True-positive rate (TPR), 247
Trustworthy AI, 229—239
  combining multiple dimensions of trustworthy AI, 250—252
    assessment across multiple dimensions, 251
    mitigation across multiple dimensions, 251—252
  explainability, 235—236
    metrics, 236

relevance, 235–236
fairness, 234–235
methodology, 238–239
  AI privacy, 246–250
  elicitation of requirements, 239–240
  technical solutions within pillar of trustworthy AI,
    240–246
model risk assessment, 237–239
pillars of, 232–237
  privacy, 232–233
qualitative *vs.* quantitative assessment, 238–239
quality, 237
relation to ethics, 231–232
relevant regulations, 230–232
  AI regulations, 231–232
  data protection regulations, 230–231
relevant use cases, 237–238
security and reliability, 234
transparency and governance, 236
Turnitin plagiarism, 200
Tweets, 354
Twitter, 356–357, 365–366

U

UAV. *See* Unmanned aerial vehicles (UAV)
UMAP. *See* User modeling, adaptation, and personalization
    (UMAP)
Uncertainty quantification (UQ), 245–246
Unintended consequences, 366
Unmanned aerial vehicles (UAV), 155
UQ. *See* Uncertainty quantification (UQ)
US Federal Bureau of Investigations (FBI), 162
US Federal Trade Commission (FTC), 230
User modeling, adaptation, and personalization (UMAP),
    26–27
Utilitarian ethics, 138
Utopian imaginaries, 185

V

VA. *See* Virtual assistants (VA)
  robots, 283–284
Validation process, 245
Verbal behavior, 11–12
Violation of copyright rights and privacy, 200–201

Violence, 365
Virtual assistants (VA), 283–284, 290–291
Virtual personalities, 357
Virtual reality (VR), 211–212, 271, 284
  brain and perception of reality in, 213–214
  design, 218–219
  ethics, empathy, and emotion, 219–222
  methods, 216–218
    study design, 217–218
    walking in small shoes, 216–217
  proteus effect, 214–215
  reflections, 223
  as empathy machine, 215–216
Virtual world (VW), 214
Virtue ethics, 137–138
Visual imagery, 213
Visualizations, 236
VR. *See* Virtual reality (VR)
VW. *See* Virtual world (VW)

W

Water quality parameters, 159
Watermark detection, 204
Watermarking systems, 272
Web crawling, 112
Web-based traceability system, 160
Websites, 131, 370–371, 375
Weed management, 155–157
WhatsApp, 367, 373
Whole assessment process, 61
Wind turbines, 285
Wireless sensor networks, 160
World Commission on Ethics of Scientific Knowledge and
    Technology (COMEST), 75

X

XAI. *See* Explainable artificial intelligence (XAI)
Xenophobia, 360–361

Y

YOLOv3. *See* You Only Look Once version 3 (YOLOv3)
You Only Look Once; version 3 (YOLOv3), 157
YouTube, 356–357

Printed in the United States
by Baker & Taylor Publisher Services